VOLUME FIVE HUNDRED AND EIGHTY FIVE

METHODS IN ENZYMOLOGY

Proteomics in Biology, Part A

METHODS IN ENZYMOLOGY

Editors-in-Chief

ANNA MARIE PYLE
Departments of Molecular, Cellular and Developmental Biology and Department of Chemistry
Investigator, Howard Hughes Medical Institute
Yale University

DAVID W. CHRISTIANSON
Roy and Diana Vagelos Laboratories
Department of Chemistry
University of Pennsylvania
Philadelphia, PA

Founding Editors

SIDNEY P. COLOWICK and NATHAN O. KAPLAN

VOLUME FIVE HUNDRED AND EIGHTY FIVE

METHODS IN ENZYMOLOGY

Proteomics in Biology, Part A

Edited by

ARUN K. SHUKLA
Indian Institute of Technology,
Kanpur, India

ACADEMIC PRESS
An imprint of Elsevier
elsevier.com

Academic Press is an imprint of Elsevier
50 Hampshire Street, 5th Floor, Cambridge, MA 02139, United States
525 B Street, Suite 1800, San Diego, CA 92101–4495, United States
The Boulevard, Langford Lane, Kidlington, Oxford OX5 1GB, United Kingdom
125 London Wall, London, EC2Y 5AS, United Kingdom

First edition 2017

ISBN: 978-0-12-809742-7
ISSN: 0076-6879

Publisher: Zoe Kruze
Acquisition Editor: Zoe Kruze
Editorial Project Manager: Magesh Mahalingam
Production Project Manager: Helene Kabes
Cover Designer: Mark Rogers

Typeset by SPi Global, India

CONTENTS

CONTRIBUTORS

J. Armengaud
CEA, DRF, IBiTec-S, SPI, Li2D, Laboratory "Innovative Technologies for Detection and Diagnostics", Bagnols-sur-Cèze, France

U. auf dem Keller
ETH Zurich, Institute of Molecular Health Sciences, Zurich, Switzerland

B. Bai
St. Jude Children's Research Hospital, Memphis, TN, United States

P.G. Barr-Gillespie
Oregon Hearing Research Center; Vollum Institute, Oregon Health & Science University; Oregon Health & Science University, Portland, OR, United States

D. Becher
Institute for Microbiology, Ernst Moritz Arndt University Greifswald, Greifswald, Germany

M. Benhar
Rappaport Institute for Research in the Medical Sciences, Faculty of Medicine, Technion-Israel Institute of Technology, Haifa, Israel

S.M. Bidlingmaier
UCSF Helen Diller Family Comprehensive Cancer Center, University of California at San Francisco, San Francisco, CA, United States

F. Bonn
Institute for Microbiology, Ernst Moritz Arndt University Greifswald, Greifswald, Germany

K. Büttner
Institute for Microbiology, Ernst Moritz Arndt University Greifswald, Greifswald, Germany

S. Cambridge
Institute of Cell Biology and Anatomy, University of Heidelberg, Heidelberg, Germany

L. Chen
Laboratory of Synthetic Microbiology, School of Chemical Engineering & Technology; Key Laboratory of Systems Bioengineering (Ministry of Education), Tianjin University; SynBio Research Platform, Collaborative Innovation Center of Chemical Science and Engineering (Tianjin), Tianjin, PR China

P.R. Cutillas
Barts Cancer Institute, Queen Mary University of London, John Vane Science Centre, London, United Kingdom

L.L. David
Oregon Health & Science University, Portland, OR, United States

M.M. Ferreira Amaral
School of Life and Health Sciences, Aston University, Birmingham, United Kingdom

L.J. Foster
Michael Smith Laboratories, University of British Columbia, Vancouver, BC, Canada

L. Frigotto
Isogenica Ltd., The Mansion, Chesterford Research Park, Essex, United Kingdom

J. Gebert
Institute of Pathology, University Hospital Heidelberg; Cancer Early Detection, German Cancer Research Center (DKFZ), Heidelberg, Germany

K.D. Ha
UCSF Helen Diller Family Comprehensive Cancer Center, University of California at San Francisco, San Francisco, CA, United States

L. Hendershot
St. Jude Children's Research Hospital, Memphis, TN, United States

O. Hermes
ETH Zurich, Institute of Molecular Health Sciences, Zurich, Switzerland

A.A. High
St. Jude Proteomics Facility, St. Jude Children's Research Hospital, Memphis, TN, United States

A.V. Hine
School of Life and Health Sciences, Aston University, Birmingham, United Kingdom

V.P. Ichhaporia
St. Jude Children's Research Hospital, Memphis, TN, United States

T. Ju
Emory University School of Medicine, Atlanta, GA, United States

J. Kast
The Biomedical Research Centre; Centre for Blood Research, University of British Columbia; University of British Columbia, Vancouver, BC, Canada

J. Kopitz
Institute of Pathology, University Hospital Heidelberg; Cancer Early Detection, German Cancer Research Center (DKFZ), Heidelberg, Germany

J.F. Krey
Oregon Hearing Research Center; Vollum Institute, Oregon Health & Science University, Portland, OR, United States

N.-K. Lee
UCSF Helen Diller Family Comprehensive Cancer Center, University of California at San Francisco, San Francisco, CA, United States

S. Lehoux
Beth Israel Deaconess Medical Center, Harvard Medical School, Boston, MA, United States

R. Li
The Biomedical Research Centre, University of British Columbia; University of British Columbia, Vancouver, BC, Canada

B. Liu
UCSF Helen Diller Family Comprehensive Cancer Center, University of California at San Francisco, San Francisco, CA, United States

S. Maaß
Institute for Microbiology, Ernst Moritz Arndt University Greifswald, Greifswald, Germany

A. McAfee
Michael Smith Laboratories, University of British Columbia, Vancouver, BC, Canada

Y. Mechref
Texas Tech University, Lubbock, TX, United States

M. Miyagi
Center for Proteomics and Bioinformatics; Case Western Reserve University, Cleveland, OH, United States

P. Ostasiewicz
Wrocław Medical University, Wrocław, Poland

A. Otto
Institute for Microbiology, Ernst Moritz Arndt University Greifswald, Greifswald, Germany

V.R. Pagala
St. Jude Proteomics Facility, St. Jude Children's Research Hospital, Memphis, TN, United States

G. Pei
Laboratory of Synthetic Microbiology, School of Chemical Engineering & Technology; Key Laboratory of Systems Bioengineering (Ministry of Education), Tianjin University; SynBio Research Platform, Collaborative Innovation Center of Chemical Science and Engineering (Tianjin), Tianjin, PR China

J. Peng
St. Jude Children's Research Hospital; St. Jude Proteomics Facility, St. Jude Children's Research Hospital, Memphis, TN, United States

W. Peng
Texas Tech University, Lubbock, TX, United States

F. Sabino
ETH Zurich, Institute of Molecular Health Sciences, Zurich, Switzerland

M. Schnölzer
Functional Proteome Analysis, German Cancer Research Center (DKFZ), Heidelberg, Germany

T. Sun
Laboratory of Synthetic Microbiology, School of Chemical Engineering & Technology; Key Laboratory of Systems Bioengineering (Ministry of Education), Tianjin University; SynBio Research Platform, Collaborative Innovation Center of Chemical Science and Engineering (Tianjin), Tianjin, PR China

Y. Su
UCSF Helen Diller Family Comprehensive Cancer Center, University of California at San Francisco, San Francisco, CA, United States

H. Tan
St. Jude Proteomics Facility, St. Jude Children's Research Hospital, Memphis, TN, United States

F. Terzi
Institute of Cell Biology and Anatomy, University of Heidelberg, Heidelberg, Germany

L. Veillon
Texas Tech University, Lubbock, TX, United States

U. Warnken
Functional Proteome Analysis, German Cancer Research Center (DKFZ), Heidelberg, Germany

P.A. Wilmarth
Oregon Health & Science University, Portland, OR, United States

J.R. Wiśniewski
Biochemical Proteomics Group, Max-Planck-Institute of Biochemistry, Martinsried, Germany

K.M. Wooding
Texas Tech University, Lubbock, TX, United States

L. Zacharias
Texas Tech University, Lubbock, TX, United States

W. Zhang
Laboratory of Synthetic Microbiology, School of Chemical Engineering & Technology; Key Laboratory of Systems Bioengineering (Ministry of Education), Tianjin University; SynBio Research Platform, Collaborative Innovation Center of Chemical Science and Engineering (Tianjin), Tianjin, PR China

S. Zhou
Texas Tech University, Lubbock, TX, United States

R. Zhu
Texas Tech University, Lubbock, TX, United States

PREFACE

Biological and cellular processes are incredibly complex and result from a large number of exceptionally well fine-tuned and interconnected individual events. Although investigation of these events at individualistic level and in isolated context typically provides important functional insights, their study at systems level and in global context is absolutely essential to decipher the inner workings of cellular and biological phenomena. This has been one of the key challenges in the modern era of biology; however, recent surge of proteomics strategies and approaches have made it possible. A combination of well-established and continuously evolving proteomics methodologies has started not only to reveal the molecular basis of a whole range of biological phenomena and regulation, but it has also emerged as a major engine to open new translational and therapeutic opportunities. The field of proteomics has evolved exceptionally rapidly in the last decade or so, and it is a monumental task and near impractical to cover every aspect in a couple of book volumes. Therefore, we have made the selection of chapters to cover a range of conventional approaches and well-established platforms, a few recent breakthroughs with respect to discovery of novel experimental framework, and finally, a few chapters highlighting the untapped potential of proteomics approaches with a futuristic outlook.

Many of the proteomics protocols have been standardized across different laboratories and published in many primary research articles, and therefore, they are mostly accessible to any interested researcher in the field. We have included a few chapters in these two volumes of *Methods in Enzymology* that briefly describe routine sample preparation strategies and then focus more on recent modifications that have been tested and validated experimentally. One of the key areas that has immensely benefited from the recent applications of proteomics is the identification, quantitation, and analysis of post-translational modifications such as glycosylation, phosphorylation, and acetylation. Here, we have collected a number of chapters describing the application of proteomics approaches in studying a wide range of post-translational modifications, either in the form of generic protocols or with a particular biological target as a case example. Proteomics-based studies typically generate large body of data which on one hand represent goldmine for extracting novel functional insights but on the other pose a major challenge in terms of unbiased, careful, and rigorous analysis. These large proteomics data sets offer unique possibilities as a hypothesis generation platform to design subsequent experimental strategies to gain novel functional insights into biological processes. Considering that data handling and analysis represents an absolutely critical component in proteomics pipeline, we have included a number of chapters in these two volumes that specifically focus on recent trends in proteomics data analysis and interpretation.

One of the powers of proteomics approaches has been the possibility of studying the biological processes at organellar and organism levels. This has opened up the possibilities of obtaining a global and holistic view of biological processes in physiologically relevant context instead of in isolation. A number of chapters in these two volumes cover this aspect, for example, analysis of body fluid degradomics, global assessment of protein synthesis in *C. elegans*, comprehensive analysis of *E. coli* membrane proteins, and a few others. These chapters underline the robust platform that proteomics can provide for generating important systems-wide clues in biology. Another important application of proteomics, in addition to being a discovery engine for basic biology, is that it can open up new possibilities in disease etiology and therapeutics, for example, by identifying novel biomarkers or by presenting a comparative proteome-wide view of pathological vs normal tissues. We have incorporated a number of chapters that specifically highlight the power of proteomics-based approaches in disease models such as identification of novel biomarkers, quantitative proteome profiling of tumor tissues, and identifying key aspects of host–pathogen interactions. These chapters categorically underscore the emerging utility of proteomics approaches in translational biology and highlight a frontier where a major focus of the field is headed currently.

I take this opportunity to sincerely thank all the authors for their excellent contributions which have made these two volumes possible. I also express my gratitude to the editorial and production team of *Methods in Enzymology* for beautifully coordinating this project. On behalf of the entire *Methods in Enzymology* team, I present two volumes entitled "Proteomics in Biology: Part A" and "Proteomics in Biology: Part B." I sincerely hope that you find the content of these two volumes extremely useful and I welcome your feedback and comments.

With best compliments

Arun K. Shukla
Indian Institute of Technology, Kanpur, India

CHAPTER ONE

An Easy and Fast Protocol for Affinity Bead-Based Protein Enrichment and Storage of Proteome Samples

A. Otto, S. Maaß, F. Bonn[2], K. Büttner, D. Becher[1]
Institute for Microbiology, Ernst Moritz Arndt University Greifswald, Greifswald, Germany
[1]Corresponding author: e-mail address: dbecher@uni-greifswald.de

Contents

Abstract

Analysis of dilute protein samples is a challenging task for scientific and industrial labs all over the world. Although there are different methods available that allow for protein enrichment from various biological sources, all of them have serious limitations apart

[2] Current address: Institute of Biochemistry II, Goethe University School of Medicine, Frankfurt am Main, Germany.

Methods in Enzymology, Volume 585
ISSN 0076-6879
http://dx.doi.org/10.1016/bs.mie.2016.09.012

from their advantages. In order to perform highly reproducible and sensitive protein analysis of lowest concentrated samples, we optimized a method to enrich proteins on affinity beads (StrataClean) recently. This chapter describes the general protocol of this strategy, thereby discussing the power as well as the limits of this technique for qualitative and quantitative proteomic studies. Moreover, additional application and protocol variants will be discussed, expanding the number of compatible up- and downstream processing techniques compared to the originally published method. Hence, we evaluated the reduction of time for sample preparation by use of preprimed affinity beads and shorter incubation durations as well as the influence of high concentration of salts or urea in the sample buffer.

1. INTRODUCTION

Enrichment of proteins is one of the most critical steps in sample preparation for proteome analyses. Hence, a variety of methods have been developed to fulfill this goal. Here, the enrichment of proteins from the extracellular compartment is of particular interest as it allows to investigate protein secretion during different (growth) conditions, to monitor protein secretion in industrial processes, or to study host–pathogen interactions mediated by secreted proteins. Moreover, successful protein enrichment is a prerequisite for the comprehensive analysis of biomarkers from body fluids. Although there are a number of scientific questions relying on efficient protein enrichment, the general method of choice is hardly to recognize. The selection of a suitable method is largely dependent on the techniques available in the respective laboratories and on the specific workflow used in the general sample preparation procedure. However, sample handling, the equipment needed, and the disposal of wastes generated during protein enrichment are central aspects during the decision process.

Several classical techniques for concentration of diluted or secreted proteins are available including approaches based on (i) protein precipitation, (ii) ultrafiltration with microfiltration units, and (iii) solid-phase extraction (SPE). Protein precipitation workflows are based either on reduced solvation of proteins in aqueous solutions with following hydrophobic aggregation by salting out or by treatment with organic solvents or on protein denaturation and subsequent precipitation by acids. Using these approaches allows for high efficiencies of precipitation without specific instrumentation. However, resolubilization of denatured and aggregated proteins can be extremely difficult. Moreover, the need for hazardous reagents like trichloroacetic acid (TCA) or high volumes of volatile organic solvents (e.g., chloroform and methanol) is a serious disadvantage of protein precipitation methods.

An alternative approach for concentration of protein-containing solutions is to use ultrafiltration based on microfiltration units. These methods can be applied to virtually all samples without the need for additional chemicals. However, users frequently experience that theoretical size-exclusion parameters do not match the size of actually passing protein species. Additionally, proteins sticking to or aggregating on the membranes during the filtration process can increase losses of several other proteins.

The third option for protein enrichment is SPE, which is a variant of affinity enrichment. These one-pot protocols require minimal use of hazardous chemicals thereby increasing sample volume only to a small extent. Moreover, SPE is compatible to most of the methods to prefractionate samples for proteomics, which expands the range of its applications.

A particular method is a commercial system of SPE, affinity beads named "StrataClean" (Ziegler, Vogt, Miersch, & Strack, 1997). It has been applied in a number of proteomic studies (Pasztor et al., 2010; Wickman et al., 2013; Zanivan et al., 2013) and, therefore, will be in the focus of this chapter.

In addition to the application of StrataClean beads for protein enrichment, it has been shown recently that these beads can be used to store and ship proteins at ambient temperatures without increased protein degradation rates (Bonn et al., 2014). The shipping of protein samples without cooling needs will save the costs of elaborate logistics and will thus provide laboratories with a strong emphasis on wet-lab work with a well-suited logistic interface to their respective analytical counterparts.

2. StrataClean BEADS FOR PROTEIN ENRICHMENT

2.1 A General Overview of the Method

The optimized protocol for reproducible protein enrichment from complex samples by StrataClean beads includes priming of the beads, bead washing prior to protein binding, washing to remove unbound proteins, and drying of the sample for storage, shipping, or further experimental procedures (Fig. 1).

The initial protocol for protein enrichment with StrataClean beads was published in 1997 (Ziegler et al., 1997). At this time, the use of StrataClean

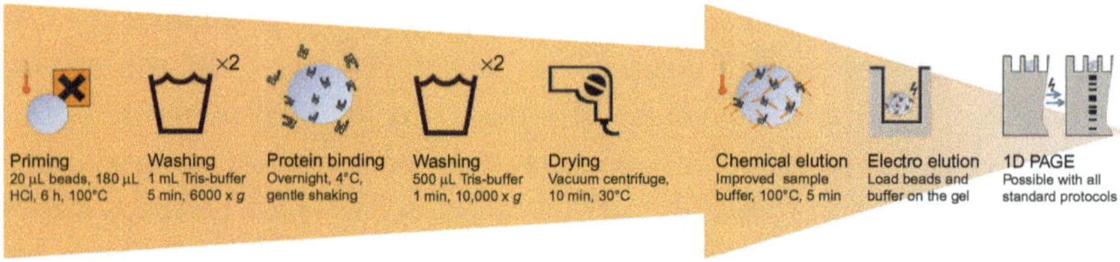

Fig. 1 General overview of the workflow for sample preparation using StrataClean beads. Proteins are enriched by priming of the beads, bead washing prior to protein binding, washing to remove unbound proteins, and drying of the sample. Elution of the protein from the beads is performed using SDS-PAGE.

beads has been prone to batch variances with respect to unwanted and nonspecific protein loads. As these contaminations were interfering with highly sensitive mass spectrometry, they biased identification of proteins in respect to reproducibility and sensitivity. Moreover, reliable protein quantification is not possible using such unstable protocols. However, a reproducible and contamination-free protein enrichment with StrataClean beads can be achieved if beads are primed by incubation in 12 *M* hydrochloric acid for at least 5 h prior to protein binding. After removal of unspecific contaminations by several washing steps and drying of the beads with sample bound, proteins can be stored at ambient temperatures for a prolonged period.

For the complete elution of proteins bound to StrataClean beads, a combined process, based on detergents and electroelution, is necessary to yield a maximum of proteins in GeLC-MS analyses (SDS-PAGE with subsequent LC–MS/MS analysis of trypsinized proteins in cut gel pieces) (Bonn et al., 2014). Therefore, loaded beads are incubated with SDS-PAGE sample buffer and are directly applied to the SDS-PAGE (Fig. 1). In order to avoid aggregation and loss of proteins, sufficient concentrations of strong reducing agents like mercaptoethanol in the elution buffer are necessary.

Nevertheless, there are studies where quantitative binding of all proteins in a sample is not central for the experiment. Indeed, cases may exist where the reproducible identification of a subset of proteins or even only very few target proteins is sufficient to answer the analytical question of interest. Occasionally, it can be beneficial to adapt and vary the existing optimized protocol to enhance the sensitivity toward the proteomic targets, to be able to combine protein enrichment with other up- or downstream methods, or to save significant analysis time. Therefore, this chapter includes variations of the general protocol examining the performance of protein enrichment by affinity beads from challenging sample buffers and for screening purposes in biotechnological studies (see Section 5).

3. GENERAL PROTOCOL FOR THE USE OF StrataClean BEADS IN PROTEOMICS

StrataClean beads are available in two formats, 3 mL (#400714, Agilent) and 9 mL (#400715, Agilent). Prior to use, vortex thoroughly to ensure an even distribution of the beads in the slurry. *Note*: It may be advisable to aliquot the beads after ordering to warrant constant bead concentration throughout consecutive experimental work.

3.1 Priming and Washing of the Beads

Priming of the affinity beads requires a prolonged time of heating in a very acidic environment. *Note*: Please take care as in the initial step of priming StrataClean, handling with highly concentrated acid is necessary. Please ensure that all operations follow local bylaws and hazardous regulations.

1. Transfer an aliquot of StrataClean resin, approximately 20 μL, in suspension to a new 1.5-mL reaction tube.
2. Add 180 μL of concentrated hydrochloric acid. *Note*: Take caution with handling of the acid. Use goggles and gloves for your safety.
3. Incubate the reaction tubes at 100°C for about 6 h. *Note*: To avoid opening of the lid and evaporation of the acid provide secure lid closure by commercial snap-on devices or by placing a heat-resistant item like a glass plate on top of the reaction tubes.
4. Let cool down to room temperature and subsequently centrifuge at low speed to sediment the resin (3500 × *g*, standard table top centrifuge).
5. Discard the supernatant.
6. Add 1 mL of washing buffer (e.g., Tris–HCl Buffer; see Table 1) and resuspend beads.
7. Centrifuge at low speed to sediment the resin (3500 × *g*, standard table top centrifuge).
8. Repeat steps 6 and 7.
9. Discard the supernatant.

The affinity beads are now ready for use (see Section 5.1).

3.2 Affinity Binding of Proteins in Solution

Primed beads are devoid from proteinaceous contaminations and can be loaded with target proteins from biological fluids or from dilute solutions. Samples containing high concentrations of interfering substances, including SDS, TCA, and organic solvents, are to dilute with a compatible buffer (suitable solvents given in Table 1). *Note*: Avoid dilution with distilled water unless the final strength of the buffering solution will sustain the solubility and buffer capacity for the proteins of interest.

1. Incubate in a 360 degree rotating shaker at 4°C for at least 2 h or overnight (*Note*: See Section 5.2).
2. Centrifuge at 10,000 × *g* at 4°C for 45 min.
3. Remove the supernatant. *Note*: In case of insufficient binding, the supernatant can be used for further extractions or protein precipitation by TCA.
4. Resuspend the resin in 500 μL of distilled water or washing buffer.
5. Centrifuge at 20,000 × *g* at 4°C for 5 min and discard the supernatant.
6. Dry the beads in a vacuum centrifuge to complete dryness.

The sample is in condition now that allows for facilitated storage and shipment. For direct analysis, both SDS-PAGE and digestion in solution are possible.

Table 1 Tested Buffer/Solvents

Buffer/Media	Composition
TE buffer	Tris–HCl 50 m*M*/EDTA 10 m*M*; pH 8.0
Tris–HCl buffer	Tris–HCl 50 m*M*; pH 8.0
Ammonium bicarbonate (ABC)	Ammonium bicarbonate 50 m*M*
Phosphate-buffered saline	10 m*M* Na_2HPO_4, 1.8 m*M* KH_2PO_4, 137 m*M* NaCl, 2.7 m*M* KCl; pH 7.4
MOPS buffer	3-(*N*-Morpholino) propanesulfonic acid 50 m*M*; pH 7.5
HEPES buffer	4-(2-Hydroxyethyl)-1-piperazineethanesulfonic acid 50 m*M*; pH 7.5

3.3 Elution of the Purified Proteins

In order to analyze the purified proteins by visualization, immunodetection of specific proteins by Western Blotting or GeLC-MS analysis, the loaded resin has to undergo electroelution in a specific loading buffer for samples in conventional SDS-PAGE. For this purpose, both self-cast gels and commercially available precast gels are usable. Apart from the buffer formulations used in the respective gel system, the StrataClean loading buffer should have the following composition: 125 m*M* Tris–HCl, pH 6.8, 20% (v/v) glycerol, 4% (w/v) SDS, 3.75% (v/v) β-mercaptoethanol, 20 m*M* DTT, 0.04% (w/w) bromophenol blue. *Note*: For direct trypsination of proteins on the beads and gel-free analysis by LC–MS/MS, please refer to the original (Bonn et al., 2014).

1. Set up the SDS-PAGE according to your laboratory standard. *Note*: Please provide a gel system that includes both resolving and loading gel. Sample loading pockets in the gel should have a dimension to hold the volume of resin in sample loading buffer.
2. Add 20 μL of sample loading buffer to the pellet. *Note*: It is crucial for the elution process that sample loading buffer contains elevated concentrations of the reducing agent as in StrataClean loading buffer.
3. Incubate at 98°C for 10 min and let cool down to room temperature.
4. Spin down the sample in a standard table top centrifuge at room temperature.
5. Load the slurry into the sample pockets of the loading gel of the SDS-PAGE.
6. Run the SDS-PAGE according to standard procedures.

After finishing the SDS-PAGE, common downstream analysis is possible with the resolved samples. *Note*: Some SDS-PAGE seems to allow affinity beads to enter the loading gel to a certain extent. Despite this, no adverse effect for downstream analytics is detectable so far.

4. APPLICATION OF StrataClean BEADS IN PROTEIN SAMPLE PREPARATION AND LIFE SCIENCE RESEARCH

4.1 Highly Diluted Protein Solutions

The affinity enrichment protocol based on StrataClean beads can be applied universally upstream of modern proteomics techniques, except for 2D-PAGE-based studies. Starting with a complex mixture of proteins of different origin, protein recovery is quantitative (Bonn et al., 2014). Sources for protein solutions include, e.g., cell extracts, growth media with extracellular proteins, and biological fluids. The StrataClean protocol has already been used in studies targeting the extracellular proteome of pathogenic microorganisms (Bonn et al., 2016) and in research elucidating the proteins secreted by actinoplanaceae during the production of the antidiabetic drug acarbose (Wendler et al., 2016). Recovery of proteins from solutions containing as little as 100 ng/mL of a complex protein extract is possible (Bonn et al., 2014).

4.2 Optimization and Monitoring of Biotechnological Processes/Screening of Biological Samples

A further application of the StrataClean protocol is to control different steps in biotechnological processes or various protein separation techniques. The advantage of a one-pot reaction avoiding the use of harsh chemicals, except from bead priming, makes this workflow suitable for large-scale screening purposes, e.g., for analysis of protein elution patterns of consecutive fractions

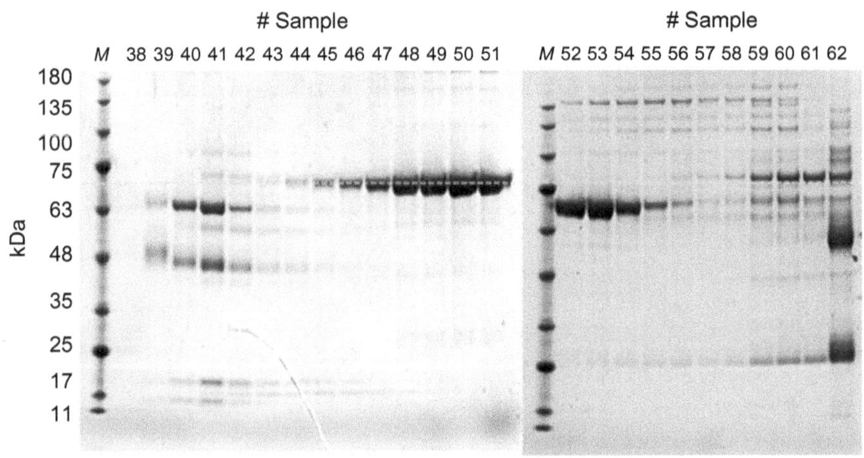

Fig. 2 Screening for proteins after resolution of a complex protein sample in a free-flow electrophoresis. Human serum runs in a field step protocol. Fractions 38–62 represent fractions of decreasing acidity. The sampling in fractions of 4 mL took place in cooled deep well plates. Fractions were incubated overnight with primed StrataClean beads. After resolution of electroeluted proteins by SDS-PAGE, affinity-enriched proteins were visualized by Coomassie staining. *M* in lane 1 depicts the molecular weight marker.

after size-exclusion chromatography or separation in a free-flow electrophoresis (Fig. 2).

4.3 Storage and Shipping of Protein Samples

In addition to the application of the StrataClean protocol to cleanup and enrich highly dilute protein samples of different origin in a one-pot process, another aspect of sample handling profits from the use of StrataClean beads. Protein samples, which have been loaded onto the beads can be desiccated together with the beads and those samples can be stored at room temperature for prolonged periods. This has been tested for storage times of up to 11 days without compromising the quality of the sample (Bonn et al., 2014). Based on these results, it is possible to decouple sample generation on one hand and sample analysis by, e.g., GeLC-MS or by gel-free proteomics in central service laboratories on the other hand.

5. PROTOCOL VARIATIONS

5.1 Storage of Primed Beads

The rationale of bead priming lies in the reduction of the overall proteinaceous contaminations that may have been introduced during the production process or storage of the beads prior to use. In contrast to the original publication, it is possible to prime and wash the beads and to store these in tightly closed reaction tubes at least over a period of a month (Fig. 3).

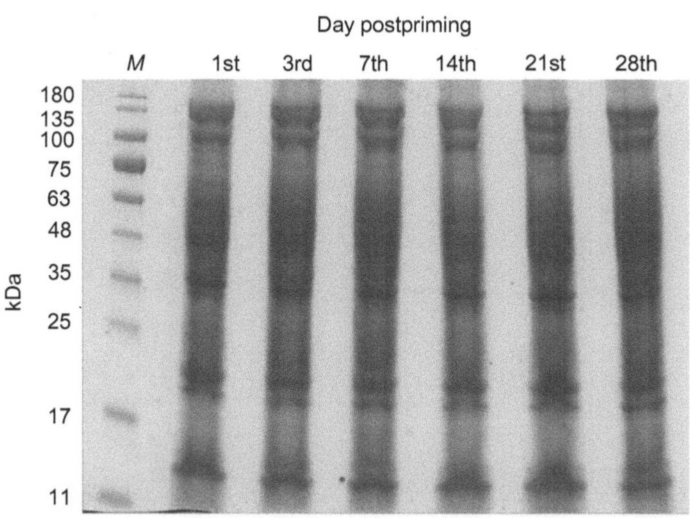

Fig. 3 Testing of capacity after different storage times of primed affinity beads. Sterile-filtered growth medium of *Peptoclostridium difficile* was incubated after cell harvest with affinity beads to capture extracellular proteins secreted into the growth medium. Affinity beads were primed at day 0 and stored for the time indicated (1, 3, 7, 14, 21, or 28 days). After the respective storage time, an aliquot of the supernatant was incubated with the beads and finally all samples were resolved on SDS-PAGE. No decrease in affinity capacity or specificity is visible. *M* in lane 1 depicts the molecular weight marker.

1. Prime the affinity beads according to the protocol "priming and washing of the affinity beads." Stop the protocol at step 6.
2. Resuspend the beads thoroughly and place an aliquot of the slurry into 2.5 mL reaction tubes with screw caps.
3. Centrifuge at low speed to sediment the resin ($3500 \times g$, standard table top centrifuge).
4. Discard the supernatant.
5. Close the reaction tubes with the screw caps and store unopened until use.

5.2 Duration of Bead Incubation

Exact quantitative analyses are possible by using affinity purification of dilute samples with the StrataClean protocol, as described earlier. Despite the advantage of complete binding of dilute proteins followed by SDS-PAGE or by direct trypsination for LC–MS/MS analyses, in experimental settings to be optimized for swift results, incubation times of several hours to overnight may be adverse to the speed of analysis intended. The time for

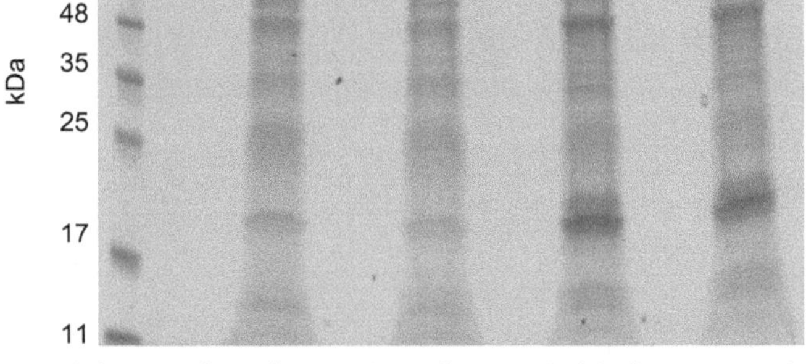

Fig. 4 Testing of shortened incubation times for protein binding to preprimed, stored affinity beads (PP), or freshly primed affinity beads (FP). Sterile-filtered growth medium of *Peptoclostridium difficile* was incubated after cell harvest with affinity beads to capture extracellular proteins secreted into the growth medium. Affinity beads were used either preprimed (PP) or primed the same day (freshly primed, FP). Supernatant samples incubated either for 30 min or overnight, according to the protocol described in this chapter. After resolution of electroeluted proteins by SDS-PAGE, affinity-enriched proteins were visualized by Coomassie staining. *M* in lane 1 depicts the molecular weight marker.

incubation of freshly primed or primed and stored beads may be shortened to 30 min without compromising the qualitative results of the purification. Although differences for single bands can be seen using the extremely shortened protocol (Fig. 4), this would not affect results in, e.g., differential proteomics approaches comparing protein abundances in a relative manner after constant sample preparation procedures. *Note*: Quantitative studies should follow the incubation time in the basic protocol.

5.3 Use of Buffers Containing Urea

Certain protocols, which may be used upstream of the affinity-binding protocol, require buffers with elevated concentrations of urea. This includes electrophoretic separation of complex protein mixtures, cell extracts, or

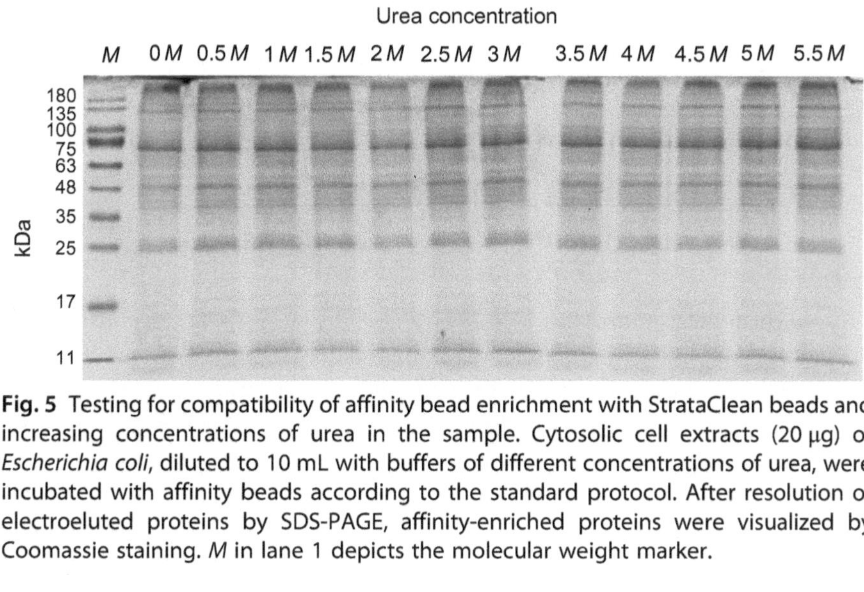

Fig. 5 Testing for compatibility of affinity bead enrichment with StrataClean beads and increasing concentrations of urea in the sample. Cytosolic cell extracts (20 μg) of *Escherichia coli*, diluted to 10 mL with buffers of different concentrations of urea, were incubated with affinity beads according to the standard protocol. After resolution of electroeluted proteins by SDS-PAGE, affinity-enriched proteins were visualized by Coomassie staining. *M* in lane 1 depicts the molecular weight marker.

dissolution of protein complexes (Hoffmann et al., 2001; Pietsch et al., 2010). Comparable protein separation patterns in SDS-PAGE show that affinity bead purification of proteins is possible from samples with up to 5.5 *M* urea (Fig. 5).

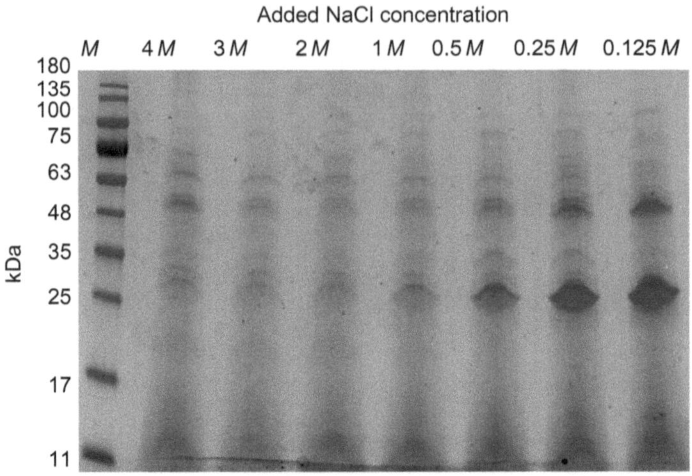

Fig. 6 Testing for compatibility of affinity bead enrichment with StrataClean beads and increasing concentrations of salt in the sample. Sterile-filtered growth medium of *Bacillus subtilis* was incubated after cell harvest with affinity beads to capture extracellular proteins secreted into the growth medium. Prior to affinity bead purification, salt (NaCl) was added with the final concentrations indicated. No protein precipitation was visible upon adding of salt to the samples. After resolution of electroeluted proteins by SDS-PAGE, affinity-enriched proteins were visualized by Coomassie staining. *M* in lane 1 depicts the molecular weight marker.

5.4 Influence of the Salt Concentration

Protein purification methods in biotechnology that are using the procedure of "salting out" proteins of interest, or which are based on media compositions of microbes in need for elevated salt concentrations, are a challenge for downstream proteomics techniques for affinity purification or analysis. With the StrataClean protocol, it is possible to analyze the qualitative composition of diluted protein samples even at high salt concentrations without affecting, e.g., the performance of SDS-PAGE resolution (Fig. 6). However, it is to note that quantitative information may be affected and protocols need to be evaluated, although qualitative analyses could be performed.

6. CONCLUSIONS

SPE based on affinity beads is a versatile method in proteomics. In the past, we were able to demonstrate reproducible and exceedingly effective enrichment of highly dilute protein by StrataClean beads. Here, we propose the applicability of the method in novel applications including control and optimization of biotechnological process as well as in sample processing downstream of specific analytical methods like free-flow electrophoresis. Furthermore, sample enrichment by affinity beads can be used to clean-up proteins from samples carrying high concentrations of urea or salt. Shortening of sample binding times allows for a quick screening method based on affinity-based protein enrichment. In summary, this method allows for a robust sample processing leading to enriched and cleaned-up protein samples.

REFERENCES

Bonn, F., Bartel, J., Büttner, K., Hecker, M., Otto, A., & Becher, D. (2014). Picking vanished proteins from the void: How to collect and ship/share extremely dilute proteins in a reproducible and highly efficient manner. *Analytical Chemistry, 86*, 7421–7427.

Bonn, F., Pané-Farré, J., Schlüter, R., Schaffer, M., Fuchs, S., Bernhardt, J., et al. (2016). Global analysis of the impact of linezolid onto virulence factor production in *S. aureus* USA300. *International Journal of Medical Microbiology, 306*, 131–140. http://dx.doi.org/10.1016/j.ijmm.2016.02.004.

Hoffmann, P., Ji, H., Moritz, R. L., Connolly, L. M., Frecklington, D. F., Layton, M. J., et al. (2001). Continuous free-flow electrophoresis separation of cytosolic proteins from the human colon carcinoma cell line LIM 1215: A non two-dimensional gel electrophoresis-based proteome analysis strategy. *Proteomics, 1*, 807–818.

Pasztor, L., Ziebandt, A.-K., Nega, M., Schlag, M., Haase, S., Franz-Wachtel, M., et al. (2010). Staphylococcal major autolysin (Atl) is involved in excretion of cytoplasmic proteins. *The Journal of Biological Chemistry, 285*, 36794–36803.

Pietsch, J., Kussian, R., Sickmann, A., Bauer, J., Weber, G., Nissum, M., et al. (2010). Application of free-flow IEF to identify protein candidates changing under microgravity conditions. *Proteomics, 10*, 904–913.

Wendler, S., Otto, A., Ortseifen, V., Bonn, F., Neshat, A., Schneiker-Bekel, S., et al. (2016). Comparative proteome analysis of Actinoplanes sp. SE50/110 grown with maltose or glucose shows minor differences for acarbose biosynthesis proteins but major differences for saccharide transporters. *Journal of Proteomics, 131*, 140–148.

Wickman, G. R., Julian, L., Mardilovich, K., Schumacher, S., Munro, J., Rath, N., et al. (2013). Blebs produced by actin-myosin contraction during apoptosis release damage-associated molecular pattern proteins before secondary necrosis occurs. *Cell Death and Differentiation, 20*, 1293–1305.

Zanivan, S., Maione, F., Hein, M. Y., Hernández-Fernaud, J. R., Ostasiewicz, P., Giraudo, E., et al. (2013). SILAC-based proteomics of human primary endothelial cell morphogenesis unveils tumor angiogenic markers. *Molecular & Cellular Proteomics, 12*, 3599–3611.

Ziegler, J., Vogt, T., Miersch, O., & Strack, D. (1997). Concentration of dilute protein solutions prior to sodium dodecyl sulfate-polyacrylamide gel electrophoresis. *Analytical Biochemistry, 250*, 257–260.

CHAPTER TWO

Filter-Aided Sample Preparation: The Versatile and Efficient Method for Proteomic Analysis

J.R. Wiśniewski[1]

Biochemical Proteomics Group, Max-Planck-Institute of Biochemistry, Martinsried, Germany
[1]Corresponding author: e-mail address: jwisniew@biochem.mpg.de

Contents

Abstract

Filter-aided sample preparation (FASP) is a versatile and efficient way of processing protein extracts for bottom-up proteomic analysis. The method repurposes centrifugal ultrafiltration concentrators for removal of detergents, protein cleavage, and isolation of pure peptide fractions. FASP can be used for protein cleavage with different proteinases either with single enzymes or in a mode of successive multienzyme digestion (MED)-FASP. The FASP methods are useful for processing of samples ranging in their sizes from submicrogram to several milligram amounts of total protein. They also allow peptide fractionation, and isolation and quantitation of total RNA and DNA acid contents. This chapter describes principles, limitations, and applications of FASP. Additionally detailed FASP and MED-FASP protocols are provided.

Methods in Enzymology, Volume 585
ISSN 0076-6879
http://dx.doi.org/10.1016/bs.mie.2016.09.013

1. INTRODUCTION

Sample preparation is the key step in the bottom-up proteomic analysis. Over the past two decades of mass spectrometry-based proteomics, a large number of methods for protein extraction, protein digestion, and peptide purification have been proposed. In general these sample preparation methods can be broadly classified by the way, how proteins are digested, in either "in-gel," "in-solution," or chemical reactor based ones. To the latter group belongs the filter-aided sample preparation (FASP) method (Wisniewski, Zougman, Nagaraj, & Mann, 2009). This technique utilizes centrifugal ultrafiltration devices for protein purification and digestion. The major advantage of FASP over other sample preparation methods is the high purity of generated peptides, which is the prerequisite for their successful liquid chromatography fractionation, mass spectrometric sequencing, and protein identification.

The initial step in proteomic sample preparation is lysis of biological material. Although this can be performed in many ways using different regents for protein extraction, the most efficient and almost quantitative procedures involve using sodium dodecyl sulfate (SDS). In contrast other detergents, including Na-deoxycholate and Na-laurate often do not enable complete protein extraction (Masuda, Tomita, & Ishihama, 2008; Rakus, Gizak, Deshmukh, & Wisniewski, 2015). Since SDS have to be removed from the sample before mass spectrometric analysis several procedures for their depletion have been proposed, such as protein precipitation or solid-phase extraction. These approaches are frequently accompanied with substantial sample losses, whereas in contrast, FASP offers an easy and straightforward way of detergent depletion (Wisniewski, 2016).

2. THE FASP PROCEDURES

2.1 FASP Methods Overview

An overview of the FASP procedures is given in Fig. 1. Beside the standard FASP procedure the multienzyme digestion (MED)-FASP and the thiol-activated polyethylene glycol (TAPEG)-FASP are shown. The first step of the FASP procedures is the removal of large excess of detergent and other low-molecular weight substances used for sample lysis (Fig. 1B). The depletion of the interfering substances is facilitated by washing with a buffer containing 8 *M* urea. The highly concentrated denaturant shrinks sizes of the

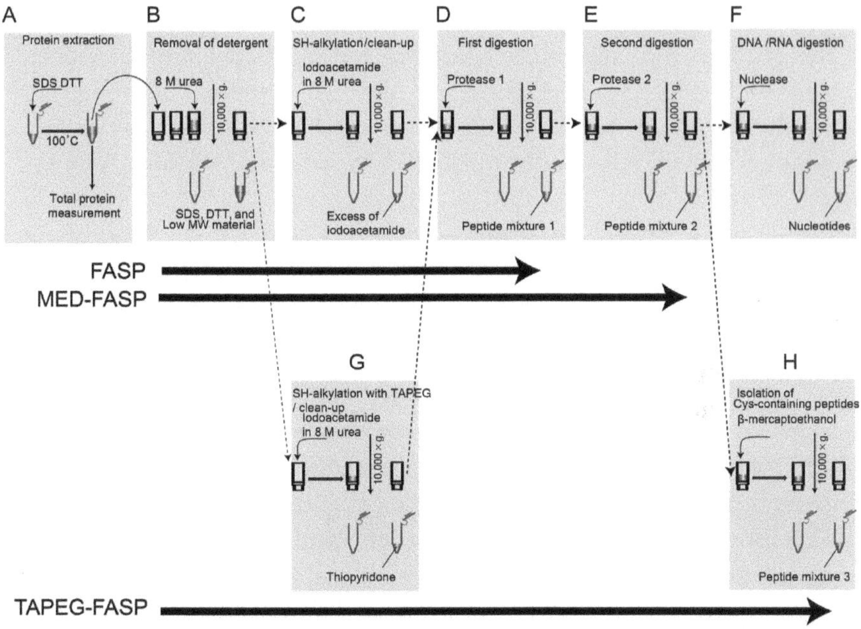

Fig. 1 Workflow of the filter-aided sample preparation (FASP) methods: FASP, multienzyme digestion (MED)-FASP, and thiol-activated polyethylene glycol (TAPEG)-FASP. (A) SDS-based protein extraction. (B) Removal of the detergent. (C) Thiol alkylation and protein clean-up. (D) Protein digestion with the first enzyme and isolation of peptides. (E) Digestion with the second enzyme and peptide elution. (F) Optional digestion of nucleic acids for quantitation of DNA and RNA. (G and H) Steps specific to the TAPEG-FASP procedure. (G) Thiol alkylation with PEG-dithio-2′pirydine for homogeneous phase immobilization of cysteine-containing peptides. (H) Disulfide bond reduction and isolation of cysteine-containing peptides.

detergent micelles and enables dissociation of protein–detergents complexes (Wisniewski, Zielinska, & Mann, 2011). In addition urea keeps proteins largely unfolded, which allows their derivatization and inhibits passage of small polypeptides through the filter pores (see also later). Denatured and at cysteinyl residues reduced proteins are treated with iodoacetamide or, in the TAPEG-FASP procedure, with PEG-dithio-2′pirydine (Fig. 1C and G). After extensive washes and buffer exchange proteins are cleaved with one (in FASP), two, or more (in MED-FASP) proteinases (Fig. 1D and E) and the resulting peptides are filtered out. In the TAPEG-FASP method the on-filter immobilized cysteine-containing peptides are released by disulfide bridge reduction (Fig. 1H). TAPEG-FASP significantly increases the number of identified peptides and proteins by 10–30%. The on-filter remaining material can be further processed. For example, cleavage of the residual with unspecific nucleases can be used for quantitation of nucleic acids (Fig. 1F). Determination of DNA content can be useful for calculation of total protein per cell and assessing protein copy numbers (Wisniewski, Hein, Cox, & Mann, 2014; Wisniewski & Rakus, 2014).

2.2 Selection of the Ultrafiltration Device

Selection of the appropriate ultrafiltration device is the prerequisite for efficient sample processing. The ultrafiltration units used in FASP had been developed for the purpose of protein concentration. Respectively to porousness of the membrane they are classified by cutoff values for proteins of different sizes. Since these cutoff values were assessed for tightly folded proteins they do not reflect retention of unstructured or denatured proteins, which are sterically expanded. For this reason, even small proteins such as core histones or insulin can be concentrated using ultrafiltration membranes with nominal cutoffs of 30 or 50 kDa (Wisniewski et al., 2011 #27). Using of ultrafiltration units with cutoffs $\leq$10 kDa is impracticable because of an increased retention of larger peptides and slower filtration flow increasing several-fold the sample preparation time. Typically the devices with horizontally arranged ultrafiltration membranes, such as "Microcon" units perform better compared to "Amicon Ultra" spin-filters with ultrafiltration membranes arranged at the sides of the units (Lipecka et al., 2016; Wisniewski, 2016). Apparently, this reflects a larger "void volume" of these ultrafiltration devices, which slows detergent depletion. In addition the inner surface of the Amicon Ultra devices is greater compared to "Microcons." Thus, irreversible sample adsorption may result in lower yields of the conversion of proteins to peptides.

2.3 Limitations in Sample Size

Efficient depletion of SDS from samples is essential for high yield conversion of the protein to peptides. Highly diluted protein lysates, containing high amounts of detergent cannot by directly processed by the protocols described in this chapter. The lowest acceptable total protein concentration in a lysate in 2% SDS is about 2 mg/mL (Fig. 2). Processing of lysates with lower total protein concentration requires additional washes with urea-containing buffer. However, this can be time consuming and further optimized.

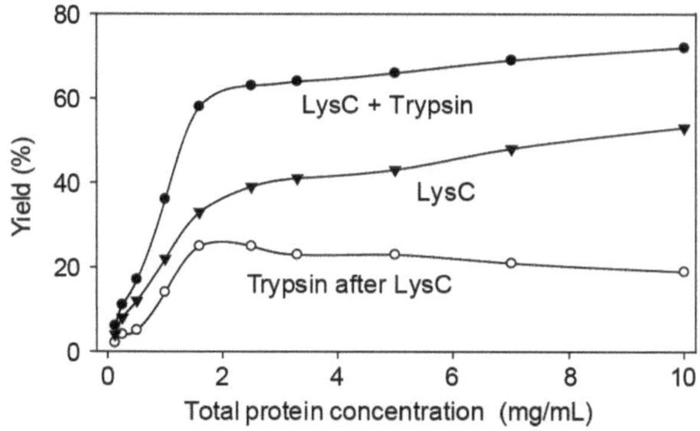

Fig. 2 Sample dilution is a critical parameter. Mouse liver lysates were prepared in the presence of 2% SDS (Wisniewski, 2016). Aliquots containing 50 μg of total protein at sample dilution varying from 0.125 to 10 mg/mL were processed with MED-FASP. The protein and peptide concentrations were determined using the WF-assay (Wisniewski & Gaugaz, 2015).

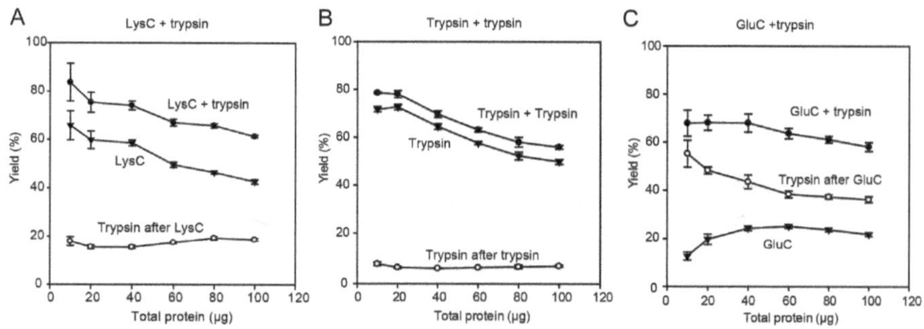

Fig. 3 MED-FASP efficiency of protein to peptide conversion at varying amounts of processed protein. Mouse liver lysates were prepared as described previously (Wisniewski, 2016). MED-FASP was carried according to the herein described protocol. Digestions were carried out using endoproteinase LysC (A), trypsin (B), and endoproteinase GluC (C) in the first cleavage step, and always trypsin for the second digestion. The protein and peptide concentrations were determined using the WF-assay (Wisniewski & Gaugaz, 2015). Each data point shows average of triplicate measurements. Bars are ±standard deviation.

The total protein amount that can be processed by FASP is critical. Using the "Microcon" ultrafiltration units an efficient protein to peptide conversion rate can be achieved within a 10–100 μg range of total protein with on average 60–70% yields (Fig. 3). The yields of peptide vary between enzymes used for the cleavage. Using single enzyme for digestion, trypsin gives slightly higher yields compared to cleavage efficiency of endoproteinase LysC (Fig. 3A and B). In contrast digestion with endoproteinase GluC allows generation of peptides at 20–30% yield (Fig. 3C). The digestion efficiency also varies between sample types. Typically protein extracts of cultured cells are easier in digestion compared to tissue lysates (Wisniewski, 2016 #61).

Since the efficiency of protein digestion depends on the staring amount of total protein and sample concentration, accurate determination of total protein in the sample is important. However, often it is not an easy task. For example, tissue lysates often contain detergents and reducing agents as DTT, which are not compatible with commonly used colorimetric protein assays, such as Bradford and BCA. In this regard the "WF"-assay offers a useful aid. It allows direct protein determination by measurement of tryptophan fluorescence in the presence of reagents used for lysis and can be used in the 96-well format (Wisniewski & Gaugaz, 2015).

Several groups have demonstrated that FASP is particularly useful for processing of very small total protein amounts and has been superior compared to other methods. Good examples are the analyzes of bacteria (Sharma et al., 2012), cultured cells (Maurer et al., 2013), mouse cochlear sensory epithelium (Darville & Sokolowski, 2013), and pancreatic islets (Lipecka et al., 2016; Schmudlach et al., 2016). In addition FASP is a valuable tool for analysis of microdissected formalin fixed and paraffin-embedded tissue (Wisniewski, Ostasiewicz, & Mann, 2011; Wisniewski et al., 2012, 2015) (see also chapter "A protocol for large-scale proteomic analysis of microdissected formalin fixed and paraffin embedded tissue" by Ostasiewicz and Wisniewski).

Samples that are subject of any kind of enrichment, such as phosphorylation or N-glycosylation, require processing of larger amounts of extracts or lysates. This can be carried out using several Microcon devices in parallel or by using larger ultrafiltration units. In large-scale phosphoproteomics studies large Amicon ultra-15 units were successfully applied (Sharma et al., 2014; Wisniewski, Nagaraj, Zougman, Gnad, & Mann, 2010).

2.4 FASP Allows Consecutive Protein Digestion With Different Proteases

In contrast to "in-gel" digestion and many other methods employing harsh digestion conditions, FASP allows protein digestion using variety of proteinases (Wisniewski et al., 2009). Furthermore, FASP enables consecutive digestion using a combination of enzymes. This protocol (Fig. 1) generates populations of different peptides, which in consequence increases the number of identified proteins (Wisniewski & Mann, 2012). MED-FASP leads to more efficient processing of analyzed material compared to FASP (Fig. 3). Higher total peptide yields are accompanied by a reduction of an amount of peptides with missed cleavages (Wisniewski, 2016). Consecutive digestion and peptide collection with endoproteinase LysC and trypsin was used in a number of large-scale proteomic studies including analyzes of human native cells isolated from human colonic mucosa (Wisniewski et al., 2015) and hepatocytes (Wisniewski, Vildhede, Noren, & Artursson, 2016). A recent study on human erythrocytes involves also digestions GluC and ArgC combined with trypsin (Brumirska-Bryk & Wisniewski, in preparation).

2.5 Multipurpose Applications of FASP

The chemical reactor nature of FASP also allows sample processing beyond detergent depletion and protein digestion (Fig. 1). It enables determination of DNA and RNA contents (Fig. 1F), an information that can used for normalization purposes and estimation of cell number in samples (Wisniewski & Rakus, 2014). This procedure can be useful for analysis of material available in limited amount, such as laser-microdissected tissue. A case where parallel quantification nucleic acids and protein digestion by common methods were impracticable. Another unique application of FASP is a reversible derivatization of proteins with polyethylene glycol that can be used for isolation of cysteine-containing peptide fractions (Fig. 1G and H). This strategy has been proven as useful for peptide fractionation (Wisniewski & Prus, 2015).

An important advantage FASP is its compatibility with different detergents (Wisniewski et al., 2011). Several studies demonstrated applicability of FASP to process extracts containing, sodium deoxycholate (Nel, Garnett, Blackburn, & Soares, 2015) and NP40 (Lipecka et al., 2016; Peuchen, Sun, & Dovichi, 2016). Such approaches can be useful when only partial extraction of proteins is intended.

In FASP, such as in the majority of other sample preparation procedures of the bottom-up proteomics protein cleavage the most time-consuming step. Studies have demonstrated that the cleavage time in the filter can be largely reduced by accelerating the enzyme activity using microwave irradiation (Yu et al., 2012).

3. FILTER-AIDED SAMPLE PREPARATION PROTOCOLS

3.1 Preparation of Lysates from Tissues and Cells

1. Homogenize tissue using a blender, Dounce, or Potter-Elvehjem-type device in about 5- to 10-fold excess (volume) of lysis buffer containing

50 m*M* DTT in 100 m*M* Tris–HCl, pH 7.6. To achieve the best results of lysis, it is essential to use enough excess of the lysis buffer over the sample, but preparation of lysates with total protein concentrations below 1 mg/mL should be avoided as too large amounts of the detergent require additional detergent depletion steps in the FASP procedure. Lysis of cultured cells, organelles, or unicellular organisms this homogenization step can be omitted.

2. After addition of SDS to a final concentration of 2% (w/v) sonicate the homogenate sonicated in a Branson type instrument, Sonifier 250 (Heinemann, Schwäbisch Gmünd), (operating at 20% duty cycle and 3–4 output for 1 min). Note that water bath-type sonicators often are inefficient for this purpose.
3. Place the tubes with the homogenate in a bath with boiling water and incubate for 3–5 min. Cool the sample to room temperature.
4. Clarify the lysate by centrifugation at 10,000 × *g* for 5 min.

3.2 Determination of Total Protein in Lysates by WF-Assay Using Fluorescence Microtiter Plate Reader

Determination of total protein content in lysates containing high concentrations of SDS and DTT using Bradford or BCA assay is unreliable. In contrast, tryptophan fluorescence-based WF-assay is an easy method allowing total protein measurement in the presence of high concentrations of detergents and disulfide-reducing reagents (Wisniewski & Gaugaz, 2015).

1. Mix 2 μL of the sample or 1 μL of the tryptophan standard (0.1 mg tryptophan/mL water) with 200 μL assay buffer (8 *M* urea, 10 m*M* Tris–HCl, pH 7.5) per well.
2. Set the excitation wavelength to 295 nm with a 5 nm bandwidth and the emission to 350 nm with a 20 nm bandwidth. Individual measurements should consist of at least 10 reads each with 50 μs integration time.
3. Read the fluorescence and calculate the protein concentration: $C = 8.56$ [sample fluorescence]/[tryptophan fluorescence] (mg/mL).

3.3 MED-FASP Protocol

1. Mix up to 40 μL of protein extract per 200 μL of UA-solution containing 8 *M* urea in 0.1 *M* Tris–HCl, pH 8.5, in the ultrafiltration units and centrifuge at 10,000 × *g* until less than 10 μL of sample remains above the filter. The centrifugation step has to be continued until less than 5% of the initial solution remains above the filter. This usually

requires a centrifugation time of 10–15 min. The time needed to achieve this varies between the filter types, lysate properties, and the total protein concentration.

2. Add 200 μL of UA to the ultrafiltration units and repeat the centrifugation.
3. Discard the flow-through from the collection tube.
4. Add 100 μL 50 m*M* iodoacetamide in UA the concentrated sample and mix at 600 rpm in a thermomixer at room temperature for 1 min.
5. Centrifuge the ultrafiltration units at 10,000 × *g* for 10 min.
6. Add 150 μL of UA to the ultrafiltration units and centrifuge at 10,000 × *g* for 15 min. Repeat this step twice.
7. Add 150 μL of Digestion Buffer (DB) (0.05 *M* Tris–HCl, pH 8.5) to the filtration units and centrifuge at 10,000 × *g* for 10 min. Repeat this step twice.
8. Add 60 μL DB with endoproteinase LysC (enzyme to protein ratio 1:100) and mix at 600 rpm in thermomixer at room temperature for 1 min.
9. Incubate the units in a humid chamber at 37°C overnight.
10. Transfer the ultrafiltration units to new collection tubes.
11. Centrifuge the ultrafiltration units at 10,000 × *g* until the solution completely passed the filter membrane (about 5 min).
12. Add 100 μL of DB and centrifuge the ultrafiltration units at 10,000 × *g* until the solution completely passed the filter membrane (about 5 min). The flow-through contains peptides obtained by LysC digestion. Repeat this step and combine the eluates.
13. Transfer the ultrafiltration units to a new tube.
14. Add 60 μL DB with trypsin (enzyme to protein ratio 1:100) and mix at 600 rpm in thermomixer at room temperature for 1 min.
15. Incubate the units in a humid chamber at 37°C for 4 h.
16. Centrifuge the ultrafiltration units at 10,000 × *g* until the solution entirely passed the filter membrane (about 5 min).
17. Add 100 μL of DB and centrifuge the ultrafiltration units at 10,000 × *g* until the solution entirely passed the filter membrane (about 5 min). The flow-through contains peptides obtained by trypsin digestion. Repeat this step and combine the eluates.
18. Determine concentrations of the LysC and tryptic peptides by A_{280} measurement in a spectrophotometer or by the fluorometric WF-assay (Wisniewski & Gaugaz, 2015) as described in Section 2.2 using DB buffer instead the urea assay buffer. Concentration of the peptides can be estimated by UV spectrometer assuming that 0.1%

solution of vertebrate proteins has at 280 nm an extinction of 1.1 units (1 mg/mL solution has an A_{280} of 1.1). Always record a spectrum from 240 to 340 nm. A distinct peak should be observed at 270–280 nm and the extinction at 320 nm should be 0. Measurements of peptide concentrations are reliable when the absorbance value at A_{280} is above 0.1 in a cuvette with a 10 mm path. Extinction values below 0.1 can be used only for rough protein estimation due to increasing contribution of light scattering. Thus, be aware that measurements carried out using "Nanodrop"-type instruments can be unreliable.

19. Desalt the peptide solutions using C18 material and store frozen at −20°C for at least 1 month or −80°C for several months.

3.4 FASP Protocol

For FASP with a single enzyme, commonly with trypsin, follow steps 1–8 and 13–19 as described in Section 3.3.

3.5 TAPEG-FASP Protocol

3.5.1 Synthesis of TAPEG (Optional)

Preparation of the TAPEG is easy does not require any specific laboratory equipment. However, the reagent can also be purchase from "Creative PEGWorks," Chapel Hill, USA or "RAPP Polymere," Tübingen, Germany.

1. Dissolve 200 mg of CH3O–PEG–SH (RAPP Polymere, Tubingen) in 4 mL of Tris–HCl, pH 7.8.
2. Add 0.25 mL of 50% (v/v) CH_3CN in H_2O containing 10 mg of 2,2′-dithiodipyridine (DTDP; Aldrithiol-2, Sigma-Aldrich) and mix thoroughly.
3. Dialyze overnight the mixture against several changes of water using dialysis membrane with a molecular weight cutoff ≤15,000.
4. Assay the concentration of the TAPEG reagent photometrically by treating with a 100-fold molar excess of 2-mecaptoethanol and measuring the absorbance of the reaction product, pyridone thione, at 343 nm. For this purpose mix 10 μL aliquots of the reagents with 0.2 mL of 0.1 *M* Tris–HCl, pH 7.8, and then add 1 μL of 2-mercaptoethanol. Calculate the concentration of PT using $\varepsilon_{343}=8000$. Reagent solutions can be stored at 4°C for several weeks or frozen at −20°C for more than 1 year without observable deterioration of activity.

3.5.2 Workflow

1. Mix aliquots of lysate containing 100 μg of total protein with 200 μL of UCE buffer containing 8 *M* urea and 1 m*M* EDTA in 0.1 *M* Tris–HCl, pH 7.8.
2. Concentrate the mixture to less than 10 μL in Microcon-30 units using centrifugation at 10,000 × *g*.
3. Add of 100 μL of UCE buffer and concentrate the solution again. Repeat this step twice.
4. Mix the concentrate with 100 μL of UCE containing 0.05 μmol of the TAPEG reagent and incubate for 30 min.
5. Transfer the ultrafiltration unit to a new tube and concentrate the mixture at 10,000 × *g*.
6. Add another 100 μL aliquot of UCE and concentrate the sample again.
7. Measure the concentration of the piridone thione in the combined filtrates (steps 5 and 6) at 343 nm using $\varepsilon_{343} = 8000$.
8. Quench unreacted excess of TAPEG in the retentate by addition of 0.05 μmol of cysteine in 100 μL of UCE for 30 min.
9. Remove excess of cysteine by adding DB (0.05 *M* Tris–HCl, pH 8.5) and filtration at 10,000 × *g*. Repeat this step once.
10. Digest proteins as described in steps 9–18 of the MED-FASP protocol (Section 3.3).
11. Add to the retentate 50 μL of 10 m*M* 2-mercaptoethanol in UCE.
12. Collect the cysteine-containing peptides by centrifugation.
13. Add 150 μL water to the unit and centrifuge again.
14. Desalt the combined eluates from steps 12 and 13 using C18 material.

ACKNOWLEDGMENTS

This work was supported by the Max-Planck Society for the Advancement of Science and by the German Research Foundation (DFG/Gottfried Wilhelm Leibniz Prize).

REFERENCES

Darville, L. N., & Sokolowski, B. H. (2013). In-depth proteomic analysis of mouse cochlear sensory epithelium by mass spectrometry. *Journal of Proteome Research*, *12*(8), 3620–3630. http://dx.doi.org/10.1021/pr4001338.

Lipecka, J., Chhuon, C., Bourderioux, M., Bessard, M. A., van Endert, P., Edelman, A., & Guerrera, I. C. (2016). Sensitivity of mass spectrometry analysis depends on the shape of the filtration unit used for filter aided sample preparation (FASP). *Proteomics*, *16*(13), 1852–1857. http://dx.doi.org/10.1002/pmic.201600103.

Masuda, T., Tomita, M., & Ishihama, Y. (2008). Phase transfer surfactant-aided trypsin digestion for membrane proteome analysis. *Journal of Proteome Research*, 7(2), 731–740. http://dx.doi.org/10.1021/pr700658q.

Maurer, M., Muller, A. C., Wagner, C., Huber, M. L., Rudashevskaya, E. L., Wagner, S. N., & Bennett, K. L. (2013). Combining filter-aided sample preparation and pseudoshotgun technology to profile the proteome of a low number of early passage human melanoma cells. *Journal of Proteome Research*, *12*(2), 1040–1048. http://dx.doi.org/10.1021/pr301009u.

Nel, A. J., Garnett, S., Blackburn, J. M., & Soares, N. C. (2015). Comparative reevaluation of FASP and enhanced FASP methods by LC-MS/MS. *Journal of Proteome Research*, *14*(3), 1637–1642. http://dx.doi.org/10.1021/pr501266c.

Peuchen, E. H., Sun, L., & Dovichi, N. J. (2016). Optimization and comparison of bottom-up proteomic sample preparation for early-stage Xenopus laevis embryos. *Analytical and Bioanalytical Chemistry*, *408*(17), 4743–4749. http://dx.doi.org/10.1007/s00216-016-9564-2.

Rakus, D., Gizak, A., Deshmukh, A., & Wisniewski, J. R. (2015). Absolute quantitative profiling of the key metabolic pathways in slow and fast skeletal muscle. *Journal of Proteome Research*, *14*(3), 1400–1411. http://dx.doi.org/10.1021/pr5010357.

Schmudlach, A., Felton, J., Cipolla, C., Sun, L., Kennedy, R. T., & Dovichi, N. J. (2016). Sample preparation protocol for bottom-up proteomic analysis of the secretome of the islets of Langerhans. *Analyst*, *141*(5), 1700–1706. http://dx.doi.org/10.1039/c5an02265g.

Sharma, R., Dill, B. D., Chourey, K., Shah, M., VerBerkmoes, N. C., & Hettich, R. L. (2012). Coupling a detergent lysis/cleanup methodology with intact protein fractionation for enhanced proteome characterization. *Journal of Proteome Research*, *11*(12), 6008–6018. http://dx.doi.org/10.1021/pr300709k.

Sharma, K., D'Souza, R. C., Tyanova, S., Schaab, C., Wisniewski, J. R., Cox, J., & Mann, M. (2014). Ultradeep human phosphoproteome reveals a distinct regulatory nature of Tyr and Ser/Thr-based signaling. *Cell Reports*, *8*(5), 1583–1594. http://dx.doi.org/10.1016/j.celrep.2014.07.036.

Wisniewski, J. R. (2016). Quantitative evaluation of filter aided sample preparation (FASP) and multienzyme digestion FASP protocols. *Analytical Chemistry*, *88*(10), 5438–5443. http://dx.doi.org/10.1021/acs.analchem.6b00859.

Wisniewski, J. R., Dus-Szachniewicz, K., Ostasiewicz, P., Ziolkowski, P., Rakus, D., & Mann, M. (2015). Absolute proteome analysis of colorectal mucosa, adenoma, and cancer reveals drastic changes in fatty acid metabolism and plasma membrane transporters. *Journal of Proteome Research*, *14*(9), 4005–4018. http://dx.doi.org/10.1021/acs.jproteome.5b00523.

Wisniewski, J. R., & Gaugaz, F. Z. (2015). Fast and sensitive total protein and peptide assays for proteomic analysis. *Analytical Chemistry*, *87*(8), 4110–4116. http://dx.doi.org/10.1021/ac504689z.

Wisniewski, J. R., Hein, M. Y., Cox, J., & Mann, M. (2014). A "proteomic ruler" for protein copy number and concentration estimation without spike-in standards. *Molecular & Cellular Proteomics*, *13*(12), 3497–3506. http://dx.doi.org/10.1074/mcp.M113.037309.

Wisniewski, J. R., & Mann, M. (2012). Consecutive proteolytic digestion in an enzyme reactor increases depth of proteomic and phosphoproteomic analysis. *Analytical Chemistry*, *84*(6), 2631–2637. http://dx.doi.org/10.1021/ac300006b.

Wisniewski, J. R., Nagaraj, N., Zougman, A., Gnad, F., & Mann, M. (2010). Brain phosphoproteome obtained by a FASP-based method reveals plasma membrane protein topology. *Journal of Proteome Research*, *9*(6), 3280–3289. http://dx.doi.org/10.1021/pr1002214.

Wisniewski, J. R., Ostasiewicz, P., Dus, K., Zielinska, D. F., Gnad, F., & Mann, M. (2012). Extensive quantitative remodeling of the proteome between normal colon tissue and adenocarcinoma. *Molecular Systems Biology*, *8*, 611. http://dx.doi.org/10.1038/msb.2012.44.

Wisniewski, J. R., Ostasiewicz, P., & Mann, M. (2011). High recovery FASP applied to the proteomic analysis of microdissected formalin fixed paraffin embedded cancer tissues retrieves known colon cancer markers. *Journal of Proteome Research*, *10*(7), 3040–3049. http://dx.doi.org/10.1021/pr200019m.
Wisniewski, J. R., & Prus, G. (2015). Homogenous phase enrichment of cysteine-containing peptides for improved proteome coverage. *Analytical Chemistry*, *87*(13), 6861–6867. http://dx.doi.org/10.1021/acs.analchem.5b01215.
Wisniewski, J. R., & Rakus, D. (2014). Multi-enzyme digestion FASP and the 'Total Protein Approach'-based absolute quantification of the *Escherichia coli* proteome. *Journal of Proteomics*, *109*, 322–331. http://dx.doi.org/10.1016/j.jprot.2014.07.012.
Wisniewski, J. R., Vildhede, A., Noren, A., & Artursson, P. (2016). In-depth quantitative analysis and comparison of the human hepatocyte and hepatoma cell line HepG2 proteomes. *Journal of Proteomics*, *136*, 234–247. http://dx.doi.org/10.1016/j.jprot.2016.01.016.
Wisniewski, J. R., Zielinska, D. F., & Mann, M. (2011). Comparison of ultrafiltration units for proteomic and N-glycoproteomic analysis by the filter-aided sample preparation method. *Analytical Biochemistry*, *410*(2), 307–309. http://dx.doi.org/10.1016/j.ab.2010.12.004.
Wisniewski, J. R., Zougman, A., Nagaraj, N., & Mann, M. (2009). Universal sample preparation method for proteome analysis. *Nature Methods*, *6*(5), 359–362. http://dx.doi.org/10.1038/nmeth.1322.
Yu, Y., Xie, L., Gunawardena, H. P., Khatun, J., Maier, C., Spitzer, W., … Chen, X. (2012). GOFAST: An integrated approach for efficient and comprehensive membrane proteome analysis. *Analytical Chemistry*, *84*(21), 9008–9014. http://dx.doi.org/10.1021/ac300134e.

CHAPTER THREE

An Overview of Advanced SILAC-Labeling Strategies for Quantitative Proteomics

F. Terzi, S. Cambridge[1]
Institute of Cell Biology and Anatomy, University of Heidelberg, Heidelberg, Germany
[1]Corresponding author: e-mail address: cambridge@ana.uni-heidelberg.de

Contents

Abstract

Comparative, quantitative mass spectrometry of proteins provides great insight to protein abundance and function, but some molecular characteristics related to protein dynamics are not so easily obtained. Because the metabolic incorporation of stable amino acid isotopes allows the extraction of distinct temporal and spatial aspects of protein dynamics, the SILAC methodology is uniquely suited to be adapted for advanced labeling strategies. New SILAC strategies have emerged that allow deeper foraging into the complexity of cellular proteomes. Here, we review a few advanced SILAC-labeling strategies that have been published during last the years. Among them, different subsaturating-labeling as well as dual-labeling schemes are most prominent for a range of analyses including those of neuronal proteomes,

Methods in Enzymology, Volume 585
ISSN 0076-6879
http://dx.doi.org/10.1016/bs.mie.2016.09.014

secretion, or cell–cell-induced stimulations. These recent developments suggest that much more information can be gained from proteomic analyses if the labeling strategies are specifically tailored toward the experimental design.

ABBREVIATIONS

Arg0/R0 standard unlabeled arginine
Arg4/R4 $^{13}C4$
Arg6/R6 $^{13}C6$
Arg10/R10 $^{13}C6$ $^{15}N4$
Arg17/R17 $^{13}C4$ $^{15}N4$ $^{2}H7$
Lys0/K0 standard unlabeled lysine
Lys4/K4 D4
Lys8/K8 $^{13}C6$ $^{15}N2$
Tyr0/Y0 standard unlabeled tyrosine
Tyr6/Y6 $^{13}C6$
Tyr10/Y10 $^{13}C9$ $^{15}N1$

1. INTRODUCTION

The field of mass spectrometry (MS)-based proteomics is advancing at an amazing pace. We are approaching the detection and identification of all proteins from individual proteomes because of recent developments in the sensitivity and dynamic range of mass spectrometers (Mann, Kulak, Nagaraj, & Cox, 2013). Consequently, the description of which proteins are present in cell lysates has been very successful for many different cell types. However, in analogy to the human genome project, the mere illustration of all genes (or proteins in this case) is certainly a great accomplishment, but there are additional and more important layers of biological complexity which are now under intense experimental investigation. Understanding which proteins are present in a certain type of cell is one basic data set, but of course, the function of each protein in time and space as it is interacting with other proteins holds more relevant information. Clearly, this kind of information is not readily extracted from MS data, and thus researchers have developed more sophisticated experimental designs including enriching, stimulating, or specifically labeling subsets of proteins. The latter is the topic of this review as several new labeling strategies have been published that allowed better characterization of cellular proteomes. In particular, the focus will be on metabolic SILAC (stable isotope labeling in cell culture) labeling, which has enjoyed considerable success (Mann, 2006; Ong et al., 2002) and which has been modified often from its originally published protocol. Compared to other popular-labeling strategies such as iTRAQ (isobaric tags for relative and absolute quantitation) (Evans et al., 2012) or TMT (tandem mass tags) (Thompson et al., 2003), the metabolic incorporation of labeled amino acids during normal protein synthesis provides a direct handle to fundamental cellular processes and proteomes. In standard SILAC-labeling protocols, stable isotopes of lysine and arginine are typically used that have a defined increase in molecular weight compared to the "normal," unmodified amino acids. For example, the weight of the stable isotope Arg10 is increased by +10 Da. This is achieved by substituting six ^{12}C carbon atoms with ^{13}C and four ^{14}N nitrogen atoms with ^{15}N to yield $^{13}C6$ $^{15}N4$ arginine. The exact substitution patterns of all the other isotopes mentioned in this chapter are listed in the abbreviation list.

Traditionally, metabolic labeling was performed with unstable, radioactive (e.g., with ^{14}C or ^{3}H) amino acids beginning in the 1960s and 1970s with important contributions from Garlick and Waterlow (1969) and Dice and Goldberg (1975a, 1975b), just to name a few. Subsaturating, pulse labeling of animals (Dice & Goldberg, 1975a, 1975b) and humans (Fern, Garlick, McNurlan, & Waterlow, 1981), was used to investigate protein turnover in many different tissues and organisms. Because proteins with high turnover incorporate more exogenous label in a certain time window compared to stable proteins, the ratio of labeled/unlabeled protein is an indicator of protein turnover. Of course, because of technical limitations, these early analyses characterized only a few proteins or biochemically purified protein fractions. Nevertheless, even almost half a century ago, researchers attempted to devise advanced labeling schemes to extract more relevant data from their experiments. For example, Arias and coworkers administered first ^{14}C and then ^{3}H isotopes of the same amino acid at two different time points and used the ratio of $^{3}H/^{14}C$ label to identify protein fractions with high turnover (Arias, Doyle, & Schimke, 1969). The second ^{3}H time point was only a few hours before cell lysis, while the ^{14}C administration occurred a few days earlier. A high $^{3}H/^{14}C$ ratio therefore indicated higher turnover and, more importantly, was independent of absolute protein amounts in the scintillation counter. In addition, the two labels provided two independent time points for calculating the decay time curve. Thus, the research community today can build upon a long history of previously established sophisticated metabolic-labeling schemes that used unstable isotopes of amino acids. With the introduction of the SILAC technology and modern MS, these early studies can eventually be extended to examining the turnover of every single protein within a proteome. Moreover, in this decade, several publications have reported advanced SILAC-labeling strategies that allow better characterization of additional protein dynamics. We here present an overview of these efforts, some from different laboratories and some reported by us, with the hope that more researchers make use of these simple adjustments for improved data collection.

2. DUAL LABELING TO EXCLUDE UNLABELED PROTEINS

Triple labeling with light, medium, and heavy stable isotopes of lysine and arginine is routinely used in MS SILAC experiments where three conditions are to be compared (Olsen et al., 2006). For two conditions, researchers traditionally have been using light and heavy isotopes to have the largest possible weight difference which in turn will improve the bioinformatics processing of the data. However, with the increased sensitivity of today's MS instruments (Michalski et al., 2011, 2012), having a large weight difference is not as crucial anymore. Thus, one can use medium and heavy stable isotopes for labeling two different conditions. The advantage that emerges from such a protocol is that unlabeled proteins can be readily excluded in silico with the appropriate settings in analysis softwares such as MaxQuant (Cox & Mann, 2008). But which MS experiments could benefit from exclusion of unlabeled proteins and limiting the analysis to labeled proteins? There are two main biological research areas where such an approach is uniquely pertinent but further applications will also be discussed.

2.1 Cell–Cell Contact Analyses

First, exclusion of unlabeled proteins is vital for quantitative analyses of cell–cell contact-dependent stimulation of protein expression. Of course, there are almost unlimited numbers of different cell–cell contacts existing and it is safe to assume that each particular cell–cell contact leads to a specific change in proteome composition, for instance because of changes in gene expression. A typical example would be the investigation of proteome changes that occur when cancer cells come in contact with endothelial cells of the vasculature. Clearly, there will be changes in either cell type. By prior labeling one of the two cell types (e.g., the cancer cells) once with medium and once with heavy stable isotopes, the medium-labeled cancer cells could be stimulated with the unlabeled endothelial cells, while the heavy-labeled cancer cells serve as unstimulated controls (or vice versa) (Fig. 1). We have used this approach to analyze the response of labeled primary astrocytes from the

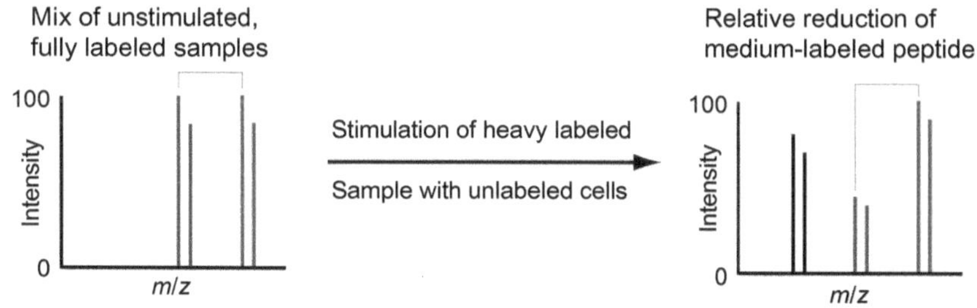

Fig. 1 Scheme for SILAC-based MS analysis of cell–cell-stimulated proteomes. Cells that will be stimulated and characterized are labeled in parallel with medium (*blue*) or heavy (*red*) stable isotopes. Without stimulation, the peptide peaks would be identical (*left*). Stimulating the heavy-labeled sample by addition of unlabeled cells from a different cell type then modifies the abundance of proteins relative within both labeled samples. Proteins from the unlabeled cells used for stimulation (*black*) do not affect MS quantification.

forebrain (cortex/hippocampus) to unlabeled neurons from either the forebrain or the hindbrain (cerebellum) (Stiess et al., 2015). To our surprise, we found that the astrocytes "remembered" their tissue origin after 2 weeks in culture as their proteomic response was different for hindbrain vs forebrain neurons. Forebrain neurons, i.e., from the same origin as the astrocytes, elicited a moderate response. However, the forebrain astrocytes mounted a massive repulsive protein response toward newly added cerebellar neurons, a cell type that these astrocytes normally do not encounter. For example, soluble CD44, which was reported to be cytotoxic to retinal ganglion cells (Choi et al., 2005), remained unchanged after addition of forebrain neurons, but increased 80-fold after addition of hindbrain neurons. In addition, these protein alterations were also compared to unstimulated astrocytes. Overall, we found three distinct proteomic responses for the three different conditions which in turn allowed extraction of the different types of protein contributions to astrocyte-related tissue homeostasis.

While such dual-labeling experiments are straightforward in cell culture, it is very well conceivable to conduct similar experiments in vivo. For example, injection of labeled cells such as iPS (induced pluripotent stem) cells into unlabeled animals. Of course, there is tremendous medical interest in identifying the key cellular and genetic processes that govern the integration of stem cells into healthy or pathological tissues in vivo (Angelos & Kaufman, 2015). With such an approach, different stem cells or different iPS cell conditions prior to injection can be directly compared in the same animal. The advantage is of course that a mixture of iPS cells, half-labeled with medium, the other half-labeled with heavy SILAC isotopes, can be injected at once, thereby eliminating potential differences from animal-to-animal variation. The proteomic responses of the iPS cells can then be readily detected after isolation and MS processing of labeled and unlabeled cells. Subsequent proteomic quantification should thus reveal how different iPS cells integrate into tissues.

2.2 Secretion Analyses

The second biological research area where one would want to exclude unlabeled proteins is in secretion analyses. For quantitative analyses of secretion, it is essential to be able to exclude a whole range of sometimes highly abundant serum proteins, many of whom are usually secreted proteins that originate from cells in the serum donor organism. Secretion studies are typically conducted in cell culture by simply collecting and examining the supernatant. Thus, it is important that serum proteins in the supernatant do not skew the data. Some cells tolerate zero or low concentrations of serum during culturing for at least a short period of time. Thus, in principle, one could grow cells in normal serum-containing medium, wash cells, and conduct the secretion studies in serum-free medium. However, we found that it is almost impossible to collect supernatants free of serum proteins even after washing cells in serum-free medium. Notably, despite washing the cells twice in serum-free medium, we detected substantial amounts of serum-derived, unlabeled secreted proteins in the MS spectra (Stiess et al., 2015). Some unlabeled serum protein peaks exceeded the peaks of the labeled peptides which would have substantially altered the quantitative analysis, for example, for Biglycan-1 (Fig. 2). Therefore, to completely eliminate serum contributions to the signals, we saturatingly labeled astrocytes with medium and heavy stable isotopes prior to collecting supernatants, so that unlabeled proteins must unequivocally have derived from serum. By restricting bioinformatics analyses with MaxQuant only to labeled peptide peaks, any peptides from unlabeled proteins did not contribute to the data. As a step further, supernatants from medium-labeled astrocytes were mixed with lysates from heavy-labeled astrocytes (and vice versa). This allowed us to quantify the percentage of extracellular vs total protein abundance for each constitutively secreted protein. The notion has existed in the secretion research community that truly secreted proteins exhibit values of roughly at least 1% of the protein being extracellular. Analysis of more than 2000 proteins strongly supported this approximation and provided the first

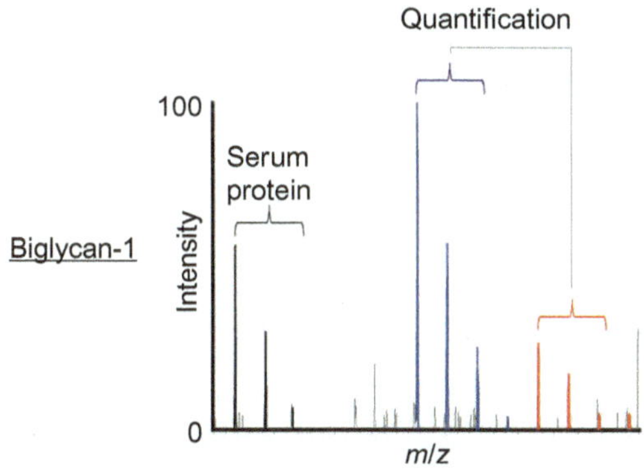

Fig. 2 Prior SILAC labeling of cells with medium or heavy isotopes allows accurate quantification of secreted proteins in supernatants in the presence of serum proteins. Raw spectrum of a peptide derived from Biglycan-1, which is secreted from astrocytes but is also present in serum. (Unlabeled) *black peaks* are derived from serum, *blue* from "medium," and *red peaks* from the "heavy"-labeled supernatants. Substantial amounts of serum-derived, unlabeled Biglycan-1 were detected despite washing the cells twice in serum-free medium. *Reprinted with permission from Stiess, M., Wegehingel, S., Nguyen, C., Nickel, W., Bradke, F., & Cambridge, S. B. (2015). A dual silac proteomic labeling strategy for quantifying constitutive and cell-cell induced protein secretion.* Journal of Proteome Research, 14, *3229–3238. Copyright 2015 American Chemical Society.*

large-scale data set of the relative extra- to intracellular protein abundancies for any cell type.

2.3 Analyses of Postmitotic Neurons

Besides secretion studies or research of cell–cell-induced protein changes, there are other applications of quantitative proteomics where this simple dual-labeling strategy will be beneficial, for example, in cells that cannot be fully labeled. In general, SILAC labeling in culture is considered to be near 100% after five cell divisions (Ong & Mann, 2006). While the cell divisions probably do not directly enhance incorporation of stable isotope amino acids, the protein turnover of rapidly dividing HeLa cells is roughly twice as high compared to arrest HeLa cells (Cambridge et al., 2011). Consequently, the exchange of unlabeled amino acids with labeled ones should occur about twice as fast. For nondividing, arrested or differentiated postmitotic cells such as neurons, achieving complete labeling is therefore more difficult, especially since (primary) differentiated cells can typically not be maintained in culture longer than a few weeks. In line of this, Neubert and colleagues demonstrated that after 10 days in culture, less than half of all proteins of primary neurons were fully labeled (Zhang, Deinhardt, & Neubert, 2014).

Therefore, to circumvent a possible scenario where a 100% unlabeled/light sample is compared with, e.g., a 80% heavy-labeled sample, researchers adopted the dual-labeling strategy in such experiments as well. Separate neuronal cultures were treated in parallel with either medium or heavy stable isotopes, and quantitative proteomic analyses were limited to labeled peptide peaks. Because the parallel processing should in principle lead to identical-labeling efficiencies, changes between the two labels should be a true reflection of the postlabeling experimental treatment. Neubert and colleagues used such an approach to study NT-3 (neurotrophin 3) signaling in dissociated primary cortical neurons and confirmed the known upregulation of many proteins of the NT-3 signaling cascade, such as the NT-3 receptor TrkC (Zhang et al., 2014). Thus, these types of dual-labeling strategies are very good for cells in culture that cannot be labeled quantitatively.

2.4 Use of Nondialyzed Serum

Such a subsaturating, dual-labeling strategy was also adopted by Ishihama and colleagues for their proteome and phosphoproteome analyses of different cell types, including HeLa cells and the hormone-sensitive breast cancer MCF-7 cells (Imami, Sugiyama, Tomita, & Ishihama, 2010). The authors found that the type of serum used for culturing these cells had a significant impact on the proteome and phosphoproteome data. SILAC labeling is commonly achieved by using dialyzed serum with a cutoff at 10,000 Da. The dialysis is meant to remove unlabeled amino acids, i.e., lysine and arginine, which are present in the serum, so that full labeling can be achieved with the added stable isotope versions of lysine and arginine. Not surprisingly, dialysis also removes other small molecules in the serum, including hormones such as estrogens, which in turn can lead to a reduction in growth rates (Gehrmann, Hathout, & Fenselau, 2004). Thus, while some cell types may not even grow in medium containing dialyzed serum, those cells that do potentially have altered proteome and phosphoproteomes compared to medium with standard serum. As a simple workaround to this problem, MCF-7 cells were grown for 6 days in medium with standard, nondialyzed serum containing either heavy or medium isotope-labeled amino acids. The authors found that only 10% or less of the peptides contained unlabeled arginines or lysines. This in turn suggests that dialysis is indeed necessary and sufficient to remove unwanted amino acids from serum formulations. Nevertheless, parallel labeling of cells in nondialyzed serum with either medium or heavy amino acids followed by in silico analyses of stable isotope peaks allowed accurate quantification under these conditions. Although the samples were not 100% labeled in nondialyzed serum, the comparison between heavy and medium peptide peaks permitted quantitative proteomics (Fig. 3). The authors used their approach to highlight differences in protein phosphorylation after application of lapatinib, an epidermal growth factor

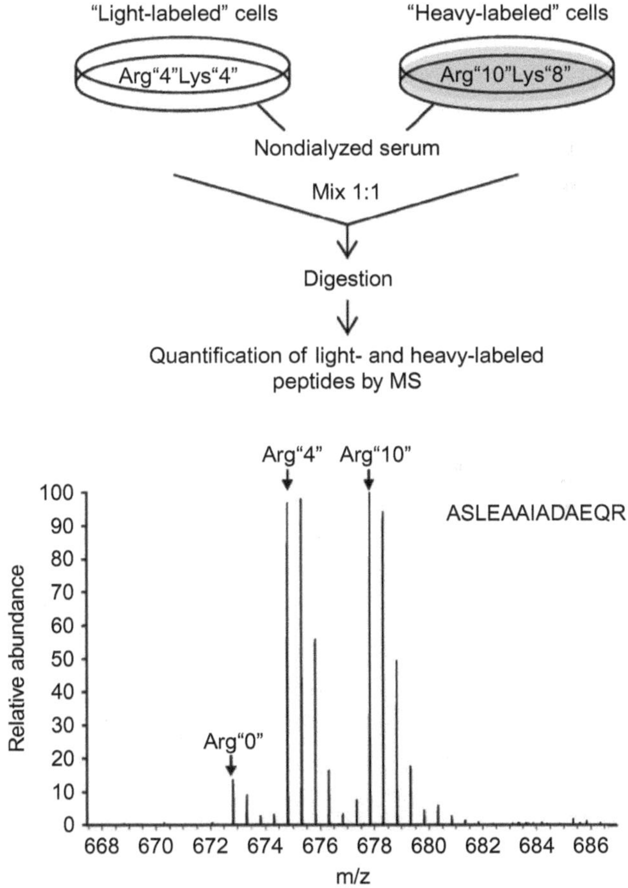

Fig. 3 SILAC with nondialyzed serum workflow. Cells are labeled in parallel with medium and heavy stable isotope amino acids using nondialyzed serum. The labeling efficiency cannot be 100% as there are unlabeled amino acids (Arg0) in the nondialyzed serum. Subsequent separate treatment of medium- and heavy-labeled samples permits quantitative proteomics using conventional SILAC procedures. *Reprinted with permission from Imami, K., Sugiyama, N., Tomita, M., & Ishihama, Y. (2010). Quantitative proteome and phosphoproteome analyses of cultured cells based on silac labeling without requirement of serum dialysis.* Molecular BioSystems, 6, *594–602. Copyright 2010 Royal Society of Chemistry.*

receptor 1 (EGFR1)/Her2 dual kinase inhibitor, and either dialyzed or nondialyzed serum (Imami et al., 2010). Of note, ERK1 phosphorylation was unchanged in nondialyzed serum despite the presence of lapatinib, while the same concentration of this kinase inhibitor in dialyzed serum reduced ERK1 phosphorylation to about 1/3. Conversely, the levels of phosphorylation of the membrane-associated progesterone receptor 2 (PGRMC2) were unaffected in dialyzed serum + lapatinib, while phosphorylation more than doubled at two separate sites in nondialyzed serum + lapatinib. Although it is no surprise that different culturing conditions produce different phosphoproteomes, these experiments should serve as a stark reminder to the MS community to consider the effects of dialyzed medium on their results, especially for drug-screening efforts. The dual-labeling SILAC strategy from Ishihama and colleagues therefore constitutes a simple solution to this dilemma.

3. SUBSATURATING LABELING

3.1 Increasing Measurement Accuracy for Medium to Slow Turnover Proteins

Pulse or pulse-chase experiments are very common in biological research. The idea is of course that the (metabolic) incorporation of exogenously supplied molecules can be controlled in time and space to allow high-resolution analyses of the cellular pathways under investigation. Such an approach has been extensively used in DNA and RNA research, for example, to label dividing cells with BrdU, and in protein research for characterization of protein turnover and dynamics. Protein turnover studies date back several decades when researchers first used unstable, radioactive isotopes of amino acids, and also then, the focus almost always was on exposing a completely unlabeled sample to 100% labeled amino acids or vice versa. Especially for short-labeling periods, such an approach undoubtedly preferentially identifies high turnover proteins, because these would incorporate the most labels in a particular time window. Another, often underappreciated drawback of these protocols is the fact that the accuracy of MS measurements can be impaired if the two SILAC peptide peaks (with and without label) are substantially different, for example, by two or more magnitudes. The difficulty of accurately quantifying the smaller, minute peptide peak then has profound consequences on the ratio computation and thus the extracted protein turnover. Imagine two peptide peaks, one with a peak height of 150 units and one with a height of 3 units, which were actually measured to be 148 and 2, respectively. The ratio would dramatically change from a true ratio value of 50 (150/3) to an experimentally determined ratio value of 74 (148/2), although the absolute measurement errors of 1 and 2 units would be considerably minute. Consequently, such skewed ratio values could lead to dramatically false values in the corresponding protein turnover. This is particularly a problem for proteins with a slow to intermediate turnover, which constitutes the majority in mammalian proteome (Cambridge et al., 2011). The reason is that after switching to a new isotope label in a pulse experiment, proteins with a slow turnover will inevitably have a very small peak because not much label could be incorporated within the given time window.

To overcome this problem, we devised a new labeling strategy by exposing cells to a 1:1 mixture of light and heavy labels for five cell divisions so that the ratio of the heavy and light peaks for every protein at time zero was 1 (Cambridge et al., 2011). A subsequent shift to 100% heavy label in a pulse experiment then allowed measuring slow turnover proteins with high accuracy, because both peptide peaks were almost of equal intensity (α-crystallin, Fig. 4). Overall, more than 99% of all proteins had a heavy vs light ratio of

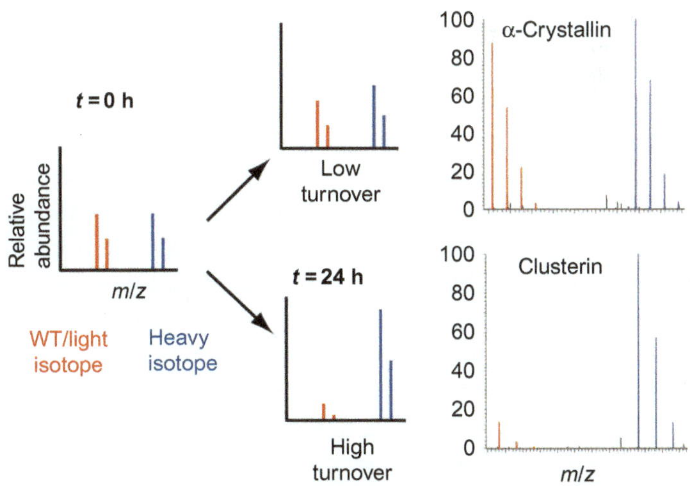

Fig. 4 Experimental SILAC scheme for improved analysis of proteins with slow or medium turnover. Prelabeled cells incubated with a 1:1 mix of unlabeled/light and heavy SILAC amino acids produce MS peptide peaks of equal intensity. Shifting these cells to 100% labeled medium leads to small or large peak differences for low (α-crystallin) or high (clusterin) turnover proteins, respectively. *Reprinted with permission from Cambridge, S. B., Gnad, F., Nguyen, C., Bermejo, J. L., Kruger, M., & Mann, M. (2011). Systems-wide proteomic analysis in mammalian cells reveals conserved, functional protein turnover.* Journal of Proteome Research, 10, *5275–5284. Copyright 2011 American Chemical Society.*

less than 10-fold, indicating that high turnover proteins were clearly a minority. Of course, just like conventional SILAC pulse experiments are suboptimal for slow turnover proteins, our strategy leads to less accurately measured ratios for high turnover proteins. Ideally, one should measure high and slow turnover proteins in separate, complementary experiments to ensure reliable, high-quality data.

We conducted these experiments with arrested HeLa cells as well as differentiated C2C12 mouse muscle cells to characterize for the first time proteome dynamics in nondividing cells. Surprisingly, both proteomes were quite dynamic with overall half-lives of 35.5 h (HeLa) or 43.2 h (C2C12). For comparison, the protein half-lives in actively dividing HeLa cells were reported to be about ~20 h (Boisvert et al., 2012). Since most cells in adult mammals are arrested in G0 phase, analyses of nondividing cells provided very pertinent information to this quiescent cell status and additionally revealed a high correlation of individual protein turnover values across both species.

3.2 Quadruple Subsaturating Labeling to Separately Measure Decay and Synthesis After Cellular Stimulation

Protein turnover is determined by the rates of synthesis and degradation, and cells can therefore govern protein abundance by controlling these opposing cellular processes. At hypothetical steady state, i.e., in the absence of external stimuli and after completion of the developmental program, protein synthesis and degradation are at equilibrium to maintain the status quo of the proteome. In a living organism with abundant cell–cell contacts and the cells being constantly exposed to a plethora of stimuli including hormones, inflammation, or changes in energy status, it is safe to assume that cells are rarely at steady state. Consequently, there is a need for strategies that can analyze protein turnover in cells that are not at steady state, while also separating the contributions of protein synthesis and decay to changes in protein abundance. To achieve this, the Quadroni laboratory has devised a complex, yet elegant strategy that allows extracting synthesis and decay rates of control and stimulated cells in a single experiment (Fierro-Monti et al., 2013). This strategy was designated pulse-chase SILAC and was based on using stable isotopes of arginine for pulse labeling, and stable isotopes of lysine for chase labeling on the same cells (Fig. 5). In this case, pulse labeling actually meant complete labeling of cells before switching to chase medium. More specifically, one sample of Jurkat cells

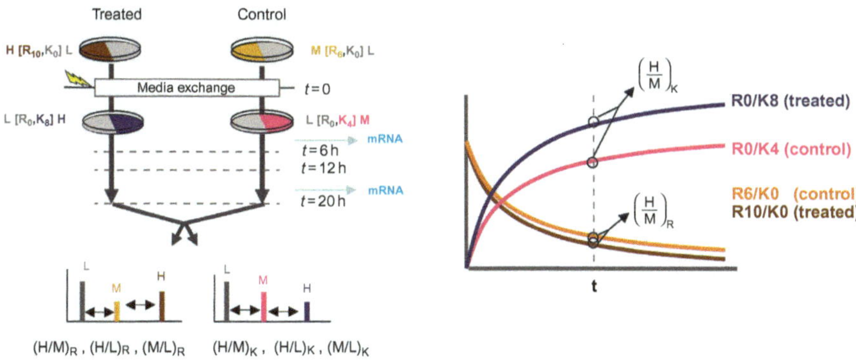

Fig. 5 *Left*: Protein-labeling scheme. Two-cell cultures were fully labeled only on Arg residues before cell treatments. For example, cells to be treated later with the HSP90 inhibitor were fully labeled with Arg10/Lys0 ("heavy" cells), while cells to be used as control were fully labeled with Arg6/Lys0 ("medium" cells). At the start of the experiment, "heavy" cells were transferred into a medium containing light Arg0 and "heavy" Lys8 (R0/K8 medium), while "medium" cells were transferred to a medium containing light Arg0 and "medium" Lys4 (R0/K4 medium). *Right*: Conceptual view of the levels of a hypothetical protein in a mixture of two (control and treated) SILAC-labeled samples. Preexisting protein is fully labeled R6/K0 and R10/K0 (*light* and *dark brown*), respectively, for the control and treated sample. Newly synthesized protein is labeled R0/K4 (control, *pink*) and R0/K8 (treated, *blue*). The SILAC (H/M) ratios for R- and K-containing peptides therefore measure the ratios of preexisting and newly synthesized proteins at time *t*. *Reprinted with permission from Fierro-Monti, I., Racle, J., Hernandez, C., Waridel, P., Hatzimanikatis, V., & Quadroni, M. (2013). A novel pulse-chase silac strategy measures changes in protein decay and synthesis rates induced by perturbation of proteostasis with an hsp90 inhibitor.* PloS One, 8, *e80423. Copyright 2013 Public Library of Science.*

was first pulsed with heavy arginine and then chased with heavy lysine, while the other Jurkat cells were pulsed with medium arginine and chased with medium lysine. One of the samples served as control, while the other was treated with the HSP90 inhibitor geldanamycin. As for the experiment, fully arginine-labeled cells (heavy or medium) were then switched to chase medium with labeled lysines (heavy or medium). At the time of the switching, geldanamycin was applied to one of cell samples for treatment. The decay of pulse arginine label in peptides of treated and control cells was used to compute protein degradation as the chase medium contained only unlabeled arginine (and labeled lysine). Conversely, the increase in peptides with labeled lysines was used to compute protein synthesis as the earlier pulse labeling was only with labeled arginine (in addition to unlabeled lysine). While the decrease of arginine label was a direct function of protein degradation, the increase in lysine label reflected the combined action of protein synthesis and protein degradation. With sophisticated mathematical computation, the authors were then able to extract a variety of different parameters and values for individual proteins, protein families, and the global proteome (Fierro-Monti et al., 2013). In short, Quadroni and colleagues could show that drug-mediated inhibition of the HSP90 chaperone induced a global decrease in protein synthesis but a specific increase in expression of stress response proteins. Likewise, they observed a global increase in decay rates and a more specific degradation of HSP90 "client proteins." These and other interesting findings suggest that this pulse-chase SILAC-labeling strategy should be adopted when attempting to separate protein synthesis from protein decay. The authors also propose to use this strategy for investigating the correlation between ubiquitination and protein degradation. The dependence of protein degradation on the level of ubiquitination has been an ongoing question in the field, and it seems that this approach is uniquely suited to decouple the ubiquitin contribution to proteolysis. Immunoenrichment of ubiquitylated proteins (chase) while simultaneously monitoring protein decay (pulse) should provide straightforward protein-specific data about the exact involvement of ubiquitin in degradation.

3.3 Determining the Contribution of Extra- vs Intracellular Amino Acids to Protein Synthesis

Pulse(-chase) experiments with labeled amino acids are used to analyze incorporation of these amino acids into newly synthesized proteins. The general assumption in many of these experiments is that the contribution of recycled, internal amino acids from recently degraded proteins is negligible compared to the large pool of labeled external amino acids. However, it could very well be that hypothetically, recycled amino acids are very effectively channeled back into the protein synthesis machinery and that external amino acids only play a minor role. If this were the case, then the dynamics of labeled amino acid incorporation from the external medium would not adhere to first-order kinetics as researchers generally assume. Therefore, attempts have been made to mathematically extract the relative contribution of internal vs external amino acids to protein synthesis from various experimental settings. We decided instead to directly measure the relative contributions with a novel SILAC-based strategy (Cambridge et al., 2011). In a simple pulse experiment with labeled arginines and lysines, analysis was limited in silico to peptides that contained both amino acids. Protein digestion was performed with LysC (cleaves only after lysines) instead of trypsin (cleaves after lysines and arginines). Thus, all peptides that contained exactly one labeled lysine and one single arginine (unlabeled or labeled) were then used for further analysis. The idea was that at early time points after the onset of the pulse experiments, peptides with a labeled lysine must be from newly synthesized proteins. In those very same peptides, the ratio of labeled vs unlabeled arginine is then a true reflection of external vs internal amino acid incorporation. Thus, the relative abundance of heavy arginine (+ heavy lysine)-containing peptides compared light arginine (+ heavy lysine)-containing peptides directly identified the contribution of externally supplied labeled amino acids to protein synthesis.

The number of peptides that fulfilled the above-mentioned criteria and which were measured at least three times was clearly limited and derived all from quite abundant proteins. However, assuming that the patterns of amino acid incorporation are the same for all proteins, the obtained data should be representative for the synthesis all proteins, i.e., with low or high abundance. Moreover, since we analyzed considerable short pulse-labeling times of 1, 2, 4, 8, or 24 h, peptides from proteins with a low turnover were rarely detected. For each time point, at least six quantifiable peptides from six different proteins could be detected. For short-labeling times of 1, 2, or 4 h, after which the label incorporation within the entire proteome was less than 10%, we found that the ratio of labeled external vs unlabeled internal amino acids was about 17 to 1 (Cambridge et al., 2011). Thus, for protein synthesis, the contribution of recycled amino acids from protein degradation is negligible, at least under the saturating concentrations used in standard SILAC-labeling experiments. Most likely, the external vs internal ratio is adjusted by the cells depending on the availability of amino acids and the growth conditions, but it is remarkable that cells appear to rely so heavily on outside supplied metabolites. Of course, lysine and arginine are essential amino acids, but at least theoretically, cells should be able to complete protein synthesis at steady state with recycled amino acids only.

4. 5-PLEXED SILAC LABELING

The majority of SILAC experiments are conducted as duplex or triplex analyses, which allows investigating protein differences between two or three different conditions, respectively. For example, a triplex experiment involving RNAi knockdown would have untreated control cells, cells transfected with a mismatch or scrambled RNAi, and cells transfected with the active RNAi. Alternatively, triplex experiments would allow investigation of three different time points. Of course, time-course experiments are preferably performed with more than three time points, while some experiments often require analysis of more than three conditions. To overcome this limitation, researchers have devised different strategies for increasing the complexity of SILAC experiments. Soon after SILAC was introduced by Matthias Mann and colleagues (Ong et al., 2002), these researchers combined two triplex experiments by having one identical condition in both for "normalization." With this approach, five different time points of phosphotyrosine signaling could be temporally resolved (Blagoev, Ong, Kratchmarova, & Mann, 2004).

Since then, two different "true" 5-plex SILAC-labeling strategies have been described. They were used to profile the temporal dynamics of adipocyte differentiation (Molina et al., 2009) or tyrosine phosphorylation (Tzouros et al., 2013). In both studies, the researchers employed five different types of stable isotope labels. Pandey and colleagues employed four different types of stable arginine isotopes: Arg4, Arg6, Arg10, and Arg17 (Molina et al., 2009) (Fig. 6). Since the tryptic digest produced peptides with either a single arginine or a lysine, only those containing an arginine could be used for quantification. Potential limitations with the small mass difference between Arg4 and Arg6 or a chromatographic shift in the elution profile of the deuterium-labeled isoform Arg17 were not observed by the authors.

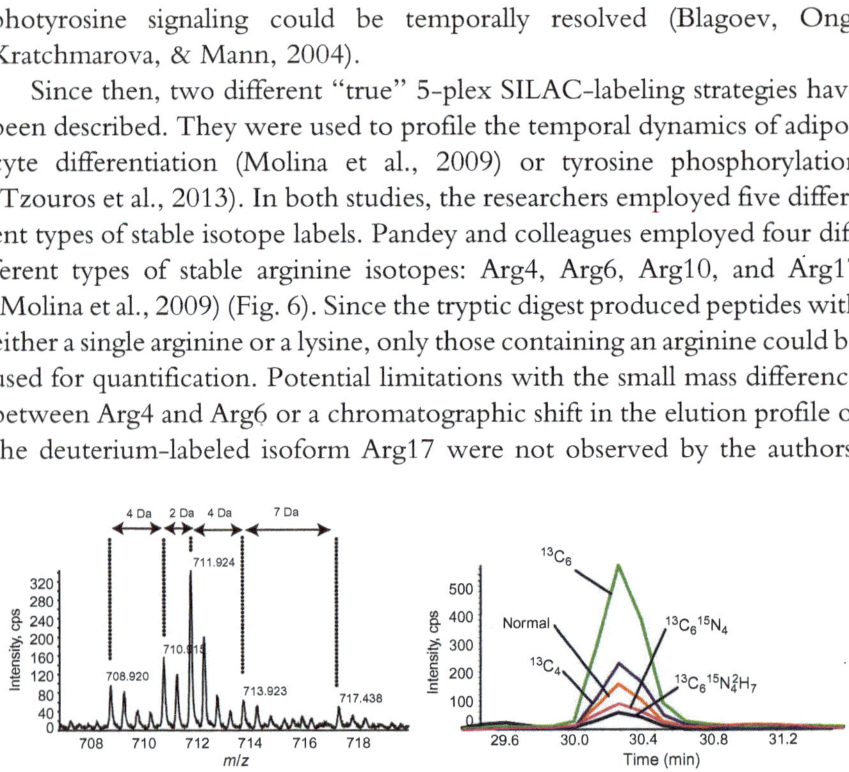

Fig. 6 Mass spectra and elution profiles for SILAC-labeled peptide sets of a 5 day time-course experiment. *Left panel* shows the mass spectra and mass differences between the isotopic-labeled peptides. *Right panel* shows elution profile. The color code of the peptide elution profiles are *red*, Day 0; *blue*, Day 1; *green*, Day 3; *violet*, Day 5; *black*, Day 7, represented by the isotopes: Arg0, Arg4, Arg6, Arg10, and Arg17, respectively. The depicted triply charged peptide derived from the fatty acid-binding protein 2. *Reprinted with permission from Molina, H., Yang, Y., Ruch, T., Kim, J. W., Mortensen, P., Otto, T., et al. (2009). Temporal profiling of the adipocyte proteome during differentiation using a five-plex silac based strategy.* Journal of Proteome Research, 8, *48–58. Copyright 2009 American Chemical Society.*

Several hundred proteins involved in adipogenesis were identified in secretome or nuclear proteome fractions over the time course of 7 days. Of those, several proteins such as SNF2 alpha, a chromatin-remodeling protein, had thus far not been reported to be involved in adipocyte differentiation.

A different 5-plex SILAC-labeling strategy was developed by Augustin and colleagues to describe the temporal dynamics of EGFR1 tyrosine phosphorylation patterns over the time course of 1 h (Tzouros et al., 2013). The authors administered different combinations of the standard stable isotopes Lys4/Lys8, Arg6/Arg10, and Tyr6/Tyr10, so that each of the five conditions/time points had a unique set of labeled amino acids and consequently unique tryptic peptides (see later).

	Lys0	Lys4	Lys8	Lys8	Lsy8
	Arg0	Arg6	Arg10	Arg10	Arg10
	Tyr0	Tyr0	Tyr0	Tyr6	Tyr10
5 μM erlotinib	–	+	+	+	+
Time	0 min	5 min	10 min	30 min	60 min

The requirement for a third stable isotope, in this case tyrosine, confines this approach in the number of possible peptides that can be analyzed as the tyrosine-containing peptides had to be immunoenriched with phosphotyrosine antibodies. Nevertheless, the authors could identify and characterize tyrosine phosphorylation patterns of more than two hundred proteins with high-temporal resolution following inhibition of EGFR1 with the selective inhibitor erlotinib.

5. CONCLUSION

Quantitative proteomics is a rapidly evolving research field both in respect to mass spectrometers and because of improved sample processing, including new labeling strategies or protein isolation procedures such as filter-aided sample preparation (Wisniewski, Zougman, Nagaraj, & Mann, 2009). We have attempted to highlight some of the recent advancements in SILAC-labeling strategies with a particular emphasis on strategies that can be readily implemented by laboratories experienced in standard SILAC technology. Of course, increasingly more efforts have been published to combine SILAC with other several methods, such as BONCAT (bioorthogonal noncanonical amino acid tagging) for selective labeling and isolation of newly synthesized proteins (Bowling et al., 2016). Some of these combinations are very powerful but require the additional introduction of an entirely new experimental procedure.

Overall, it is exciting to see how strongly proteomics has contributed to improved biomedical analyses and our understanding of cellular processes. And thus, we expect quantitative proteomics to play an even bigger role in life science research the upcoming years.

REFERENCES

Angelos, M. G., & Kaufman, D. S. (2015). Pluripotent stem cell applications for regenerative medicine. *Current Opinion in Organ Transplantation, 20*, 663–670.

Arias, I. M., Doyle, D., & Schimke, R. T. (1969). Studies on the synthesis and degradation of proteins of the endoplasmic reticulum of rat liver. *The Journal of Biological Chemistry, 244*, 3303–3315.

Blagoev, B., Ong, S. E., Kratchmarova, I., & Mann, M. (2004). Temporal analysis of phosphotyrosine-dependent signaling networks by quantitative proteomics. *Nature Biotechnology, 22*, 1139–1145.

Boisvert, F. M., Ahmad, Y., Gierlinski, M., Charriere, F., Lamont, D., Scott, M., et al. (2012). A quantitative spatial proteomics analysis of proteome turnover in human cells. *Molecular & Cellular Proteomics, 11*. M111 011429.

Bowling, H., Bhattacharya, A., Zhang, G., Lebowitz, J. Z., Alam, D., Smith, P. T., et al. (2016). Bonlac: A combinatorial proteomic technique to measure stimulus-induced translational profiles in brain slices. *Neuropharmacology, 100*, 76–89.

Cambridge, S. B., Gnad, F., Nguyen, C., Bermejo, J. L., Kruger, M., & Mann, M. (2011). Systems-wide proteomic analysis in mammalian cells reveals conserved, functional protein turnover. *Journal of Proteome Research, 10*, 5275–5284.

Choi, J., Miller, A. M., Nolan, M. J., Yue, B. Y., Thotz, S. T., Clark, A. F., et al. (2005). Soluble cd44 is cytotoxic to trabecular meshwork and retinal ganglion cells in vitro. *Investigative Ophthalmology & Visual Science, 46*, 214–222.

Cox, J., & Mann, M. (2008). MaxQuant enables high peptide identification rates, individualized p.p.b.-range mass accuracies and proteome-wide protein quantification. *Nature Biotechnology, 26*, 1367–1372.

Dice, J. F., & Goldberg, A. L. (1975a). Relationship between in vivo degradative rates and isoelectric points of proteins. *Proceedings of the National Academy of Sciences of the United States of America, 72*, 3893–3897.

Dice, J. F., & Goldberg, A. L. (1975b). A statistical analysis of the relationship between degradative rates and molecular weights of proteins. *Archives of Biochemistry and Biophysics, 170*, 213–219.

Evans, C., Noirel, J., Ow, S. Y., Salim, M., Pereira-Medrano, A. G., Couto, N., et al. (2012). An insight into iTRAQ: Where do we stand now? *Analytical and Bioanalytical Chemistry, 404*, 1011–1027.

Fern, E. B., Garlick, P. J., McNurlan, M. A., & Waterlow, J. C. (1981). The excretion of isotope in urea and ammonia for estimating protein turnover in man with [15n]glycine. *Clinical Science (London, England), 61*, 217–228.

Fierro-Monti, I., Racle, J., Hernandez, C., Waridel, P., Hatzimanikatis, V., & Quadroni, M. (2013). A novel pulse-chase silac strategy measures changes in protein decay and synthesis rates induced by perturbation of proteostasis with an hsp90 inhibitor. *PloS One, 8*, e80423.

Garlick, P. J., & Waterlow, J. C. (1969). Measurement of muscle protein turnover by constant intravenous infusion of [14c]glycine. *The Biochemical Journal, 113*, 7P–8P.

Gehrmann, M. L., Hathout, Y., & Fenselau, C. (2004). Evaluation of metabolic labeling for comparative proteomics in breast cancer cells. *Journal of Proteome Research, 3*, 1063–1068.

Imami, K., Sugiyama, N., Tomita, M., & Ishihama, Y. (2010). Quantitative proteome and phosphoproteome analyses of cultured cells based on silac labeling without requirement of serum dialysis. *Molecular BioSystems, 6*, 594–602.

Mann, M. (2006). Functional and quantitative proteomics using silac. *Nature Reviews. Molecular Cell Biology, 7*, 952–958.

Mann, M., Kulak, N. A., Nagaraj, N., & Cox, J. (2013). The coming age of complete, accurate, and ubiquitous proteomes. *Molecular Cell, 49*, 583–590.

Michalski, A., Damoc, E., Hauschild, J. P., Lange, O., Wieghaus, A., Makarov, A., et al. (2011). Mass spectrometry-based proteomics using Q Exactive, a high-performance benchtop quadrupole orbitrap mass spectrometer. *Molecular & Cellular Proteomics, 10*. M111 011015.

Michalski, A., Damoc, E., Lange, O., Denisov, E., Nolting, D., Muller, M., et al. (2012). Ultra high resolution linear ion trap orbitrap mass spectrometer (orbitrap elite) facilitates top down lc ms/ms and versatile peptide fragmentation modes. *Molecular & Cellular Proteomics, 11*. O111 013698.

Molina, H., Yang, Y., Ruch, T., Kim, J. W., Mortensen, P., Otto, T., et al. (2009). Temporal profiling of the adipocyte proteome during differentiation using a five-plex silac based strategy. *Journal of Proteome Research, 8*, 48–58.

Olsen, J. V., Blagoev, B., Gnad, F., Macek, B., Kumar, C., Mortensen, P., et al. (2006). Global, in vivo, and site-specific phosphorylation dynamics in signaling networks. *Cell, 127*, 635–648.

Ong, S. E., Blagoev, B., Kratchmarova, I., Kristensen, D. B., Steen, H., Pandey, A., et al. (2002). Stable isotope labeling by amino acids in cell culture, silac, as a simple and accurate approach to expression proteomics. *Molecular & Cellular Proteomics, 1*, 376–386.

Ong, S. E., & Mann, M. (2006). A practical recipe for stable isotope labeling by amino acids in cell culture (silac). *Nature Protocols, 1*, 2650–2660.

Stiess, M., Wegehingel, S., Nguyen, C., Nickel, W., Bradke, F., & Cambridge, S. B. (2015). A dual silac proteomic labeling strategy for quantifying constitutive and cell-cell induced protein secretion. *Journal of Proteome Research, 14*, 3229–3238.

Thompson, A., Schafer, J., Kuhn, K., Kienle, S., Schwarz, J., Schmidt, G., et al. (2003). Tandem mass tags: A novel quantification strategy for comparative analysis of complex protein mixtures by ms/ms. *Analytical Chemistry, 75*, 1895–1904.

Tzouros, M., Golling, S., Avila, D., Lamerz, J., Berrera, M., Ebeling, M., et al. (2013). Development of a 5-plex silac method tuned for the quantitation of tyrosine phosphorylation dynamics. *Molecular & Cellular Proteomics, 12*, 3339–3349.

Wisniewski, J. R., Zougman, A., Nagaraj, N., & Mann, M. (2009). Universal sample preparation method for proteome analysis. *Nature Methods, 6*, 359–362.

Zhang, G., Deinhardt, K., & Neubert, T. A. (2014). Stable isotope labeling by amino acids in cultured primary neurons. *Methods in Molecular Biology, 1188*, 57–64.

CHAPTER FOUR

Label-Free and Standard-Free Absolute Quantitative Proteomics Using the "Total Protein" and "Proteomic Ruler" Approaches

J.R. Wiśniewski[1]

Biochemical Proteomics Group, Max-Planck-Institute of Biochemistry, Martinsried, Germany
[1]Corresponding author: e-mail address: jwisniew@biochem.mpg.de

Contents

Abstract

Understanding biological systems and their variation upon stimuli requires knowledge on their composition, primarily including information on organization and dynamics of proteomes. The total protein approach (TPA) is a label- and standard-free method for absolute protein quantitation of proteins using large-scale proteomic data. The method relies on the assumption that the total MS signal from all identified proteins in the dataset reflects—in a biochemical sense—the total protein and the MS signal from a single protein corresponds its abundance in the studied sample. The method offers an easy way to quantify thousands of protein per sample. A related method, the "Proteomic Ruler," enables conversion of the protein

Methods in Enzymology, Volume 585
ISSN 0076-6879
http://dx.doi.org/10.1016/bs.mie.2016.10.002

abundance data calculated by TPA to compute numbers of protein copies per cell. TPA and the Proteomic Ruler are powerful tools for studying dynamics of cell architecture.

1. INTRODUCTION

In the bottom-up proteomics, there are two general approaches allowing identification of changes or differences between biological systems. The first one provides relative information on alteration between samples. This is most frequently assessed by differential labeling of samples with stable isotopes, which allows comparison of abundances of peptides and in consequence proteins between samples. Typical examples of such approaches are metabolic labeling (SILAC and Super SILAC) and chemical labeling involving iTRAQ and TMT technologies. In addition, there are several label-free relative quantitation approaches. Relative changes between independent analyses are difficult to compare and the relative data cannot be directly correlated to a physiological status of a biological system. In contrast this is possible using "absolute" quantitation methods. In these approaches stable isotope-labeled standards, usually peptides at known concentrations, are spiked into samples before MS analysis. Measurement of the proportion between the abundances of peptide originating from the sample and the standard allows calculation of the peptide concentration. The concentration of single or averaged concentrations of different peptides is used for assessing protein titers. This approach is also known as the "targeted," because the standards have to be designed before proteomic analyses and therefore such analyses can cover only a limited number of proteins. To circumvent this technological constrain several label-free computational methods for absolute quantification of proteins have been proposed.

Label-free methods comprise approaches using intensities of MS^1 spectra, such as iBAQ (Schwanhausser et al., 2011, 2013) and Top3 (Silva, Gorenstein, Li, Vissers, & Geromanos, 2006) and those based on counting of MS^2 spectra, the emPAI (Ishihama et al., 2005) and APEX (Braisted et al., 2008) methods (Fig. 1). Each of these label-free methods requires a biochemical input for calculation of the protein abundances. This is either determination of the total amount of the analyzed sample or a use of protein standards with defined concentrations. A direct comparison of these

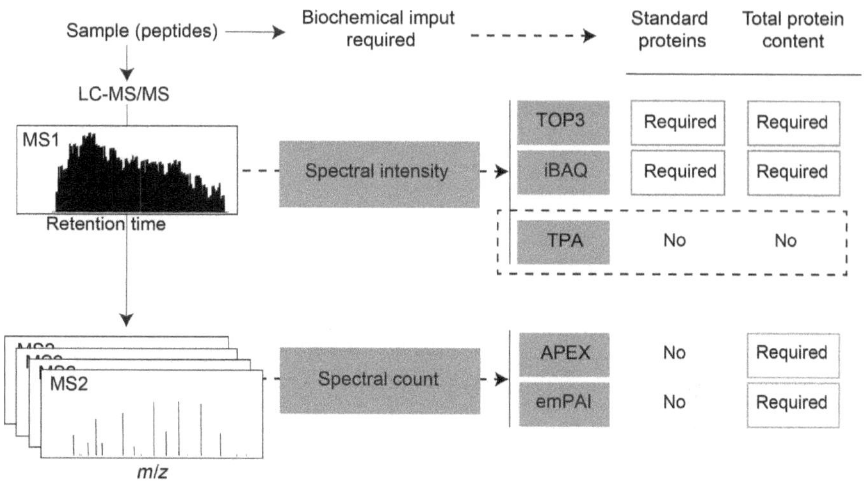

Fig. 1 Comparison of the total protein approach (TPA) with other label-free methods for absolute protein quantification. TPA, TOP3, and iBAQ used the MS^1-spectral intensity, whereas enPAI and APEX are MS^2-spectra based. TPA is a standard-free method, whereas the other label-free methods require biochemical input: measurement of the protein concentration in the sample and/or standard proteins with determined concentration.

methods showed that the TOP3 methods are the most reliable one (Ahrne, Molzahn, Glatter, & Schmidt, 2013).

In contrast to these methods, another computational procedure, the "total protein approach" (TPA), allows absolute protein quantitation without any biochemical input. In the TPA method calculation of protein abundance is based on spectral intensities acquired in the large-scale proteomic analyses (Wisniewski et al., 2012; Wisniewski & Rakus, 2014). The method does not require any specific knowledge on the sample and is standard free (Fig. 1). Therefore it can be applied to any large dataset including archival data. In large-scale analyses a related method, the "Proteomic Ruler," allows conversion of TPA abundance values into protein copy numbers per cell (Wisniewski, Hein, Cox, & Mann, 2014) (Fig. 2).

This chapter describes principles and applications of TPA and the Proteomic Ruler for absolute quantitative proteomics (Fig. 2). Combining both categories allows studying proteomes beyond the almost trivial "differential display" type analyses. In addition, consistency of the absolute proteomic data with other biochemical parameters such as total protein, nucleic acid content, or enzymatic activity provides novel research tools for studying biological systems.

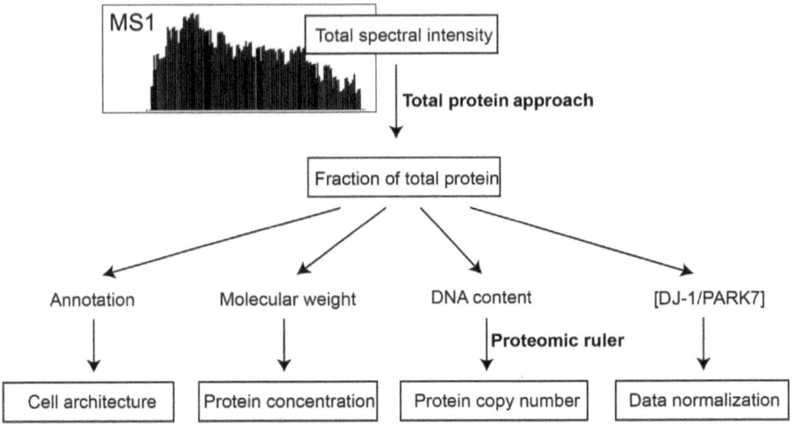

Fig. 2 Upon using additional information, such as bioinformatics annotation, molecular weight, or size of genome (DNA per cell) total protein approach data can be used for calculation of content of architectural cell components, protein concentration, and the cell size. In addition, proteins with invariable concentration can be used for normalization of proteomic data. In particular the deglycase DJ-1 is a useful normalization standard in vertebrate cells.

2. TPA: FOR DETERMINATION OF PROTEIN CONTENTS AND CONCENTRATIONS

The TPA based on the assumption that the total MS signal from all proteins in the sample reflects, in a biochemical sense, the total protein, and the total MS signal from a given protein (i) corresponds its partial abundance in the whole sample. Thus the portion of a protein in the sample is given by the following equation:

$$\text{Total protein}\,(i) = \frac{\text{MS}-\text{signal}\,(i)}{\text{Total MS}-\text{signal}}$$

and also can be expressed as percentage of the protein (i) in the sample:

$$\%\text{Total protein}\,(i) = \frac{\text{MS}-\text{signal}\,(i)}{\text{Total MS}-\text{signal}} \times 100\%$$

Further the total protein (i) is easily convertible into protein concentration per a unit of total protein that is known as specific concentration:

$$\text{Protein concentration}\,(i) = \frac{\text{MS}-\text{signal}\,(i)}{\text{Total MS}-\text{signal} \times \text{MW}(i)}\left[\frac{\text{mol}}{g\ \text{total protein}}\right]$$

Taking into account that proteins constitute 20–30% of whole cell weight, the cellular concentration per volume can be calculated. For other sample types such body fluids the content of total protein has to be assayed or retrieved from references.

The TPA method was initially validated using a mixture of standard proteins with defined concentrations (Wisniewski et al., 2012). Later, several studies revealed accuracy of the TPA values by comparing to literature data. For example, TPA-bases concentrations of plasma proteins were found to be within with the reference data intervals (Wisniewski et al., 2014). Finally, the analytical value of TPA has been strengthened by studies showing correct stoichiometry of well-known heteromeric complexes, such as the plasma membrane Na^+/K^+-ATP-ase, mitochondrial pyruvate dehydrogenase, or nuclear pore complexes (Wisniewski, Dus-Szachniewicz, et al., 2015; Wisniewski, Friedrich, Keller, Mann, & Koepsell, 2015), which were determined using TPA and the "Proteomic Ruler."

3. CELL SIZE AND PROTEIN COPY NUMBERS CAN BE ASSESSED BY THE PROTEOMIC RULER

Already in the 1970s biochemical analyses revealed that across different eukaryotic cells the ratio of the weight histones to DNA is invariably close to one. Thus a mononuclear, $2n$ human cell contains 6.5 pg histones, whereas, for example, the total amount of histones in a yeast cell is 0.025 pg. Since histones are abundant their summed content in the whole total protein is easy to determine. The total protein contents of single cell can be calculated using the relationship:

$$\text{Total protein/cell} = \frac{\text{weight of } 2n \text{ DNA[pg]} \times \text{Total MS} - \text{signal}}{\text{MS} - \text{signal (Total histone)}} [\text{pg}]$$

Knowing the total protein content per cell value the number of protein copies per cell can be calculated from:

$$\text{Protein copy number}(i) = N_A \times \text{Total protein/cell} \times \text{Protein concentration}(i)$$

Or

$$\text{Protein copy number}(i) = N_A \times \frac{\text{weight of } 2n \text{ DNA[pg]} \times \text{MS} - \text{signal}(i)}{\text{MS} - \text{signal(Total histone)} \times \text{MW}(i)}$$

where N_A is the Avogadro number.

The accuracy of the Proteomic Ruler approach has been validated by reanalysis of data from targeted approaches, which used isotopically labeled standard proteins (Wisniewski et al., 2014). This study revealed on average 1.5-fold deviation of the "Proteomic Ruler" compared to data assessed by targeted proteomics. In addition the copy numbers calculated by the Proteomic Ruler showed excellent correlation with the literature data. Notably, analysis of *Escherichia coli* proteome revealed that the estimates of copy numbers were correct for proteins at different concentrations spanning from abundant ribosomal proteins occurring at more than 7000 copies down to few copies of proteins involved in the DNA replication (Wisniewski & Rakus, 2014).

4. PROTEIN CONCENTRATIONS AND COPY NUMBERS PROVIDE DIFFERENT LAYERS OF INFORMATION

In molecular biology that primarily considers the cell as a product of gene expression, the abundances of messenger RNA are described in numbers. Almost these values are far from direct translation into biological meaning. This way of conceiving of a biological system has been copied by proteomics and is widely used as the relative protein quantitation. There is no doubt that this type of comparative protein quantitation has a potential to provide insights in changes in the cell composition, but per se it overlooks quantitative differences between varying states of cells.

To illustrate the potential bias of the relative proteomics, let us compare two cells being at the same cell cycle stage but differing in size. In this case it is to expect that the larger cells constitute more cytoplasmic protein compared to the smaller cell, whereas the content of histones and other protein constitutively organizing DNA into chromatin remains constant. A comparison of both proteomes using the relative methods will result in measuring a decrease of the nuclear proteins abundance in the larger cell relative to the smaller one, thus leading to a wrong interpretation of differences between the cells. A good illustration of this problem is a comparison of proteomic data between the native human hepatocyte and the cultured hepatoma HepG2 cell (Wisniewski, Vildhede, Noren, & Artursson, 2016). This study revealed drastic differences between proteomes of these cells, which were well reflected in statistical analyses of their proteomes. Statistical *t*-tests using either protein concentration or copy numbers led to two distinct sets of proteins that were significantly changed between the two cell types (Fig. 3). These

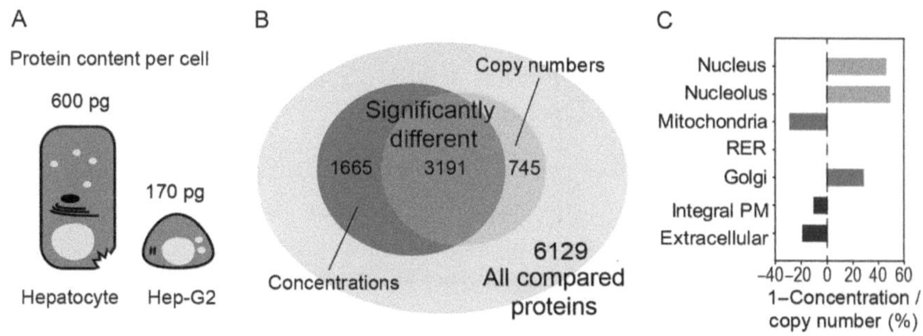

Fig. 3 TPA-derived protein concentration and copy number values allow discrimination between physiologically relevant and structurally related changes in cells. Comparative analysis of human hepatocytes and hepatoma cells revealed different sets of significant differences between both biological systems. (A) Cell sizes calculated using the Proteomic Ruler. (B) Multivariate statistical analysis revealed distinct sets of differences in protein concentrations and copy numbers between two cell types. (C) The analysis of the concentrations revealed more differences in nucleus and Golgi body compared to the copy number analysis. In turn the latter showed more differences in mitochondria and plasma membrane. *Data are from Wisniewski, J. R., Vildhede, A., Noren, A., & Artursson, P. (2016). In-depth quantitative analysis and comparison of the human hepatocyte and hepatoma cell line HepG2 proteomes.* Journal of Proteomics, 136, *234–247. http://dx.doi.org/:10.1016/j.jprot.2016.01.016.*

differences clearly show that proteomic comparisons of cells that are different in their sizes should be performed with a particular care and the proportion of proteins organized in specific organelles should be considered. Fig. 3C shows that the number of significantly changed proteins in relation to their cellular location varies between the two different calculation methods applied. Compared to the analysis of differences in protein concentrations, the analysis of protein differences in copy numbers decreased the number of significant changes for nuclear and nucleolar proteins by about 40%. The opposite happened with the numbers of significant alterations of plasma membrane and mitochondrial proteins, which rates raised by 10% and 30%, respectively.

In conclusion, the absolute protein quantitation using TPA and Proteomic Ruler allows discrimination between changes reflecting architectural and physiologically relevant alterations in cells. Using different layers of absolute proteomic quantification is essential in many situations, because cells often undergo changes in their sizes or the ratio of nucleus to cytoplasm. These alterations are difficult to observe using the relative proteomic quantitation.

5. CONSISTENCY OF TPA VALUES WITH BIOCHEMICAL DATA

Analysis of a biological system requires combining of various structural and catalytic parameters. Consistency of different parameters can provide novel insights in the cell organization and is also a prerequisite for data validation. The TPA method is a powerful tool for integrating of proteomic data with different biochemical measurements.

5.1 DNA and RNA Content

Since prokaryotic DNA is not organized into nucleosomes cell sizes of bacterial cells cannot be assessed by the Proteomic Ruler. But alternatively, determination of DNA content in the lysates of prokaryotes can provide the necessary information for calculation of the total protein content per cell. This strategy was initially tested using human cell lines (Wisniewski et al., 2014). It allowed calculation of cell sizes of *E. coli* (Wisniewski & Rakus, 2014). In addition, quantitative analysis of *E. coli* proteome using TPA revealed tight correlation of the titers of ribosomal proteins and the RNA content (Wisniewski & Rakus, 2014). A proteomic sample preparation protocol allowing a parallel determination of nucleic acids is described in Chapter "Filter-aided sample preparation: the versatile and efficient method for proteomic analysis" by Wiśniewski.

5.2 Enzymatic Activities

Metabolic capacity of enzymes is given by their abundance and kinetic properties. At substrate concentrations exceeding several fold Michaelis–Menten constant (K_m), the maximal activity V_{max} is the kinetic measure of the impact of an enzyme. Comparison of the V_{max} values of 12 glycolytic/gluconoegenic pathway enzymes revealed a high degree of correlation between enzyme concentration and the V_{max}, suggesting that enzyme concentrations provide a good approximation of their activities in living cells (Rakus, Gizak, Deshmukh, & Wisniewski, 2015; Wisniewski, Gizak, & Rakus, 2015).

Dividing the V_{max} values by the enzyme concentrations allows calculation of the catalytic constants of the enzymes, k_{cat}, which is the fundamental kinetic parameter. It represents the number of reactions catalyzed by an enzyme during a time unit. So far, determination of k_{cat}, required purification of enzymes prior to kinetic measurements. Since the purification is usually time-consuming and laborious and often accompanied with partial inactivation of enzymes, k_{cat} values have been determined only for a limited number of enzymes, mostly in nonmodel organisms. The majority of the k_{cat} values determined in by this approach from mouse organs was in the range of those determined for purified enzymes (Wisniewski, Gizak, et al., 2015).

5.3 Subcellular Fractionation

Traditionally, subcellular fractionation of tissue was accompanied with determination of total protein content and enzymatic activities across all fractions and the homogenate. This information enabled assessment of fraction yields and provided information on purity of the isolated subcellular fractions. Over the time, many of the laborious biochemical assays were replaced by Western blot analyses. At the best both the biochemical and the Western blot analyses enable insights in the subcellular distribution of a limited number of proteins. In contrast, large-scale proteomics can provide a system wide picture based on quantitative analysis of all obtained fractions. The TPA method allows quantitative analyses of thousands of proteins, providing their content and concentrations across generated fractions. Combining the TPA data with biochemical total protein measurements enables calculation of the protein contents per individual fraction (Wisniewski, Wegler, & Artursson, 2016). This information provides quantitative insights into the subcellular distribution of proteins and can be used as a guide for development of fractionation procedures.

6. NORMALIZATION AND COMPARISON OF DATASETS USING DJ-1/PARK7 TITER

Data normalization is a prerequisite for comparing quantitative data generated in various experiments. Analyzing proteins by Western blotting the total protein loads per electrophoretic lane are routinely checked using housekeeping marker proteins such as actin, tubulin, or glyceraldehyde dehydrogenase. But this approach is valid only for comparisons between similar samples, such as rapidly growing cultured cells or tissue of the same type. Across different tissues the abundances of the arbitrary selected marker proteins are variable, which disqualifies their use for normalization purposes. Thus, identification of other markers, in particular those applicable to proteomic data is demanding.

Reanalysis of a number of large-scale datasets revealed deglycase DJ-1 (PARK7) as the most invariable protein across tissues and cell lines of vertebrate organisms (Wisniewski & Mann, 2016). The protein constitutes $0.069 \pm 0.017\%$ of total cellular protein and occurs at concentration of 34.6 ± 8.7 pmol/mg total protein. Because DJ-1 is abundant and therefore detectable with several peptides, it can helpful in normalization of proteomic datasets. Moreover, DJ-1 appear as advantageous loading control for Western blot, superior to those used commonly used, and allowing comparing tissues and cells.

7. LIMITATIONS

There are two essential limitations in the use of the TPA method. The first, the approach is applicable only to large-scale proteomic analyses and the second are the structural properties of some proteins. The first limitation is of a technical nature. For human cultured cells the required minimal depth for a robust readout lies by about 12,000 peptides per dataset (Wisniewski et al., 2014). In a case of tissues and body fluids, in which often few most abundant protein contribute to 90% of the mass of the cells, the minimal number of peptides required for a correct assessment of protein abundances by TPA is usually lower. For example, an analysis of human plasma that based on identification of 5400 unique peptides resulted in correct determination of protein concentrations (Wisniewski et al., 2014).

The second limitation of the TPA method relies on nature of some proteins, which considering their primary structure have a nonproportionally low number of peptide-bound cleavage sites for enzymes used in bottom-up proteomics. But this limitation affects other label-free approaches as well as targeted proteomics too. Notable, the latter approaches are also biased by incomplete digestion of proteins that commonly accompanies proteomic analyses. In contrast, TPA does not make the attempt to use the proteomic data in a classical analytical way. TPA does not rely on comparing with standard(s) such as other quantitative methods do. Combining intensities of all peptides of dataset mitigates the influence of partial digestion on the protein titer determination.

ACKNOWLEDGMENTS

This work was supported by the Max-Planck Society for the Advancement of Science and the German Research Foundation (DFG/Gottfried Wilhelm Leibniz Prize).

REFERENCES

Ahrne, E., Molzahn, L., Glatter, T., & Schmidt, A. (2013). Critical assessment of proteome-wide label-free absolute abundance estimation strategies. *Proteomics*, *13*(17), 2567–2578. http://dx.doi.org/10.1002/pmic.201300135.

Braisted, J. C., Kuntumalla, S., Vogel, C., Marcotte, E. M., Rodrigues, A. R., Wang, R., … Pieper, R. (2008). The APEX quantitative proteomics tool: Generating protein quantitation estimates from LC-MS/MS proteomics results. *BMC Bioinformatics*, *9*, 529. http://dx.doi.org/10.1186/1471-2105-9-529.

Ishihama, Y., Oda, Y., Tabata, T., Sato, T., Nagasu, T., Rappsilber, J., & Mann, M. (2005). Exponentially modified protein abundance index (emPAI) for estimation of absolute protein amount in proteomics by the number of sequenced peptides per protein. *Molecular & Cellular Proteomics*, *4*(9), 1265–1272. http://dx.doi.org/10.1074/mcp.M500061-MCP200.

Rakus, D., Gizak, A., Deshmukh, A., & Wisniewski, J. R. (2015). Absolute quantitative profiling of the key metabolic pathways in slow and fast skeletal muscle. *Journal of Proteome Research*, *14*(3), 1400–1411. http://dx.doi.org/10.1021/pr5010357.

Schwanhausser, B., Busse, D., Li, N., Dittmar, G., Schuchhardt, J., Wolf, J., … Selbach, M. (2011). Global quantification of mammalian gene expression control. *Nature*, *473*(7347), 337–342. http://dx.doi.org/10.1038/nature10098.

Schwanhausser, B., Busse, D., Li, N., Dittmar, G., Schuchhardt, J., Wolf, J., … Selbach, M. (2013). Corrigendum: Global quantification of mammalian gene expression control. *Nature*, *495*(7439), 126–127. http://dx.doi.org/10.1038/nature11848.

Silva, J. C., Gorenstein, M. V., Li, G. Z., Vissers, J. P., & Geromanos, S. J. (2006). Absolute quantification of proteins by LCMSE: A virtue of parallel MS acquisition. *Molecular & Cellular Proteomics*, *5*(1), 144–156. http://dx.doi.org/10.1074/mcp.M500230-MCP200.

Wisniewski, J. R., Dus-Szachniewicz, K., Ostasiewicz, P., Ziolkowski, P., Rakus, D., & Mann, M. (2015). Absolute proteome analysis of colorectal mucosa, adenoma, and cancer reveals drastic changes in fatty acid metabolism and plasma membrane transporters. *Journal of Proteome Research*, *14*(9), 4005–4018. http://dx.doi.org/10.1021/acs.jproteome.5b00523.

Wisniewski, J. R., Friedrich, A., Keller, T., Mann, M., & Koepsell, H. (2015). The impact of high-fat diet on metabolism and immune defense in small intestine mucosa. *Journal of Proteome Research*, *14*(1), 353–365. http://dx.doi.org/10.1021/pr500833v.

Wisniewski, J. R., Gizak, A., & Rakus, D. (2015). Integrating proteomics and enzyme kinetics reveals tissue-specific types of the glycolytic and gluconeogenic pathways. *Journal of Proteome Research*, *14*(8), 3263–3273. http://dx.doi.org/10.1021/acs.jproteome.5b00276.

Wisniewski, J. R., Hein, M. Y., Cox, J., & Mann, M. (2014). A "proteomic ruler" for protein copy number and concentration estimation without spike-in standards. *Molecular & Cellular Proteomics*, *13*(12), 3497–3506. http://dx.doi.org/10.1074/mcp.M113.037309.

Wisniewski, J. R., & Mann, M. (2016). A proteomics approach to the protein normalization problem: Selection of unvarying proteins for MS-based proteomics and western blotting. *Journal of Proteome Research*, *15*, 2321–2326. http://dx.doi.org/10.1021/acs.jproteome.6b00403.

Wisniewski, J. R., Ostasiewicz, P., Dus, K., Zielinska, D. F., Gnad, F., & Mann, M. (2012). Extensive quantitative remodeling of the proteome between normal colon tissue and adenocarcinoma. *Molecular Systems Biology*, *8*, 611. http://dx.doi.org/10.1038/msb.2012.44.

Wisniewski, J. R., & Rakus, D. (2014). Multi-enzyme digestion FASP and the 'total protein approach'-based absolute quantification of the *Escherichia coli* proteome. *Journal of Proteomics*, *109*, 322–331. http://dx.doi.org/10.1016/j.jprot.2014.07.012.

Wisniewski, J. R., Vildhede, A., Noren, A., & Artursson, P. (2016). In-depth quantitative analysis and comparison of the human hepatocyte and hepatoma cell line HepG2 proteomes. *Journal of Proteomics*, *136*, 234–247. http://dx.doi.org/10.1016/j.jprot.2016.01.016.

Wisniewski, J. R., Wegler, C., & Artursson, P. (2016). Subcellular fractionation of human liver reveals limits in global proteomic quantification from isolated fractions. *Analytical Biochemistry*, *509*, 82–88. http://dx.doi.org/10.1016/j.ab.2016.06.006.

CHAPTER FIVE

Separation of Two Distinct O-Glycoforms of Human IgA1 by Serial Lectin Chromatography Followed by Mass Spectrometry O-Glycan Analysis

S. Lehoux[*,1], T. Ju[†,1]

[*]Beth Israel Deaconess Medical Center, Harvard Medical School, Boston, MA, United States
[†]Emory University School of Medicine, Atlanta, GA, United States
[1]Corresponding authors: e-mail address: slehoux@bidmc.harvard.edu; tju@emory.edu

Contents

Abstract

Human immunoglobulin A1 (IgA1), which carries four to six mucin-type O-glycans (O-glycans) on its hinge region (HR), is the most abundant O-glycoprotein in plasma or serum. While normal O-glycans from hematopoietic-originated cells are core 1-based complex structures, many reports showed that the IgA1 from patients with IgA nephropathy (IgAN) carries undergalactosylated or truncated O-glycans such as the Tn antigen and its sialylated version the SialylTn (STn) antigen on the HR. Yet, there is still a debate whether Tn/STn on the HR of IgA1 is specific to the IgA1 from patients with IgAN since these antigens have also been seen in serum IgA1 of healthy individuals. An additional question is whether the O-glycans at all sites on the two HRs of one IgA1 molecule are homogeneous (either all normal or all Tn/STn) or heterogeneous (both normal and Tn/STn O-glycans). To address these questions, we conducted a systematic study on the O-glycans of plasma IgA1 from both IgAN patients and healthy controls using serial HPA and PNA lectin chromatography followed by western blotting and further analysis of O-glycans from HPA-bound and PNA-bound IgA1 fractions by mass spectrometry. Unexpectedly, we found that a variable minor fraction of IgA1 from both IgAN patients and healthy controls had Tn/STn antigens, and that the O-glycoprotein

Methods in Enzymology, Volume 585
ISSN 0076-6879
http://dx.doi.org/10.1016/bs.mie.2016.10.003

IgA1 molecules from most samples had only two distinct O-glycoforms: one major glycoform with homogeneous normal core 1-based O-glycans and one minor glycoform with homogeneous Tn/STn antigens. These results raised a serious question about the role of Tn/STn antigens on IgA1 in pathogenesis of IgAN, and there is a demand for a practical methodology that any laboratory can utilize to analyze the O-glycans of IgA1. Herein, we describe the methodology we developed in more detail. The method could also be applied to the analysis of any other O-glycosylated proteins.

1. INTRODUCTION

Immunoglobulin (Ig) A (IgA) is a major Ig in biological fluids of animals, is the most produced antibody (Ab) each day, and is the most abundant Ab in mucosal. In blood, IgA is the second most abundant Ab after the IgG (Kerr, 1990; Mestecky, Russell, Jackson, & Brown, 1986; Papista, Berthelot, & Monteiro, 2011). Unlike mice who only have one type of IgA, human IgA has two subtypes: IgA1 and IgA2. The major structural difference between IgA1 and IgA2 is that IgA1 contains an extend hinge region (HR) with many Serine (Ser), Threonine (Thr), and Proline (Pro) residues which constitute a typical motif for mucin-type O-glycosylation (O-glycosylation) (Kerr, 1990). IgA1 is the major O-glycosylated proteins in human plasma with up to six of the nine O-glycosylation sites occupied in each HR (Baenziger & Kornfeld, 1974; Kerr, 1990; Tarelli, Smith, Hendry, Challacombe, & Pouria, 2004). Normal O-glycans of IgA1 are commonly described as mono- and disialylated core 1-based O-glycans (Mattu et al., 1998) since the Ig-secreting plasma cells are originally derived from hematopoietic cells.

Mucin-type O-glycosylation is the most common protein glycosylations of Ser/Thr residues of secreted and transmembrane glycoproteins and can regulate many aspects of their functions and recognition properties (Ju, Otto, & Cummings, 2011; McEver, Moore, & Cummings, 1995). O-Glycans are synthesized by a series of glycosyltransferases orchestrated mainly in the Golgi apparatus (Ju et al., 2011). In humans, the polypeptide-GalNAc-transferases form a large family of glycosyltransferases that initiate O-glycosylation by transferring a GalNAc from UDP-GalNAc to Ser or Thr residues in proteins traversing through the secretory pathway to form the Tn antigen (GalNAcα-Ser/Thr) (Fig. 1) (Schjoldager & Clausen, 2012). Typically, the Tn antigen is subsequently efficiently converted to the core 1 structure Galβ1,3GalNAcα-Ser/Thr (or T antigen) by a single gene-encoded glycosyltransferase: the Core 1 β3galactosyltransferase (Core 1 β3GalT,

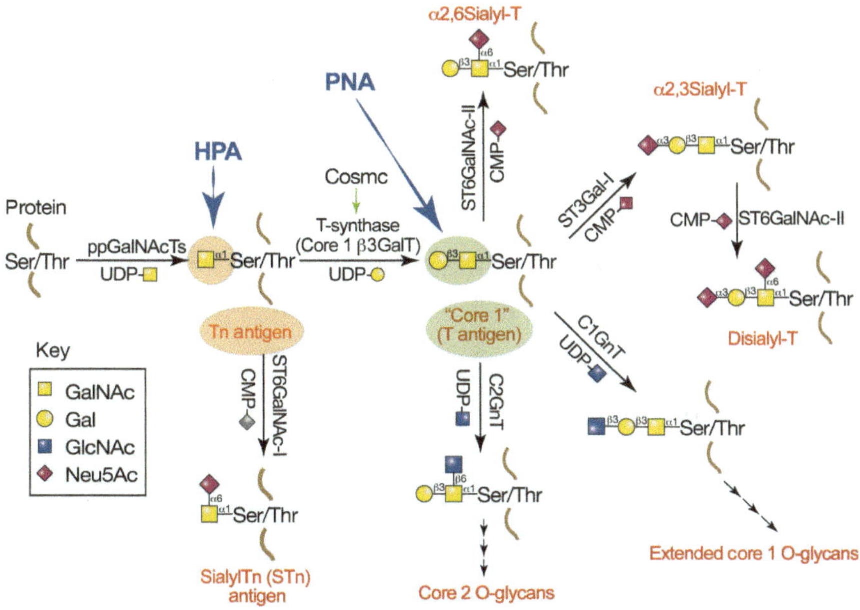

Fig. 1 The O-glycosylation of proteins in hematopoietic cells. Mucin-type O-glycosylation of proteins initiates mainly in the Golgi when polypeptide α-GalNAc-transferases (ppGalNAcTs) transfer GalNAc from UDP-GalNAc to Ser/Thr on a protein going through the secretory pathway to form Tn antigen (GalNAcα-Ser/Thr). Tn antigen is normally further modified by the T-synthase (Core 1 β3GalT) which transfers Gal from UDP-Gal to GalNAcα-Ser/Thr in a protein to form core 1 (T antigen) in all cell types. Cosmc is the specific molecular chaperone for the T-synthase. In all cell types, the T antigen is usually sialylated by ST3Gal-I which transfers neuraminic acid (sialic acid) from CMP-Neu5Ac to form α3sialyl Core 1 (Neu5Acα2,3Galβ1,3GalNAcα-Ser/Thr, α3sialyl-T), which can be further sialylated by ST6GalNAc-II to form disialyl Core 1 [Neu5Acα2,3Galβ1,3(Neu5Acα2,6)GalNAcα-Ser/Thr, Disialyl-T] structures. The T antigen can also be sialylated by ST6GalNAc-II directly to form α6sialyl Core 1 (α6sialyl-T). In hematopoietic cells, besides synthesis of sialyl core 1 structures, the T antigen is usually further modified by core 2 β6-*N*-acetylglucosaminyltransferase (C2GnT) and/or by core 1 β3-*N*-acetylglucosaminyltransferase (C1GnT) to form Core 2 [Galβ1,3(GlcNAcβ1,6)GalNAcα-Ser/Thr], and/or extended core 1-based O-glycans. Mutations in *Cosmc* or loss of T-synthase activity result in pathological expression of Tn and/or STn antigens. The STn antigen (Neu5Acα2,6GalNAcα-Ser/Thr) results from the actions of ST6GalNAc-I, which transfers Neu5Ac from CMP-Neu5Ac to the Tn antigen. Due to poor efficiency of ST6GalNAc-I, it is likely that high expression of ST6GalNAc-I could outcompete functional T-synthase to result in pathologic STn expression. Lectins, HPA can bind to the Tn antigen (*highlighted*), while PNA recognizes the core 1 structure or T antigen (*highlighted*).

T-synthase) (Ju, Brewer, D'Souza, Cummings, & Canfield, 2002; Ju, Cummings, & Canfield, 2002). T-synthase utilizes UDP-Gal as the donor and GalNAcα-Ser/Thr in proteins as the acceptor to form the T antigen. Notably, the *C*ore 1 β3GalT-*s*pecific *m*olecular *c*haperon, Cosmc, is required to assist T-synthase's correct folding (Ju & Cummings, 2002; Wang et al., 2010). Moreover, the T antigen is usually further modified, branched, and/or elongated to form complex normal O-glycans, including mono-sialyl Core 1 (or 2,3sialylT, 2,6sialylT), disialyl core 1 (disialylT), Core 2 O-glycans, and extended Core 1 O-glycans on surface glycoproteins of hematopoietic cells. The Tn antigen can also been directly sialylated by ST6GalNAc-I which is restrictedly expressed or poorly expressed at low level in most cells to form the STn antigen. Thus, the Tn and STn antigens do not normally occur in significant amounts in cellular glycoproteins (Fig. 1).

IgA nephropathy (IgAN), also called Berger's disease, was first described by Dr. Jean Berger in 1968 (Berger & Hinglais, 1968) and is the most common primary glomerulonephritis worldwide (Barratt, Feehally, & Smith, 2004). IgAN is characterized and primarily diagnosed by the deposition of IgA1 in the mesangium. Many studies found a deficiency in galactosylation of O-glycans and concomitant expression of the Tn/STn in IgA1 from patients with IgAN, suggesting that this might be responsible for the pathogenesis (Barratt et al., 2004; Giannakakis, Feriozzi, Perez, Faraggiana, & Muda, 2007; Hiki, 2009; Mestecky et al., 2008). Yet, a few studies also reported that both IgAN patients and healthy individuals had the serum IgA1 containing Tn/STn in their O-glycans (Takahashi et al., 2012; Wada, Tajiri, & Ohshima, 2010), which raised the question of the role of aberrant O-glycosylation of IgA1 in IgAN. Furthermore, the homogeneity or heterogeneity of those four to six O-glycans on each IgA1 molecules was not thoroughly investigated. With our expertise in O-glycosylation, we developed a practical lectin serial chromatography method relying on *Helix pomatia* agglutinin (HPA) and peanut agglutinin (PNA) to analyze the O-glycosylation on IgA1 (Lehoux et al., 2014). This chromatographic method can also be adapted and used for any O-glycosylated proteins in general.

2. METHODOLOGY

2.1 Reagents

2.1.1 Serum Samples

Human blood samples are collected and processed to separate the plasma from the blood. Erythrocytes and leukocytes are separated using Lymphoprep (Stemcell™ Technologies, Vancouver, Canada). The plasma samples should be aliquoted and immediately stored at −80°C or processed for experiments and kept on ice at all time.

2.1.2 Preparation of Lectin-Conjugated Beads

Among a small subset of lectins specific to O-glycans, we decided to use PNA and HPA. PNA is relatively specific to the unmasked β3Galactose on either pure Core 1 or complex Core 2 O-glycans, while HPA only recognizes terminal α-GalNAc on either Ser/Thr, which is the Tn antigen or in the blood group A (Wang et al., 2010). For chromatography purposes, we used these two lectins coupled to beads. Both PNA and HPA beads can be commercially purchased or prepared in house as described later.

Free PNA and HPA lectins (Sigma-Aldrich, St. Louis, MO, USA) are resuspended at a concentration of 2 mg/mL in a 50 m*M* lactose solution for PNA and 50 *mM* GalNAc solution for HPA. Incubation with a ligand is used to protect the lectin binding site during the coupling procedure. Five hundred microliters of the resuspended lectin are mixed with 1 mL of Affi-gel 10 beads (Bio-Rad), washed beforehand with 10 mL of ice-cold water, and incubated 1 h at RT in rotation to ensure proper mix. After a brief centrifugation ($1000 \times g$, 3′), the supernatant is removed and the beads are incubated with 1 *M* glycine at RT for 1 h in rotation in order to block any residual coupling active site on the beads. The glycine solution is then removed after centrifugation, and the beads are washed three times with 10 mL of PBS and are finally resuspended in 2 mL of PBS and stored at 4°C.

2.1.3 Choosing an Appropriate Method—Batch and Column Chromatography

Lectin chromatography can be performed in centrifuge tubes (0.5 or 1.5 mL)-batched method for small plasma volumes or on plastic columns (Poly-Prep Chromatography Column, Bio-Rad) for larger volumes of samples.

2.2 Method/Protocol

Many glycoproteins, such as IgA1, contain both N-glycans and O-glycans, mainly sialylated, and the truncated O-glycans Tn and STn are very rare in normal cells. Therefore, O-glycoproteins very rarely carry both the Tn-antigen and Core 1 structures. Affinity chromatography separation by HPA and PNA should theoretically yield to only two types of materials—HPA$^+$ only and PNA$^+$ only fractions. A material found in the HPA$^+$/PNA$^+$ fraction would indicate that both the Tn-antigen and core 1 structures are simultaneously expressed in a single O-glycoprotein molecule which has multiple O-glycosylation sites like human IgA1.

2.2.1 Serial HPA/PNA Chromatography

Both HPA and PNA bind to nonsialylated structures. Therefore, prior to HPA or PNA chromatography, O-glycoproteins should be desialylated. This can be easily achieved by incubation of glycoproteins with a general neuraminidase (from *Arthrobacter ureafaciens*) (Roche).

- Desialylation

  Dilute 50 μL of serum (or plasma) in 50 μL of PBS and incubate with 5 mU of Neuraminidase at 37°C under gentle rotation overnight (12–16 h). The efficacy of sialic acid removal can be easily tested by lectin blot using *Sambucus nigra* Agglutinin (SNA) or *Maackia amurensis* Lectin I (MAL-I or MAA1). Any signal from SNA or MAL-I should be removed upon efficient treatment with neuraminidase. Before incubating with lectin beads, the desialylated samples are centrifuged at 5000 × *g* for 10 min to remove possible precipitated proteins, and 10 μL samples are saved as the prechromatography sample.

- HPA chromatography
  1. *Binding*: incubate the desialylated plasma samples with 50 μL HPA-conjugated agarose beads (HPA beads) which are previously washed twice with 500 μL PBS. The mixture can be incubated in a 0.5-mL microtube either at room temperature for 1 h with gentle rotation or, preferably, overnight (12–16 h) at 4°C with gentle rotation.
  2. *Washing*: After incubation, the tubes containing HPA beads and samples are centrifuged at 500 × *g* for 3 min and the supernatant is collected and constitutes the HPA-unbound (HPA^-) material (Fig. 2). The beads are then washed three times with 500 μL of PBS by resuspension/centrifugation and the washed PBS is discarded.
  3. *Elution*: After washing, the HPA-bound (HPA^+) material is eluted from the beads by incubation of the beads with 45 μL of 50 m*M* GalNAc in PBS for 10–15 min at room temperature with gentle rotation. After centrifugation at 500 × *g* for 3 min, the supernatant is collected. This elution step is repeated one more time. The two eluted fractions (90 μL total) are pooled and labeled HPA^+ material (Fig. 2).

- PNA chromatography

  Both HPA-bound (HPA^+) and HPA-unbound (HPA^-) fractions are chromatographed on PNA beads, and 10 μL of each fraction is saved for western blot analysis.

  1. *Binding*: Both HPA-bound (HPA^+) and HPA-unbound (HPA^-) fractions are incubated with 50 μL of PNA-conjugated beads

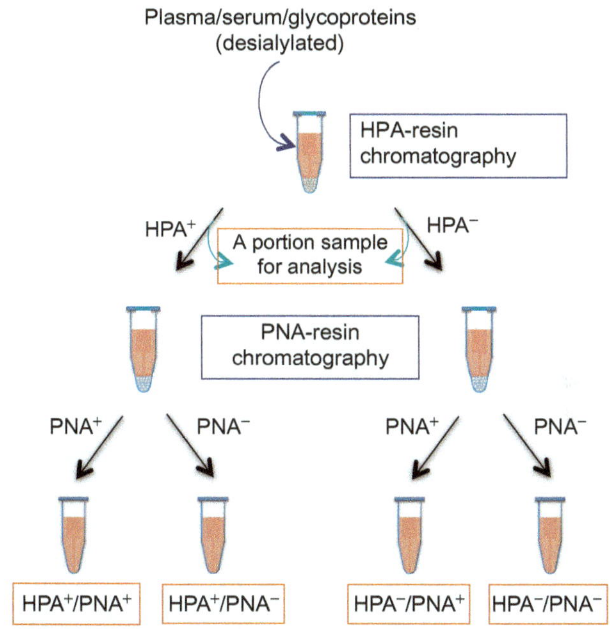

Fig. 2 Serial chromatographic separation by HPA and PNA lectins. A diagram describes the chromatography method in this chapter. The plasma sample after desialylation is first chromatographied on HPA resin. Both unbound (HPA^-) and bound (HPA^+) materials, which are eluted with 50 m*M* GalNAc, are collected and further chromatographied on PNA resin, and bound fraction (PNA^+) is eluted with 200 m*M* lactose. Thus, four fractions are finally collected (HPA^+/PNA^+; HPA^+/PNA^-; HPA^-/PNA^+; and HPA^-/PNA^-) and the material in each of these fractions can be further analyzed.

(PNA beads), which are previously washed twice with 500 μL of PBS. The mixture can be incubated at room temperature for 1 h with gentle rotation or, preferably, overnight (12–16 h) at 4°C with gentle rotation.

2. *Washing*: After incubation, the tubes containing PNA beads and samples are centrifuged at 500 × *g* for 3 min and the supernatants are collected and constitute the PNA-unbound (PNA^-) fractions. At this point, the PNA-unbound materials from HPA^+ and HPA^- correspond then to HPA^+/PNA^- or HPA^-/PNA^- materials, respectively (Fig. 2). The beads are then resuspended and washed as described earlier for HPA chromatography.
3. *Elution*: The PNA-bound (PNA^+) fractions are eluted from the beads with 40 μL of 200 m*M* lactose in PBS by incubation 10–15 min at room temperature with gentle rotation. After centrifugation at

500 × *g* for 3 min the supernatants are collected. This elution step is also repeated one more time. For each HPA^+ and HPA^- material the two PNA eluted fractions (80 μL total) are pooled and labeled as HPA^+/PNA^+ and HPA^-/PNA^+ materials, respectively (Fig. 2).

Alternatively the HPA and PNA chromatography can be performed in columns which may be more suitable for larger volumes of samples.

2.2.2 Applications

The separation of IgA1 HPA^+ and PNA^+ fractions in serum (or plasma) revealed to be a practical technique to separate and isolate different glycoforms of the human IgA1. Theoretically, not only IgA1 but all other O-glycosylated proteins can be analyzed with this approach to overall profile their O-glycans according to HPA^+ or PNA^+ only and even HPA^+/PNA^+ or HPA^-/PNA^-.

2.2.3 Identification of the O-Glycoprotein IgA1 in the Different Fractions by Immunoblotting

Following the separation of the four HPA^-/PNA^-, HPA^-/PNA^+, HPA^+/PNA^-, and HPA^+/PNA^+ fractions, immunoblotting can be performed next to detect the presence of glycoproteins of interest in each fraction with their specific antibodies.

Here, immunoblotting of human IgA1 is performed as an example. After the samples were separated in 4–20% SDS-PAGE and transferred on a nitrocellulose membrane with a standard procedure. The membrane is blocked by incubation with a TBS solution containing 0.1% Tween 20 (TBST) and 5% (w/v) nonfat-dried milk for 1 h at room temperature. The membrane is next incubated with a mouse antihuman IgA1 FITC-conjugated (Southern Biotech, Birmingham, AL, USA) at 1 μg/mL in 5% (w/v) milk (TBST) overnight at 4°C. This incubation is followed by four washes with TBST, 5 min each, prior to incubation with a rabbit anti-FITC antibody HRP-conjugated (Dako, Denmark) at 1 μg/mL in 5% (w/v) nonfat-dried milk (TBST) for 1 h at room temperature. After washed with TBST four times, the IgA1 signal is detected using enhanced chemiluminescence substrate of HRP (SuperSignal® West Pico Chemiluminescent Substrate, Thermo Scientific).

Fig. 3 illustrates results that can be obtained with the serial HPA/PNA chromatography with IgA1 and IgA2 (Lehoux et al., 2014). The immunoblot **1** shows a serum sample with both HPA^-/PNA^+ and HPA^+/PNA^- IgA1 glycoforms. The immunoblot **2** shows a serum sample with almost

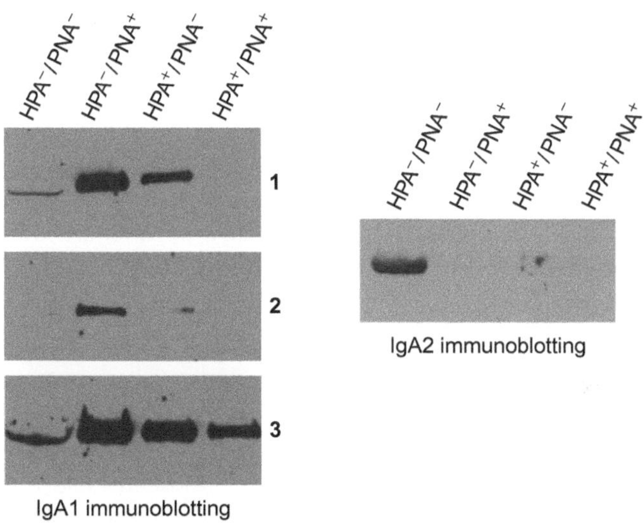

Fig. 3 IgA1 and IgA2 immunoblot analysis after serial HPA/PNA chromatography. *Left panels*: Following our lectin chromatographic separation, the samples were analyzed by western blot with antihuman IgA1 antibody, and the results from three plasma samples (**1**, **2**, and **3**) were shown to have different IgA1 profiles in the different HPA/PNA fractions. *Right panel*: The western blot analysis with antihuman IgA2 antibody only revealed a staining in the HPA^-/PNA^- fraction due to the absence of O-glycans on the human IgA2, demonstrating the specific binding of HPA and PNA to IgA1, O-glycoproteins. *This research was originally published* in Molecular & Cellular Proteomics (MCP) *by Lehoux, S., Mi, R., Aryal, R.P., Wang, Y., Schjoldager, K.T., Clausen, H., et al. (2014). Identification of distinct glycoforms of IgA1 in plasma from patients with immunoglobulin A (IgA) nephropathy and healthy individuals.* Molecular & Cellular Proteomics: MCP, 13*(11), 3097–3113. © the American Society for Biochemistry and Molecular Biology.*

exclusively HPA^-/PNA^+ IgA1 material. The immunoblot **3** shows a serum sample that contains both HPA^-/PNA^+ and HPA^+/PNA^- IgA1 materials but also more interestingly some HPA^+/PNA^+ IgA1 glycoform. More surprisingly, this sample 3 also contains a significant amount of apparent HPA^-/PNA^- IgA1 material. This was not caused by overloading since the HPA^-/PNA^- material was reloaded onto the HPA and PNA beads, respectively, and nothing was detected in the eluted materials. A similar immunoblot using a mouse antihuman IgA2 antibody (α2 chain specific; FITC-conjugated, from Dako) in the same conditions as the anti-IgA1 shows that IgA2 is, as expected since IgA2 does not have any O-glycans, exclusively found in the HPA^-/PNA^- fraction, confirming the specific detection.

2.2.4 O-Glycan Analysis by Mass Spectrometry

Serial chromatography separation of serum glycoforms of human IgA1 can also be applied prior to mass spectrometry (MS) O-glycans analysis. To have enough material of IgA to analyze, the starting material 200–500 μL of serum (or plasma) is first desialylated by incubation with 5–10 mU neuraminidase. Total serum IgA is then immunopurified with antihuman IgA (α-chain specific)-agarose beads (Sigma-Aldrich), which are washed twice with 500 μL of PBS prior to incubation with the desialylated serum sample for an overnight incubation at 4°C under gentle rotation. Next, the serum and bead mixture is transferred into a 5-mL Poly-Prep Chromatography Column (Bio-Rad, Hercules, CA, USA) and washed four times with 1 mL of PBS. The beads are then eluted three times with 1 mL of 0.2 *M* glycine–HCl pH 2.2. These three fractions are pooled, neutralized with 1.0 *M* Tris–HCl (pH 8.0) to ~pH 7.5, concentrated, and washed with PBS using a Centricon unit (Millipore, Billerica, MA, USA) to a final volume of 100 μL.

This 100 μL of IgA is used as a starting material for the serial HPA/PNA chromatography as described earlier. The presence of IgA/IgA1 in the desired fractions, e.g., mainly in the HPA^-/PNA^+ and HPA^+/PNA^- fractions, can be tested and confirmed by measurement of absorbance at 280 nm or by immunoblotting. Fractions containing at least 25 μg of material can be preceded further for O-glycan analysis by MS.

The release of O-glycans from IgA1 is achieved by pronase digestion directly followed by permethylation of the digested material (Goetz, Novotny, & Mechref, 2009; Lehoux et al., 2014). Up to 50 μg of IgA1 material is resuspended in MilliQ water to a final volume of 100 μL. Pronase (Millipore, Billerica, MA, USA) is then added to the sample to a final concentration of 0.2 mg/mL and incubated at 55°C for 48 h. The pronase-digested material is then lyophilized. The dried material is next permethylated by the addition of 200 μL of DMSO/NaOH slurry solution and 50 μL of room temperature methyl iodide (Sigma-Aldrich) followed by vigorous shaking for 30 min at room temperature. Afterward, the mixture is centrifuged at 5000 × *g* for 1 min and the supernatant collected. Five hundred microliters of MilliQ water and 500 μL of chloroform are then added and the mixture briefly mixed before centrifugation at 5000 × *g* for 1 min. The upper layer of water is discarded and 500 μL of MilliQ water is added for two more wash cycles. The resulting chloroform fraction is evaporated by Centrivap, resuspended in 200 μL of 50% methanol, and loaded onto a C18 Sep-Pak column (Waters, Milford, MA, USA)

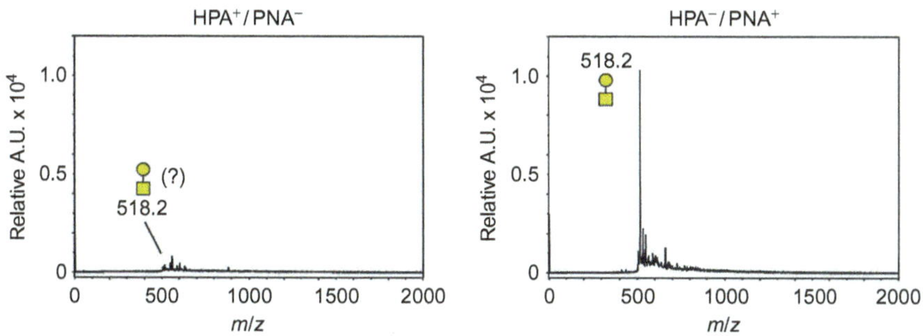

Fig. 4 Mass spectrometry O-glycan analysis of human IgA1 eluted from HPA (HPA^+) or PNA (PNA^+) chromatography. *Left panel*: No significant signals corresponding to O-glycans could be detected in the IgA1 material isolated by HPA (HPA^+), while a peak corresponding to the most abundant structure in IgA1 (the core 1 O-glycan, GalNAc-Gal) at the expected *m/z* of 518.2 (*right panel*) was clearly visible in the IgA1 material isolated by PNA (PNA^+). *This research was originally published* in Molecular & Cellular Proteomics (MCP) *by Lehoux, S., Mi, R., Aryal, R.P., Wang, Y., Schjoldager, K.T., Clausen, H., et al. (2014). Identification of distinct glycoforms of IgA1 in plasma from patients with immunoglobulin A (IgA) nephropathy and healthy individuals.* Molecular & Cellular Proteomics: MCP, 13*(11), 3097–3113. © the American Society for Biochemistry and Molecular Biology.*

preequilibrated as follows: 1 column volume (CV) of methanol, 1 CV of MilliQ water, 1 CV of acetonitrile, and 1 CV of MilliQ water. The C18 column is then washed with 3 CVs of water before being eluted with 1 CV of 50% acetonitrile. The eluted fraction is lyophilized and resuspended in 20 μL of 50% methanol before being spotted for MS analysis. 1 μL of permethylated glycans and 1 μL of 2,5-dihydroxybenzoic acid (DHB) (Sigma-Aldrich) matrix (10 mg/mL in 50:50 ACN:H_2O + 0.1% TFA) are mixed on a MALDI target plate and the data acquired.

Fig. 4 spectra were acquired using an Ultraflex-II TOF/TOF system from Bruker Daltonics in positive reflector mode (Lehoux et al., 2014). These spectra show the absence of any detectable O-glycan structures in permethylated material from the HPA^+/PNA^- fraction since the mass of GalNAc alone is too small to be detected by MS in these conditions and might be buried into the matrix signal. In contrast, a clear mass corresponding to the core-1 structure in the permethylated material from the HPA^-/PNA^+ fraction was detected.

3. DISCUSSION

With a carefully designed strategy, we have developed a workable methodology to analyze the O-glycosylations on IgA1 from human plasma (or serum) by a serial HPA, PNA chromatography followed by immunoblotting and/or MS. With this method, our results showed that a minor IgA1 fraction, not only from patients with IgAN but also from healthy individuals, carried truncated O-glycans, Tn and/or STn. Importantly, human IgA1 is readily differentiated into two distinct O-glycoforms: one with normal core 1-based complex O-glycans and another with exclusive Tn/STn antigens. The majority of the samples we examined (15 of total 16 cases) had these two distinct O-glycoforms of IgA1 and a minority of samples (1 in 16 cases) had the third O-glycoform, e.g., heterogeneous with both Tn/STn and core 1-based complex structures (HPA^+/PNA^+) in addition to those two distinct O-glycoforms: Tn/STn and sialyl core 1 O-glycans (Lehoux et al., 2014), which were often identified in glycopeptides by MS (Franc et al., 2013; Odani et al., 2010; Takahashi et al., 2010, 2012). Our results also suggested that a fraction of IgA1 was not O-glycosylated. Our data did not only raise a question about the role of Tn/STn carrying IgA1 in IgAN but also demonstrate for the first time the two distinct O-glycoforms of IgA1 in human serum. The data generated by our method are conclusive because (1) the lectin beads have sufficient capacity (not overloaded) to bind all glycoproteins that have corresponding structures; (2) the binding is specific to the O-glycans since IgA2, which is not O-glycosylated, is not in either HPA^+ or PNA^+ fractions; (3) lectin-bound and -unbound fractions of IgA1 are stained correspondingly on the lectin blotting; and (4) finally, the anti-IgA1 and IgA2 antibodies are specific.

Within human immunoglobulins, IgA1 and IgD (Mellis & Baenziger, 1983; Smith, de Wolff, Molyneux, Feehally, & Barratt, 2006) are most commonly O-glycosylated in their HR domains, yet a recent study also showed that IgG3 can be O-glycosylated in the hinge region (Plomp et al., 2015), although IgG3 is less efficiently O-glycosylated than IgA1 and IgD. It would be very interesting to investigate whether IgD and IgG3 also exist in two distinct O-glycoforms or not, since they can carry multiple O-glycans, similar to IgA1.

Modern technologies such as MS provide us unprecedented opportunities to Glycomics. Yet analysis of the truncated O-glycans, such as Tn and STn antigens, is challenging because of their size. Most studies (Franc et al., 2013; Odani et al., 2010; Takahashi et al., 2010, 2012) showed that the O-glycans on each IgA1 molecule from IgAN patients were heterogeneous, with both Tn/STn and core 1-based structures using MS to analyze the protease-generated glycopeptides of HR from IgA1, but rarely detect the Tn/STn only glycopeptide. The reason for Tn-glycopeptides of HR were discriminated in the MS analysis is unknown. Using our serial HPA-, PNA-chromatography approach, we identified that for the first time, two distinct O-glycoforms of IgA1 exist in serum of both healthy individuals and IgAN patients. While the mechanisms underlying the Tn/STn expression on the HR of IgA1 are not yet clear, we have developed a practical methodology to profile the overall O-glycan structures on glycoproteins and their potential content in Tn/STn and core 1 complex O-glycans. Are those structures homogeneous or heterogeneous on cell level? If a glycoprotein has multiple O-glycosylation sites, are those structures homogeneous or heterogeneous? It should be pointed out that the glycan profile of an individual glycoprotein needs to be carefully interpreted if the cell extracts or mixed glycoproteins are analyzed by serial lectin chromatography, since there is a possibility of protein–protein interaction and co-pull down by the lectin beads. To avoid this potential issue, denaturing of the glycoproteins prior to lectin chromatography may help. Nevertheless, although MS is the modern technology for analyzing glycan structures, it also has limitations. Our method has its merits, including its practicality for any ordinary lab to perform the analysis, and furthermore, the serial lectin chromatography can provide basic but important overall O-glycan profile of cells or a glycoprotein, which assists in the interpretation of MS data if MS analysis is followed.

ACKNOWLEDGMENT

The study here was supported by a National Institutes of Health Grant R01DK80876 (to T.J.).

REFERENCES

Baenziger, J., & Kornfeld, S. (1974). Structure of the carbohydrate units of IgA1 immunoglobulin. II. Structure of the O-glycosidically linked oligosaccharide units. *Journal of Biological Chemistry*, *249*, 7270–7281.

Barratt, J., Feehally, J., & Smith, A. C. (2004). Pathogenesis of IgA nephropathy. *Seminars in Nephrology*, *24*, 197–217.

Berger, J., & Hinglais, N. (1968). Intercapillary deposits of IgA-IgG. *Journal d'urologie et de néphrologie*, *74*, 694–695.

Franc, V., Rehulka, P., Raus, M., Stulik, J., Novak, J., Renfrow, M. B., et al. (2013). Elucidating heterogeneity of IgA1 hinge-region O-glycosylation by use of MALDI-TOF/TOF mass spectrometry: Role of cysteine alkylation during sample processing. *Journal of Proteomics*, *92*, 299–312.

Giannakakis, K., Feriozzi, S., Perez, M., Faraggiana, T., & Muda, A. O. (2007). Aberrantly glycosylated IgA1 in glomerular immune deposits of IgA nephropathy. *Journal of the American Society of Nephrology: JASN*, *18*, 3139–3146.

Goetz, J. A., Novotny, M. V., & Mechref, Y. (2009). Enzymatic/chemical release of O-glycans allowing MS analysis at high sensitivity. *Analytical Chemistry*, *81*, 9546–9552.

Hiki, Y. (2009). O-linked oligosaccharides of the IgA1 hinge region: Roles of its aberrant structure in the occurrence and/or progression of IgA nephropathy. *Clinical and Experimental Nephrology*, *13*, 415–423.

Ju, T., Brewer, K., D'Souza, A., Cummings, R. D., & Canfield, W. M. (2002). Cloning and expression of human core 1 beta1,3-galactosyltransferase. *Journal of Biological Chemistry*, *277*, 178–186.

Ju, T., & Cummings, R. D. (2002). A unique molecular chaperone Cosmc required for activity of the mammalian core 1 beta 3-galactosyltransferase. *Proceedings of the National Academy of Sciences of the United States of America*, *99*, 16613–16618.

Ju, T., Cummings, R. D., & Canfield, W. M. (2002). Purification, characterization, and subunit structure of rat core 1 beta1,3-galactosyltransferase. *Journal of Biological Chemistry*, *277*, 169–177.

Ju, T., Otto, V. I., & Cummings, R. D. (2011). The Tn antigen-structural simplicity and biological complexity. *Angewandte Chemie (International Ed in English)*, *50*, 1770–1791.

Kerr, M. A. (1990). The structure and function of human IgA. *The Biochemical Journal*, *271*, 285–296.

Lehoux, S., Mi, R., Aryal, R. P., Wang, Y., Schjoldager, K. T., Clausen, H., et al. (2014). Identification of distinct glycoforms of IgA1 in plasma from patients with immunoglobulin A (IgA) nephropathy and healthy individuals. *Molecular & Cellular Proteomics: MCP*, *13*, 3097–3113.

Mattu, T. S., Pleass, R. J., Willis, A. C., Kilian, M., Wormald, M. R., Lellouch, A. C., et al. (1998). The glycosylation and structure of human serum IgA1, Fab, and Fc regions and the role of N-glycosylation on Fc alpha receptor interactions. *Journal of Biological Chemistry*, *273*, 2260–2272.

McEver, R. P., Moore, K. L., & Cummings, R. D. (1995). Leukocyte trafficking mediated by selectin-carbohydrate interactions. *Journal of Biological Chemistry*, *270*, 11025–11028.

Mellis, S. J., & Baenziger, J. U. (1983). Structures of the O-glycosidically linked oligosaccharides of human IgD. *Journal of Biological Chemistry*, *258*, 11557–11563.

Mestecky, J., Russell, M. W., Jackson, S., & Brown, T. A. (1986). The human IgA system: A reassessment. *Clinical Immunology and Immunopathology*, *40*, 105–114.

Mestecky, J., Tomana, M., Moldoveanu, Z., Julian, B. A., Suzuki, H., Matousovic, K., et al. (2008). Role of aberrant glycosylation of IgA1 molecules in the pathogenesis of IgA nephropathy. *Kidney & Blood Pressure Research*, *31*, 29–37.

Odani, H., Yamamoto, K., Iwayama, S., Iwase, H., Takasaki, A., Takahashi, K., et al. (2010). Evaluation of the specific structures of IgA1 hinge glycopeptide in 30 IgA nephropathy patients by mass spectrometry. *Journal of Nephrology*, *23*, 70–76.

Papista, C., Berthelot, L., & Monteiro, R. C. (2011). Dysfunctions of the Iga system: A common link between intestinal and renal diseases. *Cellular & Molecular Immunology*, *8*, 126–134.

Plomp, R., Dekkers, G., Rombouts, Y., Visser, R., Koeleman, C. A., Kammeijer, G. S., et al. (2015). Hinge-region O-glycosylation of human immunoglobulin G3 (IgG3). *Molecular & Cellular Proteomics: MCP*, *14*, 1373–1384.

Schjoldager, K. T., & Clausen, H. (2012). Site-specific protein O-glycosylation modulates proprotein processing—Deciphering specific functions of the large polypeptide GalNAc-transferase gene family. *Biochimica et Biophysica Acta*, *1820*, 2079–2094.

Smith, A. C., de Wolff, J. F., Molyneux, K., Feehally, J., & Barratt, J. (2006). O-glycosylation of serum IgD in IgA nephropathy. *Journal of the American Society of Nephrology: JASN*, *17*, 1192–1199.

Takahashi, K., Smith, A. D., Poulsen, K., Kilian, M., Julian, B. A., Mestecky, J., et al. (2012). Naturally occurring structural isomers in serum IgA1 o-glycosylation. *Journal of Proteome Research*, *11*, 692–702.

Takahashi, K., Wall, S. B., Suzuki, H., Smith, A. D., 4th, Hall, S., Poulsen, K., et al. (2010). Clustered O-glycans of IgA1: Defining macro- and microheterogeneity by use of electron capture/transfer dissociation. *Molecular & Cellular Proteomics: MCP*, *9*, 2545–2557.

Tarelli, E., Smith, A. C., Hendry, B. M., Challacombe, S. J., & Pouria, S. (2004). Human serum IgA1 is substituted with up to six O-glycans as shown by matrix assisted laser desorption ionisation time-of-flight mass spectrometry. *Carbohydrate Research*, *339*, 2329–2335.

Wada, Y., Tajiri, M., & Ohshima, S. (2010). Quantitation of saccharide compositions of O-glycans by mass spectrometry of glycopeptides and its application to rheumatoid arthritis. *Journal of Proteome Research*, *9*, 1367–1373.

Wang, Y., Ju, T., Ding, X., Xia, B., Wang, W., Xia, L., et al. (2010). Cosmc is an essential chaperone for correct protein O-glycosylation. *Proceedings of the National Academy of Sciences of the United States of America*, *107*, 9228–9233.

CHAPTER SIX

Monitoring Protein Synthesis in *Caenorhabditis elegans* Using SILAC

M. Miyagi[1]
Center for Proteomics and Bioinformatics, Cleveland, OH, United States
Case Western Reserve University, Cleveland, OH, United States
[1]Corresponding author: e-mail address: masaru.miyagi@case.edu

Contents

Abstract

The static levels of proteins are the net results of their production and clearance regulated by the activities of proteins involved in their synthesis, degradation, and transportation. Therefore, the information on the rates of protein synthesis and clearance is needed to understand the underlying dynamic nature of a proteome. In this chapter, the experimental technique, we use in our laboratory for monitoring the synthesis of individual proteins in *Caenorhabditis elegans* (*C. elegans*) is described. The technique utilizes a preisotopically labeled amino acid ($^{13}C_6$-Lys) as a precursor for protein synthesis and monitors the kinetics of the precursor incorporation into the newly synthesized proteins. *C. elegans* is a powerful animal model in various fields of biomedical science such as aging, developmental biology, and neurobiology. The experimental technique would, therefore, be useful for research laboratories using *C. elegans* as an animal model.

Methods in Enzymology, Volume 585
ISSN 0076-6879
http://dx.doi.org/10.1016/bs.mie.2016.09.015

1. INTRODUCTION

Biological systems are highly dynamic. The controlled synthesis/clearance of proteins is critical for most cellular processes such as the activation of signaling pathways and the responses to external stimuli as well as maintaining the cellular homeostasis. It is important to recognize that proteins are continually synthesized and cleared in a living organism even under the steady-state conditions, during which the levels of individual proteins remain unchanged because the rates of synthesis and clearance for the proteins are balanced (equal). It is, therefore, possible that even when the abundance of a protein remains unchanged, the rate of protein synthesis is actually increased, but the clearance rate is also increased, resulting in counterbalancing the total protein abundance. Thus, the information on the synthesis and clearance of proteins is needed to understand the dynamic nature of physiological processes. Such information is, however, largely missing for many biological systems whose proteome have been characterized. Therefore, measuring the synthesis rates of proteins at a global scale in those biological systems will likely yield novel insights into the underlying physiological processes.

In contrast to the static measurement of protein concentration, for which there are many useful proteomic methods are already available (Twyman, 2014), the synthesis of proteins is not directly observable and requires the measurement of temporal changes of newly synthesized protein molecules. This can be done by administering a stable isotope-labeled precursor and monitoring the kinetics of isotope transfer into the newly synthesized protein products. There are three different strategies that have been employed for this purpose (Claydon & Beynon, 2012). The first strategy involves the uniform labeling of proteins with ^{15}N (Price, Guan, et al., 2010) or ^{13}C (Vogt, Hunzinger, et al., 2005), which can be achieved by giving animals ubiquitously ^{15}N- or ^{13}C-labeled diet. The second strategy involves the labeling of proteins with deuterium (^{2}H) or heavy oxygen (^{18}O) by administering living organisms heavy water ($^{2}H_2O$ or $H_2^{18}O$) (Rachdaoui, Austin, et al., 2009). The third strategy involves the labeling of proteins with a prelabeled amino acid(s) (e.g., $^{13}C_6$-Lys) by giving animals a diet containing preisotope-labeled amino acid(s). This latter strategy is often referred to as "stable-isotope labeling by amino acids in cell culture" (SILAC) because it was originally developed for in vitro use (Ong, Blagoev, et al., 2002). A major advantage of using SILAC over other two strategies is that the difference in mass between the labeled and unlabeled peptide species is constant for all the peptides in a sample; thus, the peptide identification and quantification is straightforward. Mass spectrometry has been the most widely used detection technique in all the three strategies because it is capable of tracking individual proteins in a complex mixture and offers high quantitative accuracy and detection sensitivity.

Caenorhabditis elegans (*C. elegans*) is a small (about 1 mm in length) and free-living (nonpathogenic and nonparasitic) organism. *C. elegans* lives in the soil and feeds primarily on microbes such as bacteria. In a laboratory setting, it can be readily raised on agar plates or liquid cultures on a diet of *Escherichia coli* (*E. coli*). *C. elegans* was the first multicellular organism whose whole genome was sequenced (C. elegans Sequencing Consortium, 1998), thus suitable for proteomic study. It is an important animal model in various fields such as aging, developmental biology, and neurobiology (http://www.wormbook.org/) due to evolutionary conservation of gene function between the worm and mammal (Burglin, Lobos, et al., 1998). Many of its genes are similar to those in humans, and many useful mutants have already been developed in *C. elegans* that mimic important human disorders. Thus, setting up an experimental platform to monitor the protein synthesis in this important model organism would be worthwhile for the laboratories conducting research using this animal.

2. PRINCIPLE OF SILAC-BASED MEASUREMENT OF PROTEIN SYNTHESIS IN *C. ELEGANS*

In order to adapt the SILAC strategy for monitoring protein synthesis in *C. elegans*, diet for the worms that contains one or more of the essential amino acids prelabeled with stable isotopes (usually ^{13}C and/or ^{15}N) needs to be prepared. *E. coli* in which all the proteins are fully labeled with $^{13}C_6$-Lys has been used as the food for worms (Yuan, Kadiyala, et al., 2012). $^{13}C_6$-Lys has been the preferred amino acid because no metabolic exchange of the isotopically labeled elements in $^{13}C_6$-Lys with other amino acids occurs (Yuan et al., 2012). An important thing to note is that both the proteome (Yuan et al., 2012) and the lifespan (Vukoti, Yu, et al., 2015) of *C. elegans* that has been fully labeled with $^{13}C_6$-Lys are essentially the same as those of unlabeled animal. This implies that no significant physiological changes occur by the isotope labeling and therefore justifying the use of SILAC strategy for biological studies on this animal.

A general experimental workflow of SILAC-based measurement of protein synthesis in *C. elegans* is shown in Fig. 1A. Unlabeled worms that have been grown on light-Lys ($^{12}C_6$-Lys)-labeled *E. coli* are transferred to heavy-Lys ($^{13}C_6$-Lys)-labeled *E. coli* plates and cultured for different durations and harvested. Since lysine is an essential amino acid for *C. elegans* all the proteins synthesized during the incubation incorporate the heavy-Lys. This allows us to distinguish the newly synthesized proteins (labeled with the heavy-Lys) from the original proteins (labeled with the light-Lys) that had existed before switching the medium using mass spectrometry. Proteins are extracted from the worms, digested by Lys-C, and the digest is analyzed by LC–MS/MS. Lys-C is the preferred protease for the digestion of proteins in this case

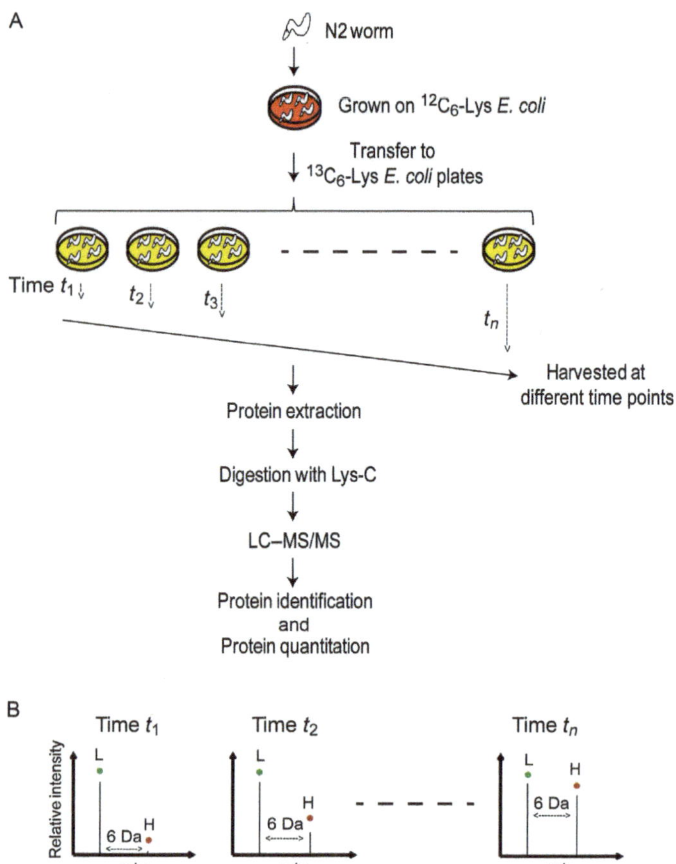

Fig. 1 SILAC-based strategy for monitoring the newly synthesized proteins in *C. elegans*.

because the complete digestion results in producing peptides that contain a single lysine residue (except the C-terminal peptides of proteins) with a constant 6 Da mass difference between the heavy and light peptide pair. The light and heavy forms of a peptide coelute from the LC column and therefore appear in the same mass spectrum, which makes the subsequent identification and quantification of the peptide straightforward and reliable. Fig. 1B shows the hypothetical mass spectra of a peptide at different time points. The $^{13}C_6$-Lys labeled "heavy" peak (H) increases relative to the $^{12}C_6$-Lys labeled "light" peak (L) as the duration of incubation increases, thus indicating an increase of the newly synthesized protein from which the peptide was originated. It should be noted that the newly synthesized proteins are labeled with heavy-Lys in this example; however, the experiment can be done reversely as well, in which fully heavy-Lys-labeled worms are transferred to light-Lys-labeled *E. coli* plates and cultured for different durations (Vukoti et al., 2015).

3. EXPERIMENTAL PROCEDURES

3.1 Preparation of Light ($^{12}C_6$)-Lys and Heavy ($^{13}C_6$)-Lys-Labeled Bacteria

1. Streak the arginine and lysine auxotrophic *E. coli* strain AT713 (obtained from the *E. coli* Genetic Stock Center at Yale University) on a lysogeny broth (LB) plate and culture overnight at 37°C.
2. A single bacterial colony is then inoculated into 10 mL of LB and incubated overnight in an incubator shaker (37°C, 180 rpm).
3. A portion of bacterial culture (~100 μL) is then inoculated into 50 mL of M9 minimal medium (50 m*M* Na_2HPO_4, 20 m*M* KH_2PO_4, 10 m*M* NaCl, 20 m*M* NH_4Cl, 2 m*M* $MgSO_4$, 0.1 m*M* $CaCl_2$, and 0.2% glucose) supplemented with arginine (100 μg/mL), cysteine (100 μg/mL), and lysine (100 μg/mL, either light ($^{12}C_6$)- or heavy ($^{13}C_6$, 99 atom%)-Lys) and continuously cultured in an incubator shaker (37°C, 200 rpm) until the absorbance of the culture at 600 nm (A600) reached 1.0.
4. Then, 10 mL of the resulting labeled bacteria is inoculated into 1000 mL of M9 basal medium, with amino acid supplementation, and cultured in an incubator shaker (37°C, 200 rpm) until A600 reached 2.0.
5. Harvest the labeled bacteria by pelleting by centrifugation (8000 × *g*, 10 min).

6. Resuspend the bacteria in 15 mL sterile water and spread onto a peptone-free nematode growth media (NGM) plate (500 μL for each 10-cm plate).
7. Expose the plates to 1000 mJ/cm^2 of ultraviolet light (SpectroLinker XL-1500, Spectronics Corp, Westbury, NY) to kill the bacteria. The plates with the light- and heavy-Lys containing bacteria can be stored for 1–2 weeks at 4°C. The labeling yield should be checked by analyzing the bacterial proteins by mass spectrometry. This can be done by extracting proteins with 100 m*M* ammonium bicarbonate (only soluble proteins are extracted), digesting the proteins by Lys-C, and analyzing the digest by LC–MS/MS. The labeling yield of proteins can be estimated by comparing the peak heights of the $^{13}C_6$-Lys labeled "heavy" peak to the unlabeled "light" peak (L) from any of the proteins observed. Almost complete isotopic labeling (>98%) should be achieved for all the peptides except for the peptides that do not contain a lysine residue (mostly the C-terminal peptides of proteins). A representative spectrum of a peptide derived from a protein expressed in the $^{13}C_6$-Lys labeled-*E. coli* is shown in Fig. 2. As can be seen in the figure, the intensity of $^{13}C_6$-Lys labeled "heavy" peak is much higher than that of unlabeled peak, indicating that almost complete isotope labeling is achieved.

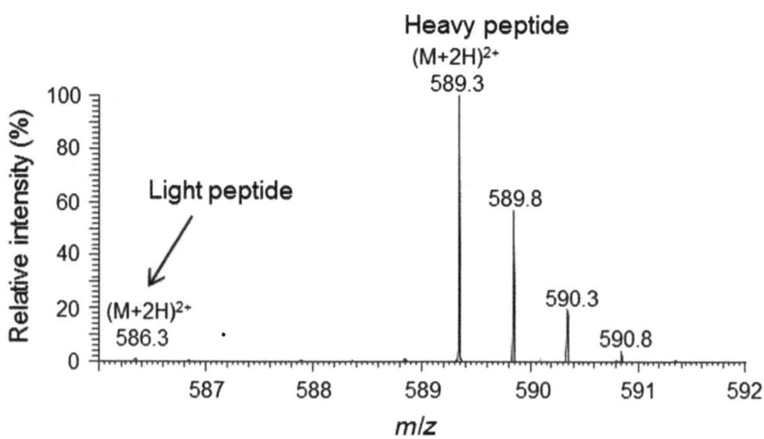

Fig. 2 Labeling efficiency of *E. coli* proteins by heavy lysine. Proteins extracted from $^{13}C_6$-Lys-labeled *C. elegans* were digested by Lys-C and analyzed by LC–MS/MS. The mass spectrum of a representative peptide shows near complete labeling of the peptide with $^{13}C_6$-Lys.

3.2 *C. elegans* Maintenance and Age Synchronization

1. Culture worms (WT Bristol N2 strain) on the peptone-free NGM plates (51 m*M* NaCl, 25 mm K_3PO_4, 5 μg/mL cholesterol, 1 m*M* $CaCl_2$, 1 m*M* $MgSO_4$) seeded with the *E. coli* strain AT713 raised on the $^{12}C_6$-Lys (light-Lys) media by transferring 8–12 single L4 larvae to the plates and growing for 3–5 days to achieve several hundred (300–500) fully gravid nematodes. WT Bristol N2 strain is the strain that has its genome sequenced (C. elegans Sequencing Consortium, 1998).
2. Wash plate several times with sterile distilled water to loosen worms and eggs stuck in bacteria. To synchronize the worms' age, treat gravid nematodes with bleach solution (5% sodium hypochlorite solution, 0.25 *M* NaOH) for 3–9 min, with occasional vortexing. The bleaching results in the death of worms, while the eggshell protects the embryos. Longer bleaching times will lead to loss of viable eggs. Wash worms several times with sterile distilled water to remove the remaining bleach solution.
3. Incubate the surviving axenic eggs in liquid M9 buffer (22 m*M* KH_2PO_4, 42 m*M* Na_2HPO_4, 86 m*M* NaCl) with shaking overnight without food and allow to hatch as starved L1 larvae.
4. Aspirate liquid (to remove dauer pheromone accumulated during starvation) and transfer all eggs onto peptone-free NGM plates seeded with light ($^{12}C_6$)-Lys *E. coli*.
5. After hatching, culture the age-synchronized animals up to L4 larvae stage before transferring to several (depending on the number of experimental groups or replicates) peptone-free NGM plates seeded with light ($^{12}C_6$)-Lys *E. coli* plates containing 25 mg/L 5-fluoro-2-deoxyuridine and growing worms to desired ages for experiment. 5-Fluoro-2-deoxyuridine is used to prevent egg hatching, but should be avoided during larval stages prior to adulthood, as it leads to developmental defects.

3.3 Label-Chase Experiment

1. Transfer the light-Lys-labeled worms from each peptone-free NGM plate seeded with light-Lys-labeled *E. coli* onto a peptone-free NGM plate previously seeded with heavy-Lys-labeled *E. coli* containing 25 mg/L 5-fluoro-2-deoxyuridine. Each plate should contain approximately 500–1000 worms.

2. Harvest the worms at different time points from each 10-cm diameter plate with 40 mL of sterile distilled water.
3. Place the worms in a 50-mL Falcon tube and centrifuge at 2000 × *g* for 2 min.
4. Collect the worms and carefully overlay onto chilled 30% sucrose and centrifuge at 2000 × *g* for 5 min.
5. Collect the upper layer of worms (these worms are alive) in a 15-mL Falcon tube, wash several times with sterile distilled water, centrifuge at 2000 × *g* for 2 min.
6. Flash freeze and store the worms at −80°C until analyzed.

3.4 Preparation of Protein Samples for Mass Spectrometry Analysis

1. To the worms harvested from the 10-cm plates, add 250 μL of 100 m*M* ammonium bicarbonate buffer containing 4% perfluorooctanoic acid (*w/v*) and protease inhibitor mixture (1 m*M* EDTA, 0.2 m*M* PMSF, 0.7 μg/μL leupeptin, and 0.5 μg/μL pepstatin A), and extract proteins by ultrasonication (4.5 kHz three times for 9 s with a 3-min pause on ice between the pulses). Note that perfluorooctanoic acid has been demonstrated to be capable of extracting membrane proteins (Kadiyala, Tomechko, et al., 2010). Therefore, this procedure extracts not only the soluble proteins but also the membrane proteins.
2. Centrifuge at 15,000 × *g* for 10 min and collect the supernatant.
3. Add dithiothreitol to be 10 m*M* and incubate at 37°C for 30 min, and then add iodoacetamide to be 25 m*M* and incubate at 25°C for 45 min.
4. After the reaction, add a ninefold excess volume of ice-cold acetone, and leave for 2 h at −20°C to precipitate the proteins.
5. Centrifuge at 2400 × *g* for 10 min and collect the pellet.
6. Wash the pellet with ice-cold 90% acetone, centrifuge at 2400 × *g* for 1 min, remove the supernatant. Repeat this two to three times.
7. Dissolve the pellet in 50 μL of 100 m*M* ammonium bicarbonate containing 8 *M* urea, and then slowly dilute with 450 μL of 100 m*M* ammonium bicarbonate.
8. Estimate protein concentration. Various methods can be used. We typically use a DC protein assay kit from Bio-Rad (Hercules, CA).
9. Digest the proteins by Lys-C (1:50 Lys-C to protein ratio [*w/w*]) at 37°C overnight.

3.5 LC–MS/MS Analysis

LC–MS/MS analysis of the Lys-C digest can be done in a variety of ways depending on the availability of HPLC and mass spectrometer. We recommend using a reverse-phase nano-HPLC column for the separation of the Lys-C peptides and a high-resolution mass spectrometer such as time-of-flight or Fourier transform ion cyclotron (FT-ICR) instrument for the peptides detection. If the HPLC system used is not equipped with a trapping column for removing urea in the digest, the urea must be removed prior to the LC–MS/MS analysis because it likely interferes the analysis. This can be done using a disposable small reverse-phase C18 column. The use of low-resolution mass spectrometer such as the quadrupole instrument likely complicates the downstream spectrum analysis because its resolution is not sufficient to achieve the baseline separation between adjacent isotopic peaks. The detailed LC–MS/MS protocol we use can be found in our recent report (Vukoti et al., 2015).

3.6 Protein Identification

The identification of peptide, hence protein, can be done by subjecting the acquired LC-MS/MS data against a protein sequence database using a database search engine. We use Wormpep (can be obtained from: ftp://ftp.sanger.ac.uk/pub/wormbase/releases/), which contains only *C. elegans* proteins, as the protein sequence database and Mascot (Matrix Science, London, UK) as the database search engine. Other search engines such as SEQUEST (Eng, McCormack, et al., 1994) can also be used. To account for the *S*-alkylation of protein samples, carbamidomethylation of cysteine should be set as a fixed modification, whereas replacement of C-terminal Lys with heavy-Lys should be set as variable modification. Strict Lys-C specificity is usually applied, and missed cleavages are not allowed; therefore, all the identified peptides contain (except the C-terminal peptides of proteins) only one lysine residues at the C-termini.

3.7 Quantification of Light- and Heavy-Lys-Labeled Peptides and Proteins

We use the SILAC Quantification Suite in ProteomicsTools software (freely available from: https://github.com/shengqh/RCPA.Tools/releases) (Guo, Miyagi, et al., 2014; Sheng, Dai, et al., 2012) to determine the abundances of light- and heavy-Lys-labeled proteins. Other protein quantification software such as MaxQuant (Cox & Mann, 2008) can also be used.

The SILAC Quantification Suite extracts the precursor ion of a peptide to be quantified from a raw LC–MS/MS file based on the information on the MS/MS identification of the peptide. Five scans around the identified MS/MS spectrum are included in quantification scan window. The standard deviations (SD) for the m/z values of both light- and heavy-Lys-labeled precursors are calculated based on those scans. Then, the quantification scan window is extended at both ends until the m/z offset of either light- or heavy-Lys-labeled precursor becomes larger than threefold of their corresponding SD. Once the quantification scan window is determined, the light- and heavy-Lys-labeled precursor intensities are extracted, and the ratio is calculated by applying nonnegative least-square fitting. Proteins identified with multiple peptides are also quantified by the nonnegative least-square method using the light- and heavy-Lys-labeled precursor intensities of all peptides identified from the same protein. The R^2 values of nonnegative least square at both peptide and protein level are used to indicate the quantification quality. All the quantified proteins can be manually validated at both peptide and scan levels with the user-friendly graphic interface (Fig. 3). The final quantification results can then be exported as a spreadsheet. Peptides with the R^2 values greater than 0.8 are usually accepted as quantified peptides.

3.8 Data Presentation

Once the abundance of light- and heavy-Lys-labeled protein is obtained using ProteomicsTools software, the fraction of the heavy-Lys-labeled protein [(heavy protein)/(light protein + heavy protein)], which is the fraction of newly synthesized protein, can be calculated for each protein. The value at each time point is then plotted against time as shown in Fig. 4. The hypothetical plot shows that how the fraction of newly synthesized protein increases over the time. From the plot, it is obvious that the turnover rate of Protein A is fastest and then followed by Protein B and C. When steady state can be assumed for the conducted experiment, the rate constant for the synthesis of each protein can be calculated by fitting the data to the first-order kinetic curve: $y = 1 - e^{-kt}$, where y is the fraction of newly synthesized protein at time $= t$, k is the rate constant, and t is the time.

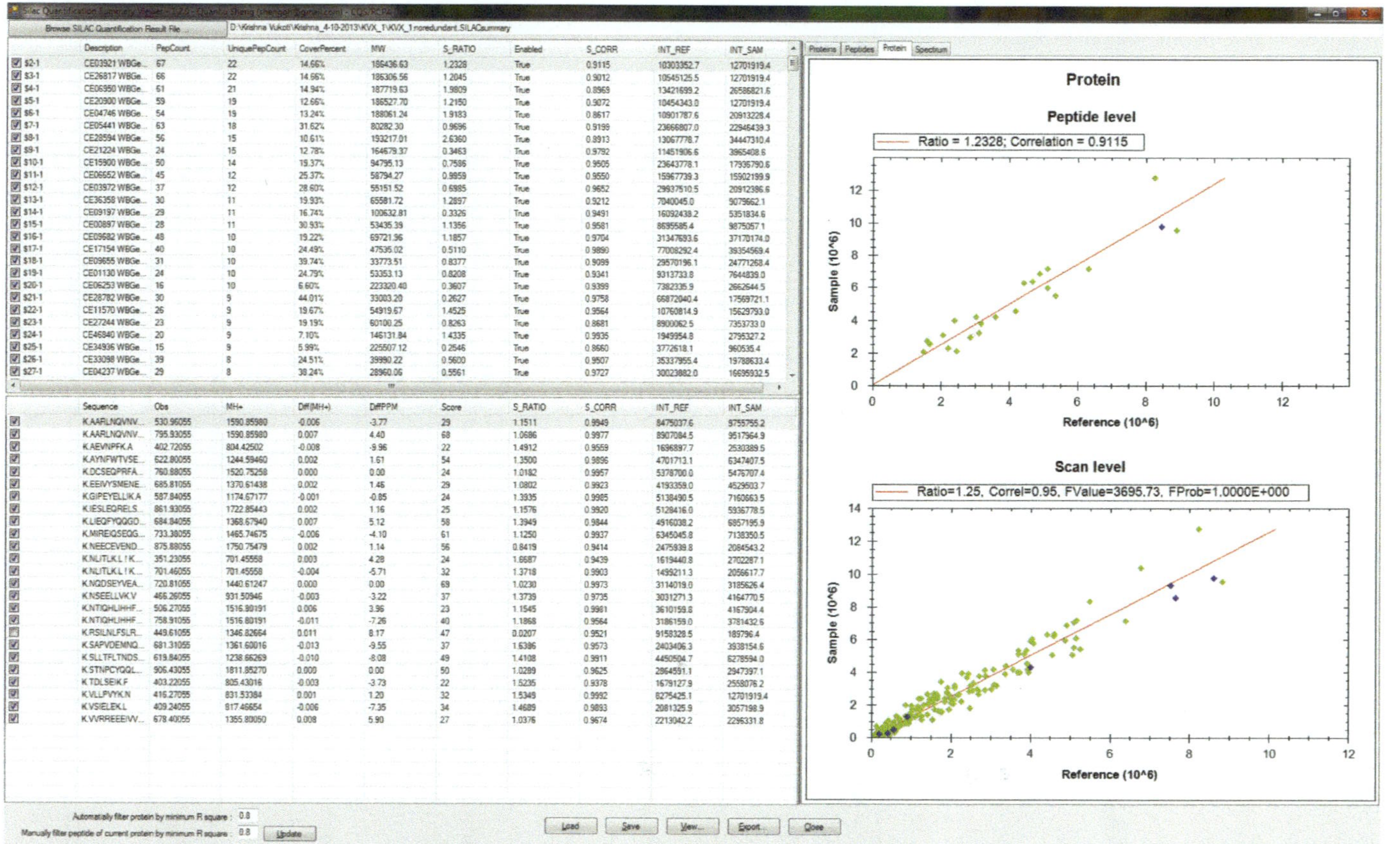

Fig. 3 SILAC quantification summary viewer in ProteomicsTool Suite (version 4.0.13.1). The protein (*upper panel* on the *left*) and peptide (*bottom panel* on the *left*) information are displayed in a *spread sheet*. The *scatter plots* (peak intensities of light- vs heavy-Lys-labeled peptides) at the peptide (*upper panel* on the *right*) and the scan (*bottom panel* on the *right*) level and the robust linear model fitted line are visualized. This graphical user interface allows users to perform visual quality control and manually exclude outliers.

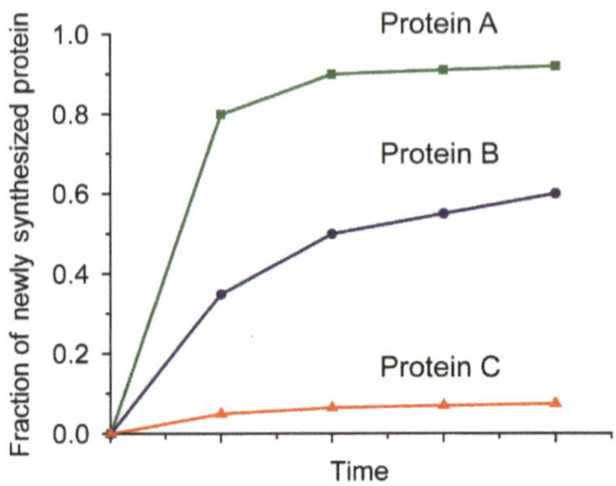

Fig. 4 The appearance of newly synthesized proteins during the label-chase experiment.

4. CONCLUSION

Experimental procedures for monitoring the protein synthesis in *C. elegans* have been described. Since most of the proteomic studies have measured only the static concentrations of proteins expressed in this animal, it is likely that monitoring the synthesis of proteins provide new information. Although we have used the technique only for the wild-type animals, many interesting experiments could be designed using the described experimental platform, for example, comparing the rates of synthesis of individual proteins between wild-type and long-lived nematode (*daf-2* mutant). The results from such studies will have an immediate impact in the field of aging research.

In addition to monitoring protein synthesis, the experimental platform can be used for various comparative proteomic experiments such as for comparing protein expressions in $^{13}C_6$-Lys-labeled and -unlabeled worms. We have used the platform to compare the proteome of wild-type and *eat-2* mutant nematodes, a genetic model that manifests chronic caloric restriction. In this experiment, an equal weight of the fully $^{13}C_6$-Lys-labeled wild-type and unlabeled *eat-2* mutant nematodes were mixed, and their proteins were subjected to proteome analysis (Yuan et al., 2012). Since all peptides show $^{13}C_6$-Lys-labeled and -unlabeled peaks originated from the wild-type and the *eat-2* mutant worms, respectively, we were able to determine the differences in the expression levels of individual proteins between the two animal groups. Similarly, $^{13}C_6$-Lys-labeled and -unlabeled worms will be useful for studying protein modifications, protein–protein interactions, and protein–drug interaction.

ACKNOWLEDGMENT

I would like to thank Zhaoyang Feng, Ao-Lin Hsu, Yiyuan Yuan, and Krishna Vukoti for providing expertise in nematode maintenance, Quanhu Sheng for developing the software for LC-MS/MS data analysis, and Nelson Hsieh for his critical comments on this chapter.

REFERENCES

Burglin, T. R., Lobos, E., et al. (1998). *Caenorhabditis elegans* as a model for parasitic nematodes. *International Journal for Parasitology*, *28*(3), 395–411.

C. elegans Sequencing Consortium. (1998). Genome sequence of the nematode *C. elegans*: A platform for investigating biology. *Science*, *282*(5396), 2012–2018.

Claydon, A. J., & Beynon, R. (2012). Proteome dynamics: Revisiting turnover with a global perspective. *Molecular & Cellular Proteomics*, *11*(12), 1551–1565.

Cox, J., & Mann, M. (2008). MaxQuant enables high peptide identification rates, individualized p.p.b.-range mass accuracies and proteome-wide protein quantification. *Nature Biotechnology*, *26*(12), 1367–1372.

Eng, J. K., McCormack, A. L., et al. (1994). An approach to correlate tandem mass spectral data of peptides with amino acid sequences in a protein database. *Journal of the American Society for Mass Spectrometry*, *5*(11), 976–989.

Guo, Y., Miyagi, M., et al. (2014). O18Quant: A semiautomatic strategy for quantitative analysis of high-resolution 16O/18O labeled data. *BioMed Research International*, *2014*, 971857.

Kadiyala, C. S., Tomechko, S. E., et al. (2010). Perfluorooctanoic acid for shotgun proteomics. *PLoS One*, *5*(12), e15332.

Ong, S. E., Blagoev, B., et al. (2002). Stable isotope labeling by amino acids in cell culture, SILAC, as a simple and accurate approach to expression proteomics. *Molecular & Cellular Proteomics*, *1*(5), 376–386.

Price, J. C., Guan, S., et al. (2010). Analysis of proteome dynamics in the mouse brain. *Proceedings of the National Academy of Sciences of the United States of America*, *107*(32), 14508–14513.

Rachdaoui, N., Austin, L., et al. (2009). Measuring proteome dynamics in vivo: As easy as adding water? *Molecular & Cellular Proteomics*, *8*(12), 2653–2663.

Sheng, Q., Dai, J., et al. (2012). BuildSummary: Using a group-based approach to improve the sensitivity of peptide/protein identification in shotgun proteomics. *Journal of Proteome Research*, *11*(3), 1494–1502.

Twyman, R. M. (2014). *Principles of proteomics*. New York: Garland Science.

Vogt, J. A., Hunzinger, C., et al. (2005). Determination of fractional synthesis rates of mouse hepatic proteins via metabolic 13C-labeling, MALDI-TOF MS and analysis of relative isotopologue abundances using average masses. *Analytical Chemistry*, 77(7), 2034–2042.

Vukoti, K., Yu, X., et al. (2015). Monitoring newly synthesized proteins over the adult life span of *Caenorhabditis elegans*. *Journal of Proteome Research*, *14*(3), 1483–1494.

Yuan, Y., Kadiyala, C. S., et al. (2012). Enhanced energy metabolism contributes to the extended life span of calorie-restricted Caenorhabditis elegans. *The Journal of Biological Chemistry*, *287*(37), 31414–31426.

CHAPTER SEVEN

Identification of Novel Macropinocytosing Human Antibodies by Phage Display and High-Content Analysis

K.D. Ha, S.M. Bidlingmaier, Y. Su, N.-K. Lee, B. Liu[1]
UCSF Helen Diller Family Comprehensive Cancer Center, University of California at San Francisco, San Francisco, CA, United States
[1]Corresponding author: e-mail address: bin.liu@ucsf.edu

Contents

Abstract

Internalizing antibodies have great potential for the development of targeted therapeutics. Antibodies that internalize via the macropinocytosis pathway are particularly promising since macropinocytosis is capable of mediating rapid, bulk uptake and is selectively upregulated in many cancers. We hereby describe a method for identifying antibodies that internalize via macropinocytosis by screening phage-displayed single-chain antibody selection outputs with an automated fluorescent microscopy-based high-content analysis platform. Furthermore, this method can be similarly applied to other endocytic pathways if other fluorescent, pathway-specific, soluble markers are available.

Methods in Enzymology, Volume 585
ISSN 0076-6879
http://dx.doi.org/10.1016/bs.mie.2016.10.004

1. INTRODUCTION

There is currently much interest in utilizing internalizing antibodies as the basis for developing targeted therapeutics including antibody–drug conjugates (Behrens & Liu, 2014; Carter & Senter, 2008; Scott et al., 2012; Sherbenou et al., 2015). The uptake efficiency and intracellular fate of delivered agents will affect therapeutic potency and depends on the internalization pathway utilized. Macropinocytosis is a form of bulk uptake that permits rapid and efficient internalization of antibody-delivered cargos (Ha et al., 2014). Additionally, macropinocytosis is selectively upregulated in many cancer types. For example, Ras-transformed pancreatic cancer cells upregulate macropinocytosis to increase amino acid uptake (Commisso et al., 2013). Thus, antibodies capable of efficient internalization via macropinocytosis are intriguing candidates for development into targeted therapeutics, particularly against cancer.

Given the existence of multiple pathways of internalization, robust and efficient methods for identifying macropinocytosing antibodies will greatly aid in their discovery and translation into therapeutics. To address this need, we developed a high-throughput, high-content analysis (HT-HCA) screening protocol that employs automated fluorescent microscopy-based analysis to identify phage antibodies that colocalize with Texas Red-conjugated 70 kDa neutral dextran (ND70-TR), a macropinocytosis marker (Ha et al., 2014). We hereby provide a detailed description of this protocol, including phage antibody identification, validation of macropinocytosis with full-length human IgG, functional internalization based on payload delivery, and identification of receptors bound by novel macropinocytosing antibodies.

2. HT-HCA SCREENING OF MACROPINOCYTIC ANTIBODIES: OVERVIEW OF THE STRATEGY AND METHOD

This HT-HCA of novel antibodies from phage libraries largely depends on fluorescent colocalization. Using either standard or confocal fluorescent microscopy techniques, two standard correlation coefficients, called the Pearson colocalization coefficient (PCC) or the Mander's overlap coefficient (MOC), can be used to quantify the degree of colocalization between two distinct, fluorescent sources both in vitro and in vivo (Adler & Parmryd, 2010; Huang et al., 2015). PCC largely relies on degrees of spatial overlap, while MOC additionally considers fluorescent intensities (Dunn et al., 2011). One critical limitation to utilizing PCC and MOC for determining colocalization through fluorescent microscopy is that the typical optical resolutions of confocal microscopes tend to range between 200 and 300 nm, potentially leading to false-positive colocalization results (Xu et al., 2016). However, in this HT-HCA method, ND70 is used to label macropinosomes that range in size from 200 to 2500 nm in diameter (Hewlett et al., 1994; Kerr & Teasdale, 2009). The large size of macropinosomes therefore obviates a need for high optical resolution.

The experiment utilizes phage-displayed, single-chain variable fragment (scFv) antibody libraries as a source for novel antibody clones. Multiple methods exist in generating phage-displayed antibody libraries (Andris-Widhopf et al., 2011; Marks et al., 1991; O'Connell et al., 2002; Sheets et al., 1998; Weber et al., 2014). Phage-display libraries need to be preselected against target tissues or cell types of interest to generate polyclonal outputs that are greatly enriched for binding clones; otherwise, the frequency of binding clones in the nonselected libraries is too low to allow HT-HCA screening (Ha et al., 2014). Multiple methods have been published that describe ways to select for high-affinity phage antibodies from cell and tissue specimen. For cell-based selection, phage antibody display libraries are incubated with live cells to enrich both surface-bound and internalized antibodies (An et al., 2008; Liu et al., 2004; Poul et al., 2000; Zhu et al., 2010). For tissue-based selection, we previously developed a novel method that involves selecting phage antibody libraries on cancer tissues with the aid of laser capture microdissection to specifically identify phage antibody binding to tumor cells in situ residing in their tissue microenvironment (Ruan et al., 2006). Antibodies identified by this method bind to clinically relevant tumor epitopes and show excellent tumor targeting in vivo (He et al., 2010; Iyer et al., 2011a, 2011b; McCabe et al., 2012). Alternatively, directed ultraviolet light can introduce genomic cross-linking in phage antibody clones bound to undesired regions of tissues, thus rendering them nonreplicable (Larsen et al., 2015). Another method involves formalin-fixed, paraffin-embedded tissue biopsies on glass slides (Ten Haaf et al., 2015).

The HT–HCA method begins with polyclonal phage-displayed antibody library selection outputs enriched for binding to target tissues or cell lines of interest. Colonies from the polyclonal selection outputs are picked and arrayed into 96-well plates for monoclonal culturing. Monoclonal phage antibodies are incubated with the cell line of interest and analyzed for binding by FACS. Positive binders are then rearrayed and tested using

HCA to further select for phage antibodies that colocalize with ND70-TR (Ha et al., 2014). Colocalization is quantified by PCC (or MOC) score, and clones with high PCC (or MOC) scores are further characterized by fluorescence confocal microscopy to quantify colocalization with the lysosomal marker lysosomal-associated membrane protein 1 (LAMP1). Clones that exhibit colocalization are then converted into full-length human antibodies, retested to confirm for both high-affinity binding and macropinocytosis-specific endocytosis, and then used to identify the corresponding receptor bound by the antibody. The method is summarized in a flowchart, depicted in Fig. 1 (Ha et al., 2014).

3. HT-HCA SCREENING PROTOCOL FOR MACROPINOCYTIC PHAGE ANTIBODIES

3.1 Monoclonal Phage Antibody Array From a Polyclonal Selection Output

1. Inoculate 1–5 mL of appropriate bacterial culture media with frozen glycerol stock of bacteria previously infected with the preselected phage antibody display library, containing appropriate antibiotic(s) (e.g., TG1 bacteria infected with fd phage displaying scFv antibody fragments are grown in 2× YT media containing 12.5 μg/mL tetracycline).
2. Measure A_{600}, where 1 O.D. is equivalent to about 6.8×10^8 cells/mL.
3. Dilute bacterial culture to yield ~90 cells/μL.
4. Plate 200–5000 cells onto 10 cm agar media plates containing the appropriate antibiotic(s). Plated cell counts will depend on bacterial cell viability within the glycerol stock.
5. Incubate plate(s) at 37°C for 18 h.
6. After 18 h, the incubated plates should yield physically visible colonies with respectable spacing in between colonies to permit facile plucking using pipette tips or sterilized toothpicks.
7. Add 100–150 μL of bacterial culture media with appropriate antibiotic(s) into each well of a sterile, round-bottom, polystyrene, or polypropylene 96-well plate with lid (BD Biosciences).
8. Inoculate each well with individual colonies using sterile pipette tips or toothpicks, leaving at least one well mock inoculated with a bacteria-free pipette tip or toothpick. This well will serve as a negative control to test for potential contamination.

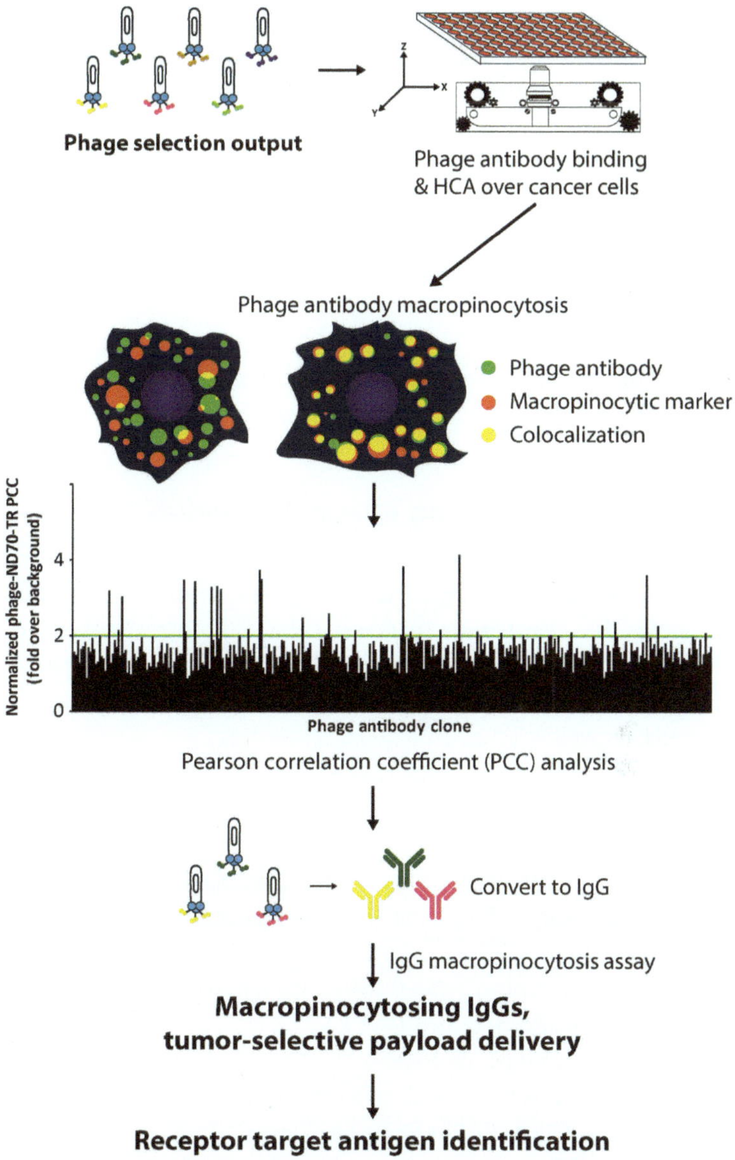

Fig. 1 Outline of HCA screening strategy for identification of macropinocytosing phage antibody. Sublibraries generated previously by us from laser capture microdissection-based phage antibody library selection (Ruan et al., 2006) are used as the starting material for screening. HCA instruments enable automated high-throughput detection of antibody colocalization with a macropinocytosis marker, which is quantified by PCC scores. Phage antibodies with PCC scores above two standard deviations of the mean are identified and converted to full-length human IgGs for further characterization,

(Continued)

9. Cover each 96-well plate with its accompanied lid and then cover each plate with aluminum foil.
10. Incubate 96-well plate(s) at 37°C with shaking at 150–225 RPM for 18 h.
11. After 18 h, transfer 50 μL of bacterial/phage culture with a multichannel pipette into wells of a new, sterile 96-well round-bottom plate.
12. Add 50 μL of bacterial media containing 50% glycerol to each well and pipette up/down five times to mix. Label and freeze plate at −80°C as a glycerol stock plate.
13. The remaining 50–100 μL of bacterial/phage culture can be sealed with paraffin wax tape and stored at 4°C for up to 24 h, to be used in high-affinity phage clone selection with tissue culture cells.

3.2 Identification of Binding Phage Antibodies

1. Seed cell line of interest into sterile, optical flat-bottom 96-well plates with lids (BD Biosciences) to yield 70–90% confluence the next day.
2. The next day, centrifuge bacterial/phage culture 96-well plate at 3900 × *g* for 5 min and add 20 μL of supernatant containing phage particles into the respective wells of the 96-well plate containing tissue culture cells.
3. Incubate phage particles over tissue culture cells at 37°C for 90 min, with standard carbon dioxide (CO_2, 5%) and humidity requirements for the cell line of interest.
4. After 90 min, gently wash unbound phage particles away from tissue culture cells using 150 μL of 1 × phosphate-buffered saline (PBS) twice.
5. Fix cells and any bound phages using 4% paraformaldehyde in PBS for 10 min at room temperature (RT).
6. Wash 3 × with 150 μL of PBS.
7. Permeabilize cells with 150 μL of PBS containing 1% fraction V bovine serum albumin (BSA; Fisher Scientific) and 0.1% Triton X-100 (Sigma) for 15 min at RT.

Fig. 1—Cont'd including macropinocytosis, tumor-selective payload delivery, and target antigen identification. *Adapted from our original publication in Ha, K. D., et al. (2014). High-content analysis of antibody phage-display library selection outputs identifies tumor-selective macropinocytosis-dependent rapidly internalizing antibodies.* Molecular & Cellular Proteomics, 13*(12), 3320–3331.*

8. Add 50 μL of biotinylated rabbit antibacteriophage fd antibody (Sigma, diluted 1:1000 in PBS) to each well and incubate at RT for 1–2 h.
9. Wash twice with PBS.
10. Incubate wells with 50 μL of fluorophore-conjugated secondary agent (e.g., phycoerythrin-conjugated streptavidin (Life Technologies/ Invitrogen) or Alexa Fluor® 488-conjugated streptavidin (Jackson ImmunoResearch), final concentration 1 μg/mL) in PBS buffer containing Hoechst 33342 dye at 1 μg/mL for 15–30 min at RT. *Note*: If the antibacteriophage antibody from Section 3.2, step 8 is already conjugated with a fluorophore, incubate cells with just Hoechst dye in PBS.
11. Wash wells thrice with PBS.
12. Image 96-well plates on automated, fluorescent plate imager such as CellInsight™ NXT HCS (Thermo Scientific), imaging a minimum of 300 cells per well. Cell count can be determined through Hoechst dye fluorescence.
13. For each phage-displayed antibody clone, clones yielding high antibacteriophage fluorescence values per cell are identified (typically two standard deviations above the mean).
14. Consolidate binding phage clones from their respective glycerol stock plates into fresh, sterile 96-well round-bottom plates, using 100–150 μL of bacterial media with appropriate antibiotic(s) (e.g., for fd phage, 2 × YT with 12.5 μg/mL tetracycline).
15. Culture 96-well plates with at 37°C with 150–225 RPM shaking for 18 h.
16. After 18 h, prepare glycerol stocks of the consolidated phage clones as described in Section 3.1, step 12, storing the remaining cultures at 4°C for up to 24 h until further testing with HT-HCA.

3.3 HT-HCA Screening of Macropinocytosing Phage Antibodies

1. Seed tissue culture cell line of interest into flat-bottom 96-well plates with lids to yield 70–90% confluence the next day.
2. The next day, aspirate media from all wells and add 100 μL of media containing 70 kDa neutral dextran conjugated with Texas Red (ND70-TR, ThermoFisher/Life Technologies/Invitrogen) at a final concentration of 50 μg/mL.
3. Add 20 μL of supernatant from Section 3.2, step 16 to each well. *Note*: At this dilution, most of the tissue culture cell lines are viable for at least

24 h. If, however, viability is affected by phage-containing supernatant, reduce the amount added (e.g., 10 or 5 μL) or shorten the time of incubation.

4. Incubate at 37°C for 18 h, with standard carbon dioxide and humidity requirements for the tissue culture cell line of interest. *Note*: Incubation time can vary between 8 and 18 h depending on the cell line.
5. After 18 h, gently wash unbound phage particles and noninternalized ND70-TR away from tissue culture cells using 150 μL of PBS twice.
6. Fix cells and any bound phages using 4% paraformaldehyde in PBS for 10 min at RT.
7. Wash with 150 μL of PBS.
8. Permeabilize cells for 15 min at RT.
9. Wash once with PBS, add 50 μL of biotinylated rabbit antibacteriophage fd antibody (Sigma, diluted 1:1000 in 1:5 PBS-diluted permeabilization buffer) to each well, and incubate at RT for 1–2 h.
10. Wash wells twice with PBS.
11. Incubate wells with 50 μL of fluorophore-conjugated secondary agent (e.g., phycoerythrin-conjugated streptavidin (Life Technologies/Invitrogen) or Alexa Fluor® 488-conjugated streptavidin (Jackson ImmunoResearch), final concentration 1 μg/mL) in PBS buffer containing Hoechst 33342 dye at 1 μg/mL for 15–30 min at RT. *Note*: If the antibacteriophage antibody from Section 3.3, step 9 is already conjugated with a fluorophore, incubate cells with just Hoechst dye in 1:5 diluted permeabilization buffer.
12. Wash wells 3 × with PBS.
13. Image the processed 96-well plate within an automated, fluorescent plate imager, e.g., CellInsight™ NXT HCS platform (Thermo Scientific) with a semiapochromat 20 × LUCPLFLN objective (Olympus) utilizing > 6 fields per well with a minimum of 300 cells per well.
14. Quantify colocalization through PCC (or MOC) score using analysis software associated with the instrument, utilizing Hoechst nuclei staining as the reference channel for cell counts (for CellInsight, use the built-in colocalization analysis). MOC analysis will require careful adjustment of fluorescent threshold settings, and this threshold should be applied across all image fields. We typically use PCC in our HCA screening.
15. Phage antibody clones yielding high PCC values (typically two standard deviations above the mean of control cells) are identified as

candidates with macropinocytosis-specific internalization qualities. Control cells should have bound phage antibodies previously identified that show no internalizing activity.

3.4 Sequencing, Purification, and Secondary Characterization of Phage Clones

1. Culture bacteria/phage clones exhibiting ND70-TR colocalization in appropriate bacterial media with antibiotic(s).
2. Extract plasmid DNA from bacterial cultures and sequence.
3. Culture phage with unique antibody sequence in 100 mL bacterial culture media with appropriate antibiotic(s) for 18 h.
4. After 18 h, centrifuge bacteria at 3900 × *g* for 20 min at 4°C.
5. Collect ~95 mL supernatant and filter-sterilize through 0.45 μm filters into sterile, conical tubes.
6. Add 25 mL 20% polyethylene glycol (PEG)-8000 (Sigma-Aldrich or VWR) in 2.5 *M* NaCl, mix, and incubate over ice for 5 h.
7. Centrifuge solution at 3900 × *g* for 20 min.
8. Carefully decant supernatant.
9. Resuspend the phage pellet in 10 mL PBS.
10. Add 2.5 mL 20% PEG-8000 in 2.5 *M* NaCl, mix, and incubate over ice for 10 min.
11. Centrifuge at 3900 × *g* for 20 min.
12. Carefully aspirate supernatant.
13. Resuspend pellets with 3 mL PBS, filter-sterilize through 0.45 μm filters into cryovials, and store at −80°C. *Note*: Alternatively, purified phages may be stored at 4°C for many months, depending on the specific clone. Binding against tissue culture cells of interest will need to be reassessed on a regular basis to determine stability and display level of purified phage stored at 4°C.
14. Seed tissue culture cell line of interest onto a sterile 8-well Lab-Tek II chambered cover glass for subsequent fluorescent confocal microscopy and incubate at 37°C for 18 h. Aim for 50–60% confluence by the next day.
15. Next day, incubate cells with complete tissue culture media containing ND70-TR at final 50 μg/mL and purified phage (1:10 dilution) for desired time period (e.g., 1 h, 4 h, 8 h, etc.).
16. Wash twice with PBS unbound phage particles and ND70-TR away from tissue culture cells.
17. Fix cells using 4% paraformaldehyde in PBS for 10 min at RT.

18. Wash twice with PBS.
19. Permeabilize cells with permeabilization buffer comprising PBS, 1% BSA, and 0.1% Triton X-100, for 15 min at RT.
20. Add biotinylated rabbit antibacteriophage fd antibody diluted at 1:1000 in 1:5 PBS-diluted permeabilization buffer for 1 h at RT.
21. Wash twice with PBS.
22. Add to chamber fluorophore-conjugated secondary agent (e.g., phycoerythrin-conjugated streptavidin (ThermoFisher/Life Technologies/Invitrogen) or Alexa Fluor® 488-conjugated streptavidin (Jackson ImmunoResearch), both at final concentrations of 1 μg/mL) in PBS buffer containing Hoechst 33342 dye at 1 μg/mL for 15–30 min at RT.
23. Wash 3 × with PBS.
24. Image on a confocal microscope such as Fluoview (Olympus), capturing slices of cells that are sufficient to cover the entire *Z*-depth of each cell.
25. Quantify the degree of colocalization using on-board analysis software between phage and ND70-TR using PCC (using confocal *Z*-slices) or MOC (using *Z*-projected confocal images). Phage antibody clones yielding at least two standard deviations greater PCC or MOC values when compared to control phage antibody clones are considered candidates with macropinocytosis-specific internalization qualities. Standard deviation thresholds can be adjusted accordingly to limit or increase the number of potential, screened phage antibody clones. The control phage antibody should bind to the plasma membrane and does not internalize into the cell.

3.5 Converting Single-Chain Antibody Fragments into Human IgG Antibodies

Phage antibodies that exhibit binding to tissue culture cells of interest and ND70-TR colocalization by HCA and confocal microscopy can then be converted into full-length human antibody for further analysis of internalization pathway and functional delivery.

1. Inoculate phage-infected TG1 bacteria in 3 mL of bacterial culture media with antibiotic(s) (e.g., 2 × YT containing 12.5 μg/mL tetracycline) and incubate at 37°C at 150–225 RPM for 18 h.
2. After 18 h, pellet bacteria at 8000 × *g* for 5 min.
3. Extract plasmid DNA from bacterial pellet using standard molecular biology techniques.

4. PCR-amplify heavy- and light-chain variable fragments separately from plasmid using appropriate primer sets (matching antibody sequences) and clone into either in-house expression vectors (Smith et al., 2009) or commercial heavy- and light-chain expression vectors such as pFUSE-CHIg-hG1 for human IgG1 heavy chain (InvivoGen) and pFUSE2-CLIg-hK (InvivoGen) for human kappa light chain or pFUSE2-CLIg-hL2 (InvivoGen) for human lambda 2 light chain.
5. Prepare plasmid DNA of both the heavy and light expression constructs and sequence to confirm successful cloning.

3.6 Transient Expression of Full-Length Human IgGs

Full-length human IgG production is accomplished by transient cotransfection of a mix of heavy- and light-chain expression plasmids into HEK293a cells. For untested antibody clones, conduct small-volume, initial transfections before scaling up. The ratio of heavy to light can be varied to achieve the best results (we typically start with heavy to light at molar ratio 1:1). Stable cell line can be generated by electroporation of CHO cells, which is not covered by this protocol.

1. Seed HEK293a (ThermoFisher/Life Technologies/Invitrogen) into sterile 10 cm tissue culture plates or T-75 flasks to yield 80–90% confluence the next day.
2. Transfect cells using polyethylenimine (PEI) (Sigma-Aldrich) in Opti-MEM (ThermoFisher/Life Technologies/Invitrogen) or other lipid-based transfection means (e.g., Lipofectamine®3000 (ThermoFisher/Life Technologies/Invitrogen)), using 10 μg of each heavy- and light-chain DNA construct for each 10 cm dish, mix according to the protocol of the transfection reagent.
3. For PEI-based transfection, incubate overnight at 37°C with 5% CO_2, aspirate the transfection complexes and media from the tissue culture vessels, and replace with serum-free media (50% DMEM/50% RPMI 1640) supplemented with penicillin–streptomycin and Nutridoma-SP (Roche).
4. Collect after 6 days antibody-containing media from tissue culture vessels. Add fresh serum-free media over tissue culture vessels and repeat media collection 3 days later.
5. Centrifuge media at 5000 × *g* for 5 min at 4°C using a bench top centrifuge and transfer supernatant into a fresh, sterile vessel. If available, quantify total IgG content within the collected media using a BLItz® system or equivalent label-free interferometry method.

6. Store antibody-containing media at 4°C for up to 3 days for future purification.
7. Before purification, FACS can be performed to confirm that the antibody can bind to the intended target cell line using procedures described later.
8. Detach target tissue culture cells from growth flasks or dishes using EDTA-based cell dissociation solution for 5–10 min at 37°C.
9. Centrifuge detached cells at $700 \times g$ for 3 min and then aspirate the supernatant.
10. Incubate cells with supernatant containing human antibodies at RT for 1 h. Aim for 10 μg/mL antibody based on the BLItz® measurement. *Note*: If antibody concentrations in unpurified supernatant fall under 10 μg/mL but above 1 μg/mL, incubate cells with undiluted media. If less than 1 μg/mL antibody is present in the unpurified supernatant, the transfection may need to be repeated with varying heavy to light ratio. Antibodies that are poorly produced despite multiple trials are generally not moved to the next stage of analysis.
11. Wash cells twice with PBS/0.5% BSA and incubate cells with secondary antibody (antihuman Fc-specific antibody conjugated with a fluorophore).
12. Wash cells twice with PBS/0.5% BSA and analyze binding by FACS.

3.7 Purification of Full-Length Human IgG Antibodies

This section largely adheres to standard, affinity-based antibody purification by protein A or G chromatography.

1. Prepare vertical drip column containing protein A/G agarose slurry, utilizing the appropriate amount of slurry for the expected, total quantity of antibodies present in the media.
2. Wash columns with five bed volumes of PBS.
3. Add antibody-containing media to column of agarose beads and permit to drip through. Save the flow-through, stored on ice.
4. Wash columns thrice with five bed volumes of PBS.
5. Elute antibodies using 1 × bed volume of 0.1 *M* glycine, pH 2.8.
6. Immediately neutralize with 1 *M* Tris–HCl, pH 8.0. Repeat elution and neutralization.
7. Quantify IgG in eluate fractions and pool main antibody-containing fractions based on BLItz® measurement or equivalent label-free interferometry method.

8. Concentrate pooled fractions using an Amicon® Ultra-4 centrifuge filter with 30,000 Da molecular weight cutoff (MilliPore) per manufacturer's recommendation. Buffer exchange is also accomplished at this step using either PBS or HEPES (pH 7.0).
9. Analyze purified IgG by 10% reducing SDS-PAGE. *Note*: We also perform HPLC (Agilent Technologies, 1220 infinity LC) analysis to determine aggregation status using the size-exclusion column. IgGs that show more than 10% aggregation are not moved to the next stage of analysis as internalization is sensitive to cross-linking by aggregated antibodies. Those aggregation-prone antibodies may be "rescued" by mutagenesis or chain shuffling, which is beyond the scope of this protocol.

3.8 Confirming Human IgG Antibody Internalization Into Cells Through Macropinocytosis

It is important to confirm that the full-length human antibody is indeed internalized by macropinocytosis as indicated by the phage antibody because (1) internalization pathway could be different between single-chain and full-length antibodies, and (2) phage antibody can nonspecifically interact with cell membrane to yield internalization patterns/properties that are not truly associated with the target receptor.

1. Seed tissue culture cell line of interest onto a sterile 8-well Lab-Tek II chambered cover glass and aim for 50–60% confluence by the next day.
2. Next day, incubate cells with tissue culture media containing ND70-TR (final concentration at 50 μg/mL) and purified IgG (final concentration at 10 μg/mL) for various time points.
3. Wash cells twice with PBS to remove any unbound IgG antibodies and ND70-TR.
4. Fix cells using 4% paraformaldehyde in PBS for 10 min at RT.
5. Wash 3 × with PBS.
6. Permeabilize cells with permeabilization buffer comprising PBS, 1% BSA, and 0.1% Triton X-100, for 15 min at RT.
7. Incubate cells with antibodies against various organelles such as lysosomes (LAMP1) and early endosomes (early endosomal antigen 1 or EEA1).
8. Wash 3 × with PBS.
9. Incubate cells with fluorescently conjugated antihuman Fc secondary antibody and fluorescently conjugated secondary antibodies against antiorganelle antibodies, diluted in 1:5 PBS-diluted permeabilization buffer and Hoechst dye for 30 min.

10. Wash twice with PBS and fix cells with 4% paraformaldehyde in PBS for 10 min RT.
11. Image on confocal microscope as described in Section 3.4, step 24.
12. Quantify the degree of colocalization between IgG antibody and ND70-TR (and organelles) using either PCC (using confocal *Z*-slices) or MOC (using *Z*-projected confocal images).

3.9 Identification of Target Receptor Bound by Macropinocytosing Antibody

Once a macropinocytosing antibody is identified and characterized, it can be used to identify the bound antigen. This procedure depends on antibody affinity capture of membrane-bound receptors that are labeled by cell surface biotinylation (Goenaga et al., 2007; Liu et al., 2007; Ruan et al., 2006). Following immunoprecipitation of cell lysates using the target antibody of interest and a control antibody binding to a known antigen against the same cell line, the immunoprecipitate is analyzed by SDS-PAGE and immunoblot that reveal protein bands unique to the immunoprecipitate pulled down by antibody of interest (Ha et al., 2014). The corresponding protein bands can then be excised from an SDS-PAGE and analyzed by tandem mass spectrometry (MS/MS) (Andersen et al., 2010; Liu et al., 2002; Lund et al., 2007). Other novel methods for antigen identification, such as screening yeast surface cDNA display libraries, can also be used to identify antigens bound by either phage or full-length antibodies (Bidlingmaier & Liu, 2006; Bidlingmaier et al., 2009). These methods are described in details elsewhere (Bidlingmaier & Liu, 2011; Bidlingmaier et al., 2015) and are beyond the scope of this chapter.

1. Seed target tissue culture cells of interest to yield roughly 9×10^6 cells the next day for each antibody (target and control antibodies).
2. The next day, harvest and pellet cells at $1000 \times g$ for 5 min at 4°C.
3. Wash cell pellet 3× with PBS (pH 8.0), centrifuging cells at $1000 \times g$ for 5 min after each wash.
4. Resuspend cells at a concentration of 2.5×10^7 cells/mL with PBS (pH 8.0).
5. Equilibrate EZ-Link Sulfo-NHS-LC-Biotin (Thermo Pierce #21335) to RT prior to opening the reagent vessel.
6. Prepare 10 m*M* biotin-labeling reagent in PBS.
7. Immediately add 200 μL of reagent per mL of resuspended cells and incubate on ice for 20 min with occasional agitation.

8. Wash cell pellet 3 × with PBS (pH 8.0) containing 100 m*M* glycine to quench the reaction.
9. Lyse cell pellet with lysis buffer, such as RIPA lysis buffer (50 m*M* Tris–HCl, pH 7–8, 150 m*M* NaCl, 0.1% SDS, 0.5% sodium deoxycholate, 1% NP-40).
10. Centrifuge lysates in a refrigerated microcentrifuge (Eppendorf) at 14,000 × *g* at 4°C for 10 min.
11. Collect supernatant and incubate with 150 μL of protein A agarose (ThermoFisher/Pierce) slurry per mL of lysate.
12. Preclear lysate by incubating with protein A agarose slurry overnight at 4°C with gentle rotation or at RT for 3 h with gentle rotation.
13. Spin down beads at 5000 × *g* for 1 min and collect precleared, biotinylated lysates.
14. Store at −80°C or use immediately for immunoprecipitation.
15. Conjugate target antibody onto protein A/G agarose beads. Additionally conjugate a separate control antibody that binds to a known antigen expressed on the cell line of interest. This can be performed with chemical cross-linking by DMP (Sigma #D8388), as described in the next steps.
16. Bind roughly 2 mg of antibody per mL of wet protein A/G agarose beads in PBS (pH 7.4) and incubate for 1 h at RT with gentle rotation, or 4°C overnight.
17. Wash beads twice with 10 bead volumes of 0.2 *M* sodium borate (pH 9.0), centrifuging beads at 1000 × *g* for 1 min at 4°C.
18. Resuspend beads in 10 bead volumes of 0.2 *M* sodium borate (pH 9.0).
19. Remove roughly 10 μL of agarose beads for future SDS-PAGE analysis (as preconjugated antibody-beads sample).
20. Add DMP (Sigma #D8388) to bead slurry to yield a final DMP concentration of 13 mg/mL.
21. Incubate for 30 min at RT with gentle rotation.
22. Stop the reaction by pelleting and washing beads with 10 bed volumes of 0.2 *M* ethanolamine in PBS (pH 8.0).
23. Incubate beads with ethanolamine solution for 2 h at RT with gentle rotation.
24. Pellet beads and resuspend with PBS (pH 7.4) containing 0.02% sodium azide.
25. Remove 10 μL of agarose beads, for future SDS-PAGE analysis (as postconjugated antibody-beads sample).

26. Check the efficiency of antibody conjugation onto agarose beads by performing SDS-PAGE on the pre- and postconjugated antibody-bead samples. Efficient conjugation is indicated by heavy-chain presence (~55 kDa) in the preconjugation sample, but not in the post-conjugation sample.
27. Add precleared biotinylated cell lysates to target antibody-conjugated agarose beads and incubate at RT with gentle rotation for 1 h or 4°C overnight. Additionally, perform the same for the control antibody.
28. Pellet beads at 14,000 × *g* for 5 min at 4°C using a refrigerated microcentrifuge.
29. Aspirate remaining supernatant and wash beads 3 × with 10 bed volumes of lysis buffer used, such as the RIPA buffer.
30. Split slurry in half into two separate tubes, one will be used for SDS-PAGE and the other for immunoblotting with horseradish peroxidase (HRP)-conjugated streptavidin.
31. Add appropriate amount of 5 × SDS sample buffer to pelleted beads and boil beads for 2 min at >95°C.
32. Run two SDS-PAGE gels in parallel, where one gel is stained with GelCode (ThermoFisher), or other mass spectrometry safe colloidal stain, and the other gel is Western transferred to PVDF membrane and immunoblotted with streptavidin-HRP.
33. The band on the immunoblot that is unique to the target antibody, but not the control antibody, indicates the gel position(s) of the receptor(s) bound by the antibody.
34. Excise the corresponding protein band(s) from the GelCode-stained gel and then submit the excised gel fragment(s) for protein identification through MS/MS. In-gel trypsin digestion, sample analysis by MS/MS, and analysis of MS/MS data using Sequest (ThermoFisher) are commonly available (e.g., http://cbs.umn.edu/cmsp/home).
35. Confirm antigen identification by ectopic expression of the cDNA encoding the putative target antigen in CHO or HEK293 cells that do not express the target (not bound by the antibody), followed by FACS analysis (Ha et al., 2014). Alternatively, if antibodies against the putative target antigen are available commercially, they can be used in Western blot analysis to confirm the target antigen identification (Ruan et al., 2006).

3.10 Functional Internalization and Targeted Payload Delivery

Besides microscopic studies, a functional internalization assay can be used to confirm antibody internalization and demonstrate cell-type-specific intracellular payload delivery. The assay is based on saporin (Stirpe et al., 1983), a plant toxin that is not internalized on its own but is internalized when conjugated to an internalizing antibody. Internalization of saporin inhibits protein synthesis of within the target cell, resulting in cell death (Rahman et al., 2008; Vago et al., 2005). The following procedure is based on biotin—labeling the macropinocytosing antibody, mixing with streptavidin-conjugated toxin to form the immunotoxin, and incubating the immunotoxin with target and control cells to assess cell-type-specific loss of viability.

1. To biotin-label antibody, remove the vial of Sulfo-NHS-LC-Biotin from −20°C and warm it to RT before use.
2. Immediately before use, prepare a 10 m*M* EZ-Link Sulfo-NHS-LC-Biotin solution using ultrapure water.
3. Add 27 μL of 10 m*M* biotin solution to 1 mL of 2 mg/mL purified human IgG antibody in PBS at pH 7.4, rotate the mixture at RT for 45 min.
4. Add 20% (v/v) of 1 *M* Tris–HCl (pH 8.0) to quench the reaction, mix gently by pipetting up and down.
5. Buffer-exchange to PBS and remove nonreacted biotin using Zeba™ spin desalting columns (ThermoFisher) according to the manufacturer's instruction. Biotin-labeled antibody can be stored at −20°C for months or 4°C for a couple of weeks until use in immunotoxin assays.
6. Seed target and control cell lines (e.g., tumor vs nontumor lines) at a density of 1000–2000 cell per well in a 96-well flat-bottom plate with 50 μL growth media and grow for 16 h at 37°C with 5% CO_2.
7. Mix streptavidin-conjugated saporin (SA-ZAP, Advanced Targeting Systems) with biotinylated IgG at a molar ratio of 1:1, vortex, and incubate on ice for 30 min to form the immunotoxin.
8. Add 50 μL serially diluted immunotoxin in PBS to each well and incubate for 96 h at 37°C with 5% CO_2. For initial assessment, 1:10 serial dilution is often used to find the range. For assessment of the half-maximal effective concentration (EC_{50}), 1:3 serial dilution is often used to improve accuracy.

9. Remove the cell growth media from each well, add 100 μL of diluted CCK-8 solution (Dojindo, 10 μL CCK-8 solution mixed with 90 μL PBS) to each well in the 96-well plates, and incubate for 1–4 h at 37°C in 5% CO_2. *Note*: Remove any air bubble in the well, as it interferes with absorbance reading.
10. Measure the absorbance at 450 nm using a microtiter plate reader (e.g., Synergy HT from Bioteck).
11. Determine the EC_{50} value by curve fitting using Prism (GraphPad).

4. CONCLUSION

The method described enables identification of macropinocytosis-specific internalizing antibodies through a high-content analysis-based screening of phage antibody display libraries. Novel phage antibodies are identified by colocalization with macropinocytosis marker, converted into full-length human antibodies, and further characterized with regard to cell binding, pathway of internalization, and intracellular payload delivery. Target antigen can be identified by immunoprecipitation and mass spectrometry analysis. This HT-HCA screening platform is generally applicable and should allow systematic discovery, evaluation, and exploitation of macropinocytosis, a subset of the internalizing space, whose activity is significantly upregulated in many types of cancer.

ACKNOWLEDGMENTS

Work in our laboratory is supported by grants from the National Institutes of Health/National Cancer Institute (R01 CA171315, R01 CA118919, and R01 CA129491).

REFERENCES

Adler, J., & Parmryd, I. (2010). Quantifying colocalization by correlation: The Pearson correlation coefficient is superior to the Mander's overlap coefficient. *Cytometry. Part A*, 77(8), 733–742.

An, F., et al. (2008). Targeted drug delivery to mesothelioma cells using functionally selected internalizing human single-chain antibodies. *Molecular Cancer Therapeutics*, 7(3), 569–578.

Andersen, J. D., et al. (2010). Identification of candidate biomarkers in ovarian cancer serum by depletion of highly abundant proteins and differential in-gel electrophoresis. *Electrophoresis*, *31*(4), 599–610.

Andris-Widhopf, J., et al. (2011). Generation of human scFv antibody libraries: PCR amplification and assembly of light- and heavy-chain coding sequences. *Cold Spring Harbor Protocols*, *2011*(9), 1139–1150.

Behrens, C. R., & Liu, B. (2014). Methods for site-specific drug conjugation to antibodies. *MAbs*, *6*(1), 46–53.

Bidlingmaier, S., & Liu, B. (2006). Construction and application of a yeast surface-displayed human cDNA library to identify post-translational modification-dependent protein-protein interactions. *Molecular & Cellular Proteomics*, *5*(3), 533–540.

Bidlingmaier, S., & Liu, B. (2011). Identification of protein/target molecule interactions using yeast surface-displayed cDNA libraries. *Methods in Molecular Biology*, *729*, 211–223.

Bidlingmaier, S., et al. (2009). Identification of MCAM/CD146 as the target antigen of a human monoclonal antibody that recognizes both epithelioid and sarcomatoid types of mesothelioma. *Cancer Research*, *69*(4), 1570–1577.

Bidlingmaier, S., et al. (2015). Combining phage and yeast cell surface antibody display to identify novel cell type-selective internalizing human monoclonal antibodies. *Methods in Molecular Biology*, *1319*, 51–63.

Carter, P. J., & Senter, P. D. (2008). Antibody-drug conjugates for cancer therapy. *Cancer Journal*, *14*(3), 154–169.

Commisso, C., et al. (2013). Macropinocytosis of protein is an amino acid supply route in Ras-transformed cells. *Nature*, *497*(7451), 633–637.

Dunn, K. W., et al. (2011). A practical guide to evaluating colocalization in biological microscopy. *American Journal of Physiology. Cell Physiology*, *300*(4), C723–C742.

Goenaga, A. L., et al. (2007). Identification and characterization of tumor antigens by using antibody phage display and intrabody strategies. *Molecular Immunology*, *44*(15), 3777–3788.

Ha, K. D., et al. (2014). High-content analysis of antibody phage-display library selection outputs identifies tumor selective macropinocytosis-dependent rapidly internalizing antibodies. *Molecular & Cellular Proteomics*, *13*(12), 3320–3331.

He, J., et al. (2010). Targeting prostate cancer cells in vivo using a rapidly internalizing novel human single-chain antibody fragment. *Journal of Nuclear Medicine*, *51*(3), 427–432.

Hewlett, L. J., et al. (1994). The coated pit and macropinocytic pathways serve distinct endosome populations. *The Journal of Cell Biology*, *124*(5), 689–703.

Huang, S., et al. (2015). Comparison of dextran perfusion and GSI-B4 isolectin staining in a mouse model of oxygen-induced retinopathy. *Eye Science*, *30*(2), 70–74.

Iyer, A. K., et al. (2011a). Novel human single chain antibody fragments that are rapidly internalizing effectively target epithelioid and sarcomatoid mesotheliomas. *Cancer Research*, *71*(7), 2428–2432.

Iyer, A. K., et al. (2011b). The effect of internalizing human single chain antibody fragment on liposome targeting to epithelioid and sarcomatoid mesothelioma. *Biomaterials*, *32*(10), 2605–2613.

Kerr, M. C., & Teasdale, R. D. (2009). Defining macropinocytosis. *Traffic*, *10*(4), 364–371.

Larsen, S. A., et al. (2015). Selection of cell-type specific antibodies on tissue-sections using phage display. *Journal of Cellular and Molecular Medicine*, *19*(8), 1939–1948.

Liu, B., et al. (2002). Towards proteome-wide production of monoclonal antibody by phage display. *Journal of Molecular Biology*, *315*(5), 1063–1073.

Liu, B., et al. (2004). Mapping tumor epitope space by direct selection of single-chain Fv antibody libraries on prostate cancer cells. *Cancer Research*, *64*(2), 704–710.

Liu, B., et al. (2007). Recombinant full-length human IgG1s targeting hormone-refractory prostate cancer. *Journal of Molecular Medicine*, *85*(10), 1113–1123.

Lund, T. C., et al. (2007). iTRAQ is a useful method to screen for membrane-bound proteins differentially expressed in human natural killer cell types. *Journal of Proteome Research*, *6*(2), 644–653.

Marks, J. D., et al. (1991). By-passing immunization. Human antibodies from V-gene libraries displayed on phage. *Journal of Molecular Biology*, *222*(3), 581–597.

McCabe, K. E., et al. (2012). An engineered cysteine-modified diabody for imaging activated leukocyte cell adhesion molecule (ALCAM)-positive tumors. *Molecular Imaging and Biology*, *14*(3), 336–347.

O'Connell, D., et al. (2002). Phage versus phagemid libraries for generation of human monoclonal antibodies. *Journal of Molecular Biology*, *321*(1), 49–56.
Poul, M. A., et al. (2000). Selection of tumor-specific internalizing human antibodies from phage libraries. *Journal of Molecular Biology*, *301*(5), 1149–1161.
Rahman, W., et al. (2008). Selective ablation of dorsal horn NK1 expressing cells reveals a modulation of spinal alpha2-adrenergic inhibition of dorsal horn neurones. *Neuropharmacology*, *54*(8), 1208–1214.
Ruan, W., et al. (2006). Identification of clinically significant tumor antigens by selecting phage antibody library on tumor cells in situ using laser capture microdissection. *Molecular & Cellular Proteomics*, *5*(12), 2364–2373.
Scott, A. M., et al. (2012). Antibody therapy of cancer. *Nature Reviews. Cancer*, *12*(4), 278–287.
Sheets, M. D., et al. (1998). Efficient construction of a large nonimmune phage antibody library: The production of high-affinity human single-chain antibodies to protein antigens. *Proceedings of the National Academy of Sciences of the United States of America*, *95*(11), 6157–6162.
Sherbenou, D. W., et al. (2015). The development of potential antibody-based therapies for myeloma. *Blood Reviews*, *29*(2), 81–91.
Smith, K., et al. (2009). Rapid generation of fully human monoclonal antibodies specific to a vaccinating antigen. *Nature Protocols*, *4*(3), 372–384.
Stirpe, F., et al. (1983). Ribosome-inactivating proteins from the seeds of Saponaria officinalis L. (soapwort), of Agrostemma githago L. (corn cockle) and of Asparagus officinalis L. (asparagus), and from the latex of Hura crepitans L. (sandbox tree). *The Biochemical Journal*, *216*(3), 617–625.
Ten Haaf, A., et al. (2015). Phage display-based on-slide selection of tumor-specific antibodies on formalin-fixed paraffin-embedded human tissue biopsies. *Immunology Letters*, *166*(2), 65–78.
Vago, R., et al. (2005). Saporin and ricin A chain follow different intracellular routes to enter the cytosol of intoxicated cells. *The FEBS Journal*, *272*(19), 4983–4995.
Weber, M., et al. (2014). A highly functional synthetic phage display library containing over 40 billion human antibody clones. *PLoS One*, *9*(6). e100000.
Xu, L., et al. (2016). Resolution, target density and labeling effects in colocalization studies—Suppression of false positives by nanoscopy and modified algorithms. *The FEBS Journal*, *283*(5), 882–898.
Zhu, X., et al. (2010). Identification of internalizing human single-chain antibodies targeting brain tumor sphere cells. *Molecular Cancer Therapeutics*, *9*(7), 2131–2141.

CHAPTER EIGHT

Beyond the Natural Proteome: Nondegenerate Saturation Mutagenesis—Methodologies and Advantages

M.M. Ferreira Amaral*, L. Frigotto†, A.V. Hine*,[1]

*School of Life and Health Sciences, Aston University, Birmingham, United Kingdom
†Isogenica Ltd., The Mansion, Chesterford Research Park, Essex, United Kingdom
[1]Corresponding author: e-mail address: a.v.hine@aston.ac.uk

Contents

Abstract

Beyond the natural proteome, high-throughput mutagenesis offers the protein engineer an opportunity to "tweak" the wild-type activity of a protein to create a recombinant protein with required attributes. Of the various approaches available, saturation mutagenesis is one of the core techniques employed by protein engineers, and in recent times, nondegenerate saturation mutagenesis is emerging as the approach of choice. This review compares the current methodologies available for conducting nondegenerate saturation mutagenesis with traditional, degenerate saturation and briefly outlines the options available for screening the resulting libraries, to discover a novel protein with the required activity and/or specificity.

Methods in Enzymology, Volume 585
ISSN 0076-6879
http://dx.doi.org/10.1016/bs.mie.2016.10.005

1. INTRODUCTION

Natural proteomes provide a plethora of proteins that scientists employ in widespread applications. But despite nature's bounty, there are many occasions where the natural proteins do not quite have the activity needed. Perhaps an enzyme's affinity for a key substrate is poor, or that enzyme lacks the necessary stability. Maybe a transcription factor that binds a particular DNA sequence is needed and the required specificity cannot be found naturally. More frequently, the CDR region of an antibody or antibody fragment will need to be changed to create a key therapeutic. This is where protein engineering comes to the fore, to create those novel proteins. Protein engineering itself relies on two key components: mutagenesis of the encoding gene and screening of the resulting proteins created.

The earliest methods of mutagenesis were conceptually crude. The entire genome of an organism was targeted with a mutagen (chemical or physical), high kill rates were accepted, and any surviving organisms were screened for a phenotype of interest. Notwithstanding the lack of knowledge regarding causative mutation(s), the study of such mutants led to information that was key to elucidating many biochemical pathways. However, by the late 1970s, alternative, more refined approaches had begun to emerge. For example, Shortle and Nathans (1978) were able to target such random mutagenesis to short, specific regions of DNA within a viral chromosome. Around this time, the first reports of successful site-directed mutagenesis were also published. Smith and coworkers had already established the importance of comparing mutant with wild-type sequences in localizing genetic function (Smith et al., 1977), and in 1978, Smith and coworkers published the first example of site-directed mutagenesis mediated by oligonucleotides (Hutchison et al., 1978). These two approaches: site-directed mutagenesis and targeted random mutagenesis would serve as forerunners for the high-throughput mutagenesis strategies of today.

Site-directed mutagenesis developed rapidly from 1978 onward, with protocols to improve mutagenesis efficiency such as Kunkel mutagenesis (using *Escherichia coli* dut^- ung^- strains) coming to the fore in the 1980s (Kunkel, 1985). However, even such "rapid" protocols were relatively lengthy and by the late 1980s favor had returned to simple primer extension-based methods, as originally described by Smith and coworkers. Smith shared the Nobel Prize for Chemistry in 1993 "For his fundamental contributions to the establishment of oligonucleotide-based, site-directed mutagenesis and its development for protein studies," with Kary B. Mullis, the inventor of PCR (Smith, 1994). Site-directed mutagenesis and PCR were subsequently combined to create megaprimer mutagenesis and its various modifications (Ke & Madison, 1997; and references therein). Megaprimer mutagenesis is a rapid process that generates a PCR amplicon for incorporation into the original gene via cassette mutagenesis (Wells, Vasser, & Powers, 1985). Meanwhile, a commercial approach, the Stratagene QuikChange® Site-Directed Mutagenesis Kit (Agilent Technologies, cat # 200518), and its improvements (Liu & Naismith, 2008) were yet faster since they require no further cloning steps.

Thus, by the early 1990s, introduction of a single-point mutation at a single location was a routine and rapid laboratory procedure. The next major advancement would be to introduce multiple mutations in one step, to generate a whole library of variations based on a single gene. Now, methodologies diversified, depending on the knowledge of the original gene and the likely number of mutations required. To introduce random mutations at random locations within an amplicon, error-prone PCR was first employed (Cadwell & Joyce, 1991; Leung, Chen, & Goeddel, 1989). Thereafter, error-prone PCR was itself employed within gene shuffling (Crameri, Raillard, Bermudez, & Stemmer, 1998; Stemmer, 1994), which effectively "breeds" homologous genes together on a laboratory timescale. Such techniques are immensely powerful, albeit that by their nature, they can never cover all of the theoretical sequence space. Nonetheless, gene shuffling offers an elegant solution when engineering a protein to create a desired activity, particularly when predicting the necessary locations of mutations would be impossible (e.g., Campbell et al., 2002). By contrast, saturation mutagenesis offers changes to protein residues at one or more defined locations simultaneously, also producing diverse variants but potentially within the theoretical sequence space, depending on the number of codons targeted and the methodology employed. In effect, saturation mutagenesis is conceptually a simple extension of site-directed mutagenesis, but in a high-throughput format. Instead of a single location, saturation mutagenesis targets multiple codons, making multiple substitutions in each, in a single experiment. Unsurprisingly, therefore, saturation mutagenesis is a key tool in a protein engineer's arsenal. However, the techniques used to generate such mutations can be far from straightforward and are the subject of the current review.

2. ADVANTAGES OF NONDEGENERATE SATURATION

By its very nature, the genetic code is degenerate. A codon (three bases) encodes a single amino acid. Each position within that codon can be any one of the four bases A, C, G, and T. Thus, there are a total of 64 codons (4^3 combinations of the four bases, where order is important). Of those 64 codons, 3 are termination codons, leaving the remaining 61 codons to encode the 20 naturally occurring amino acids. However, as any student of biology knows, those 61 codons are not distributed evenly. Specifically, some amino acids are encoded by six codons (Leu, Ser, Arg), some by four codons (Ala, Gly, Pro, Thr, Val), one by three codons (Ile), several by two codons (Cys, Asp, Glu, Phe, His, Lys, Asn, Gln, Tyr), while two amino acids are encoded by just one codon each (Met, Trp). Thus a conventional degenerate saturation codon (NNN, where N = a mixture of A, C, G, & T and NNN a mixture that collectively contains all four bases at each position of the codon) is actually a mixture of all 64 possible sequences that necessarily encodes the various amino acids disproportionately. More degenerate codons equate to yet more sequences/disproportionality (bias). Thus a piece of DNA containing three such codons would in practice be a mixture of 262,144 (64^3) different DNA sequences.

Aside from the huge numbers involved, the disproportionality/bias encoded by conventional saturation has major impact when screening the encoded library for the "best" protein(s), particularly where ligand-based screening technologies are involved (see Section 4.1). Thus, scientists have attempted to reduce both sheer numbers and concomitantly the encoded bias either by reducing redundancy or by eliminating it altogether.

Initial approaches to reduce redundancy involved using simple *limited* codon redundancy such as NNK or NNS saturation codons (K = T/G, S = G/C), and indeed, this became the "norm" for many years. NNK/NNS reduces from 64 to 32 codons to encode the 20 amino acids (+1 termination codon), so reducing numbers and decreasing the bias but not yet eliminating it. More recently, the 22c-trick has been described, which reduces the number of codons to just 22 per saturated position (Kille et al., 2013), resulting in two codons for Val and Leu, one codon for all other amino acids, and zero termination codons. The consequences of these methods in comparison with fully nondegenerate saturation (exactly 20 codons encoding 20 amino acids) are compared in Fig. 1.

A Diversity

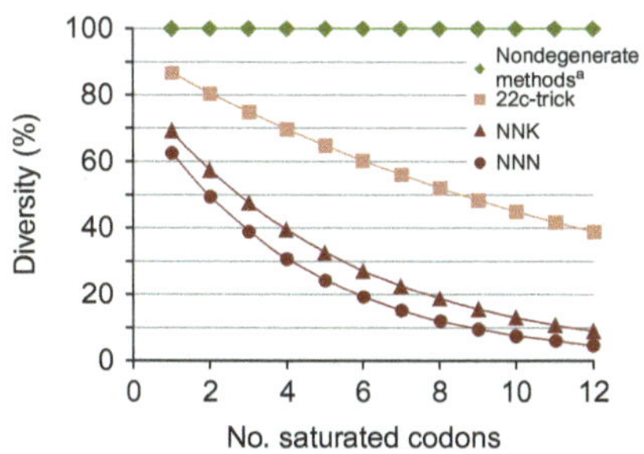

B Encoded bias

No. saturated codons	Ratio most common:rarest codon combinations[b] NNN	NNK/NNS	22c-trick	Nondegenerate methods[a]
3	216:1	27:1	8:1	1:1
6	4.7×10^4:1	729:1	64:1	
9	1.0×10^7:1	2.0×10^4:1	512:1	
12	2.2×10^9:1	5.3×10^5:1	4096:1	

C Encoded truncation

No. saturated codons	NNN	NNK/NNS	22c-trick	Nondegenerate methods[a]
3	13	9	0%	
6	25	17		
9	35	25		
12	44	32		

Fig. 1 Comparison of performance of common saturation mutagenesis techniques. *Green coloration* indicates ideal performance, *pale pink coloration* indicates tolerable performance, and *deep pink coloration* indicates unacceptable performance, where nondegenerate methods[a] may be created via various methodologies as described in Section 3. (A) Diversity was calculated using the formula $d=1/(N\sum_k p_k^2)$ (Makowski & Soares, 2003) and is in agreement for a 12-mer peptide saturated with codon NNN (Krumpe, Schumacher, McMahon, Makowski, & Mori, 2007). (B) Ratios[b] represent the theoretical relative concentrations of each individual gene combining any of the most common codons (Leu/Arg/Ser, NNN/NNK; or Leu/Val, 22c-trick) vs each individual gene containing any combination of the rarest codons (Met/Trp, NNN; Cys/Asp/Glu/Phe/His/Ile/Lys/Met/Asn/Gln/Trp/Tyr, NNK; or 18 codons (omitting Leu/Val), 22c-trick). (C) Truncation is calculated as the percentage of sequences that contain one or more termination codons within the saturated region. *Reproduced in part from Ashraf, M., Frigotto, L., Smith, M.E., Patel, S., Hughes, M.D., Poole, A.J., et al. (2013). ProxiMAX randomization: A new technology for non-degenerate saturation mutagenesis of contiguous codons.* Biochemical Society Transactions, 41, *1189–1194 under the Creative Commons Attribution Licence (CC-BY).*

As can be seen from Fig. 1, conventional methods of degenerate saturation are largely inadequate. Diversity is the gold standard when creating gene libraries. Within the context of protein engineering, diversity is a measure of the percentage of unique species within a library. Fig. 1A demonstrates that diversity is poor even when just one codon is saturated via conventional degenerate codons and even drops off rapidly with the 22c-trick. By contrast, it is maintained at a theoretical level of 100% via nondegenerate approaches, no matter how many codons are saturated. Bias is an alternative way of addressing this problem. Fig. 1B explains why diversity is so poor in conventionally constructed saturation libraries. While the numbers in Fig. 1B reflect a worst-case scenario (i.e., the ratio of the most common codon combinations to the rarest codon combinations), it is clear that there is no equality of representation between different gene sequences, with some being very populous in comparison with others. Finally, only nondegenerate methods and the 22c-trick prevent encoded truncation with the gene library (Fig. 1C). Truncation can be a problem because truncated, nonfunctional proteins can be prone to aggregation, which leads to protein precipitation.

By contrast, nondegenerate methods allow the user to include all 20 codons in approximately equal ratio. Some nondegenerate methodologies go still further and permit the user to choose exactly which amino acids are (and are *not*) encoded at a specified codon and further still, in which relative proportions (either equal or alternative, defined ratios)—see Section 3.2. In these ways, along with the removal of termination codons, nondegenerate methods permit the size of the library to be minimized while concomitantly maximizing the number of encoded variants for screening. The combination of those attributes has very positive ramifications since modern, directed evolution puts pressure on creating high-quality libraries with reduced number of variants to boost the efficiency of screening experiments (Tang et al., 2012)—see Section 4.

3. SATURATION MUTAGENESIS METHODS

3.1 Core Techniques in Saturation Mutagenesis

Whether saturation mutagenesis is degenerate, near nondegenerate (the 22c-trick, Kille et al., 2013), or wholly nondegenerate, all methodologies rely on the same core techniques, as described in this section.

3.1.1 Cassette Mutagenesis

A DNA "cassette" is simply a double-stranded piece of DNA that can be spliced into a gene of interest, so replacing the original gene sequence. Splicing usually occurs via conveniently located restriction sites that are present in the original gene and are incorporated at either end of the cassette. In its first iteration, saturation mutagenesis was achieved via cassette mutagenesis using synthetic oligonucleotide cassettes containing degenerate bases at defined locations (Wells et al., 1985). DNA cassettes may vary in length from tens to hundreds of base pairs and can be simple synthetic DNA (containing degenerate bases as required), PCR amplicons (made from primers with degenerate bases as required), or be created by nondegenerate methodologies such as MAX randomization (Section 3.2.2), ProxiMAX randomization (Section 3.2.3), and Slonomics® (Section 3.2.4).

3.1.2 Simple Primer Extension Mutagenesis

The original method of site-directed mutagenesis was based on primer extension (Hutchison et al., 1978; Smith, 1994). In essence, a primer containing a centrally located mutation was annealed to a single-stranded circular DNA template and extended around the template to create a heteroduplex. Because the parental strand of wild-type DNA remained intact, the efficiency of this mutagenesis was low. However, the methodology has been developed to high efficiency over the decades. One of the most favored strategies is now QuikChange® Mutagenesis (Agilent Technologies). Here, double-stranded plasmid template is amplified by PCR with a pair of complementary primers that each contain the required mutation(s) in a central location. After extension around each strand of the entire plasmid, the parental DNA is destroyed by *Dpn*I digestion (requires Dam methylation) and the two mutated strands are annealed together (to create a plasmid with two staggered nicks) and transformed into bacteria without further modification. In terms of saturation mutagenesis, the pair of primers need not contain a single mutation, but may instead contain a degenerate codon or indeed take the form of a DNA cassette as described in Section 3.1.1. To expand the technology further, the QuikChange® Multi Site-Directed Mutagenesis Kit has been developed to target up to five sites simultaneously (Hogrefe, Cline, Youngblood, & Allen, 2002). Target residues have to be at least five codons apart and so automatically exclude targeting contiguous sites. Although quick and simple, QuikChange® has many limitations, for example, complementary primers are prone to self-priming or give low efficiency with more than one mutated position (Liu & Naismith, 2008). The protocol of QuikChange® has been the subject of modifications by many research laboratories.

3.1.3 Overlap Extension Mutagenesis

The use of overlap extension PCR in mutagenesis was first described in 1988 (Higuchi, Krummell, & Saiki, 1988). In essence, PCR fragments (created from primers that contain mutations) which have complementary 3′ regions can be used to prime each other and so join those fragments together. This technique is used extensively in both degenerate and nondegenerate saturation mutagenesis, with fragments created as originally described by Higuchi et al. (in which the overlaps contain the mutations), or for joining cassettes together in which the mutations are contained internally.

3.2 Nondegenerate Saturation Methodologies

3.2.1 TRIM Technology: Trinucleotide Phosphoramidites

The original approach to nondegenerate saturation mutagenesis, TRIM technology, employs trinucleotide phosphoramidites (Virnekäs et al., 1994). Essentially, rather than adding one base at a time as in oligonucleotide DNA synthesis, three bases are added to a growing oligonucleotide in a single reaction. Thus to saturate a given codon, a predefined mixture of trinucleotide phosphoramidites is added to the growing DNA strand. However, single bases of DNA have differing coupling efficiencies during oligonucleotide synthesis, meaning that a biased mixture of phosphoramidites is required to effect equimolar addition (Ho et al., 1996) and this effect is amplified when coupling trinucleotides. To mitigate the problem, it is probably wise to use a commercial source of DNA produced using TRIM technology, such as Gene Art™ (Thermofisher) rather than undertaking synthesis with trinucleotide phosphoramidites in one's own laboratory. The resulting DNA can be used in any saturation protocol of choice.

3.2.2 MAX Randomization

MAX randomization (Hughes, Nagel, Santos, Sutherland, & Hine, 2003) was one of the first techniques to be published that achieves nondegenerate saturation without the need for any specialized chemicals, reagents, or equipment. In essence, MAX randomization relies on "selectional hybridization" in which a series of short selection oligonucleotides hybridize with a complementary template oligonucleotide that is conventionally saturated at the relevant codons. After the selection oligonucleotides have been ligated, asymmetric PCR ensures that only the selection strand is amplified (Fig. 2). Thus, MAX randomization generates a cassette which is typically used in cassette mutagenesis, but could be employed as a double-stranded primer in QuikChange® mutagenesis, or be joined with other sequences via overlap PCR. Note that although MAX randomization can be used to mutate multiple codons, only a maximum of two contiguous (neighboring) codons can be saturated, owing to the addressing function of the conserved part of the selection oligonucleotides. Again, MAX randomization yields a double-stranded DNA cassette that can be used in any mutagenesis protocol of choice.

3.2.3 ProxiMAX Randomization

Like other nondegenerate techniques, ProxiMAX is a nondegenerate saturation technology that uses one codon only per amino acid (Ashraf et al.,

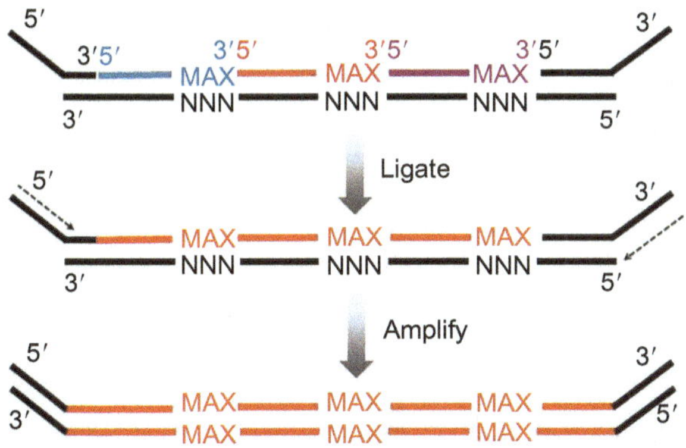

Fig. 2 Overview of the MAX randomization technique (Hughes et al., 2003). A single template oligonucleotide is synthesized that is fully degenerate at the designated, saturated codons. Meanwhile, a set of up to 20 small selection oligonucleotides are synthesized individually, for each saturated position. Each selection oligonucleotide consists of a short (typically in the order of 6 bp) addressing region that is fully complementary to the template and one MAX codon, where a MAX codon is the favored codon for a single amino acid in the organism of interest. The selection oligonucleotides are mixed as required and, alongside two terminal oligonucleotides, are hybridized with the template and ligated together. The ligated strand is then selectively amplified with primers complementary to the terminal oligonucleotides, to generate a randomization cassette.

2013). In common with MAX randomization, ProxiMAX does not require any specialized reagents, but rather relies on conventional oligonucleotides, a Type IIS restriction enzyme and blunt-ended ligation. The process involves cycles of ligation, amplification, and digestion with the consequence that one codon is added to the end of a growing DNA fragment, per cycle (Fig. 3). However, that "codon" is really a user-defined mixture of up to 20 individual, double-stranded oligonucleotides. These oligonucleotides are largely conserved sequences that can be fully or partially double-stranded, or be self-complementary hairpins, but each terminates with a unique triplet encoding just one amino acid (typically chosen for *max*imal expression in the organism of choice—hence the names MAX/ProxiMAX randomization). In practice, the mixture is often selected to comprise less than 20 such MAX codons, for example, omitting codons for cysteine and methionine or selecting just the polar residues. As such, ProxiMAX randomization allows the user to define exactly which amino acids are encoded at each individual saturated position, which can either be contiguous or separated by regions of conserved sequence. The relative proportions of codons at each saturated position may also be user defined. ProxiMAX

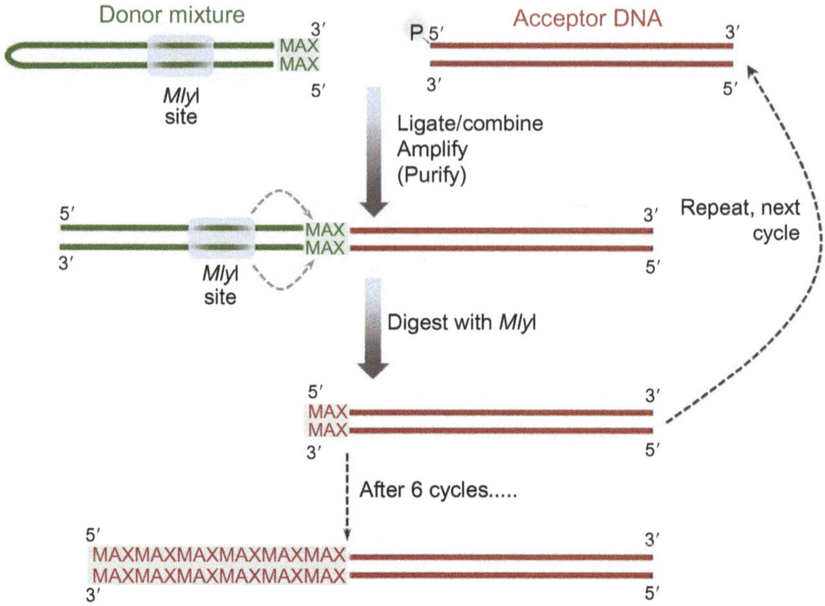

Fig. 3 Overview of the ProxiMAX randomization process. Double-stranded DNA donors, carrying the required "MAX" codons at their termini, are ligated onto a double-stranded DNA acceptor sequence (phosphorylated at the required 5′ end only). The donors can take the form of partially double-stranded DNA, fully double-stranded DNA, or hairpin oligonucleotides (as shown). After ligation, the products are amplified by PCR. Depending on whether the process is performed with automation or manually, the donor oligonucleotides can either be combined before or after ligation, with the automated process substantially reducing the number of steps involved and permitting the use of hexanucleotide donors as required (Frigotto et al., 2015). The amplified, purified product is then digested with *MlyI* and the process repeated, using the digestion product from round 1 as the acceptor for the next round of ligation. Different sets of donor oligonucleotides (up to 20 independently synthesized, double-stranded oligonucleotides) are cycled to prevent potential carryover from one round of addition to the next. *Adapted from Ashraf, M., Frigotto, L., Smith, M.E., Patel, S., Hughes, M.D., Poole, A.J., et al. (2013). ProxiMAX randomization: A new technology for non-degenerate saturation mutagenesis of contiguous codons.* Biochemical Society Transactions, 41, *1189–1194.*

can be achieved manually with good results (Poole, 2015) or via automation, which gives excellent compliance with library design (Ashraf et al., 2013). Automation also permits the addition of hexameric, two-codon units rather than one codon per cycle, though this modification requires automation, owing to the sheer number (400) of oligonucleotides (codon donors) involved (Frigotto et al., 2015). The use of two codons or hexamer nucleotides, rather than one per cycle, boosts efficiency and enhances performance of high-throughput mutagenesis reducing synthesis time. The commercial development of ProxiMAX, Colibra™ (Isogenica Ltd.), involves comprehensive quality control tests, via next-generation sequencing, which allow for careful monitoring of library fragments manufacture. ProxiMAX can be used to saturate multiple contiguous codons and the resulting DNA cassettes are typically linked together by ligation or used in cassette mutagenesis, but could also be used in overlap extension PCR or as primers in QuikChange® mutagenesis.

3.2.4 *Slonomics®/SlonoMax™*

Originally published as an automated gene synthesis technology (Schatz et al., 2010; Van den Brulle et al., 2008), Slonomics® is also a cycle-based process, though one that avoids amplification and involves sticky-ended ligation of hairpin oligonucleotide building blocks typically with three-base single-stranded overhangs, called splinkers and anchors. The process requires a total of 64 splinkers and 4096 anchors. Initially, a selected splinker is ligated to a selected anchor. The ligated product is immobilized via a biotin moiety contained within the anchor. Washing removes any unligated splinker and the resulting immobilized product is then digested with a Type IIS endonuclease that leaves a three-base, single-stranded overhang. The immobilized sequences are then discarded, so removing digested (and any unligated) anchor. Meanwhile the supernatant, containing the extended splinker, enters the next cycle. In essence, each cycle transfers six bases of DNA (a staggered three bases on each strand) from the anchor to the splinker (Fig. 4). The process is repeated to generate up to six codons (18 bp) in what is termed an "elongation fragment" and several elongation fragments can be combined via digestion with two further Type IIS restriction enzymes that each leave unique sticky ends, followed by multiple ligations. Slonomics® has been adapted to make combinatorial libraries by using mixtures of splinkers and anchors in a commercial process called SlonoMax™ (Waldmann, 2006, 2013).

3.2.5 *DC Analyzer and MDC Analyzer*

DC analyzer stands for Degenerate Codon Analyzer and is a computational approach to designing a saturated gene library (Tang et al., 2012). Specifically, DC analyzer is downloadable software that designs combinatorial degenerate primers. As a consequence of DC analyzer's programming, termination codons, codon degeneracy, and rare codons of *E. coli* are eliminated from the design. In practice, DC analyzer selects one codon for each

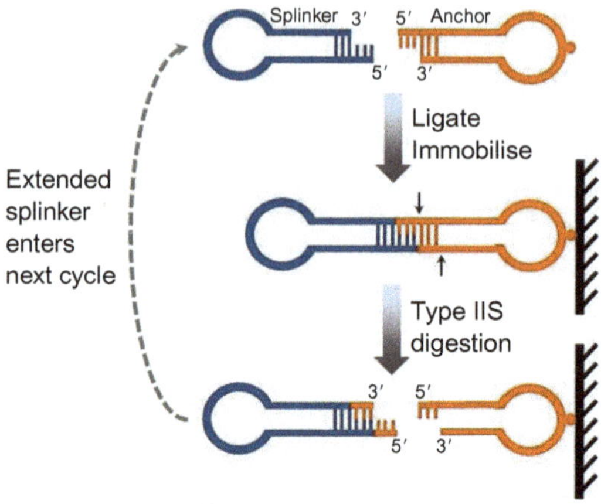

Fig. 4 Overview of the Slonomics® process (Van den Brulle et al., 2008; Waldmann, 2006, 2013). Hairpin splinker oligonucleotides are joined, via sticky-ended ligation, to a mixture of hairpin anchor oligonucleotides. The ligated product is then immobilized via a biotin moiety contained within the anchors, washed and digested with Type IIS restriction enzyme *Eam*1104I, which generates a three-base sticky-ended overhang. The resulting extended splinker, which is now free in solution, enters the next round of addition. Up to six rounds of addition are performed to generate an "elongation block" and several elongation blocks may be joined together via further Type IIS digestion and subsequent ligation (Van den Brulle et al., 2008). No PCR amplification is involved in the process.

amino acid and outputs a series of mainly limited degeneracy primers that the user must synthesize and then employ in combination within a PCR reaction or choice, to perform saturation mutagenesis. As such, DC analyzer eliminates bias and so increases the diversity of the encoded library. DC analyzer is used to design "small-intelligent libraries," as an alternative to NNS randomization. Owing to the numbers of primers involved, DC analyzer typically targets small numbers of codons (one and two codons were demonstrated by Tang et al., 2012). To target higher numbers of codons, Multi-Site Degenerate Codon Analyzer (MDC analyzer) was developed (Tang et al., 2014).

MDC analyzer allows for randomization of more than two codons that may be contiguous (three contiguous codons were demonstrated) and, rather than producing full saturation, aims to reduce library size by designing DNA sequences to encode selected subsets of amino acids at saturated positions, rather than all 20 amino acids. Using prediction programs in conjugation with mutagenesis technology gives a so-called rational random approach. Once the required subsets of amino acids have been chosen, using applications such as ConSurf-HSSP (Glaser, Rosenberg, Kessel, Pupko, & Ben-Tal, 2005) and HotSpot Wizard (Pavelka, Chovancova, & Damborsky, 2009) which analyze the favorable properties of amino acids at key regions of a protein, MDC designs multiple degenerate oligonucleotides to encode the required amino acids without termination codons or *E. coli* rare codons. By accepting that some nonrequired amino acids will also be encoded, MDC analyzer drastically reduces the number of oligonucleotides required to perform the mutagenesis, in comparison with DC analyzer (Tang et al., 2014).

3.3 Near Nondegenerate Saturation: The 22c-Trick

Conceptually similar to DC analyzer, the 22c-trick (Kille et al., 2013) uses a combination of PCR primers that contain predefined, limited degeneracy at the selected codons. Specifically, saturated codons are generated during a PCR reaction via a mixture of three primers: one contains codon NDT (A/C/G/T; A/G/T; T), another VHG (A/C/G; A/C/T; G), and the final, codon TGG. As described in Section 2, when used in combination, these primers collectively encode valine and leucine twice, the other 18 amino acids once each, and no termination codons. Note that this process cannot saturate multiple contiguous codons. Moreover, optimization of annealing temperature is essential to achieve good saturation (Kille et al., 2013), since different primers will necessarily have differing annealing temperatures and diversity would be affected quite severely if higher numbers of codons were targeted using this methodology (Fig. 1).

4. SCREENING THE LIBRARY: METHODS, LIBRARY SIZE, AND THE IMPORTANCE OF DIVERSITY

In extending the natural proteome via protein engineering, library production is only the first part of a complex process. Once the gene library has been expressed, the resulting protein library may contain from hundreds up to 10 trillion (10^{13}) different components, depending on the library design and methodology used in construction. Without a suitable screening procedure such libraries would be useless—they can hardly be examined one component at a time. In fact, the majority of libraries fall somewhere in between the two size limits and thus screening several million to billion components is nontrivial. Libraries with sizes greater than 10^{10} cannot generally be screened in vivo since transformation efficiency will limit the yield of clones. Thus the largest libraries are usually screened by using in vitro methodologies. Meanwhile, the type of protein library is also important in selecting a screening methodology. Ligand-binding libraries such as antibodies, transcription factors, etc., tend to be screened via biopanning approaches (immobilized ligand bound by solution-based protein—Section 4.1). In contrast, unless seeking an essentially irreversible enzyme inhibitor, enzyme libraries tend to require a different approach, since binding an immobilized substrate (ligand) would normally be followed by processing and release of the substrate, so preventing the immobilization of functional proteins that forms a key part of the biopanning process.

4.1 Methods for Screening Ligands

When biopanning a ligand-binding protein library, the ligand of choice is immobilized and the protein library added in solution. After binding, the majority of the library (nonbinding protein) is washed away, bound proteins are eluted, these populations are amplified, and the process is repeated, typically 4–5 times. Thereafter, individual species are isolated, sequenced, and their characteristics investigated.

Thus, biopanning is dependent on mass action, which in turn requires that the library components are all present in approximately equal concentrations for accurate discovery of the "best" proteins. Thus, good diversity is essential. If all library components are unique, i.e., 100% diversity, and expressed similarly (which is why rare codons should be avoided during saturation), all protein components will have approximately equal concentrations. In contrast, a library with low diversity will have many different genes encoding one protein and just one gene encoding another, which will lead to biased concentrations of proteins (see Fig. 1) and thus delivery of the populous proteins as the best "hits," regardless of their suitability.

The various options for biopanning strategies are described in this section.

4.1.1 Phage Display

Smith first proposed that filamentous phage can serve as carriers of foreign DNA and so provide a means to screen library variants such as those of antibodies (Scott & Smith, 1990; Smith, 1985).

This method of screening libraries relies on infecting bacterial cells with viruses called bacteriophage. Phage chromosomes have capacity to take in foreign DNA fragments and both replicate and express them from within host bacteria, usually *E. coli*. Since this foreign DNA is spliced into the phage coat protein genes, upon its expression the protein will be exposed on the outer surface of the phage particle (Smith & Petrenko, 1997). Specifically, the library of variants is cloned into phage vectors which then transfect bacterial cells. One cell will carry one variant only and expressed proteins of that variant will be exposed on the surface of phage which are then screened by biopanning. After each round of selection, the eluted phage is amplified by passage through *E. coli*. In addition to antibody fragments, phage display has been successful in screening libraries to find insulin and IRF-1 receptor agonists and antagonists (Dedova et al., 2004); small peptides that mimic erythropoietin (McConnell et al., 1998; Wrighton et al., 1996); and even an inhibitor for the enzyme pancreatic lipase (Lunder, Bratkovic, Kreft, & Strukelj, 2005). Though useful for many applications, phage display is limited by transformation efficiency and cannot therefore be used to screen libraries of greater than 10^9–10^{10} components (Odegrip et al., 2004).

4.1.2 Ribosome Display

Ribosome display is an in vitro method of selection invented by Plückthun (Hanes & Plückthun, 1997). It was the first in vitro selection method, inspired by work of Mattheakis, Bhatt, and Dower (1994) who demonstrated affinity selection using polysomes which enable the critical link between genotype and phenotype that is essential for any biopanning application. As an in vitro method, ribosome display is not limited by transformation efficiency and can therefore be used to screen very large libraries.

The *E. coli* S30 system is a coupled transcription/translation system that forms a key part of ribosome display and was also used by Mattheakis and coworkers. They displayed a library of short, 13-mer opioid peptides on *E. coli* S30 followed by affinity screening against immobilized antibodies. Because the expressed proteins were coupled with mRNA, selected hits could be reverse transcribed into cDNA and then sequenced to determine the nucleotide sequence encoding the displayed peptide (Mattheakis et al., 1994).

Ribosome display also uses the *E. coli* S30 system to create whole, correctly folded proteins that remain coupled with the S30 complex and mRNA (Hanes & Plückthun, 1997). In the first instance, PCR is employed to both amplify the library and couple it with a T7 promoter and ribosome-binding site. Following transcription to RNA, the *E. coli* S30-coupled transcription/translation system translates the mRNA in vitro. This system also includes various factors that enable correct folding of the translated proteins and stabilize produced ribosome–mRNA–protein complexes. These complexes are then screened by biopanning. After each round of selection, isolated library members are eluted from the ribosome complex and the mRNA is reverse transcribed into cDNA ready for another round of biopanning or for sequencing of individual isolates (Hanes & Plückthun, 1997).

Ribosome display is often used to improve protein affinity or stability and can be used in combination with other selection pressure mechanisms (Buchanan et al., 2012). The first exemplification of this cell-free system was used to screen single-chain fragments (scFv) of an antibody (Hanes & Plückthun, 1997). Subsequently, Minter and coworkers have used ribosome display in conjunction with three stability selective pressure factors for two therapeutic proteins that gave problems during the drug development phase (Buchanan et al., 2012). Specifically, the tendency of erythropoietin to aggregate was reduced and a 1000-fold improvement in the soluble expression of granulocyte colony-stimulating factor was gained.

In a conceptually similar approach to ribosome display, mRNA may be covalently bound to its encoded protein (Xiao, Bao, & Zhao, 2014). Because this modified approach and ribosome display more generally are performed wholly in vitro, neither method is limited by transformation efficiency.

4.1.3 CIS Display

CIS display is an in vitro screening technology that uses the RepA family of proteins of the R1 plasmid as key components. RepA is a bacterial replication initiator protein which has a special feature of high-fidelity *cis* activity: it binds only to the DNA from which it was expressed. The first part of the method involves construction of DNA consisting of N-terminal promoter, library, RepA gene (*repA*), CIS element, and C-terminal *ori*. Once these elements are joined in that order, in vitro transcription begins at the promoter and ends when it reaches the CIS element. Simultaneously, in vitro translation produces RepA protein which binds to its encoding DNA at the CIS region and *ori* sequence. As a result, the nascent polypeptide (from the library) is fused with RepA. Ultimately, through its *cis* activity, RepA protein provides the crucial physical linkage between genotype (DNA) and phenotype (protein; Fig. 5). After each round of biopanning, DNA from the eluted complex is either amplified by PCR ready for the next round of biopanning, or sequenced to identify the interacting library component (Mathonet, Ioannou, Betley, & Ullman, 2011; Odegrip et al., 2004).

CIS display has been used in antibody research (Odegrip et al., 2004), peptide maturation, ligand discovery including therapeutic peptides (Mathonet et al., 2011), and engineering of small WW scaffolds (small

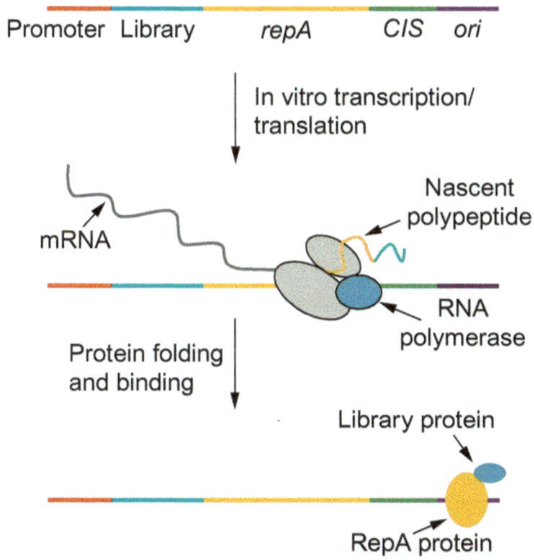

Fig. 5 Overview of CIS display. Double-stranded DNA is generated consisting of (in 5′ → 3′ order) a promoter, the saturated library fused in-frame to *repA*, the *CIS* element, and the *ori* sequence. The DNA is then subject to coupled in vitro transcription/translation using an *E. coli* S-30 extract. Transcription pauses when RNA polymerase reaches the *CIS* element. Meanwhile, translation of the newly produced mRNA produces a library component/RepA fusion protein that binds to the *ori* sequence, so linking the protein to its encoding gene. The resulting complex is screened via biopanning. *Adapted from Odegrip, R., Coomber, D., Eldridge, B., Hederer, R., Kuhlman, P.A., Ullman, C., et al. (2004). CIS display: In vitro selection of peptides from libraries of protein–DNA complexes.* Proceedings of the National Academy of Sciences of the United States of America, 101, *2806–2810.*

β-sheet motifs; Patel, Mathonet, Jaulent, & Ullman, 2013). CIS display technology has also been used to identify 12-mer peptides resistant to thrombin, chymotrypsin, and plasma proteases (Eldridge et al., 2009).

4.2 Methods for Screening Enzymes

Enzymes are often engineered to effect a change in substrate specificity, solvent tolerance, or stability. Unless an irreversible enzyme inhibitor is sought (see Section 4.1), there is much more variety in screening methodologies, since screening is necessarily based on phenotypic properties. Examples of two key strategies are given herein (Sections 4.2.1. and 4.2.2).

4.2.1 Double Selection: Positive and Negative

Functional assays are performed in in vivo systems and rely on the production of phenotypes which are different to the native molecule and, thus, recognized as mutant. Double selection is a method that has been used with different selection markers—one positive and one negative (Liu & Schultz, 1999; Pastrnak & Schultz, 2001). A well-established positive marker is based on antibiotic resistance such as the β-lactamase gene, while negative selection can be based on a toxic gene such as barnase. In further studies, β-lactamase was replaced with chloramphenicol acetyltransferase, since chloramphenicol has proved to have stronger selection pressure than other antibiotics owing to its bacteriostatic nature (Pastrnak, Magliery, & Schultz, 2000).

Schultz and coworkers use this approach to study the aminoacyl-tRNA synthetase (AARS) family of enzymes, specifically to engineer these enzymes to aminoacylate suppressor tRNA (specific to the amber stop codon) with various unnatural amino acids. During positive selection, variants of an AARS engineered in the amino acid binding pocket are selected based on aminoacylation of the suppressor tRNA with either a natural amino acid (AA) or the chosen unnatural amino acid (UAA). Thus, any enzyme that effectively aminoacylates the suppresssor tRNA (with any amino acid) will permit read-through of the amber stop codon so that antibiotic resistance can be expressed. Subsequently, during negative selection, the UAA is omitted. This time, read-through of the stop codon will permit expression of a toxic gene (barnase or uracil phosphoribosyltransferase). Negative selection is carried out in the absence of the UAA, so that any clones that express AARS enzymes that aminoacylate with native amino acids will be killed, so leaving behind enzymes that aminoacylate only with the required UAA (Melançon & Schultz, 2009).

4.2.2 FACS Screening

Fluorescence-activated cell sorting (FACS) is a cell display and activity-based selection screening procedure that employs flow cytometry. It is an ultrahigh-throughput technique, capable of screening up to 10^8 mutants per day (Yang & Withers, 2009). FACS is also characterized by high sensitivity. FACS is suitable for screening enzymatic activity where production of fluorescence is feasible. As with any another screening, FACS relies on linking the genotype with phenotype. For comparison, an example used in engineering AARSs is again described. Here, amber stop codons were engineered into genes for both T7 RNA polymerase and GFP under the control of a T7 promoter. Using genes for an orthogonal AARS/suppressor tRNA pair (i.e., the suppressor tRNA is not recognized by host AARSs and the orthogonal AARS does not recognize the host tRNA), the amino acid binding pocket of the orthogonal AARS was engineered to bind UAAs. In this instance, two positive selections (rather than a positive and then a negative selection) are employed. The first again involves chloramphenicol resistance as described previously (Section 4.2.1), while the second relies on UAA aminoacylation of the suppressor tRNA to read-through both the T7 RNA polymerase and the GFP genes, to generate a fluorescent signal detectable by FACS (Santoro, Wang, Herberich, King, & Schultz, 2002).

Alternative methods for FACS-based screening of gene/protein libraries have been reviewed recently (Xiao et al., 2014).

5. CONCLUSIONS

The transition from low- to high-throughput protein engineering has enabled the creation of novel proteins in a myriad of applications, and saturation mutagenesis plays an invaluable role within that sphere. However, the protocol selected to perform that saturation depends on several factors including the design of the library and the importance placed on the quality vs the complexity of library synthesis, as follows:

i. how many codons are being targeted,
ii. whether full or partial saturation is required,
iii. whether or not controlled ratios of codons are required,
iv. importance of the quality of the retrieved product, vs
v. the effort/expense involved in library synthesis.

The conflicting and synergistic features of these properties are examined herein.

In the preceding sections of this review, it becomes fairly clear that degenerate saturation is of very limited or indeed *no* practical utility when saturating three or more codons in a library that is to be screened by biopanning (Fig. 1), although its simplicity may be more attractive when screening enzyme libraries (Section 4.2), where life/death rather than mass action (Section 4.1) determines whether or not a protein is selected. In contrast, for small numbers of saturated codons (~1–3 codons), the near-nondegenerate 22c-trick or the fully nondegenerate DC anaylzer/MDC analyzer used in conjunction with overlap PCR or QuikChange® mutagenesis (Section 3.1) are attractive options owing to their simplicity of use and lack of relative expense. Indeed, the financial benefit of employing these methodologies in comparison with NNK/NNS has been examined recently (Acevedo-Rocha, Reetz, & Nov, 2015). However, three saturated codons are the maximum exemplified for each of these techniques as described by the inventing authors (Kille et al., 2013; Tang et al., 2012, 2014), presumably because the number of primers becomes unmanageable for higher numbers or saturated positions and also because those primers are specific to individual saturation experiments.

For higher numbers of saturated positions, a more complex strategy of nondegenerate saturation is required. ProxiMAX randomization has recently been used to target 24 consecutive codons (Frigotto et al., 2015), while the maximum number of residues that can be targeted by SlonoMAX™ has not been defined by the inventors (Van den Brulle et al., 2008), but is surely equal to that of ProxiMAX. Both of these techniques are expensive in terms of DNA synthesis, but the oligonucleotides involved may be consistently reused, because everything except the saturated codons is removed by Type IIS restriction digestion, once the saturation protocol has been completed. Moreover, both offer the ability to define not only the precise residues encoded at each saturated position but also the relative ratios of each codons at a saturated location, meaning that natural antibody libraries (for example) can be mimicked. In contrast, neither the 22c-trick (Kille et al., 2013) nor MDC analyzer (Tang et al., 2014) offers that possibility, even with low numbers of saturated positions. Meanwhile, DC analyzer (Tang et al., 2012) does allow the user to specify which amino acids are encoded, but not their relative proportions, because of the defined degenerate primer synthesis involved.

Thus in the end, the selected method of saturation will come down to a decision about where the funds and effort are to be spent. Upfront investment in library synthesis is economically advantageous even when considering small libraries (Acevedo-Rocha et al., 2015). This effect can only be amplified when screening large (>10^9 protein libraries). Here, even the expense of having a gene library synthesized commercially must pale into insignificance in comparison with screening costs of examining so many components via a biopanning strategy. More serious still is the possibility of taking forward a suboptimal candidate protein that was identified because the library from which it was identified contained poor diversity. The relative expense of such a mistake is hard to calculate.

ACKNOWLEDGMENTS

This work was supported by BBSRC Grant BB/L015633/1. The authors gratefully acknowledge Dr. Andrew J. Sutherland (Aston University) for critical reading of the manuscript.

REFERENCES

Acevedo-Rocha, C. G., Reetz, M. T., & Nov, Y. (2015). Economical analysis of saturation mutagenesis experiments. *Scientific Reports*, *5*, 10654. http://dx.doi.org/10.1038/srep10654.

Ashraf, M., Frigotto, L., Smith, M. E., Patel, S., Hughes, M. D., Poole, A. J., et al. (2013). ProxiMAX randomization: A new technology for non-degenerate saturation mutagenesis of contiguous codons. *Biochemical Society Transactions*, *41*, 1189–1194.

Buchanan, A., Ferraro, F., Rust, S., Sridharan, S., Franks, R., Dean, G., et al. (2012). Improved drug-like properties of therapeutic proteins by directed evolution. *Protein Engineering, Design & Selection*, *25*, 631–638.

Cadwell, R., & Joyce, G. (1991). Randomization of genes by PCR mutagenesis. *PCR Methods and Applications*, *2*, 28–33.

Campbell, R. E., Tour, O., Palmer, A. E., Steinbach, P. A., Baird, G. S., Zacharias, D. A., et al. (2002). A monomeric red fluorescent protein. *Proceedings of the National Academy of Sciences of the United States of America*, *99*, 7877–7882.

Crameri, A., Raillard, S. A., Bermudez, E., & Stemmer, W. P. C. (1998). DNA shuffling of a family of genes from diverse species accelerates directed evolution. *Nature*, *391*(6664), 288–291.

Dedova, O., Fletcher, P., Liu, H., Wang, P., Blume, A., Brissette, R., et al. (2004). US Patent No. US2004023887. Washington, DC: U.S. Patent and Trademark Office.

Eldridge, B., Cooley, R. N., Odegrip, R., McGregor, D. P., FitzGerald, K. J., & Ullman, C. G. (2009). An in vitro selection strategy for conferring protease resistance to ligand binding peptides. *Protein Engineering, Design & Selection*, *22*, 691–698.

Frigotto, L., Smith, M. E., Brankin, C., Sedani, A., Cooper, S. E., Kanwar, N., et al. (2015). Codon-precise, synthetic, antibody fragment libraries built using automated hexamer codon additions and validated through next generation sequencing. *Antibodies*, *4*, 88–102.

Glaser, F., Rosenberg, Y., Kessel, A., Pupko, T., & Ben-Tal, N. (2005). The ConSurf-HSSP database: The mapping of evolutionary conservation among homologs onto PDB structures. *Proteins*, *58*, 610–617.

Hanes, J., & Plückthun, A. (1997). *In vitro* selection and evolution of functional proteins by using ribosome display. *Proceedings of the National Academy of Sciences of the United States of America*, *94*, 4937–4942.

Higuchi, R., Krummell, B., & Saiki, R. K. (1988). A general method of in vitro preparation and specific mutagenesis of DNA fragments: Study of protein and DNA interactions. *Nucleic Acids Research*, *16*, 7351–7367.

Ho, S. P., Britton, D. H. O., Stone, B. A., Behrens, D. L., Leffet, L. M., Hobbs, F. W., et al. (1996). Potent antisense oligonucleotides to the human multidrug resistance-1 mRNA are rationally selected by mapping RNA-accessible sites with oligonucleotide libraries. *Nucleic Acids Research*, *24*, 1901–1907.

Hogrefe, H. H., Cline, J., Youngblood, G. L., & Allen, R. M. (2002). Creating randomized amino acid libraries with the QuikChange® multi site-directed mutagenesis kit. *Biotechniques*, *33*, 1158–1160.

Hughes, M. D., Nagel, D. A., Santos, A. F., Sutherland, A. J., & Hine, A. V. (2003). Removing the redundancy from randomised gene libraries. *Journal of Molecular Biology*, *331*, 967–972.

Hutchison, C. A., III, Phillips, S., Edgell, M. H., Gillam, S., Jahnke, P., & Smith, M. (1978). Mutagenesis at a specific position in a DNA sequence. *The Journal of Biological Chemistry*, *253*(18), 6551–6560.

Ke, S.-H., & Madison, E. L. (1997). Rapid and efficient site-directed mutagenesis by single-tube 'megaprimer' PCR method. *Nucleic Acids Research*, *25*, 3371–3372.

Kille, S., Acevedo-Rocha, C. G., Parra, L. P., Zhang, Z. G., Opperman, D. J., Reetz, M. T., et al. (2013). Reducing codon redundancy and screening effort of combinatorial protein libraries created by saturation mutagenesis. *ACS Synthetic Biology*, *2*, 83–92.

Krumpe, L. R. H., Schumacher, K. M., McMahon, J. B., Makowski, L., & Mori, T. (2007). Trinucleotide cassettes increase diversity of T7 phage-displayed peptide library. *BMC Biotechnology*, *7*, 65.

Kunkel, T. A. (1985). Rapid and efficient site-specific mutagenesis without phenotypic selection. *Proceedings of the National Academy of Sciences of the United States of America*, *82*, 488–492.

Leung, D. W., Chen, E., & Goeddel, D. V. (1989). A method for random mutagenesis of a defined DNA segment using a modified polymerase chain reaction. *Technique*, *1*, 11–15.

Liu, H., & Naismith, J. H. (2008). An efficient one-step site-directed deletion, insertion, single and multiple-site plasmid mutagenesis protocol. *BMC Biotechnologies*, *8*, 91.

Liu, D. R., & Schultz, P. G. (1999). Progress toward the evolution of an organism with an expanded genetic code. *Proceedings of the National Academy of Sciences of the United States of America*, *96*, 4780–4785.

Lunder, M., Bratkovic, T., Kreft, S., & Strukelj, B. (2005). Peptide inhibitor of pancreatic lipase selected by phage display using different elution strategies. *Journal of Lipid Research*, *46*, 1512–1516.

Makowski, L., & Soares, A. (2003). Estimating the diversity of peptide populations from limited sequence data. *Bioinformatics*, *19*, 483–489.

Mathonet, P., Ioannou, A., Betley, J., & Ullman, C. (2011). CIS display, a DNA-based in vitro selection technology for therapeutic peptides. *Chimica Oggi/Chemistry Today*, *29*, 10–12.

Mattheakis, L. C., Bhatt, R. R., & Dower, W. J. (1994). An *in vitro* polysome display system for identifying ligands from very large peptide libraries. *Proceedings of the National Academy of Sciences of the United States of America*, *91*, 9022–9026.

McConnell, S. J., Dinh, T., Le, M.-H., Brown, S. J., Becherer, K., Blumeyer, K., et al. (1998). Isolation of erythropoietin receptor agonist peptides using evolved phage libraries. *Biological Chemistry*, *379*, 1279–1286.

Melançon, C. E., III, & Schultz, P. G. (2009). One plasmid selection system for the rapid evolution of aminoacyl-tRNA synthetases. *Bioorganic & Medicinal Chemistry Letters*, *19*, 3845–3847.

Odegrip, R., Coomber, D., Eldridge, B., Hederer, R., Kuhlman, P. A., Ullman, C., et al. (2004). CIS display: In vitro selection of peptides from libraries of protein–DNA complexes. *Proceedings of the National Academy of Sciences of the United States of America*, *101*, 2806–2810.

Pastrnak, M., Magliery, T. J., & Schultz, P. G. (2000). A new orthogonal suppressor trna/aminoacyl-tRNA synthetase pair for evolving an organism with an expanded genetic code. *Helvetica Chimica Acta*, *83*, 2277–2286.

Pastrnak, M., & Schultz, P. G. (2001). Phage selection for site-specific incorporation of unnatural amino acids into proteins in vivo. *Bioorganic & Medicinal Chemistry*, *9*, 2373–2379.

Patel, S., Mathonet, P., Jaulent, A. M., & Ullman, C. G. (2013). Selection of a high-affinity WW domain against the extracellular region of VEGF receptor isoform-2 from a combinatorial library using CIS display. *Protein Engineering, Design & Selection*, *26*, 307–315.

Pavelka, A., Chovancova, E., & Damborsky, J. (2009). HotSpot Wizard: A web server for identification of hot spots in protein engineering. *Nucleic Acids Research*, *37*, W376–W383.

Poole, A. J. (2015). The integration of ProxiMAX randomisation with CIS display for the production of novel peptides. Aston University, PhD thesis.

Santoro, S. W., Wang, L., Herberich, B., King, D. S., & Schultz, P. G. (2002). An efficient system for the evolution of aminoacyl-tRNA synthetase specificity. *Nature Biotechnology*, *20*, 1044–1048.

Schatz, O., O'Connell, T., Schwer, H., & Waldmann, T. (2010). US Patent No. US 7695906. Washington, DC: U.S. Patent and Trademark Office.
Scott, J. K., & Smith, G. P. (1990). Searching for peptide ligands with an epitope library. *Science*, *249*, 386–390.
Shortle, D., & Nathans, D. (1978). Local mutagenesis: A method for generating viral mutants with base substitutions in preselected regions of the viral genome. *Proceedings of the National Academy of Sciences of the United States of America*, *75*(5), 2170–2174.
Smith, G. P. (1985). Filamentous fusion phage: Novel expression vectors that display cloned antigens on the virion surface. *Science*, *228*, 1315–1317.
Smith, M. (1994). Synthetic DNA and biology. *Bioscience Reports*, *14*(2), 52–66.
Smith, M., Brown, N. L., Air, G. M., Barrell, B. G., Coulson, A. R., Hutchison, C. A., III, et al. (1977). DNA sequence at the C termini of the overlapping genes *A* and *B* in bacteriophage ΦX174. *Nature*, *265*, 702–705.
Smith, G. P., & Petrenko, V. A. (1997). Phage display. *Chemical Reviews*, *97*, 391–410.
Stemmer, W. P. C. (1994). Rapid evolution of a protein in vitro by DNA shuffling. *Nature*, *370*(6488), 389–391.
Tang, L., Gao, H., Zhu, X., Wang, X., Zhou, M., & Jiang, R. (2012). Construction of "small-intelligent" focused mutagenesis libraries using well-designed combinatorial degenerate primers. *Biotechniques*, *52*, 149–158.
Tang, L., Wang, X., Ru, B., Sun, H., Huang, J., & Gao, H. (2014). MDC-Analyzer: A novel degenerate primer design tool for the construction of intelligent mutagenesis libraries with contiguous sites. *BioTechniques*, *56*, 301–310.
Van den Brulle, J., Fischer, M., Langmann, T., Horn, G., Waldmann, T., Arnold, S., et al. (2008). A novel solid phase technology for high-throughput gene synthesis. *Biotechniques*, *45*, 340–343.
Virnekäs, B., Ge, L. M., Plückthun, A., Schneider, K. C., Wellnhofer, G., & Moroney, S. E. (1994). Trinucleotide phosphoramidites-ideal reagents for the synthesis of mixed oligonucleotides for random mutagenesis. *Nucleic Acids Research*, *22*, 5600–5607.
Waldmann, T. (2006). *Sloning announces the development of SlonoMax™ gene variant libraries of previously unmatched quality—Creating new possibilities for the directed evolution of proteins*. Retrieved July 13, 2016, from http://www.businesswire.com/news/home/20060928005357/en/Sloning-Announces-Development-SlonoMax-TM-Gene-Variant.
Waldmann, T. (2013). *Achieving a new quality level in tailoring genetic diversity*. Retrieved July 13, 2016, from http://www.scientistlive.com/content/19548.
Wells, J. A., Vasser, M., & Powers, D. B. (1985). Cassette mutagenesis: An efficient method for generation of multiple mutations at defined sites. *Gene*, *34*, 315–323.
Wrighton, N. C., Farrell, F. X., Chang, R., Kashyap, A. K., Barbone, F. P., Mulcahy, L. S., et al. (1996). Small peptides as potent mimetics of the protein hormone erythropoietin. *Science*, *273*, 458–463.
Xiao, H., Bao, Z., & Zhao, H. (2014). High throughput screening and selection methods for directed enzyme evolution. *Industrial & Engineering Chemistry Research*, *54*, 4011–4020.
Yang, G., & Withers, S. G. (2009). Ultrahigh-throughput FACS-based screening for directed enzyme evolution. *ChemBioChem*, *10*, 2704–2715.

CHAPTER NINE

WGCNA Application to Proteomic and Metabolomic Data Analysis

G. Pei*,†,‡, L. Chen*,†,‡, W. Zhang*,†,‡,1

*Laboratory of Synthetic Microbiology, School of Chemical Engineering & Technology, Tianjin University, Tianjin, PR China

†Key Laboratory of Systems Bioengineering (Ministry of Education), Tianjin University, Tianjin, PR China

‡SynBio Research Platform, Collaborative Innovation Center of Chemical Science and Engineering (Tianjin), Tianjin, PR China

[1]Corresponding author: e-mail address: wwzhang8@tju.edu.cn

Contents

Abstract

Progresses in mass spectrometric instrumentation and bioinformatics identification algorithms made over the past decades allow quantitative measurements of relative

Methods in Enzymology, Volume 585
ISSN 0076-6879
http://dx.doi.org/10.1016/bs.mie.2016.09.016

or absolute protein/metabolite amounts in cells in a high-throughput manner, which has significantly expedited the exploration into functions and dynamics of complex biological systems. However, interpretation of high-throughput data is often restricted by the limited availability of suitable computational methods and enough statistical power. While many computational methodologies have been developed in the past decades to address the issue, it becomes clear that network-focused rather than individual gene/protein-focused strategies would be more appropriate to obtain a complete picture of cellular responses. Recently, an *R* analytical package named as weighted gene coexpression network analysis (WGCNA) was developed and applied to high-throughput microarray or RNA-seq datasets since it provides a systems-level insights, high sensitivity to low abundance, or small fold changes genes without any information loss. The approach was also recently applied to proteomic and metabolomic data analysis. However, due to the fact that low coverage of the current proteomic and metabolomic analytical technologies, causing the format of datasets are often incomplete, the method needs to be modified so that it can be properly utilized for meaningful biologically interpretation. In this chapter, we provide a detailed introduction of the modified protocol and its tutorials for applying the WGCNA approach in analyzing proteomic and metabolomic datasets.

ABBREVIATION

GO gene topology
GS gene significance
iTRAQ isobaric tags for absolute and relative quantification
KEGG Kyoto encyclopedia of genes and genomes
MAR missing at random
MCAR missing completely at random
MNAR missing not at random
MS metabolite significance
PMM predictive mean matching
PS peptide significance
TOM topological overlap measure
WGCNA weighted gene coexpression network analysis

1. INTRODUCTION

Proteomics aims to cover a complex dimension of genome expression and activity directly relevant to physiological phenotypes. Metabolomics aims to determination and quantification of small biomolecules, which are the end products of gene expression in cells (Blanchet et al., 2011; Gauguier, 2016). In the past decades, various strategies of high-throughput mass spectrometry (MS)-based proteomics and metabolomics, including stable isotopes labeled (ICAT, ICPL, IDBEST, iTRAQ, TMT, IPTL, and SILAC), label-free statistical assessment (MRM, SWATH), and absolute quantification approaches (AQUA), have been developed (Chahrour, Cobice, & Malone, 2015), which have allowed global-level quantitative measurements of proteins or metabolites for a better understanding of complex biological systems (Chahrour et al., 2015; Yin & Xu, 2014).

With advances of MS technologies in terms of detection sensitivity and target coverage, current proteomics and metabolomics analyses typically yield vast amounts of measurement data. However, it remains as a major challenge how to translate the information into a better biological understanding by conventional differential expression analysis (Ruffalo, Koyuturk, & Sharan, 2015). For example, it is difficult to identify those proteins/metabolites with low abundance or small fold changes (e.g., signal transduction proteins), which play important regulatory roles in biological responses against external environmental perturbations (Pei, Chen, Wang, Qiao, & Zhang, 2014; Ruffalo et al., 2015). In addition, due to differential expression analysis is typically constrained to paired samples analysis, it is thus unable to make systematically analysis to deal with the large datasets from heterogeneous sources simultaneously (Ruffalo et al., 2015). However, coexpression network approaches, which construct relationship networks based on pairwise or low-order conditional pairwise association measures without using any artificial cutoff, have been demonstrated to be a powerful tool in systematically deciphering cellular responses or identifying critical pathways relevant to key traits or conditions (Lehtinen et al., 2013; Zhang, Li, Xue, Kong, & Wang, 2016). Furthermore, one new method named as weighted gene coexpression network analysis (WGCNA) was recently proposed to describe correlation relationship between genes across microarray samples, and correlation relationship between clusters of highly correlated genes or modules and external conditions or sample traits, which reflects the continuous nature of the underlying coexpression information and avoids information loss by setting artificial threshold parameters compared to unweighted network (Langfelder & Horvath, 2008). Due to the earlier advantages, the application of the WGCNA method has, therefore, been extended to other types of high-throughput datasets, such as proteomic and metabolomic data in recent years (Kunowska, Rotival, Yu, Choudhary, & Dillon, 2015; Pei et al., 2014; Su et al., 2014; Yu et al., 2015).

Compared to transcriptomics, proteomic and metabolomics datasets are often incomplete due to the imperfect identification of coding sequences within a genome and the limited sensitivity of current detection technologies (Lazar, Gatto, Ferro, Bruley, & Burger, 2016). Current proteomics technologies typically allow detection of only one-half to two-third of all coded proteins (Zhang et al., 2006). It is reported that the proportion of peptides/proteins that exhibits at least one missing value is very high in proteomic datasets, ranging between 70% and 90% (Albrecht, Kniemeyer, Brakhage, & Guthke, 2010). Similarly, large fraction of missing values is also typical for metabolomic datasets (Huan & Li, 2015). To address the issue, several approaches, such as *k* nearest neighbor, least square method, and local least squares method, originally developed for missing data imputation for microarray data, have been proposed and applied to proteomic and metabolomic datasets (Nie, Wu, Culley, Scholten, & Zhang, 2007; Xia, Sinelnikov, Han, & Wishart, 2015). However, as each of the imputation methods is devoted to a specific type of missingness mechanisms (Lazar et al., 2016), it requires better understanding and precisely controlling of each missing values before a suitable analytical approach is utilized for proteomic and metabolomic data analysis. In addition, appropriate transformation of proteomic and metabolomic datasets is a prerequisite for capturing the true correlation in datasets. For example, the total signals detected can be significantly different due to sample sizes, weights, or volumes (Wu & Li, 2016). Therefore, without performing a suitable transformation to normalize protein and metabolite abundance data, coexpression correlation cannot be properly constructed. In the case of semiquantitative proteomic and metabolomic abundance data, a suitable data transformation may be even more important in stabilizing variances and capturing true correlation (Nie et al., 2007). Moreover, as isotope reagents in proteomics only allowing quantitation of a limited number of independent samples (Chahrour et al., 2015), simultaneous analysis of multiple samples obtained from different batches or analytical strategies would have to overcome issues related to high data heterogeneity. Thus, batch effect removal should be considered as a necessary step before analysis. Finally, as module enrichment analysis in the standardized protocol of WGCNA is limited only to GO (gene ontology) enrichment analysis of several model species, it is necessary to modify the protocol and extend its metabolomic datasets with only Kyoto encyclopedia of genes and genomes (KEGG) pathway information. In this chapter, we describe a modified WGCNA protocol and demonstrate its applications to both proteomic and metabolomic datasets.

2. OVERVIEW OF WGCNA PROTOCOL

A brief overview of the standard procedure of WGCNA analysis includes.

2.1 Network Construction

In weighted coexpression networks, "genes" (or "proteins," "metabolites") are referred as nodes, the node profile x_i is referred as the gene expression profile, and the node significance measure (GS_i) is referred as the gene significance. A network is fully specified by its *adjacency matrix* a_{ij}, a symmetric $n \times n$ matrix with entries in [0, 1] whose component a_{ij} encodes the network connection strength between nodes i and j. To calculate adjacency matrix, an intermediate quantity called the *coexpression similarity* s_{ij} as the absolute value of the correlation coefficient between the profiles of nodes i and j (alternative measures, e.g., biweight midcorrelation, or the *Spearman* correlation).

$$s_{ij} = |\mathrm{cor}(x_i, x_j)|$$

However, similar to unweighted networks by hard thresholds, adjacency functions for weighted networks require users to choose threshold parameters for random noise removing, for example, by applying the approximate scale-free topology criterion (Zhang & Horvath, 2005).

2.2 Module Identification

Once the weighted network constructed, densely interconnected genes clusters for module detection is conducted. To make networks less sensitive to spurious connections or to connection missing due to random noises, the topological overlap measure (TOM) is introduced to identify modules of highly coexpressed genes based on the adjacency of a_{ij} (Ravasz, Somera, & Mongru, 2002).

$$\mathrm{TOM}_{ij} = \frac{\sum_{u \neq i,j} a_{iu} a_{uj} + a_{ij}}{\min(k_i, k_j) + 1 - a_{ij}}$$

As the WGCNA approach identifies gene modules using unsupervised cluster, users could apply one of several module detection methods. The default method is hierarchical clustering using the standard *R* function hclust; branches of the hierarchical clustering dendrogram correspond to modules and can be identified using one of several branch-cutting methods available, for example, the constant-height cut or two dynamic branch-cut methods (Langfelder, Zhang, & Horvath, 2008). To determine the optimal cutting parameters or to estimate how many clusters are present in datasets, cluster stability/robustness analysis is recommended (Langfelder et al., 2008).

Several options have been implemented for summarizing gene expression profiles of a given module. For example, the function moduleEigengenes represents the module expressions of the qth module by the module eigengene $E(q)$, defined as the first principal component of the expression matrix. The eigengene (eigenprotein/eigenpeptide/eigenmetabolite) E can be considered as a weighted average expression profile (Langfelder et al., 2008). Alternatively, the intramodular connectivity measure can be used to define the most highly connected intramodular hub gene as the module representative. The intramodular hub genes are highly correlated with the module eigengene (Horvath & Dong, 2008).

2.3 Relationship of Modules and External Conditions/Traits

Identification of biologically significant modules and genes is one major goal of coexpression analysis. The WGCNA approach defines a GS that assigns a nonnegative number to each gene; the higher GS_i the more biologically significant is gene i. A microarray sample trait T can be used to define a trait-based GS as the absolute correlation between the traits and the expression profiles. A measure of module significance can be defined as average GS across genes in the module. When dealing with a sample trait T, a measure of statistical significance between the module eigengene E and the trait T can be defined, for example, using correlation or a p-value obtained from a univariate regression model between E and T. Modules with a great trait significance measure to a given trait represent a high association with the trait. Genes with high module membership (MM) in the modules highly related to a given trait are biologically important and could subject to further validation (Fuller et al., 2007; Ghazalpour et al., 2006; Horvath et al., 2006; Oldham, Horvath, & Geschwind, 2006).

$$GS_i = |\mathrm{cor}(x_i, T)|$$
$$GS_i = -\log p_i$$

2.4 Topological Property Analysis

Topological property of networks can be described using network concepts, also known as network statistics or indices, including whole network connectivity (degree), intramodular connectivity, topological overlap, clustering coefficient, density, etc. (Dong & Horvath, 2007; Horvath & Dong, 2008). Differential analysis of network concepts, such as intramodular connectivity, can be performed to reveal regulatory changes in gene expressions (Fuller et al., 2007; Oldham et al., 2006). The WGCNA package implements several functions, such as softConnectivity, intramodularConnectivity, TOMSimilarity, clusterCoef, and networkConcepts, for computing these network concepts. These functions can be used to create summary statistics of these concepts and to evaluate their differences across networks or interested modules (Langfelder et al., 2008).

2.5 Network Visualization

Visualization of module structures and network connections can be achieved by several approaches. For example, coexpression module structures can be visualized by heatmap plots of gene–gene connectivity, which can be generated by the function TOMplot. An alternative is a multidimensional scaling plot. Relationships between modules can be summarized by a hierarchical clustering dendrogram of their eigengenes, or by heatmap plots of the corresponding eigengene network (function labeledHeatmap). In addition, functions exportNetworkToVisANT and exportNetworkToCytoscape allow networks exported in a format suitable for analysis by VisANT (Hu, Snitkin, & DeLisi, 2008) and Cytoscape (Shannon et al., 2003), respectively.

3. APPLICATION IN PROTEOMIC ANALYSIS

3.1 Description of Experimental Design and Proteomics Dataset

Description of the proteomic analysis and relevant biological findings for this application example has been described previously (Pei et al., 2014). Briefly, our aim was to identify unique responses of the model cyanobacterium *Synechocystis* sp. PCC 6803 (hereafter *Synechocystis*) to biofuels stress so that the low tolerance of cyanobacterial hosts to biofuels can be understood and overcome by subsequently synthetic engineering work. Toward the goal, the growth of *Synechocystis* under ethanol, butanol, hexane, and salt stress conditions with dosages of 1.5% (v/v), 0.2% (v/v), 0.8% (v/v), 4% (w/v), and nitrogen starvation conditions, which led to ~50% growth reduction, were determined. For each batch, cells under control or treated condition were harvested at two time points (i.e., 24 and 48 h) that were corresponding to middle-exponential and exponential-stationary transition phases, and each sample with two biological replicates, respectively. The iTRAQ analysis led to identification and quantification of about 6252 ~ 8282 distinct peptides in five batch proteomics datasets, which were mapped to 1300 ~ 1703 proteins in the *Synechocystis* genome (Pei et al., 2014). Although both protein-based and peptide-based proteomic datasets can be used for further statistical analysis (Gibbs, Baratt, & Baric, 2013; Yu & Feng, 2016; Zhang, Liu, et al., 2016), to avoid issues related to multiple mapping of the same peptide, in this study peptide-based proteomic dataset was used directly for network construction and WGCNA analysis (Gibbs et al., 2013).

3.2 Data Preprocessing

Batch effects are subgroups of measurements that have qualitatively different behaviors across conditions and are unrelated to biological or scientific variables in the study. For example, batch effects may occur if two subsets of experiments were run on different days, on different incubators, or by two technicians (Leek et al., 2010). If not properly handled, batch effects can cause strong bias to the following data interpretation (Leek et al., 2010). Although batch effects are difficult to detect in low-dimensional assays, they can be detected (such as using principal component analysis or hierarchical cluster analysis) and even removed (using ComBat package in *R* software) from the high-throughput datasets (Langfelder et al., 2008).

Although the proteomic datasets were from different batches of cultivation experiments, control samples were established for each batch proteomic datasets. In order to eliminate batch effects and stabilize the variances of different peptides, relative ratios of peptides between treated and control samples were used for later analysis (log2 transformed). To facilitate the calculation, only average relative ratios in replicates samples were kept for further peptide-based WGCNA analysis.

It is worth noting that unlike transcriptome, proteomic datasets are often incomplete due to imperfect identification of coding sequences within a genome and limited sensitivity of current peptide detection technologies (Webb-Robertson et al., 2015; Wilkins et al., 1996). Missing data in proteomic datasets can be issues for WGCNA analysis, because either simply assigning undetected proteins to a concentration value of zero or excluding them from the correlation analysis would introduce significant bias or cause information loss, respectively. Therefore, it is necessary to develop methodologies to conduct imputation for missing values before network construction and module detection are performed. A recent study by Lazar et al. (2016) summarized different statistical methods to conduct imputation for missing proteomic values, and the comparative analysis showed that each imputation method is typically devoted to a specific type of missingness mechanisms (Lazar et al., 2016). Currently, three types of missing value mechanisms have been identified (Bhaskaran & Smeeth, 2014; Rubin, 1976):

Missing completely at random (MCAR) is corresponding to combination and propagation of multiple minor errors or stochastic fluctuations in a proteomics dataset. For instance, a misidentified peptide can or cannot be balanced by the alignment of the precursor maps, leading to an abundance value or, on the contrary, to a missing value. As results, each missing value cannot be directly explained by the nature of the peptide or by its measured intensity (Lazar et al., 2016).

Missing at random (MAR) is a more occurred class than MCAR, where missingness can be fully accounted for by variables where there is complete information. It means that there might be systematic differences between the missing and observed abundance values, but these can be entirely explained by other observed conditional variables (Bhaskaran & Smeeth, 2014). However, some MAR imputation methods can also be used for MCAR missing values, and thus applied to proteomics data sets (Lazar et al., 2016).

Missing not at random (MNAR), on the contrary, has a targeted effect. In MS-based analysis, chemical species whose abundances are close enough to detection limitation of instruments record a higher rate of missing values. This is why MNAR-devoted imputation methods used in proteomics focus on left-censored data, that is, the distribution of proteomic data is truncated on the left side, region depicting the lower abundances (Lazar et al., 2016).

It is critical to identify the nature of missing values in the datasets before any strategy for imputation is applied. Based on numerous experimental and theoretical arguments, it is recommended to apply MCAR/MAR imputation method in case of no enough knowledge available to determine the natures of missing values in quantitative proteomic datasets (Lazar et al., 2016; Webb-Robertson et al., 2015). For the example, proteomic datasets used in this chapter, which carry possible mechanisms of MCAR/MAR missing values (Gao, Pei, Chen, & Zhang, 2015), predictive mean matching (PMM) approach in the multiple imputations based on chained equations (MICE) was conducted using *R* Mice package (Zhang, 2016).

Even with a suitable imputation methodology, too much missing data can still be a problem for imputation accuracy. In general, methods perform well when there are fewer missing values (<10%) and become inaccurate with increasing proportion of missing values (>20%) (Gao et al., 2015; Lazar et al., 2016), it is thus necessary to remove peptides with a large fraction of missing entries. Depends on the type of missingness mechanisms and different imputation approaches applied, the threshold of data missing fraction for reasonable estimation could be different (Lazar et al., 2016; Nie et al., 2007). For example, PMM-based method achieved the best imputation accuracy when missing rate is low (<10%) (Gao et al., 2015), while a recent effort was to integrate the GO information could enhance data imputation even when the missing fraction is large (Tuikkala, Elo, Nevalainen, & Aittokallio, 2006).

Finally, for the example proteomic datasets, the number of multiple imputations was set as "$m=10$" to create 10 imputed datasets, and the random number generator was set as "seed = 1." After combining results of 10 imputed data analysis, the imputed results showed a *Pearson* correlation greater than 0.6 to the real experimental data, indicating a reliability of the imputation (data not shown) for further WGCNA analysis.

3.3 Construction of Signed Weighted Coexpression Network

After data normalization and missing value imputation, the format of proteomic datasets could be similar to that of microarray and RNA-seq data, and suitable for WGCNA analysis. One central parameter of a weighted coexpression gene network is correlation direction for signed or unsigned networks. Usually, in a signed correlation network, nodes with negative correlation are considered unconnected (their connection strength is zero or very close to zero). In contrast, in an unsigned correlation network, nodes with strong negative correlations have high connection strengths: the unsigned network adjacency is based on the absolute values of correlation, so positive and negative correlations are treated equally. In general, it is recommended to construct signed networks over unsigned networks, because unsigned networks contain no information about the sign of the underlying correlations (Langfelder et al., 2008). Therefore, the networkType = "signed" for peptides weighted coexpression network construction in function blockwiseModules was set.

Construction of a weighted coexpression network needs applying another key parameter named as soft-thresholding powers (Zhang & Horvath, 2005). Usually, the soft-thresholding power in signed networks should be twice as much as that in unsigned networks (Langfelder et al., 2008). It has been recommended to choose soft-thresholding power based on the criterion of approximate scale-free topology. As shown in Fig. 1A, the power = 29 was chosen, for the example proteomics datasets, which is the lowest power for the curve that the scale-free topology fit index flat upon reaching a high value above 0.85, and with a moderate mean connectivity (Fig. 1B).

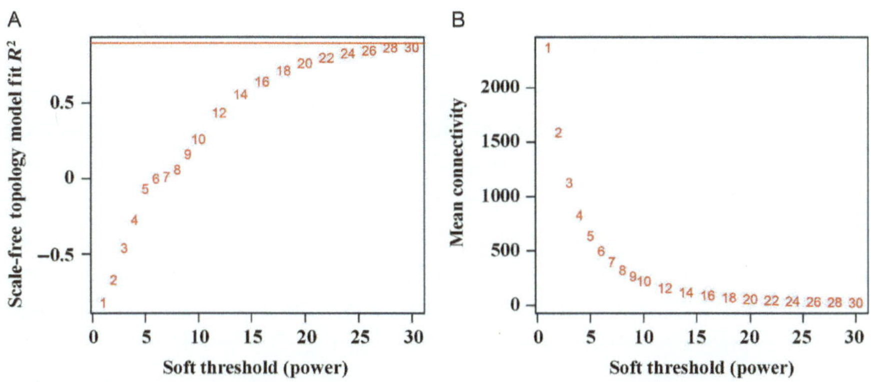

Fig. 1 Analysis of proteomic network topology for various soft-thresholding powers. (A) The scale-free fit index (*y*-axis) as a function of the soft-thresholding power (*x*-axis) and (B) the mean connectivity (degree, *y*-axis) as a function of the soft-thresholding power (*x*-axis).

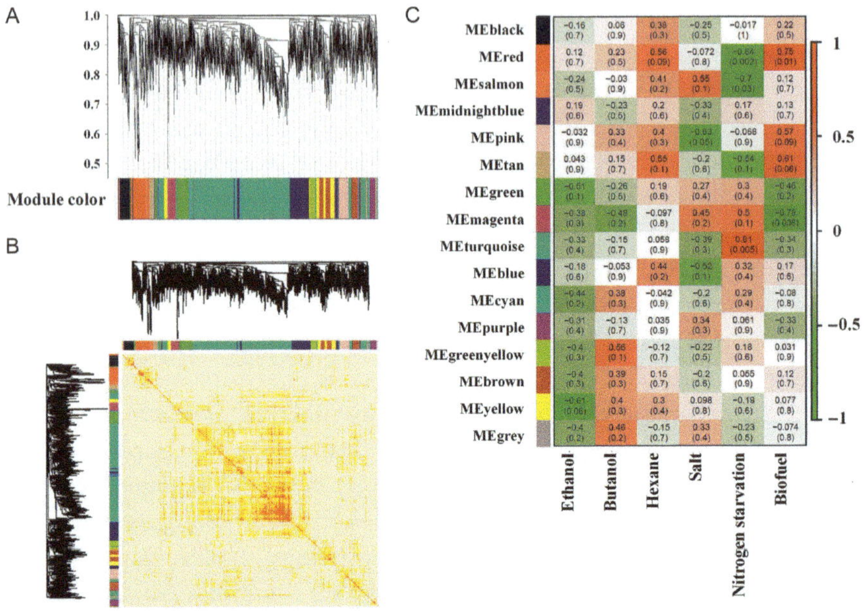

Fig. 2 Network visualization plots. (A) Clustering dendrograms of peptides, with dissimilarity based on topological overlap, together with assigned module colors. There is one peptide dendrogram per block: (B) visualizing peptide network using a heatmap plot. The heatmap depicts the topological overlap matrix (TOM) among all peptides in the analysis. *Light color* represents low overlap and progressively *darker red color* represents higher overlap. Blocks of *darker colors* along the diagonal are the modules. The peptides dendrogram and module assignment are also shown along the *left side* and the *top*: (C) module–trait associations. Each row corresponds to a module eigenpeptide, column to a trait, respectively. Each cell contains the corresponding correlation and *p*-value. The table is *color-coded* by correlation according to the *color legend*.

Many parameters in blockwiseModules need to be defined for WGCNA analysis, such as TOMType = "signed" that counts the directed connection strengths in TOM and minModuleSize = 30 that define a minimal module size of 30. For most other parameters, their default values in the package can be used. As results, a total of 16 modules in the example proteomic datasets were identified. The hierarchical clustering dendrograms and the heatmap plots of peptides networks were visualized in Fig. 2A and B.

3.4 Identification of Modules Associated With Conditions

To identify modules that are significantly associated with the biofuels treated traits, 16 eigenpeptides of modules were correlated with external traits. The color-coded table in Fig. 2C showed the full module–trait correlations.

Several significant module–trait associations were identified. For example, the module eigenpeptide of the red ($r = 0.75$, correlation p-value $= 0.01$) and magenta ($r = -0.78$, correlation p-value $= 0.008$) were positively correlated or negatively correlated with the biofuel treatment, respectively; and another two module eigenpeptides of the pink ($r = 0.57$, correlation p-value $= 0.09$) and tan ($r = 0.61$, correlation p-value $= 0.06$) were both positively correlated with the biofuel treatment.

Associations of individual peptides with a trait of interest (biofuels treatment) were quantified by defining peptide significance (*PS*) as the absolute value of the correlation between the peptides and the trait. For each module, a quantitative measure of *MM* was defined as the correlation between the module eigenpeptide and the peptides expression profile. The modules were ordered according to their *PS* for the biofuel treatment. Based on the *PS* and *MM* measures, peptides that have a higher significance for the biofuel treatment as well as high *MM* in modules of interest were identified.

As an example, the results showed that the red, pink, tan, and magenta modules were significantly associated with the biofuel treatment. Scatter plots of *PS* vs *MM* were plotted for these three modules (Fig. 2). As signed networks were conducted, clearly, *PS* and *MM* are positively correlated in red, pink, and tan modules, and negatively correlated in magenta module, respectively, illustrating a reliability of our results.

3.5 Pathway Enrichment Analysis of Condition-Responsive Modules

Four protein modules that are highly associated with the biofuel treatment were identified. To facilitate a biological interpretation, enrichment analysis was applied to determine whether any metabolic pathway or GO was significantly enriched in the four biofuel-responsive modules. Several functional enrichment analysis suites such as DAVID (Dennis et al., 2003) and KOBAS (Mao, Cai, Olyarchuk, & Wei, 2005) are available. Besides, function GoenrichmentAnalysis is also available in WGCNA package for GO enrichment analysis. However, the Goenrichment-Analysis function is available only to several eukaryotic species with organism-specific information (i.e., GO terms) already included in the package so that it cannot be used for microbial species of our interests. Alternatively, based on the improved protocol and Goseq function in *R* GOseq package (Young, Wakefield, Smyth, & Oshlack, 2010), the WGCNA was extended to whole organisms' gene/protein/metabolite

Table 1 Protein Enrichment Analysis Results for Biofuels Response Modules

Modules	Pathways	*p*-Values
Red	ko00196 Photosynthesis-antenna proteins	3.83×10^{-7}
Red	ko00860 Porphyrin and chlorophyll metabolism	0.0110
Pink	ko00620 Pyruvate metabolism	0.0338
Pink	ko00195 Photosynthesis	0.0432
Tan	ko00196 Photosynthesis-antenna proteins	7.70×10^{-5}
Tan	ko00710 Carbon fixation in photosynthetic organisms	0.0004
Tan	ko00630 Glyoxylate and dicarboxylate metabolism	0.0338
Magenta	ko00240 Pyrimidine metabolism	0.0110
Magenta	ko03018 RNA degradation	0.0120
Magenta	ko00010 Glycolysis/gluconeogenesis	0.0200

enrichment analysis in any interested module, as long as the latest mappings between gene/protein/metabolite and KEGG pathway or GO terms can be extracted.

As the modules were constructed and identified based on peptides, the WGCNA of peptides module information should be transformed to proteins module information matrix first (Details in Table S1 (http://dx.doi.org/10.1016/bs.mie.2016.09.016)). Briefly, a protein Sll1214 in column "red" was labeled as "1," indicating at least one peptide of Sll1214 was detected in red module; otherwise, it should be set as "0." By setting a threshold of hypergeometric test p-value < 0.05, the analysis showed that peptides in red module were significantly enriched in pathway "Photosynthesis-antenna proteins" (p-value $= 3.83 \times 10^{-7}$) and "Porphyrin and chlorophyll metabolism" (p-value $= 0.011$) (Table 1). Peptides in pink and tan module were significantly enriched in pathway "Photosynthesis" (p-value $= 0.043$), "Photosynthesis-antenna proteins" (p-value $= 7.70 \times 10^{-5}$) and "Carbon fixation in photosynthetic organisms" (p-value $= 0.0004$), respectively.

3.6 Topological Properties and Network Visualization

As hub proteins typically represent important biological functions, it will be interesting to identify centrally located intramodular hub peptides in the

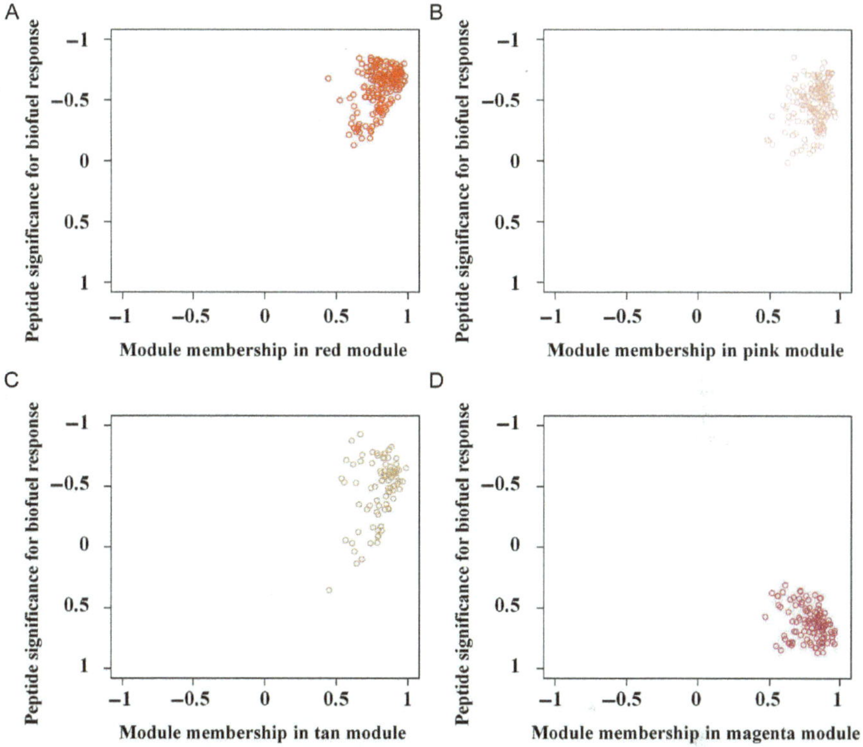

Fig. 3 The scatter plot of peptide significance (*PS*) for biofuels treatments vs module membership (*MM*) in (A) *red*, (B) *pink*, (C) *tan*, and (D) *magenta* modules, respectively. *PS* and *MM* exhibit a very significant correlation, implying that hub peptides in these modules are highly correlated with biofuels treatments.

modules related to the biofuel treatment (Horvath & Dong, 2008). To do so, the *PS* and *MM* were used (Fig. 3). The high correlation between *PS* and *MM* suggested that hub peptides in the module might also be highly correlated with the biofuel treatment. For example, the peptide "ASGVEFPSLVDR" of the protein Slr1874 in tan module has a peptide significance of 0.928 to the biofuel treatment, significantly higher than the eigenpeptide of the tan module. Finally, the WGCNA package provides a convenient way for exporting the networks to VisANT (Hu et al., 2008) and Cytoscape (Shannon et al., 2003) for network visualization. A Cytoscape plot of the most connected peptides in these modules was shown in Fig. 4.

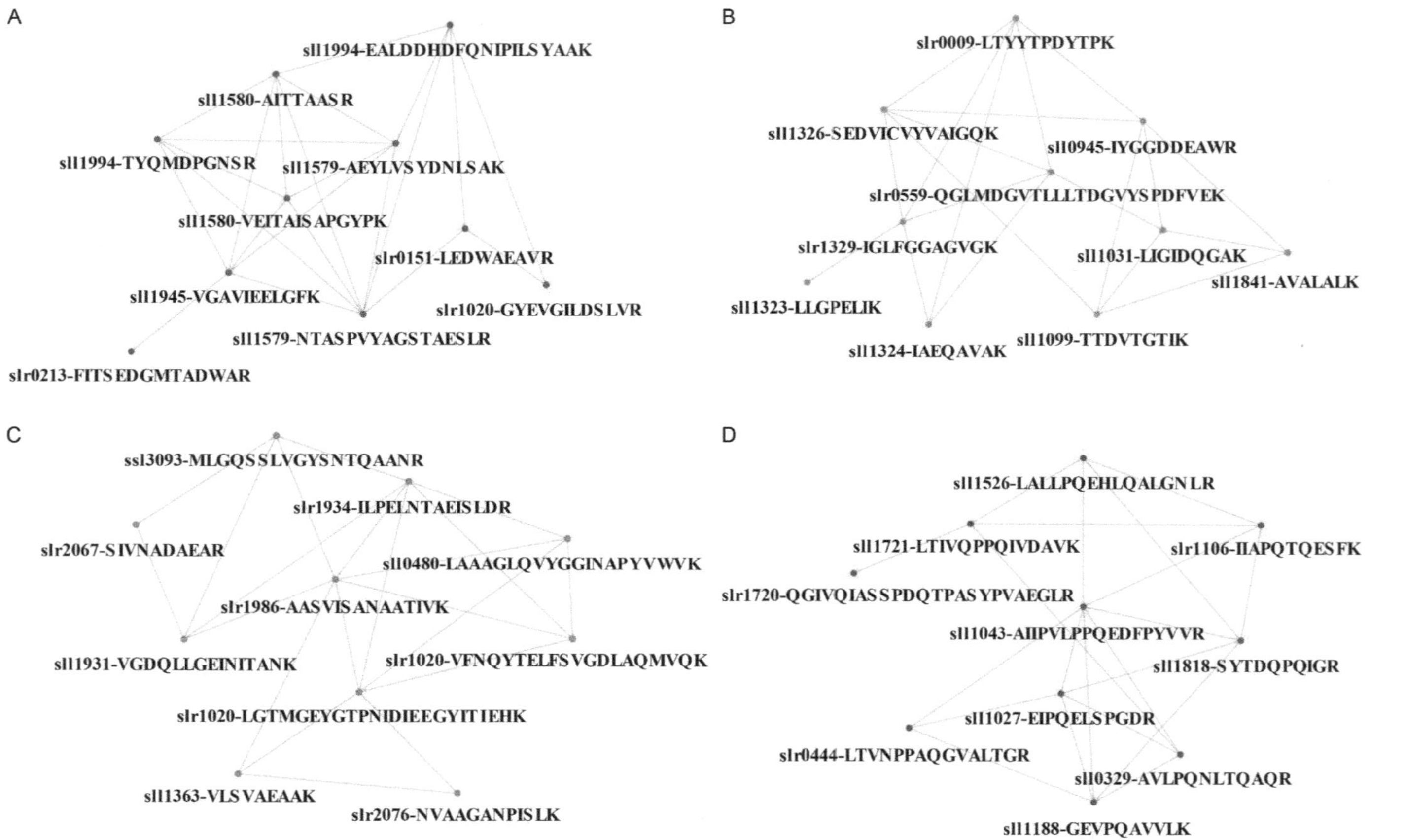

Fig. 4 Visualization of 10 most highly connected peptides in (A) *red*, (B) *pink*, (C) *tan*, and (D) *magenta* modules, generated by the Cytoscape software. The four plots showed network connections whose topological overlap is above the thresholds of 0.21, 0.11, 0.12, and 0.09, respectively.

4. APPLICATION IN METABOLOMICS ANALYSIS

4.1 Description of Experimental Design and Metabolomic Dataset

Description of the metabolomic analysis and relevant biological findings analysis related to this example metabolomic dataset has been described previously (Yu et al., 2015). Briefly, the aim was to reveal possible metabolic modules and biomarkers related to effects of several chemical modulators on astaxanthin biosynthesis. The information obtained will be valuable in deciphering regulatory mechanism of astaxanthin biosynthesis in *Haematococcus pluvialis* (here after *H. pluvialis*) and discovering or even designing new chemical modulators to further enhance astaxanthin accumulation (Yu et al., 2015). Toward the goals, the *H. pluvialis* cells under control, treatments of chemicals solvent dimethyl sulfoxide (DMSO), oxygenant including 2,2′-azo-bis(2-amidinopropane)-dihydrochloride (AAPH), methylene blue (MB), methyl viologen (MV), and signal transducers, including salicylic acid (SA), gibberellic acid (GA_3), and abscisic acid (ABA) were collected for gas chromatography-mass spectrometry (GC-MS) analysis, respectively, and each sample with three biological replicates. Typically, GS-MS analysis of the *H. pluvialis* samples led to identification and quantification of 91 ~ 94 distinct chemically classified metabolites.

4.2 Data Preprocessing

Even all metabolomic datasets were from the same batch, the coverage or concentration of metabolites can be significantly different between samples due to analytical errors; it is thus critical to reduce or eliminate effects of these variations by a proper preprocessing (Wu & Li, 2016). With the aid of mean-centered normalization approach combined with log transformation scaling, the influence of variation in concentration from different metabolites can be eliminated. Briefly, the absolute concentration of metabolites was converted into ratios between the concentration of a metabolite in a given sample and the average concentration of the metabolite in all samples, and then log2 transformed.

A possible mechanism of MNAR missing values was determined, for example, metabolomic datasets, and a modified MNAR-devoted imputation method named as deterministic minimum imputation was therefore used for imputing metabolomic missing data (Lazar et al., 2016). Briefly, missing values of a metabolite were replaced by a minimum value of the metabolite quantified. In addition, similar to that discussed in the proteomic example, to ensure the networks to be biologically meaningful, it is recommended to remove metabolites with too many missing values (Langfelder et al., 2008).

4.3 Construction of Unsigned Weighted Coexpression Networks

After data normalization and missing value imputation, the format of metabolomic could be similar to that of microarray and RNA-seq data, and suitable for WGCNA analysis. The majority of parameters for WGCNA construction have been explained earlier in the example of proteomic analysis. An unsigned weighted metabolites coexpression network was constructed by setting the networkType = "unsigned" in function blockwiseModules for the metabolomic example. In addition, as only fewer than one hundred metabolites identified, the scale-free topology fit index failed to reach values above 0.8 for a reasonable power (Fig. 5A).

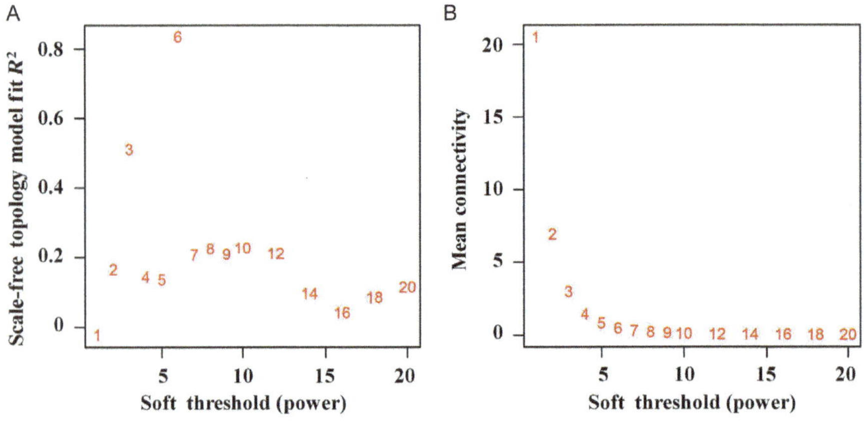

Fig. 5 Analysis of metabolomic network topology for various soft-thresholding powers. The *left panel* shows the scale-free fit index (*y*-axis) as a function of the soft-thresholding power (*x*-axis). The *right panel* displays the mean connectivity (degree, *y*-axis) as a function of the soft-thresholding power (*x*-axis).

Nevertheless, as previous reports have found that the lack of scale-free topology does not necessary invalidate the datasets, mean connectivity of the network should be examined (Langfelder et al., 2008). Taking consideration of both scale-free topology fit index and mean connectivity (Fig. 5B), the soft-thresholding power = 3 and minModuleSize = 5 were chosen for the analysis of the example metabolomic datasets. As shown in the hierarchical clustering dendrograms and the heatmap plots of metabolites, the weighted coexpression network constructed allowed identification of nine metabolic modules in the example metabolic dataset (Fig. 6).

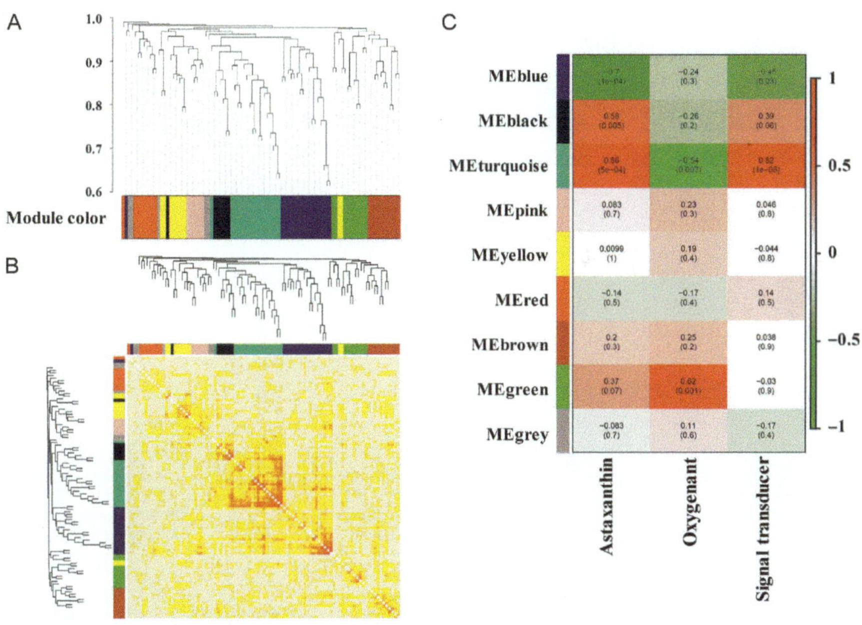

Fig. 6 Network visualization plots. (A) Clustering dendrograms of metabolites, with dissimilarity based on topological overlap, together with assigned *module colors*. There is one metabolite dendrogram per block. (B) Visualizing metabolites network using a heatmap plot. The heatmap depicts the topological overlap matrix (TOM) among all metabolites in the analysis. *Light color* represents low overlap and progressively *darker red color* represents higher overlap. Blocks of *darker colors* along the diagonal are the modules. The metabolites dendrogram and module assignment are also shown along the left side and the top. (C) Module–trait associations. Each row corresponds to a module eigenmetabolite, column to a trait. Each cell contains the corresponding correlation and *p*-value. The table is *color coded* by correlation according to the *color legend*.

4.4 Identification of Modules Associated With Multiple Traits

To identify modules that are significantly associated with the astaxanthin accumulation and further decipher the regulatory mechanism of *H. pluvialis* to treatments of different chemical modulators, nine eigenmetabolites of the modules were correlated with external traits and the full module–trait correlations were shown in Fig. 6C. It was found that the eigenmetabolites in blue module (18 metabolites with astaxanthin $r = -0.7$, correlation p-value $= 1 \times 10^{-4}$) were negatively correlated with astaxanthin accumulation, and the eigenmetabolites in the turquoise module ($r = 0.66$, correlation p-value $= 5 \times 10^{-4}$) and the black module ($r = 0.56$, correlation p-value $=0.005$) were positively correlated with astaxanthin accumulation. Meantime, the eigenmetabolite of the green module ($r = 0.62$, correlation p-value $= 0.001$) was positively correlated with oxygenant treatments, and the eigenmetabolite of the turquoise module ($r = 0.82$, correlation p-value $= 1 \times 10^{-6}$) was positively correlated with treatments of signal transducers. In addition, the scatter plot of metabolite significance (*MS*) vs *MM* (Fig. 7A) show a significant correlation.

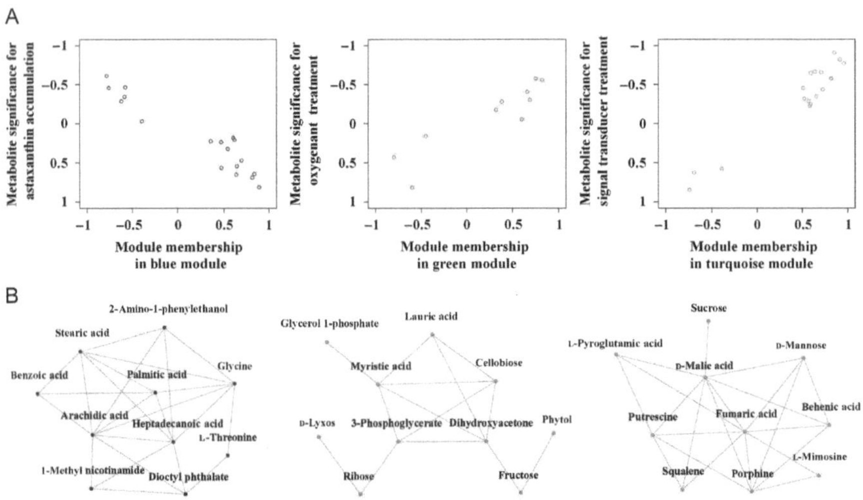

Fig. 7 Example of several modules associated with multiple traits. (A) The scatter plot of metabolite significance (*MS*) for astaxanthin accumulation vs *blue* module membership (*MM*), Oxygenant response vs *green* module membership (*MM*), and signal transducer response vs *turquoise* module membership (*MM*); (B) Visualization of 10 most highly connected metabolites network connections in *blue*, *green*, and *turquoise* modules, whose topological overlap is above the thresholds of 0.12, 0.06, and 0.2, respectively.

Table 2 Metabolite Enrichment Analysis Results for Biofuels Responsive Modules

Modules	Pathways	*p*-Values
Blue	map01040 Biosynthesis of unsaturated fatty acids	0.0069
Green	map00561 Glycerolipid metabolism	0.0084
Green	map00564 Glycerophospholipid metabolism	0.0108
Turquoise	map00650 Butanoate metabolism	0.0309

4.5 Pathway Enrichment Analysis of Traits Related Modules

Several modules with significant correlation with traits of interest were chosen for pathway enrichment analysis. After collecting information of all metabolites participated in metabolic pathways in *H. pluvialis* from the KEGG database and defining a threshold of hypergeometric test p-value < 0.05, enrichment analysis was conducted, and the results showed that metabolites in the blue module were significantly enriched in pathway "biosynthesis of unsaturated fatty acids" (p-value $= 6.9 \times 10^{-3}$), while metabolites in the green module were significantly enriched in pathway "glycerolipid metabolism" (p-value $= 8.43 \times 10^{-3}$) and "glycerophospholipid metabolism" (p-value $= 8.43 \times 10^{-3}$) (Table 2), suggesting these pathways could be biologically relevant to the respective modules.

4.6 Topological Properties and Network Visualization

Hub metabolites are with a high degree of connectivity in biological interaction networks and are thus supposed to be biological important, it will thus be interesting to identify centrally located intramodular hub metabolites in the modules related astaxanthin accumulation or treatments of chemical modulators (Horvath & Dong, 2008). To do so, the *MS* and *MM* were calculated and correlated (Fig. 7B), the high correlation between *MS* and *MM* suggested that hub metabolites in the modules could also be highly correlated with the external traits. Since the procedures of topological property analysis and network visualization are identical to the analysis of example proteomic datasets described earlier, technical details were omitted here.

5. CONCLUSIONS

WGCNA is a powerful and comprehensive approach for high-dimensional data analysis. In view of the limitations in its application to proteomic and metabolomic datasets, we introduced a modified WGCNA protocol as a complement to the standardized protocol of WGCNA, and provided detailed tutorial to its application to proteomic and metabolomic example datasets. The results demonstrated that the modified WGCNA protocol can be successfully extended to the application of high-throughput proteomic and metabolomic data for meaningful biological interpretation.

ACKNOWLEDGMENTS

The research was supported by grants from National Basic Research Program of China ("973" program, project no. 2011CBA00803, 2012CB721101, and 2014CB745101), National High-tech R&D Program ("863" program, project no. 2012AA02A707), and Nature Science Foundation of China (no. 31370115 and 31470217).

DATA AVAILABILITY

The raw data in proteomic and metabolomic examples, and WGCNA code used in this chapter are available upon request (http://dx.doi.org/10.1016/bs.mie.2016.09.016).

REFERENCES

Albrecht, D., Kniemeyer, O., Brakhage, A. A., & Guthke, R. (2010). Missing values in gel-based proteomics. *Proteomics*, *10*(6), 1202–1211.

Bhaskaran, K., & Smeeth, L. (2014). What is the difference between missing completely at random and missing at random? *International Journal of Epidemiology*, *43*(4), 1336–1339.

Blanchet, L., Smolinska, A., Attali, A., Stoop, M. P., Ampt, K. A., Aken, H. V., et al. (2011). Fusion of metabolomics and proteomics data for biomarkers discovery: Case study on the experimental autoimmune encephalomyelitis. *BMC Bioinformatics*, *12*, 254.

Chahrour, O., Cobice, D., & Malone, J. (2015). Stable isotope labelling methods in mass spectrometry-based quantitative proteomics. *Journal of Pharmaceutical and Biomedical Analysis*, *113*, 2–20.

Dennis, G., Jr., Sherman, B. T., Hosack, D. A., Yang, J., Gao, W., Lane, H. C., et al. (2003). DAVID: Database for annotation, visualization, and integrated discovery. *Genome Biology*, *4*(5), P3.

Dong, J., & Horvath, S. (2007). Understanding network concepts in modules. *BMC Systems Biology*, *1*, 24.

Fuller, T. F., Ghazalpour, A., Aten, J. E., Drake, T. A., Lusis, A. J., & Horvath, S. (2007). Weighted gene coexpression network analysis strategies applied to mouse weight. *Mammalian Genome*, *18*(6–7), 463–472.

Gao, L., Pei, G., Chen, L., & Zhang, W. (2015). A global network-based protocol for functional inference of hypothetical proteins in *Synechocystis* sp. PCC 6803. *Journal of Microbiological Methods*, *116*, 44–52.

Gauguier, D. (2016). Application of quantitative metabolomics in systems genetics in rodent models of complex phenotypes. *Archives of Biochemistry and Biophysics*, *589*, 158–167.

Ghazalpour, A., Doss, S., Zhang, B., Wang, S., Plaisier, C., Castellanos, R., et al. (2006). Integrating genetic and network analysis to characterize genes related to mouse weight. *PLoS Genetics*, *2*(8), e130.

Gibbs, D., Baratt, A., & Baric, R. (2013). Protein co-expression network analysis (ProCoNA). *Journal of Clinical Bioinformatics*, *3*(1), 11.

Horvath, S., & Dong, J. (2008). Geometric interpretation of gene coexpression network analysis. *PLoS Computational Biology*, *4*(8), e1000117.

Horvath, S., Zhang, B., Carlson, M., Lu, K. V., Zhu, S., Felciano, R. M., et al. (2006). Analysis of oncogenic signaling networks in glioblastoma identifies ASPM as a molecular target. *Proceedings of the National Academy of Sciences of the United States of America*, *103*(46), 17402–17407.

Hu, Z., Snitkin, E. S., & DeLisi, C. (2008). VisANT: An integrative framework for networks in systems biology. *Briefings in Bioinformatics*, *9*(4), 317–325.

Huan, T., & Li, L. (2015). Counting missing values in a metabolite-intensity data set for measuring the analytical performance of a metabolomics platform. *Analytical Chemistry*, *87*(2), 1306–1313.

Kunowska, N., Rotival, M., Yu, L., Choudhary, J., & Dillon, N. (2015). Identification of protein complexes that bind to histone H3 combinatorial modifications using super-SILAC and weighted correlation network analysis. *Nucleic Acids Research*, *43*(3), 1418–1432.

Langfelder, P., & Horvath, S. (2008). WGCNA: An R package for weighted correlation network analysis. *BMC Bioinformatics*, *9*, 559.

Langfelder, P., Zhang, B., & Horvath, S. (2008). Defining clusters from a hierarchical cluster tree: The dynamic tree cut package for R. *Bioinformatics*, *24*(5), 719–720.

Lazar, C., Gatto, L., Ferro, M., Bruley, C., & Burger, T. (2016). Accounting for the multiple natures of missing values in label-free quantitative proteomics data sets to compare imputation strategies. *Journal of Proteome Research*, *15*(4), 1116–1125.

Leek, J. T., Scharpf, R. B., Bravo, H. C., Simcha, D., Langmead, B., Johnson, W. E., et al. (2010). Tackling the widespread and critical impact of batch effects in high-throughput data. *Nature Reviews. Genetics*, *11*(10), 733–739.

Lehtinen, S., Marsellach, F. X., Codlin, S., Schmidt, A., Clement-Ziza, M., Beyer, A., et al. (2013). Stress induces remodelling of yeast interaction and co-expression networks. *Molecular BioSystems*, *9*(7), 1697–1707.

Mao, X., Cai, T., Olyarchuk, J. G., & Wei, L. (2005). Automated genome annotation and pathway identification using the KEGG Orthology (KO) as a controlled vocabulary. *Bioinformatics*, *21*(19), 3787–3793.

Nie, L., Wu, G., Culley, D. E., Scholten, J. C., & Zhang, W. (2007). Integrative analysis of transcriptomic and proteomic data: Challenges, solutions and applications. *Critical Reviews in Biotechnology*, *27*(2), 63–75.

Oldham, M. C., Horvath, S., & Geschwind, D. H. (2006). Conservation and evolution of gene coexpression networks in human and chimpanzee brains. *Proceedings of the National Academy of Sciences of the United States of America*, *103*(47), 17973–17978.

Pei, G., Chen, L., Wang, J., Qiao, J., & Zhang, W. (2014). Protein network signatures associated with exogenous biofuels treatments in cyanobacterium *Synechocystis* sp. PCC 6803. *Frontiers in Bioengineering and Biotechnology*, *2*, 48.

Ravasz, E., Somera, A. L., & Mongru, D. A. (2002). Hierarchical organization of modularity in metabolic networks. *Science*, *297*(30), 1551–1555.

Rubin, D. B. (1976). Inference and missing data. *Biometrika*, *63*(3), 581–592.

Ruffalo, M., Koyuturk, M., & Sharan, R. (2015). Network-based integration of disparate omic data to identify "silent players" in cancer. *PLoS Computational Biology*, *11*(12), e1004595.

Shannon, P., Markiel, A., Ozier, O., Baliga, N., Wang, J., Ramage, D., et al. (2003). Cytoscape: A software environment for integrated models of biomolecular interaction networks. *Genome Research*, *13*(11), 2498–2504.

Su, Y., Wang, J., Shi, M., Niu, X., Yu, X., Gao, L., et al. (2014). Metabolomic and network analysis of astaxanthin-producing *Haematococcus pluvialis* under various stress conditions. *Bioresource Technology*, *170*, 522–529.

Tuikkala, J., Elo, L., Nevalainen, O. S., & Aittokallio, T. (2006). Improving missing value estimation in microarray data with gene ontology. *Bioinformatics*, *22*(5), 566–572.
Webb-Robertson, B. J., Wiberg, H. K., Matzke, M. M., Brown, J. N., Wang, J., McDermott, J. E., et al. (2015). Review, evaluation, and discussion of the challenges of missing value imputation for mass spectrometry-based label-free global proteomics. *Journal of Proteome Research*, *14*(5), 1993–2001.
Wilkins, M. R., Pasquali, C., Appel, R. D., Ou, K., Golaz, O., Sanchez, J. C., et al. (1996). From proteins to proteomes: Large scale protein identification by two-dimensional electrophoresis and amino acid analysis. *Biotechnology (NY)*, *14*(1), 61–65.
Wu, Y., & Li, L. (2016). Sample normalization methods in quantitative metabolomics. *Journal of Chromatography A*, *1430*, 80–95.
Xia, J., Sinelnikov, I. V., Han, B., & Wishart, D. S. (2015). MetaboAnalyst 3.0—Making metabolomics more meaningful. *Nucleic Acids Research*, *43*(W1), W251–W257.
Yin, P., & Xu, G. (2014). Current state-of-the-art of nontargeted metabolomics based on liquid chromatography-mass spectrometry with special emphasis in clinical applications. *Journal of Chromatography. A*, *1374*, 1–13.
Young, M. D., Wakefield, M. J., Smyth, G. K., & Oshlack, A. (2010). Gene ontology analysis for RNA-seq: Accounting for selection bias. *Genome Biology*, *11*(2), R14.
Yu, X., & Feng, L. (2016). Quantitative proteomics reveals the novel co-expression signatures in early brain development for prognosis of glioblastoma multiforme. *Oncotarget*, *7*(12), 14161–14171.
Yu, X., Niu, X., Zhang, X., Pei, G., Liu, J., Chen, L., et al. (2015). Identification and mechanism analysis of chemical modulators enhancing astaxanthin accumulation in *Haematococcus pluvialis*. *Algal Research*, *11*, 284–293.
Zhang, Z. (2016). Multiple imputation with multivariate imputation by chained equation (MICE) package. *Annals of Translational Medicine*, *4*(2), 30.
Zhang, W., Gritsenko, M. A., Moore, R. J., Culley, D. E., Nie, L., Petritis, K., et al. (2006). A proteomic view of *Desulfovibrio vulgaris* metabolism as determined by liquid chromatography coupled with tandem mass spectrometry. *Proteomics*, *6*(15), 4286–4299.
Zhang, B., & Horvath, S. (2005). A general framework for weighted gene co-expression network analysis. *Statistical Applications in Genetics and Molecular Biology*, *4*. Article 17.
Zhang, Q., Li, J., Xue, H., Kong, L., & Wang, Y. (2016). Network-based methods for identifying critical pathways of complex diseases: A survey. *Molecular BioSystems*, *12*(4), 1082–1089.
Zhang, L., Liu, Y. Z., Zeng, Y., Zhu, W., Zhao, Y. C., Zhang, J. G., et al. (2016). Network-based proteomic analysis for postmenopausal osteoporosis in Caucasian females. *Proteomics*, *16*(1), 12–28.

CHAPTER TEN

A Protocol for Large-Scale Proteomic Analysis of Microdissected Formalin Fixed and Paraffin Embedded Tissue

P. Ostasiewicz*, J.R. Wiśniewski[†,1]
*Wrocław Medical University, Wrocław, Poland
[†]Biochemical Proteomics Group, Max-Planck-Institute of Biochemistry, Martinsried, Germany
[1]Corresponding author: e-mail address: jwisniew@biochem.mpg.de

Contents

Abstract

Extensive use of biomarkers in diagnostic and therapeutic procedures is probably the hallmark of future medicine. Biomarkers can be of different chemical nature—most frequently DNA, RNA, and proteins are considered as molecules of interest. As the most rational approach appears analysis of proteins, because these are the "effector" molecules directly involved for cell homeostasis. There is no commonly accepted standard procedure in proteomic-based biomarker search. In this chapter, we described a protocol for proteomic analysis of formalin-fixed and paraffin-embedded tissue, which has been proven to be highly effective for proteomics-based biomarker discovery.

1. INTRODUCTION

Extensive use of biomarkers in diagnostic and therapeutic procedures is probably the hallmark of future in medicine. Biomarkers can be of different chemical nature—most frequently DNA, RNA, and proteins are considered as molecules of interest. Despite the massive investments in genomics

Methods in Enzymology, Volume 585
ISSN 0076-6879
http://dx.doi.org/10.1016/bs.mie.2016.09.017

and transcriptomics, exceeding any investments in proteomics, all recent efforts to identify mRNA-based cancer biomarkers resulted in introduction to clinics only two commonly used microarrays—"MammaPrint" and "Oncotype" (Kondo, 2014). It is worth to note also that among biomarkers recommended by leading international societies to use in clinical practice there are only few DNA/RNA biomarkers and several proteins (Harris et al., 2007; Sturgeon et al., 2008).

The most rational approach for biomarker discovery is the analysis of proteins, as the proteins are the "effector" molecules responsible for cell function and growth (Kondo, 2014). For a long time, low sensitivity of the proteomic methods has been considered to be its major drawback of the technology; however, it has recently been demonstrated that proteomics can provide data, which depth and extent is comparable to that, which can be achieved using genetic methods (Wisniewski et al., 2012). An advantage of proteomic analysis over genetic methods is higher stability of proteins compared to nucleic acids. This allows analysis of archival, chemically fixed samples, which have been stored for many years (Balgley et al., 2009; Ostasiewicz, Zielinska, Mann, & Wisniewski, 2010).

There are two major classes of clinical material used in biomarkers search—body fluids, such as blood, plasma, urine, and solid tissues. The main advantage of analyzing body fluids is a relatively straightforward accessibility of samples. Unfortunately, biomarkers are usually present in the body fluids at extremely low concentrations (Hori & Gambhir, 2011), because their occurrence commonly is a result of so called "tumor leaking" rather than an effect of active secretion. Concentration of the most abundant proteins in a given fluid is often several orders of magnitude higher than concentrations of majority of other proteins, including potential biomarkers. These titer differences are the substantial analytical limitation because the dynamic range of commonly used mass spectrometers does not exceed 3 orders (Ray et al., 2011). Therefore, discovery of biomarkers in serum or other fluids remains extremely difficult (Kondo, 2014). As a consequence solid tissues appear to be more suitable for biomarker search, because in most cases, about 95% of proteins in solid spans 3–4 orders of magnitude dynamic range (Wisniewski et al., 2015). However, the titers of potential biomarkers have to be explored among lower abundant proteins (Simpson, Bernhard, Greening, & Moritz, 2008).

Considering solid tissue for a proteomic analysis a choice has to be made between fresh or flash frozen and formalin-fixed and paraffin-embedded (FFPE) material. The fresh or frozen tissue has for a long time been considered as the only one suitable for proteomic investigation and the chemically modified FFPE tissue as a difficult in the analysis. In the practice, however, the fresh/frozen material is only rarely available, in particular when samples from patients with known clinical outcome are needed. On the other hand, there are several important advantages of FFPE material. The most important is that formalin fixing and paraffin embedding is a standard procedure at all pathology departments resulting in almost unlimited number of available samples. Furthermore, the fact that prolonged storage of such samples does not negatively influence its quality; therefore, comparison of tissues from patients with different outcomes is possible, which allows identification prognostic markers. Such specimens are also much more resistant and are easier to handle which results in less false negative or positive results that are often due to degradation of nonfixed samples. This aspect is particularly important if tissue slices are subject of microdissection, when large surface of thin material is directly exposed to air and sometimes to the high temperature from halogen microscope light. In context of microdissection it is also worth mentioning the fact that fixed material offers better morphology preservation than the fresh one, which in turn allows easier and more precise dissection. Benefits of using FFPE material were already reviewed (Giusti & Lucacchini, 2013; Gustafsson, Arentz, & Hoffmann, 2015; Maes et al., 2013; Magdeldin & Yamamoto, 2012).

The removal of the chemical cross-links introduced by formalin fixing remained a difficult task for the long time. It was known that usage of elevated temperatures and high concentrations strong detergents, such as SDS, should facilitate protein retrieval but because of the detergent incompatibility with mass spectrometry many of the published protocols avoided its use (Jiang et al., 2007; Sprung et al., 2009). Substituting SDS by less effective reagents led to incomplete lysis of the tissue and resulted in limited depth of proteomic analysis. In other approaches, SDS was used, but the necessary depletion of the detergent, before mass spectrometry, was accompanied with significant losses of the extracted proteins (Balgley et al., 2009; Jiang et al., 2007; Xu et al., 2008). An overview of different approaches used for protein recovery from FFPE samples can be found in Table 1. Table 2 list studies using commercially available kits. Typically most of the methods allowed only partial insights in the tissue proteome covering several hundreds to 2000 proteins, and only few studies provides more detailed datasets with up to 6000 proteins. In contrast, using the multienzyme filter-aided sample preparation (MED FASP) method (see chapter "Filter aided sample preparation the versatile and efficient method for proteomic analysis"

Table 1 Overview of Proteomic Studies on FFPE Tissue

References	Number of Proteins Identified	Comments	Equivalence With Fresh Samples	Compatible With Microdissection	PTM Analysis Possible
SDS free					
Nazarian, Santi, Hathout, and Macdonald (2008)	200	–	Not tested	n.a.	n.a.
Sprung et al. (2009)	2500	–	Yes	n.a.	n.a.
Nirmalan et al. (2011)	300	RapiGest	Yes	n.a.	n.a.
Alkhas et al. (2011)	1500	–	Not tested	Yes	n.a.
Sousa et al. (2012)	2500	–	Not tested	Yes	n.a.
Sprung et al. (2012)	2000	–	Yes	Yes	n.a.
SDS based					
Protein precipitation					
Xiao et al. (2010)	700	Acetone precipitation	Not tested	n.a.	n.a.
Hammer et al. (2014)	1000	Methanol-chloroform precipitation	Yes	n.a.	n.a.
Bronsert et al. (2014)	1000	Acetone precipitation	Yes	n.a.	n.a.
Dunne et al. (2015)	300	ProteoExtract Kit	Not tested	Yes	n.a.

Yin et al. (2015)	6000	Acetone precipitation	Not tested	n.a.	n.a.
Lai et al. (2016)		Acetone precipitation	Yes	n.a.	n.a.
In-gel digestion					
Fowler, Waybright, Veenstra, O'Leary, and Mason (2012)	5000	–	About 50% overlap between frozen and FFPE samples	n.a.	n.a.
Byrum et al. (2013)	1500	–	Not tested	n.a.	n.a.
Fowler, O'Leary, and Mason (2014)	5000	Elevated pressure	Yes	n.a.	n.a.
Noberini, Uggetti, Pruneri, Minucci and Bonaldi (2016)	–	Analysis of histone posttranslational modifications	Yes	n.a.	Yes
SDS removal-based methods					
Shi, Liu, Balgley, Lee, and Taylor (2006)	2000	SDS removal via dialysis	Yes	n.a.	n.a.
Guo et al. (2007)	2500	SDS removal via dialysis	Yes	Yes	n.a.
Balgley et al. (2009)	2000	SDS removal via dialysis; study evaluating storage time effect	Not tested	n.a.	n.a.

Continued

Table 1 Overview of Proteomic Studies on FFPE Tissue—cont'd

References	Number of Proteins Identified	Comments	Equivalence With Fresh Samples	Compatible With Microdissection	PTM Analysis Possible
Ostasiewicz et al. (2010)	5000	FFPE-FASP	Yes	Yes	Yes
Wisniewski et al. (2012)	8500	FFPE-FASP			
Wisniewski et al. (2015)	10,900	FFPE-FASP			
Pozniak et al. (2016)	10,000	FFPE-FASP			
Tanca et al. (2016)	2500	Detergent removal spin columns	Not tested	n.a.	n.a.
Other					
Crockett, Lin, Vaughn, Lim, and Elenitoba-Johnson (2005)	130	RIPA buffer		n.a.	n.a.

n.a., not assayed.

Table 2 Overview of Proteomic Studies on FFPE Tissue Using Commercial Kits

References	Number of Proteins Identified	Equivalence With Fresh Samples	Compatible With Microdissection	PTM Analysis Possible
Liquid tissue (expression pathology)				
Hood et al. (2005)	1100	Partial or n.a.	Yes	Yes, according to information on manufacturer website
Prieto et al. (2005)	300			
Scicchitano, Dalmas, Boyce, Thomas, and Frazier (2009)	200			
Byrum et al. (2011)	~1000 (FDR dependent)			
Bateman et al. (2011)	9400 (at FDR of 5.84%)			
Bell et al. (2011)	360			
Jain et al. (2012)	260			
QProteome (Qiagen)				
Gamez-Pozo et al. (2011)	150	Yes	Yes	Yes
Gundisch et al. (2012)	1800			
Hammer et al. (2014)	1100			

by Wiśniewski) for processing FFPE lysates enables an in-depth proteomic exploration of human tissue. Recent studies, in which this FASP has been used, datasets more than 10,000 proteins were generated (Table 1).

In this chapter, we describe the FASP-based protocol for minute amounts of sample, of 100–200 nL of fixed tissue, which can be collected by laser capture microdissection. The sample volume corresponds to 10–20 mm^2 of a 10 μm thick slice. This tissue amount yields 10–30 μg total protein. The protocol consists of several steps including the microdissection procedure, fixation reversal, protein digestion, and a strong anion exchange-based

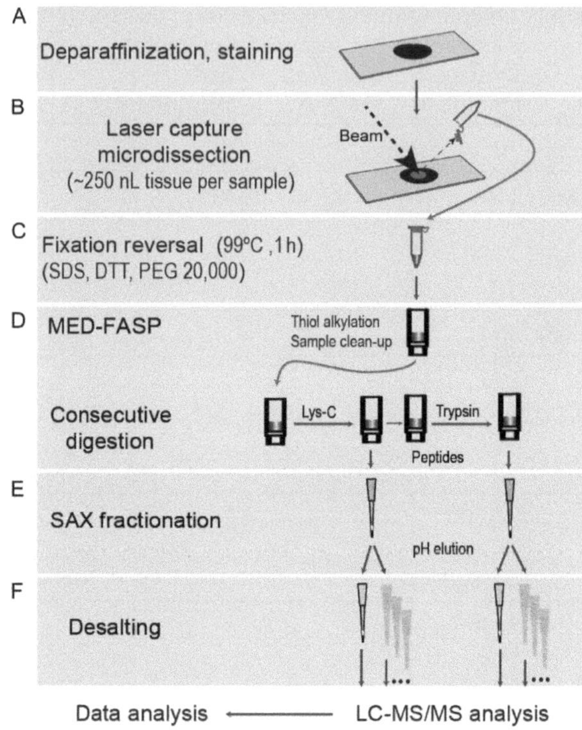

Fig. 1 Workflow of the protocol for the in-depth proteomic analysis of formalin-fixed and paraffin-embedded tissue using laser-capture microdissection. (A) Staining of sample slices for visualization of areas of interest and preparation for microdissection. (B) Microdissection. (C) Reversal of fixation. (D) Sample processing using multienzyme filter-aided sample preparation (MED FASP) method. (E) Fractionation of isolated peptides using strong anion exchanger in a cartridge format. (F) Desalting of peptide fractions before LC-MS/MS analysis.

peptide fractionation (Fig. 1). The number of SAX fractions can be optimized in respect to the total peptide amounts obtained from MED FASP-based processing. The principles, types, and recommendations for using MED FASP are described in detail in chapter "Filter aided sample preparation the versatile and efficient method for proteomic analysis" by Wiśniewski.

2. FFPE–FASP–SAX PROTOCOL

2.1 Materials

2.1.1 Dissection, Deparaffinization, and Lysis of FFPE Tissue

a. Laser microdissector (for example, Zeiss, Göttingen, Germany).

b. Membrane-coated slides for microdissection (Zeiss).
   - CRITICAL: Microdissection can be also done with regular glass slides but in our experience, membrane slides are preferable. Direct collecting of pieces of tissue from regular glass slides results in lower yields and presumably greater destruction by laser light.

c. Adhesive cap collecting tubes (Zeiss, Göttingen, Germany).
   - CRITICAL: Opaque caps enable better visualization of tissue details.

d. Thermoblock set to 99°C or boiling water bath.

e. Bench-top centrifuge.

f. Tissue lysis buffer (TLB): of 0.1 *M* Tris–HCl, pH 8.0, 0.1 *M* DTT, 0.5% (w/v) polyethylene glycol 20,000 and 2% SDS.
   - CRITICAL: Some brands of polyethylene glycol contain impurities that could affect the sample. Therefore it is advisable to prepare first a stock solution of the polyethylene glycol and to dialyze it exhaustively against water before use.

2.1.2 MED-FASP

a. Ultrafiltration units (Microcon30k, Merck Millipore Darmstadt, Germany).

b. Thermostated bench-top centrifuge, temperature set to 20°C.

c. Humid chamber (box) with a rack for Eppendorf-type disposal tubes.
   - TIP: The humid chamber consists of a closed plastic box with moist paper towels at the bottom.

d. Incubator set to 37°C.

e. UA buffer: 8 *M* urea in 0.1 *M* Tris–HCl, pH 8.5. Prepare 1 mL per 1 sample.
   - CRITICAL: The UA and IAA solutions must be freshly prepared and have to be used within a day.

f. IAA solution: 0.05 *M* iodoacetamide in UA. Prepare 0.1 mL per 1 sample.

g. Endoproteinase Lys-C, stock solution 1 μg/μL.

h. Trypsin, stock solution 0.4 μg/μL.

i. Digestion buffer (DB): 0.05 *M* Tris–HCl, pH 8.5. Prepare 0.25 mL per 1 sample.

2.1.3 SAX Pipette-Tip Peptide Fractionation

a. Bench-top centrifuge.

b. SAX "pipette-tip columns" (one "pipette-tip column" per sample).

- Stack six layers plugs of Empore-Anion membrane (Bioanalytical Technologies 3M Company, St. Paul, USA) in a standard 0.2 mL micropipette tip.
  - Place two layers/two disks on the clear, hard surface (e.g., small petri dish).
  - Use a blunt-needle (Hamilton, NDL-KF (ga16/51 mm/pst13) purchase no. 50516/00) to cut out the plug from membranes.
  - Stack it gently but firmly in the pipette tip using a plastic wire.
  - Repeat it twice.
  - CRITICAL: Try to use same pressure when stacking the membranes into all "pipette-tip columns."

c. Desalting C18-"StageTip" columns Prepare 6 Stage tips for each sample.
- The C18-"StageTips" (Ishihama, Rappsilber, Andersen, & Mann, 2002) are prepared by stacking three layers of Empore-C18 (Bioanalytical Technologies 3M Company, St. Paul, MN; cat no. 2215) in a 0.2 mL pipette tip as described above.

d. Stock solution of Britton & Robinson Universal Buffer (BRUB).
- Prepare a solution containing 0.1 *M* CH_3COOH, 0.1 *M* H_3PO_4, and 0.1 *M* H_3BO_3.
- Make 5 aliquots and adjust them with 1 *M* NaOH to 2, 4, 5, 6, and 12.
- Once prepared BRUB solutions can be stored for a long time. For better results readjust pH at least once a week.
- Before use dilute fivefold with deionized water.

e. Buffer A: 0.1% (v/v) CF_3COOH in deionized water.

f. Buffer B: 60% (v/v) CH_3CN, 0.1% (v/v) CF_3COOH in deionized water.

2.2 Protocol

2.2.1 Lysis of FFPE Tissue

TIMING: Depends on slides number. Approximately, for 10 slides about 3–4 h is required, excluding microdissection time. Microdissection time is extremely variable ranging from 1 to 12 h per sample, depending on type of tissue and required accuracy of dissection.

a. Prepare membrane-coated slides with 45 min UV irradiation.
- CRITICAL STEP: Omission of this step often results in a loss sample sections. Prepared slides have to be used within 3–5 days, than slides have to be retreated.

b. Cut the 5–10 μm slices and mount them on membrane-coated slides.

c. Dry the section for 1 h in 37°C incubator.

- PAUSE POINT: Prepared slides can be stored in room temperature for at least 2–4 days without any negative impact on tissue quality.

d. Deparaffinize the slide for 2.5 min in xylene (in histological jar).

e. Transfer the slide to the fresh xylene for another 1.5 min.

f. Transfer the slide to the absolute ethanol for 1 min.

g. Transfer the slide to the 70% ethanol for 1 min.

- TIP: The incubation times should be adjusted according to the needs. Times listed here are usually suitable for 7 μm section. When working with 10 μm section add 30–60 s to each incubation time. The longer incubations, the better morphology preservation under microscopic observations but the greater protein loses.

h. Transfer the slide to water for 1 min.

i. Transfer the slide to the Mayer's hematoxylin for 20 s.

- TIP: Exact time depends on "strength" of hematoxylin used. Be careful not to overstain the sections. Too dark section makes microscopic analysis more difficult.

j. Rinse the slide with water for 1 min.

- PAUSE POINT: The stained sections should not be stored for longer than 12–24 h.

k. Dissect required amount of tissue.

- PAUSE POINT: If dissecting of all the required material is not possible in one day, you can store dissected material in room temperature overnight. Be sure to spin down tissue fragments down into TLB.

l. Lyse the microdissected FFPE tissue in the TLB buffer 99°C in a heating block with agitation (600 rpm) for 1 h. Use 3 μL of the buffer for 10 nL of the microdissected tissue.

- CRITICAL: During the sample lysis water condenses on the tube lid. Therefore it is advisable to collect the water back in the tube cone by short centrifugation in an interval of about 15 min.

m. Clarify the crude extract by centrifugation at 16,000 $\times g$ at 18°C for 10 min.

- PAUSE POINT: The lysate can be stored for few days in room temperature or frozen at −20°C for several months.
- TIP: Do not store the samples in the fridge (4°C).

2.2.2 Sample Processing

TIMING: 3 h + 18 h incubation + 6 h (including 4 h incubation).

a. Mix up to 50 μL of the tissue of FFPE lysate with 200 μL of UA in the ultrafiltration unit.
  - CRITICAL STEP: Be careful not to scratch the surface of the filter with pipette tip.

b. Centrifuge it at 10,000 × *g* for until less than 10 μL of the solution remains in the filter. This requires about 10–15 min centrifugation.
  - TIP: The time required for the centrifugation varies between samples of different origin and their total protein content.
  - TIP: Less than 10 μL in the filter unit means that it should look empty but wet. Completely dry surface after 10–15 min suggests that the filter is broken.

c. Add 200 μL of UA to the ultrafiltration unit and centrifuge at 10,000 × *g* for 15 min. Discard the flow-through from the collection tube.

d. Add 50 μL IAA solution and mix at 600 rpm in a thermo-mixer for 1 min and centrifuge at 10,000 × *g* for 10 min.

e. Add 100 μL of UA to the ultrafiltration unit and centrifuge at 10,000 × *g* for 15 min. Repeat this step twice.

f. Add 100 μL of the DB to the ultrafiltration unit and centrifuge at 10,000 × *g* for 10 min. Repeat this step twice.

g. Add 40 μL DB with endoproteinase Lys-C (enzyme to protein ratio 1:100) and mix at 600 rpm in thermo-mixer for 1 min. Incubate the units in the humid chamber at 37°C for 18 h.
  - PAUSE POINT: Incubation.

h. Transfer the filter units to new collection tubes and centrifuge at 10,000 × *g* for 10 min.

i. Add 160 μL of deionized water and centrifuge the ultrafiltration units at 10,000 × *g* for 10 min. The combined filtrate (3.8 and 3.9) contains the peptides released by endoproteinase Lys C.

j. Transfer the ultrafiltration units to new collection tubes and add 40 μL DB with trypsin (enzyme to protein ratio 1:100). Mix at 600 rpm in thermo-mixer for 1 min. Incubate the units in a humid chamber at 37°C for 4 h.

k. Centrifuge the ultrafiltration units at 10,000 × *g* for 10 min.

l. Add 160 μL of deionized water and centrifuge the ultrafiltration units at 10,000 × *g* for 10 min. The combined filtrate (k and l) contains trypsin cleaved peptides.
  - PAUSE POINT: Peptide solutions (j, l) can be stored in 4°C for few days or frozen at −20°C for several months.

2.2.3 Fractionation of the Lys-C and Tryptic Peptides

a. Dilute the peptide solutions obtained by digestions with Lys-C and trypsin with 0.2 mL of BRUB, pH 11 and 5, respectively.

b. Assemble the SAX "pipette-tip column" in the centrifugal tube adapter lid

c. Wash and equilibrate the SAX "pipette-tip column."

- **i.** Add 0.1 mL of methanol to the column and centrifuge it at 4000 × *g* for until the column is empty but the membranes inside still wet.
  - TIP: This requires about 5 min centrifugation depending on the pressure used for column assembling.
- **ii.** Add 0.1 mL of 1 *M* NaOH and centrifuge as in Section 2.2.3 c.i
- **iii.** Add 0.1 mL of the BRUB, pH 11 for the separation of Lys-C peptides and with BRUB pH 5 for the tryptic peptides. Centrifuge as in Section 2.2.3 c.i. Repeat this step once.

d. Equilibrate the C18-"StageTips" as described in Section 2.2.3 c.i subsequently with:

- **i.** 0.05 mL of methanol
- **ii.** 0.05 mL of Buffer B
- **iii.** 0.05 mL of Buffer A
  - TIP: Prepare six C18-"StageTips"; four for the separation of the Lys-C peptides and two for the separation of the tryptic peptides.
  - TIP: The described protocol for peptide fractionation was optimized for peptides samples containing 10–20 μg of total peptide. In a case of smaller amounts of peptides the number of fractions can be reduced.

e. Assemble the SAX "pipette-tip column" in the C18-"StageTip." Load the sample solutions into the equilibrated SAX "pipette-tip columns" and centrifuge the columns at 5000 × *g* for 3 min.

f. Add 0.1 mL of the BRUB pH 11 or 5, respectively to the Lyc-C and tryptic samples, to the SAX "pipette-tip column" and centrifuge at 5000 × *g* for 3 min.

g. Transfer the SAX "pipette-tip column" to the next C18-"StageTip."

h. Continue eluting of peptides with BRUB pH 6, 4, and 2 for the Lys C sample and with BRUB pH 2 for the sample with the tryptic peptides. Collect each eluate in a separate C18-"StageTip."

i. Add 0.05 mL 0.1% (v/v) CF_3COOH in water to the all C18-"StageTips" and centrifuge the columns at 5000 × *g* for 3 min.

j. Elute peptides from the C18-"StageTips" with Buffer B in deionized water into vials used for injection of the samples into LC system.

- PAUSE POINT: Elutes can be stored frozen at −20°C for several months.

2.3 Improving Peptide Yields

a. Protein and peptide content have to be determined at the beginning and the end of the FASP-based sample processing. We recommend using the tryptophan fluorescence (WF) method (Wisniewski & Gaugaz, 2015). This method allows accurate measurement of protein content in the presence of SDS and direct determination peptide concentrations without any wasting of sample.
b. Use TLB with polyethylene glycol to prevent unspecific binding of protein to plastics. For samples containing more than 10 μg of total protein the addition of the polyethylene glycol is not required.
c. Use membrane-coated slides for microdissection.
d. When working with adhesive caps avoid dissecting more than 5 mm^2 of tissue at once. Overfilled caps have smaller adhesive surface, threshold of 5 mm is an arbitrary limit. If more material is needed, after collecting 5 mm^2 put about 10 μL of TLB on the surface of the adhesive cap, close the tube, and spin it for 30 s.
e. At the end of microdissection make sure that you collect all the tissue pieces from cap to the tube.
f. Minimize tissue/lysate handling, especially pipetting and number of tubes used. E.g., use the same tube for microdissection and lysis.

2.4 Anticipated Results

A combination of MED-FASP with a "pipette-tip" peptide fractionation enables generation of six peptide populations with less 30% of peptides detectable in more than only one fraction. Using mass spectrometers with high sequencing speed 50,000–100,000 peptides per sample can be identified. This allows identification of more than 10,000 proteins. The data generated from samples using MED FASP are suitable for label free absolute protein quantification using the "Total Protein Approach" and "Proteomic Ruler" computational approaches (Wisniewski, Hein, Cox, & Mann, 2014; Wisniewski & Rakus, 2014). These methods are described in chapter "Label free and standard free absolute quantitative proteomics using the 'Total Protein' and 'Proteomic Ruler' approaches" by Wiśniewski.

ACKNOWLEDGMENTS

This work was supported by the Max-Planck Society for the Advancement of Science and the German Research Foundation (DFG/Gottfried Wilhelm Leibniz Prize).

REFERENCES

Alkhas, A., Hood, B. L., Oliver, K., Teng, P. N., Oliver, J., Mitchell, D., ... Conrads, T. P. (2011). Standardization of a sample preparation and analytical workflow for proteomics of archival endometrial cancer tissue. *Journal of Proteome Research*, *10*(11), 5264–5271. http://dx.doi.org/10.1021/pr2007736.

Balgley, B. M., Guo, T., Zhao, K., Fang, X., Tavassoli, F. A., & Lee, C. S. (2009). Evaluation of archival time on shotgun proteomics of formalin-fixed and paraffin-embedded tissues. *Journal of Proteome Research*, *8*(2), 917–925. http://dx.doi.org/10.1021/pr800503u.

Bateman, N. W., Sun, M., Bhargava, R., Hood, B. L., Darfler, M. M., Kovatich, A. J., ... Conrads, T. P. (2011). Differential proteomic analysis of late-stage and recurrent breast cancer from formalin-fixed paraffin-embedded tissues. *Journal of Proteome Research*, *10*(3), 1323–1332. http://dx.doi.org/10.1021/pr101073s.

Bell, L. N., Saxena, R., Mattar, S. G., You, J., Wang, M., & Chalasani, N. (2011). Utility of formalin-fixed, paraffin-embedded liver biopsy specimens for global proteomic analysis in nonalcoholic steatohepatitis *Proteomics. Clinical Applications*, *5*(7–8), 397–404. http://dx.doi.org/10.1002/prca.201000144.

Bronsert, P., Weisser, J., Biniossek, M. L., Kuehs, M., Mayer, B., Drendel, V., ... Schilling, O. (2014). Impact of routinely employed procedures for tissue processing on the proteomic analysis of formalin-fixed paraffin-embedded tissue *Proteomics. Clinical Applications*, *8*(9–10), 796–804. http://dx.doi.org/10.1002/prca.201300082.

Byrum, S., Avaritt, N. L., Mackintosh, S. G., Munkberg, J. M., Badgwell, B. D., Cheung, W. L., & Tackett, A. J. (2011). A quantitative proteomic analysis of FFPE melanoma. *Journal of Cutaneous Pathology*, *38*(11), 933–936. http://dx.doi.org/10.1111/j.1600-0560.2011.01761.x.

Byrum, S. D., Larson, S. K., Avaritt, N. L., Moreland, L. E., Mackintosh, S. G., Cheung, W. L., & Tackett, A. J. (2013). Quantitative proteomics identifies activation of hallmark pathways of cancer in patient melanoma. *Journal of Proteomics & Bioinformatics*, *6*(3), 43–50. http://dx.doi.org/10.4172/jpb.1000260.

Crockett, D. K., Lin, Z., Vaughn, C. P., Lim, M. S., & Elenitoba-Johnson, K. S. (2005). Identification of proteins from formalin-fixed paraffin-embedded cells by LC-MS/MS. *Laboratory Investigation*, *85*(11), 1405–1415. http://dx.doi.org/10.1038/labinvest.3700343.

Dunne, J. C., Lamb, D. S., Delahunt, B., Murray, J., Bethwaite, P., Ferguson, P., ... Jordan, T. W. (2015). Proteins from formalin-fixed paraffin-embedded prostate cancer sections that predict the risk of metastatic disease. *Clinical Proteomics*, *12*(1), 24. http://dx.doi.org/10.1186/s12014-015-9096-3.

Fowler, C. B., O'Leary, T. J., & Mason, J. T. (2014). Improving the proteomic analysis of archival tissue by using pressure-assisted protein extraction: A mechanistic approach. *Journal of Proteomics & Bioinformatics*, 7(6), 151–157. http://dx.doi.org/10.4172/jpb.1000315.

Fowler, C. B., Waybright, T. J., Veenstra, T. D., O'Leary, T. J., & Mason, J. T. (2012). Pressure-assisted protein extraction: A novel method for recovering proteins from archival tissue for proteomic analysis. *Journal of Proteome Research*, *11*(4), 2602–2608. http://dx.doi.org/10.1021/pr201005t.

Gamez-Pozo, A., Sanchez-Navarro, I., Calvo, E., Diaz, E., Miguel-Martin, M., Lopez, R., ... Vara, J. A. (2011). Protein phosphorylation analysis in archival clinical cancer samples by shotgun and targeted proteomics approaches. *Molecular BioSystems*, 7(8), 2368–2374. http://dx.doi.org/10.1039/c1mb05113j.

Giusti, L., & Lucacchini, A. (2013). Proteomic studies of formalin-fixed paraffin-embedded tissues. *Expert Review of Proteomics*, *10*(2), 165–177. http://dx.doi.org/10.1586/epr.13.3.

Gundisch, S., Hauck, S., Sarioglu, H., Schott, C., Viertler, C., Kap, M., … Becker, K. F. (2012). Variability of protein and phosphoprotein levels in clinical tissue specimens during the preanalytical phase. *Journal of Proteome Research*, *11*(12), 5748–5762. http://dx.doi.org/10.1021/pr300560y.

Guo, T., Wang, W., Rudnick, P. A., Song, T., Li, J., Zhuang, Z., … Balgley, B. M. (2007). Proteome analysis of microdissected formalin-fixed and paraffin-embedded tissue specimens. *The Journal of Histochemistry and Cytochemistry*, *55*(7), 763–772. http://dx.doi.org/10.1369/jhc.7A7177.2007.

Gustafsson, O. J., Arentz, G., & Hoffmann, P. (2015). Proteomic developments in the analysis of formalin-fixed tissue. *Biochimica et Biophysica Acta*, *1854*(6), 559–580. http://dx.doi.org/10.1016/j.bbapap.2014.10.003.

Hammer, E., Ernst, F. D., Thiele, A., Karanam, N. K., Kujath, C., Evert, M., … Barthlen, W. (2014). Kidney protein profiling of Wilms' tumor patients by analysis of formalin-fixed paraffin-embedded tissue samples. *Clinica Chimica Acta*, *433*, 235–241. http://dx.doi.org/10.1016/j.cca.2014.03.020.

Harris, L., Fritsche, H., Mennel, R., Norton, L., Ravdin, P., Taube, S., … American Society of Clinical Oncology. (2007). American Society of Clinical Oncology 2007 update of recommendations for the use of tumor markers in breast cancer. *Journal of Clinical Oncology*, *25*(33), 5287–5312. http://dx.doi.org/10.1200/JCO.2007.14.2364.

Hood, B. L., Darfler, M. M., Guiel, T. G., Furusato, B., Lucas, D. A., Ringeisen, B. R., … Krizman, D. B. (2005). Proteomic analysis of formalin-fixed prostate cancer tissue. *Molecular & Cellular Proteomics*, *4*(11), 1741–1753. http://dx.doi.org/10.1074/mcp.M500102-MCP200.

Hori, S. S., & Gambhir, S. S. (2011). Mathematical model identifies blood biomarker-based early cancer detection strategies and limitations. *Science Translational Medicine*, *3*(109), 109ra116. http://dx.doi.org/10.1126/scitranslmed.3003110.

Ishihama, Y., Rappsilber, J., Andersen, J. S., & Mann, M. (2002). Microcolumns with self-assembled particle frits for proteomics. *Journal of Chromatography. A*, *979*(1–2), 233–239.

Jain, M. R., Li, Q., Liu, T., Rinaggio, J., Ketkar, A., Tournier, V., … Li, H. (2012). Proteomic identification of immunoproteasome accumulation in formalin-fixed rodent spinal cords with experimental autoimmune encephalomyelitis. *Journal of Proteome Research*, *11*(3), 1791–1803. http://dx.doi.org/10.1021/pr201043u.

Jiang, X., Jiang, X., Feng, S., Tian, R., Ye, M., & Zou, H. (2007). Development of efficient protein extraction methods for shotgun proteome analysis of formalin-fixed tissues. *Journal of Proteome Research*, *6*(3), 1038–1047. http://dx.doi.org/10.1021/pr0605318.

Kondo, T. (2014). Inconvenient truth: Cancer biomarker development by using proteomics. *Biochimica et Biophysica Acta*, *1844*(5), 861–865. http://dx.doi.org/10.1016/j.bbapap.2013.07.009.

Lai, Z. W., Weisser, J., Nilse, L., Costa, F., Keller, E., Tholen, M., … Schilling, O. (2016). Formalin-fixed, paraffin-embedded tissues (FFPE) as a robust source for the profiling of native and protease-generated protein amino termini. *Molecular & Cellular Proteomics*, *15*(6), 2203–2213. http://dx.doi.org/10.1074/mcp.O115.056515.

Maes, E., Broeckx, V., Mertens, I., Sagaert, X., Prenen, H., Landuyt, B., & Schoofs, L. (2013). Analysis of the formalin-fixed paraffin-embedded tissue proteome: Pitfalls, challenges, and future prospectives. *Amino Acids*, *45*(2), 205–218. http://dx.doi.org/10.1007/s00726-013-1494-0.

Magdeldin, S., & Yamamoto, T. (2012). Toward deciphering proteomes of formalin-fixed paraffin-embedded (FFPE) tissues. *Proteomics*, *12*(7), 1045–1058. http://dx.doi.org/10.1002/pmic.201100550.

Nazarian, J., Santi, M., Hathout, Y., & Macdonald, T. J. (2008). Protein profiling of formalin fixed paraffin embedded tissue: Identification of potential biomarkers for pediatric brainstem glioma. *Proteomics Clinical Applications*, *2*(6), 915–924. http://dx.doi.org/10.1002/prca.200780061.

Nirmalan, N. J., Hughes, C., Peng, J., McKenna, T., Langridge, J., Cairns, D. A., … Banks, R. E. (2011). Initial development and validation of a novel extraction method for quantitative mining of the formalin-fixed, paraffin-embedded tissue proteome for biomarker investigations. *Journal of Proteome Research*, *10*(2), 896–906. http://dx.doi.org/10.1021/pr100812d.

Noberini, R., Uggetti, A., Pruneri, G., Minucci, S., & Bonaldi, T. (2016). Pathology tissue-quantitative mass spectrometry analysis to profile histone post-translational modification patterns in patient samples. *Molecular & Cellular Proteomics*, *15*(3), 866–877. http://dx.doi.org/10.1074/mcp.M115.054510.

Ostasiewicz, P., Zielinska, D. F., Mann, M., & Wisniewski, J. R. (2010). Proteome, phosphoproteome, and N-glycoproteome are quantitatively preserved in formalin-fixed paraffin-embedded tissue and analyzable by high-resolution mass spectrometry. *Journal of Proteome Research*, *9*(7), 3688–3700. http://dx.doi.org/10.1021/pr100234w.

Pozniak, Y., Balint-Lahat, N., Rudolph, J. D., Lindskog, C., Katzir, R., Avivi, C., … Geiger, T. (2016). System-wide clinical proteomics of breast cancer reveals global remodeling of tissue homeostasis. *Cell Systems*, *2*(3), 172–184. http://dx.doi.org/10.1016/j.cels.2016.02.001.

Prieto, D. A., Hood, B. L., Darfler, M. M., Guiel, T. G., Lucas, D. A., Conrads, T. P., … Krizman, D. B. (2005). Liquid tissue: Proteomic profiling of formalin-fixed tissues. *BioTechniques*, *38*(Suppl.), 32–35.

Ray, S., Reddy, P. J., Jain, R., Gollapalli, K., Moiyadi, A., & Srivastava, S. (2011). Proteomic technologies for the identification of disease biomarkers in serum: Advances and challenges ahead. *Proteomics*, *11*(11), 2139–2161. http://dx.doi.org/10.1002/pmic.201000460.

Scicchitano, M. S., Dalmas, D. A., Boyce, R. W., Thomas, H. C., & Frazier, K. S. (2009). Protein extraction of formalin-fixed, paraffin-embedded tissue enables robust proteomic profiles by mass spectrometry. *The Journal of Histochemistry and Cytochemistry*, *57*(9), 849–860. http://dx.doi.org/10.1369/jhc.2009.953497.

Shi, S. R., Liu, C., Balgley, B. M., Lee, C., & Taylor, C. R. (2006). Protein extraction from formalin-fixed, paraffin-embedded tissue sections: Quality evaluation by mass spectrometry. *The Journal of Histochemistry and Cytochemistry*, *54*(6), 739–743. http://dx.doi.org/10.1369/jhc.5B6851.2006.

Simpson, R. J., Bernhard, O. K., Greening, D. W., & Moritz, R. L. (2008). Proteomics-driven cancer biomarker discovery: Looking to the future. *Current Opinion in Chemical Biology*, *12*(1), 72–77. http://dx.doi.org/10.1016/j.cbpa.2008.02.010.

Sousa, J. F., Ham, A. J., Whitwell, C., Nam, K. T., Lee, H. J., Yang, H. K., … Goldenring, J. R. (2012). Proteomic profiling of paraffin-embedded samples identifies metaplasia-specific and early-stage gastric cancer biomarkers. *The American Journal of Pathology*, *181*(5), 1560–1572. http://dx.doi.org/10.1016/j.ajpath.2012.07.027.

Sprung, R. W., Jr., Brock, J. W., Tanksley, J. P., Li, M., Washington, M. K., Slebos, R. J., & Liebler, D. C. (2009). Equivalence of protein inventories obtained from formalin-fixed paraffin-embedded and frozen tissue in multidimensional liquid chromatography-tandem mass spectrometry shotgun proteomic analysis. *Molecular & Cellular Proteomics*, *8*(8), 1988–1998. http://dx.doi.org/10.1074/mcp.M800518-MCP200.

Sprung, R. W., Martinez, M. A., Carpenter, K. L., Ham, A. J., Washington, M. K., Arteaga, C. L., … Liebler, D. C. (2012). Precision of multiple reaction monitoring mass spectrometry analysis of formalin-fixed, paraffin-embedded tissue. *Journal of Proteome Research*, *11*(6), 3498–3505. http://dx.doi.org/10.1021/pr300130t.

Sturgeon, C. M., Duffy, M. J., Stenman, U. H., Lilja, H., Brunner, N., Chan, D. W., … National Academy of Clinical Biochemistry (2008). National Academy of Clinical

Biochemistry laboratory medicine practice guidelines for use of tumor markers in testicular, prostate, colorectal, breast, and ovarian cancers. *Clinical Chemistry*, *54*(12), e11–e79. http://dx.doi.org/10.1373/clinchem.2008.105601.

Tanca, A., Addis, M. F., Pisanu, S., Abbondio, M., Pagnozzi, D., Eccher, A., … Fanciulli, G. (2016). Atypical carcinoid and large cell neuroendocrine carcinoma of the lung: A proteomic dataset from formalin-fixed archival samples. *Data in Brief*, 7, 529–531. http://dx.doi.org/10.1016/j.dib.2016.02.083.

Wisniewski, J. R., Dus-Szachniewicz, K., Ostasiewicz, P., Ziolkowski, P., Rakus, D., & Mann, M. (2015). Absolute proteome analysis of colorectal mucosa, adenoma, and cancer reveals drastic changes in fatty acid metabolism and plasma membrane transporters. *Journal of Proteome Research*, *14*(9), 4005–4018. http://dx.doi.org/10.1021/acs.jproteome.5b00523.

Wisniewski, J. R., & Gaugaz, F. Z. (2015). Fast and sensitive total protein and peptide assays for proteomic analysis. *Analytical Chemistry*, *87*(8), 4110–4116. http://dx.doi.org/10.1021/ac504689z.

Wisniewski, J. R., Hein, M. Y., Cox, J., & Mann, M. (2014). A "proteomic ruler" for protein copy number and concentration estimation without spike-in standards. *Molecular & Cellular Proteomics*, *13*(12), 3497–3506. http://dx.doi.org/10.1074/mcp.M113.037309.

Wisniewski, J. R., Ostasiewicz, P., Dus, K., Zielinska, D. F., Gnad, F., & Mann, M. (2012). Extensive quantitative remodeling of the proteome between normal colon tissue and adenocarcinoma. *Molecular Systems Biology*, *8*, 611. http://dx.doi.org/10.1038/msb.2012.44.

Wisniewski, J. R., & Rakus, D. (2014). Multi-enzyme digestion FASP and the 'Total Protein Approach'-based absolute quantification of the Escherichia coli proteome. *Journal of Proteomics*, *109*, 322–331. http://dx.doi.org/10.1016/j.jprot.2014.07.012.

Xiao, Z., Li, G., Chen, Y., Li, M., Peng, F., Li, C., … Chen, Z. (2010). Quantitative proteomic analysis of formalin-fixed and paraffin-embedded nasopharyngeal carcinoma using iTRAQ labeling, two-dimensional liquid chromatography, and tandem mass spectrometry. *The Journal of Histochemistry and Cytochemistry*, *58*(6), 517–527. http://dx.doi.org/10.1369/jhc.2010.955526.

Xu, H., Yang, L., Wang, W., Shi, S. R., Liu, C., Liu, Y., … Balgley, B. M. (2008). Antigen retrieval for proteomic characterization of formalin-fixed and paraffin-embedded tissues. *Journal of Proteome Research*, 7(3), 1098–1108. http://dx.doi.org/10.1021/pr7006768.

Yin, X., Zhang, Y., Guo, S., Jin, H., Wang, W., & Yang, P. (2015). Large scale systematic proteomic quantification from non-metastatic to metastatic colorectal cancer. *Scientific Reports*, *5*, 12120. http://dx.doi.org/10.1038/srep12120.

CHAPTER ELEVEN

Body Fluid Degradomics and Characterization of Basic N-Terminome

F. Sabino, O. Hermes, U. auf dem Keller[1]
ETH Zurich, Institute of Molecular Health Sciences, Zurich, Switzerland
[1]Corresponding author: e-mail address: ulrich.aufdemkeller@biol.ethz.ch

Contents

Abstract

Rapid improvements in instrumentation and data analysis make mass spectrometry-based proteomics the method of choice for global characterization of proteomes and discovery of protein-based biomarkers. On the contrary to tissue biopsies, body fluids—e.g., blood, wound fluid, urine, and saliva—are noninvasive and easy to collect and process. However, they are very complex and present high dynamic ranges of protein concentrations, rendering direct shotgun proteomics analysis as inefficient for identification of low-abundance proteins in these specimens. Sample prefractionation, immunoaffinity depletion of highly abundant proteins, and enrichment of posttranslational modifications (PTM) are common strategies for proteome simplification of body fluids. Combinatorial peptide ligand libraries (CPLL) relatively deplete high-abundance

Methods in Enzymology, Volume 585
ISSN 0076-6879
http://dx.doi.org/10.1016/bs.mie.2016.09.018

proteins by binding equimolar amounts of protein species in the sample and provide an elegant species-independent alternative to immunoaffinity-based approaches. By cleaving target proteins, proteases catalyze an irreversible PTM, whereby uncontrolled proteolysis is associated with many diseases. Thus, proteolytic events represent powerful indicators for disease progression and their specific identification in body fluids holds great promises for establishment of novel biomarkers. Quantitative N-terminal enrichment strategies, such as terminal amine isotopic labeling of substrates (TAILS) detect protease-generated neo-N-termini with high specificity and increase coverage of low-abundance proteins by inherent proteome simplification. In this chapter, we describe a protocol that combines the CPLL technology with iTRAQ-based TAILS to systematically characterize the basic N-terminome of body fluid proteomes and its alterations in disease conditions that we have successfully applied to explore the wound fluid degradome at multiple time points after skin injury.

1. INTRODUCTION

The completion of the human genome project (International Human Genome Sequencing, 2004; McPherson et al., 2001) changed the focus of life science research from DNA sequencing to identification of gene function and proteome analysis. The human Proteome Project (HPP) was launched in 2011 aiming at identifying and characterizing at least one protein isoform coded by each human gene and to identify associated posttranslational modifications (PTMs). The biology/disease centric branch of HPP focuses on proteome analysis of major human diseases and on the identification of perturbations in protein abundances associated with each particular condition. It is believed that proteins, which show stable changes in abundance in specific pathologic states may function as biomarkers for those diseases, even if the function of the proteins is unknown (Eckhard, Marino, Butler, & Overall, 2015; Falk, Ramstrom, Stahl, & Hober, 2007).

Biological tissues are challenging to analyze due to difficulties in obtaining suitable biopsies and sample heterogeneity resulting from differences in cell-type composition. Hence, body fluids such as blood (serum and plasma), wound fluid, cerebrospinal fluid, urine, and saliva are attractive alternatives for biomarker discovery based on easy collection, low invasiveness, low cost, and ease to process (Ahn & Simpson, 2007; Hu, Loo, & Wong, 2006). Among body fluids, blood is the most widely used for diagnostic purposes. A plasma sample is obtained when blood is collected in the presence of an anticoagulant, while in the absence of coagulation inhibitors a serum sample is provided. Plasma comes in contact with every organ in the body, and thus reflects the physiological or pathological state of the organism, containing proteins and chemicals that can be used for diagnostic purposes (Good et al., 2007; Hu et al., 2006; Wildes & Wells, 2010).

In recent years, mass spectrometry (MS)-based proteomics accomplished great advances in terms of accuracy and confidence in protein identification, becoming the tool of choice for biomarker discovery (Good et al., 2007; Hu et al., 2006). However, the search for biomarkers in plasma and serum mostly did not reach the initial expectations, with only a few novel biomarkers used in clinical applications (Pernemalm & Lehtio, 2014). Because of their contact with every organ in the body, plasma constituents are less specific for a particular organ or disease (Good et al., 2007). Moreover, plasma and serum present high dynamic ranges, with protein concentration spanning more than 10 orders of magnitude. This difference in concentration hampers the detection of low-abundance tissue-leakage proteins and protein species. Alternative splicing and PTMs make the analysis even more demanding and sample handling is pivotal, since some proteins are more prone to degradation than others (Falk et al., 2007; Pernemalm & Lehtio, 2014).

1.1 Simplification of Body Fluid Proteomes by Protein Depletion

Despite the improvements achieved by MS technologies in terms of sensitivity, direct analysis of plasma and serum for detection of low-abundance proteins (e.g., many proteases and their inhibitors) is still inefficient. Therefore, several strategies have been implemented to overcome these limitations (Hermes, Schlage, & auf dem Keller, 2011; Pernemalm & Lehtio, 2014). Prefractionation is a technique frequently used for reduction of sample complexity and dynamic range of protein concentration. Generally, this means the use of a particular biochemical property of a group of proteins for separation purposes, such as molecular size, hydrophobicity, isoelectric point, or enrichment of proteins with specific PTMs. However, prefractionation is not always sufficient to enable detection of low-abundance proteins (Falk et al., 2007; Nilsson et al., 2010). Since only 22 blood proteins are estimated to account for 99% of the total blood protein mass, depletion of highly abundant proteins is a strategy commonly implemented for dynamic range reduction (Wildes & Wells, 2010). Immunoaffinity has been extensively used to deplete albumins or immunoglobulins, and commercially available affinity products can deplete up to 100 highly and moderately abundant proteins from blood plasma and other body fluids (Pernemalm & Lehtio, 2014). However, protein depletion has several disadvantages, since it is associated with high costs, difficulties in reproducibility, and most importantly possible loss of information derived from unspecific reduction in abundance of other proteins (Granger, Siddiqui, Copeland, & Remick, 2005; Shen et al., 2005). Nonetheless, the improvement in detection limits compensates the possible disadvantages of the method (Falk et al., 2007).

An alternative to antibody-based depletion methods is offered by combinatorial peptide ligand libraries (CPLL), which have been commercialized in form of the ProteoMiner™ system by Bio-Rad Laboratories (Hercules, CA). This system uses a library of hexapeptides that bind equimolar amounts of each protein species present in the sample. The most abundant proteins saturate their ligands and the excess is washed away, enriching the sample with low-abundance proteins that do not saturate their peptide binding partners (Boschetti & Righetti, 2008; Pernemalm & Lehtio, 2014). This technology avoids the disadvantages of a complete depletion of certain proteins and is particularly useful, if no antibodies are available for the proteins to deplete, e.g., in samples from nonhuman origin.

1.2 Proteolysis in Biological Fluids and Positional Proteomics

Protein PTMs comprise an important mechanism of regulation of protein folding, subcellular localization, activity, interactions, and degradation (Lange & Overall, 2013). Although in many cases the development of a disease is associated with a particular PTM of a protein, this information is lost by simple analysis of protein abundances (Good et al., 2007). PTMs are commonly low in abundance and usually undersampled and overlooked without PTM-specific enrichment protocols and parameters of analysis (Lange & Overall, 2013).

With more than 560 members, proteases constitute the largest enzyme family in humans (Lange & Overall, 2013). These enzymes catalyze an irreversible PTM by hydrolyzing scissile bonds in target substrates and provide an important mode of regulation for pivotal biological processes, such as the cell cycle and apoptosis (Pernemalm & Lehtio, 2014). Most body fluid proteins have their secretory signal peptide removed in the endoplasmic reticulum, and additional processing events may also happen during secretion or by other proteases circulating in plasma or wound fluid (for example, in complement activation or blood coagulation). Due to implications in numerous important biological processes, uncontrolled proteolysis is frequently associated with pathological events and disease progression, including cardiovascular and neurodegenerative disorders, chronic inflammation, and cancer. Hence, protein termini have been proposed to reflect a mechanistic state of bioactive proteins, constituting possible biomarkers for particular conditions (Eckhard et al., 2015; Huesgen, Lange, & Overall, 2014; Plasman, Van Damme, & Gevaert, 2013). In most cases a disease will not be defined by a single proteolytic event or by the aberrant activity of a single protease, but by an ensemble of cleavage events—a proteolytic signature. Thus, the system-wide study of proteolysis gained significant attention in biomarker discovery (Lai, Petrera, & Schilling, 2015).

Despite the frequency of proteolytic events in biological systems, numbers of protein N- and C-termini are still low in comparison to internal protein peptides, and thus enrichment strategies are necessary for efficient identification of proteolytic signatures. Since primary amines are more reactive than carboxyl groups, N-termini are more reactive than C-termini, and therefore most strategies to study proteolysis focus on amine reactivity to enrich for protein N-termini (Van Damme et al., 2010). Positive enrichment specifically selects N-terminal peptides for separation from internal peptides. In general, free ε-amines are blocked and α-amines are reacted with affinity tags. After digestion with trypsin, the tagged peptides can be affinity purified. A common problem of methods that use positive selection is the inability to study natural N-terminal PTMs, such as acetylation and cyclization, which are known to affect protein structure, function, stability, targeting, and interaction with other proteins (Agard & Wells, 2009; Huesgen et al., 2014; Rogers & Overall, 2013). Moreover, discrimination between N-terminal α-amines and ε-amines of lysines is problematic (Rogers & Overall, 2013). As an alternative, negative selection strategies enrich protein N-termini by depletion of internal peptides. These methods rely on blocking of both α- and ε-amines on the protein level and subsequent removal of internal tryptic peptides based on reactivity of newly generated peptide α-amines. Negative selection approaches facilitate higher proteome coverage than positive selection, and also allow the study of natural PTMs of protein N-termini (Rogers & Overall, 2013).

By enrichment of protein termini and concomitant depletion of internal peptides, positional proteomics approaches provide an efficient method for reduction of the challenging dynamic range of protein concentrations in body fluids (Eckhard et al., 2015; Plasman et al., 2013; Wildes & Wells, 2010). Moreover, common shotgun proteomics approaches rely on the generation of peptides with suitable length, hydrophobicity and ionization properties for MS analysis. Thus, the use of only one endopeptidase (in general trypsin) for protein digestion is limiting for proteome characterization.

With N- and C-termini enrichment the analyses focus on semitryptic peptides, which possess different properties than the fully tryptic counterparts, thereby allowing complementary proteome coverage (Eckhard et al., 2015).

2. N-TERMINAL ENRICHMENT STRATEGIES

In recent years, several positional proteomics technologies have been developed that aim at enriching protein termini for identification of protease cleavage products in complex biological samples (Vizovisek, Vidmar, Fonovic, & Turk, 2016). The inherent proteome simplification predefines them as optimal approaches to mine body fluid proteomes for biomarkers of disease either by monitoring changes in protein abundance or by indicative proteolytic processing events. Some of these methods received particular attention and have been applied to assess N-terminomes and proteolysis in serum, plasma, and other body fluids from patient origin and animal models.

2.1 Positive Enrichment

2.1.1 N-Terminomics

This positive enrichment strategy introduced by Guy Salvesen's group exploits selective guanidation of lysine ε-amines over N-terminal α-amines to specifically label protein N-termini with an amine-reactive biotin tag. Upon tryptic digest, N-terminal peptides can be affinity purified with help of streptavidin beads and subjected to LC–MS/MS analysis (Timmer et al., 2007). This leads to efficient proteome simplification and helped to identify hundreds of protein N-termini in human serum as well as known and novel protease processing sites that modulate the blood proteome, e.g., by N-terminal trimming of serum proteins through aminopeptidases.

2.1.2 Subtiligase Method

Jim Wells' laboratory at the University of San Francisco developed a powerful positive selection method for selective biotinylation of free N-termini based on the enzyme subtiligase (an engineered form of subtilisin protease) and a biotin-labeled tagging peptide (Yoshihara, Mahrus, & Wells, 2008). In this workflow, subtiligase is used for biotinylation of N-terminal α-amines, the proteome is digested with trypsin, and the biotin-labeled N-termini are enriched by affinity purification with help of streptavidin resins. The peptides are finally retrieved from the column with a tobacco etch virus protease, which cleaves a consensus sequence integrated into the biotin-labeled tagging peptide. Employing this technique, the Wells group identified more than 770 N-termini in 222 blood proteins and by combination with isobaric labeling for relative quantification identified substrates for an exogenously added protease (Wildes & Wells, 2010). Moreover, they recorded biological signatures of apoptosis in plasma from tumor patients undergoing chemotherapy (Wiita, Hsu, Lu, Esensten, & Wells, 2014). As a minor disadvantage, the subtiligase approach introduces a bias by not allowing quantitative labeling of N-terminal peptides starting with a proline, isoleucine, aspartate, and glutamate, leading to false negatives (Plasman et al., 2013).

2.2 Negative Enrichment

2.2.1 Combined Fractional Diagonal Chromatography

Combined fractional diagonal chromatography (COFRADIC) was introduced by Kris Geveart in 2003 and uses two-dimensional chromatography to fractionate peptides based on different physicochemical properties. Initially, primary amines are acylated, aiding identification of free N-termini present in the sample. The mixture is digested with trypsin and treated with glutamine cyclotransferases and pyroglutamyl aminopeptidases for removal of N-terminal pyroglutamates. Next, strong cation exchange (SCX) chromatography is used to enrich for N- and C-terminal peptides. At low pH, the terminal peptides are less positively charged than fully tryptic peptides, do not bind to the resin, and are collected in the flow through, while tryptic peptides are retained on the column. The flow-through fraction is treated with 2,4,6-trinitrobenzenosulfonic acid (TNBS), which confers a very hydrophobic trinitrophenyl group to the α-amines of C-terminal and remaining fully tryptic peptides. Finally, a series of reverse-phase chromatography steps are used to remove the trinitrophenyl group containing peptides, leaving only N-termini in the sample (Plasman et al., 2013; Staes et al., 2011). COFRADIC has been demonstrated to significantly increase coverage of serum proteomes (Sandra et al., 2007) and was applied, e.g., to identify novel diagnostic biomarkers of heart failure in human plasma samples (Mebazaa et al., 2012).

2.2.2 Terminal Amine Isotopic Labeling of Substrates

Terminal amine isotopic labeling of substrates (TAILS) is a technique that was introduced by Chris Overall's group at the University of British Columbia in 2010 (Kleifeld et al., 2010) and further extended to a highly multiplexed

N-terminomics platform for protease substrate discovery and N-terminome profiling (Sabino et al., 2015; Schlage et al., 2014). In this approach, the α-amines of free protein N-termini and the ε-amines of lysine side chains are blocked at the protein level by dimethylation or isobaric mass tags (iTRAQ/TMT). The proteome is digested in most cases using trypsin, which recognizes only arginines as cleavage sites as a consequence of the labeled lysine residues. As a key step of the procedure, a high-molecular weight (~100 kDa) aldehyde-derivatized polymer (HPG-ALD) with affinity for primary amines binds the free N-terminus of newly generated internal tryptic peptides. The polymer is recovered by ultrafiltration, and the flow through corresponds to a fraction enriched for N-terminal peptides. Each of these chemically labeled N-termini is quantifiable due to the presence of an isobaric mass tag or isotopically labeled dimethyl group. Moreover, naturally modified N-termini can be discriminated from chemically labeled groups, allowing the study of acetylation and pyroglutamate formation (Eckhard et al., 2015; Kleifeld et al., 2010; Lange & Overall, 2013). TAILS has some advantages over alternative approaches, since it avoids the hurdles and reproducibility issues of cumbersome serial reverse-phase chromatography steps used by COFRADIC and the false negatives that may arise from the use of the subtiligase method.

3. BODY FLUID DEGRADOMICS BY iTRAQ-TAILS

The combination of efficient depletion of highly abundant proteins and specific enrichment of N-terminal peptides holds great promises for the comprehensive characterization of body fluid proteomes and their alterations in disease. Here, we describe a protocol that uses the CPLL (ProteoMiner™) technology and iTRAQ-based TAILS (iTRAQ-TAILS) for proteomic analysis and identification of proteolytic cleavage events in body fluids, which we successfully implemented for the recording of proteolytic signatures in porcine wound exudates (Fig. 1) (Sabino et al., 2015). The procedure exploits species-independent depletion of highly abundant proteins by CPLL and efficient proteome simplification by N-terminal enrichment by iTRAQ-TAILS and allows for quantitative comparisons between conditions with help of isobaric mass tags. Applying this workflow allowed for identification of hundreds of proteins in wound exudates spanning a concentration range of six orders of magnitude and including low-abundance proteases and inhibitors (Fig. 2).

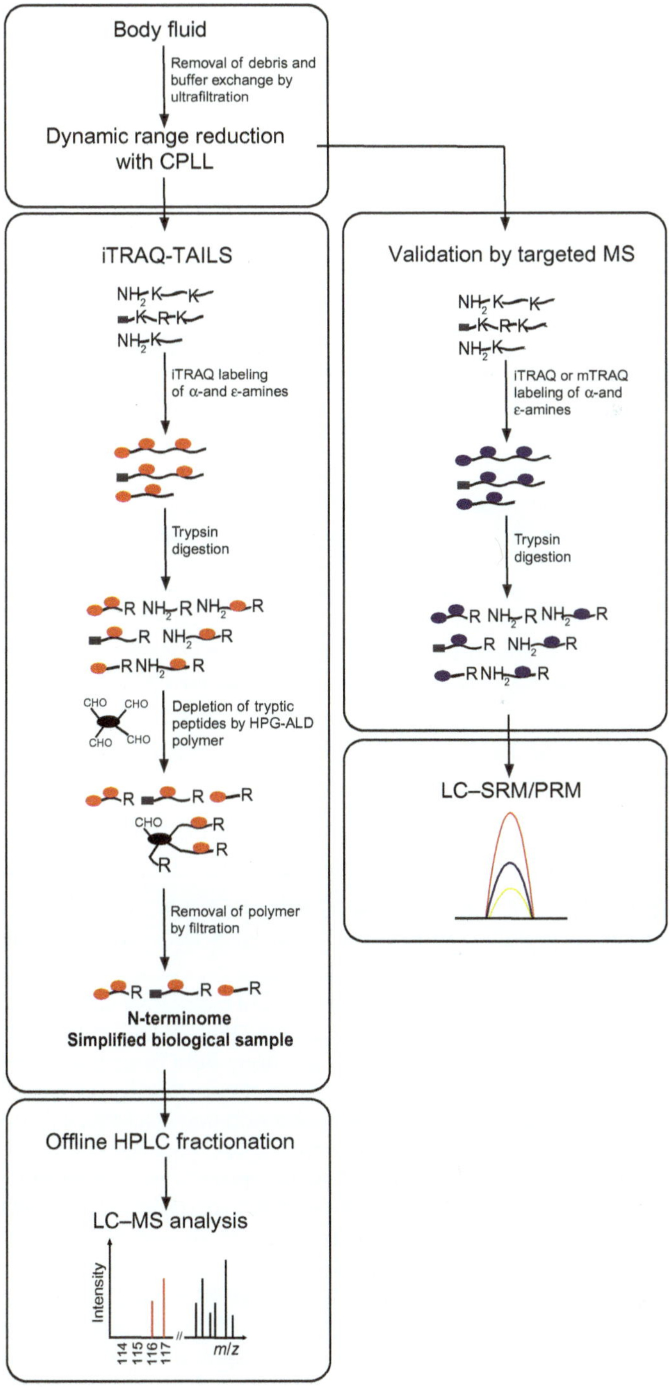

Fig. 1 See legend on next page.

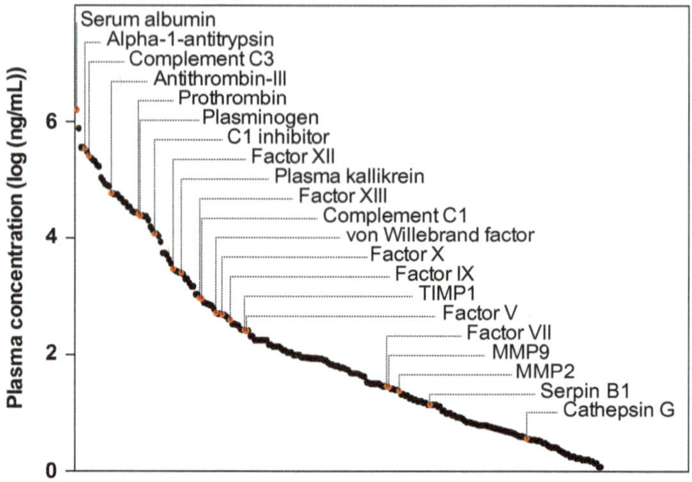

Fig. 2 Dynamic range of proteins identified in porcine wound exudates by iTRAQ-TAILS. Selected proteins were related to known blood plasma concentrations (Farrah et al., 2011). *Red* (*gray* in the print version) *dots* indicate proteases and inhibitors of major proteolytic systems. *Figure was adapted with permission from Sabino, F., Hermes, O., Egli, F. E., Kockmann, T., Schlage, P., Croizat, P., et al. (2015). In vivo assessment of protease dynamics in cutaneous wound healing by degradomics analysis of porcine wound exudates.* Molecular & Cellular Proteomics, 14*(2), 354–370. http://dx.doi.org/10.1074/mcp.M114.043414.*

3.1 Processing of Biological Fluid and Reduction of Concentration Dynamic Range

Starting point for this protocol is a biological fluid that has been obtained from human patient material or an animal model. The fluid might be provided as raw specimen (e.g., serum, plasma) or might have been extracted from a suitable receptacle, which could be also dressing material that has

Fig. 1—Cont'd Workflow for iTRAQ-TAILS analysis of biological fluids. The body fluid is harvested and applied to a resin containing a CPLL (ProteoMiner™ (BioRad, Hercules, CA)) for depletion of high-abundance proteins. Body fluid proteomes from different conditions are labeled at the protein level with amine-reactive isobaric mass tags (iTRAQ), combined and subjected to tryptic digest. Addition of a HPG-ALD polymer with affinity for primary amines removes internal tryptic peptides, and the resultant N-terminome is analyzed by LC–MS/MS. Relative quantification between conditions (e.g., healthy and diseased donor) is achieved with help of iTRAQ reporter ion intensities (*m/z* 114–117). For validation and establishment of biomarker assays, body fluid proteomes from different conditions are labeled at the protein level with a label with identical chemistry as used for iTRAQ-TAILS and digested separately. Targeted proteomics assays (selected/parallel reaction monitoring (S/PRM)) are established for specific indicative N-terminal peptides and used to reproducibly monitor their relative abundances in samples from different patient populations.

been soaked with fluid at a wound site. In our experience, material of good quality can be obtained, e.g., from polyurethane foams used in negative pressure wound therapy by incubation with a physiological buffer (e.g., 50 m*M* HEPES, pH 7.8) and subsequent centrifugation as described in Sabino et al. (2015). Immediately after harvesting or already during extraction protease inhibitors should be added to reduce proteolytic events after sample collection that would interfere with detection of genuine proteolytic modifications of the body fluid proteome, which are indicative for a biological process or disease state. Moreover, samples should be maintained on ice during the entire process, and cycles of freezing and thawing should be avoided to reduce protein degradation. In the following, we describe the necessary steps to equalize protein amounts, and thereby relatively deplete high-abundance proteins prior to N-terminal enrichment. These steps have to be performed in parallel for each sample to be compared by multiplexed iTRAQ-TAILS (e.g., biological replicates per condition).

1. (Optional): Determine total protein concentration in each sample by standard assay (Bradford assay, bicinchoninic acid (BCA) assay). This step can be omitted for raw fluids (e.g., serum, plasma) that have an inherent protein concentration of 50–70 μg/μL.
2. (Optional): If necessary, concentrate the biological fluid to 50–70 μg/μL using 5 kDa cutoff ultrafiltration concentrators (e.g., Vivaspin) at 4°C. During this step sample buffer should be exchanged to 50 m*M* HEPES, pH 7.8. Store samples at –80°C before further processing.
3. Remove solid particles by centrifugation at 13,000 × *g* for 10 min at 4°C and load 1 mL of sample onto a preequilibrated ProteoMiner™ (BioRad, Hercules, CA) Large-Capacity column (equilibrate according to manufacturer's instructions).
4. Incubate the loaded column for 2 h at room temperature with gentle rotation.
5. Wash the resin and elute bound proteins according to manufacturer's instructions.
6. (Optional): Take sample for SDS-PAGE analysis.
7. Change buffer to TAILS buffer (250 m*M* HEPES (pH 7.8), 2.5 *M* guanidine hydrochloride) by ultrafiltration using 5 kDa molecular weight cutoff ultrafiltration concentrators (e.g., Vivaspin). Important: The elution buffer provided with the ProteoMiner™ kit contains CHAPS, which is not compatible with the iTRAQ-TAILS protocol. The addition of 2.5 *M* guanidine hydrochloride is crucial to avoid protein precipitation due to the lower content of serum albumin in the eluted sample.

8. Adjust protein concentration to 1 μg/μL with TAILS buffer and store at −80°C until further processing.

Efficient depletion of very highly abundant proteins may be visualized by analysis of samples prior and after ProteoMiner™ treatment by SDS-PAGE analysis. A typical example is shown in Fig. 3.

3.2 N-Terminal Enrichment by iTRAQ-TAILS

The iTRAQ-TAILS workflow does not only allow to identify low-abundance proteins through proteome simplification by N-terminal enrichment but also to quantitatively compare up to eight samples for differential protein abundance and proteolytic processing. The general procedure does not differ for 4plex or 8plex iTRAQ reagents, which are chosen based on the specific experimental design. According to standard shotgun proteomics workflows proteomes are denatured and disulfide bridges broken by reduction and alkylation of cysteine residues. Next, the addition of isobaric labels (iTRAQ) at the protein level blocks all free N-termini (α-amines), allowing discrimination between cleavage events that happened prior to or after trypsin digestion. The side chains of lysine residues (ε-amines) are also labeled by iTRAQ, and thus skipped by trypsin. The addition of a polymer with

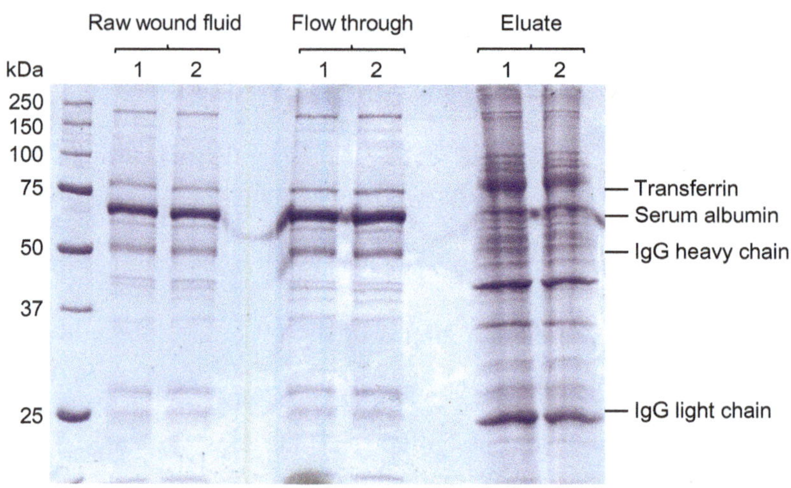

Fig. 3 Reduced dynamic range of protein concentration in wound fluids after ProteoMiner™ treatment. 1 and 2: aliquots of the same wound fluid sample. 10 μg protein of the raw wound fluid, the flow through, and the eluate were analyzed by SDS-PAGE and Coomassie staining. The reduced dynamic range of protein concentration after ProteoMiner™ (BioRad, Hercules, CA) treatment is visualized by the higher number of bands exhibiting less different intensities.

affinity for primary amines allows depletion of fully tryptic peptides and enrichment of N-termini. Samples depleted for highly abundant proteins and transferred to TAILS buffer are ready to be used for iTRAQ-TAILS analysis.

1. Start with 250 μg total protein (1 μg/μL in TAILS buffer) per biological condition to analyze.
2. Incubate the samples at 65°C for 15 min to assure total denaturation. Add 1 m*M* tris(2-carboxy-ethyl)phosphine (TCEP) and incubate at 65°C for 45 min to reduce cysteine residues.
3. Add 5 m*M* iodoacetamide and incubate for 30 min at 65°C in the dark to alkylate the reduced cysteines.
4. Resuspend content of each vial of iTRAQ reagent (1 mg) in DMSO and mix with each sample in a 1:4 protein/iTRAQ ratio (w/w) and 1:1 (v/v) ratio, i.e., 1 mg iTRAQ in ~250 μL DMSO per channel. Incubate for 30 min at room temperature. Important: The 30 min of labeling should be counted precisely for each condition to reduce quantification errors derived from different labeling times.
5. Stop the labeling reaction by adding 10 m*M* ammonium bicarbonate (NH_4HCO_3). Incubate for 30 min at room temperature. Important: Vortex samples after addition of ammonium bicarbonate to assure efficient quenching of excess iTRAQ reagent.
6. Combine samples from all conditions labeled with differential iTRAQ reagents and clean up proteome by precipitation with 6–8 volumes of ice-cold acetone and 1 volume of ice-cold methanol. Incubate at least 2 h at –80°C and precipitate by centrifugation at 10,000 × *g* for 20 min at 4°C. Wash the pellet with 5 mL ice-cold methanol, centrifuge at 10,000 × *g* for 10 min at 4°C and air dry the pellet.
7. Resuspend pellet in 100 m*M* NaOH in a 10/1 ratio protein/NaOH (w/v), i.e., 1 mg total protein in 100 μL 100 m*M* NaOH. Adjust the volume with 100 m*M* HEPES, pH 7.8 to a concentration of 1 μg/μL protein (assuming no sample losses).
8. Digest overnight with trypsin gold (Promega) at a ratio of 1:100 trypsin/protein (w/w) at 37°C.
9. (Optional): Remove 1/10 of the sample for whole-proteome analysis (preTAILS sample). *Note*: Most internal tryptic peptides will have iTRAQ-labeled lysine residues and can be used for relative quantification at the protein level.
10. Adjust pH to 6.5–7 with hydrochloric acid (HCl), add HPG-ALD polymer (available without commercial or company restriction from

Flintbox Innovation Network, The Global Intellectual Exchange and Innovation Network (http://www.flintbox.com/public/project/1948/)) in a threefold excess (polymer:protein) (w/w) together with 50 m*M* $NaBH_3CN$ solution and incubate overnight at 37°C to deplete internal tryptic peptides. *Note*: The neutral pH assures a positive charge of the primary amines and thereby binding of the tryptic peptides to the polymer.

11. Filter the sample through a 30-kDa molecular weight cutoff ultrafiltration concentrator (Vivaspin). The polymer is retained in the filtrate, while the N-terminal peptides are recovered as the flow through.
12. Wash the polymer with 100 μL of 100 m*M* ammonium bicarbonate, centrifuge again, and mix the flow through with the recovered N-terminal peptides (TAILS sample).
13. Store at –80°C before further analysis.

3.3 Offline SCX-HPLC Fractionation and Peptide Cleanup

Despite reduction of dynamic range in concentration by use of CPLL (ProteoMiner™) high-abundance proteins (serum albumin, immunoglobulins) are still more abundant in the sample than proteins of low abundance, such as chemokines, growth factors, and proteases. However, upon N-terminal enrichment, high-abundance proteins are represented by mostly a single peptide with distinct physicochemical properties—the protein N-terminus. This can be exploited by offline HPLC fractionation, which confines N-terminal peptides from high-abundance proteins to specific fractions, increasing chances of identifying low-abundance proteins by their N-termini that elute in separate fractions along the gradient. Fig. 4 demonstrates this principle by showing SCX elution profiles from porcine wound fluids prior and after N-terminal enrichment by iTRAQ-TAILS. While multiple tryptic peptides derived from the highly abundant proteins serum albumin and transferrin elute in several fractions, N-termini from serum albumin, immunoglobulins, and hemoglobin are confined to specific fractions, allowing their efficient separation from N-terminal peptides of low-abundance proteins. As a standard procedure, we describe steps for peptide fractionation by SCX chromatography, but alternative systems, such as hydrophilic interaction chromatography might also be used.

1. Adjust volumes of preTAILS and TAILS samples to 1 mL with SCX A buffer (10 m*M* KH_2PO_4, 25% acetonitrile, pH 2.7), check pH and adjust to ≤2.7 with 50% phosphoric acid.

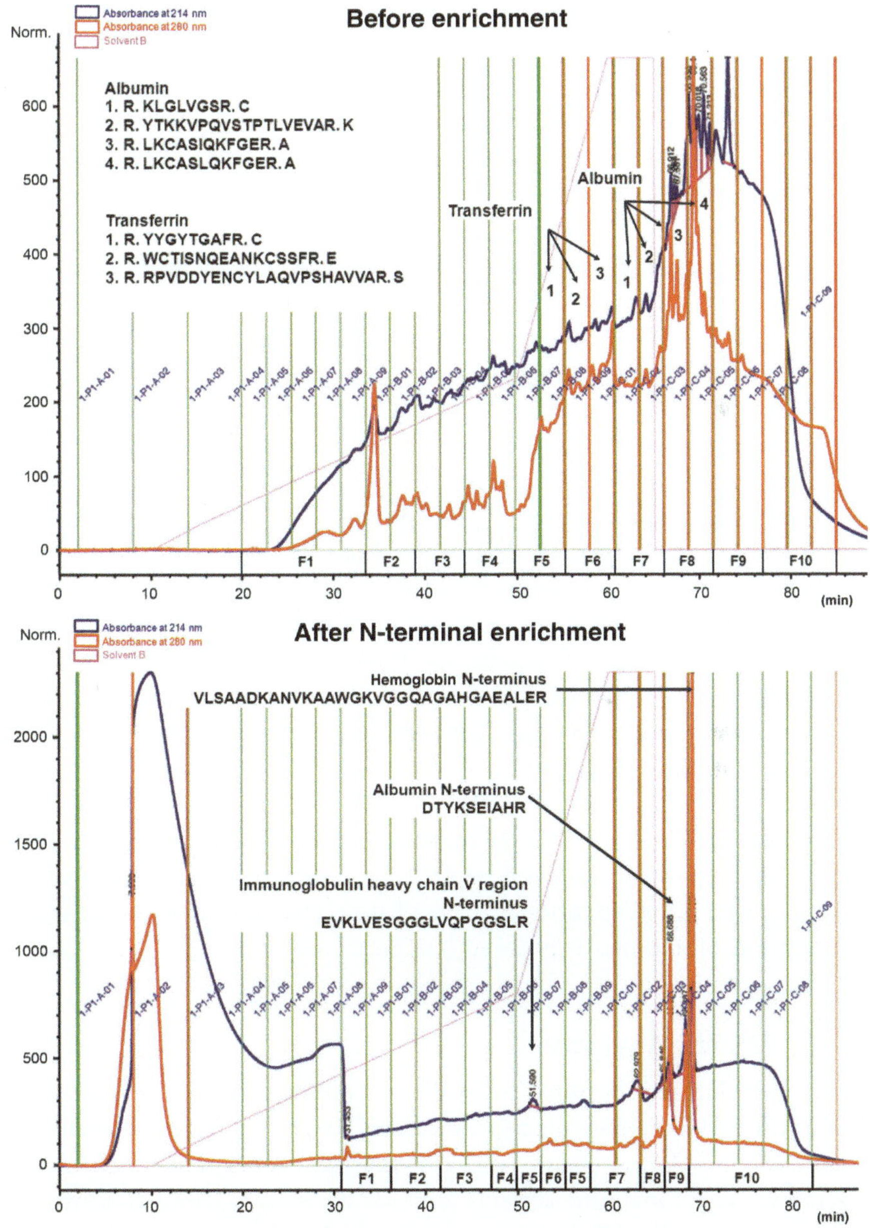

Fig. 4 SCX-HPLC UV absorption chromatograms at 214 and 280 nm of tryptic digests of wound fluids before and after N-terminal enrichment. A total of 24 SCX-HPLC fractions per run were pooled to 10 fractions (F1–F10). Each of the pooled HPLC fractions was analyzed for the most abundant peptides by MALDI-TOF/TOF. Tryptic peptides of highly abundant proteins were identified in several fractions before enrichment and N-termini of high-abundance proteins were confined to single fractions after N-terminal enrichment.

2. Remove any precipitate by centrifugation (10 min, 13,000 × *g*, RT).
3. Load samples onto SCX column (e.g., PolySULFOETHYL A column, 200 × 2.1 mm, 5 μm 300-Å (PolyLC Inc.)) and elute with a linear gradient (0 min 0% SCX buffer B (SCX buffer A + 0.5 *M* KCl); 60 min 0% SCX buffer B; 65 min 5% SCX buffer B; 100 min 35% SCX buffer B; 110 min 100% SCX buffer B) at a flow rate of 0.2 mL/min.
4. Monitor eluate by absorbance at 214 and 280 nm and collect ~25–30 fractions of 0.5 mL.
5. Reduce volume of fractions containing eluted peptides to 100 μL using a SpeedVac concentrator.
6. Wet C18 OMIX tip (Agilent Technologies) with 100 μL 50% acetonitrile and equilibrate with 100 μL washing buffer (3% acetonitrile, 0.1% trifluoroacetic acid (TFA)).
7. Load peptide solutions to C18 resin and pool multiple fractions by binding to the same C18 OMIX tip. *Note*: In our experience peptides from ~25–30 fractions can be pooled to 8–10 samples according to their absorbance intensities in the SCX elution profile (high intensities: pool 2–3 fractions; low intensities: pool 4–6 fractions).
8. Wash C18 OMIX tip five to seven times with 100 μL washing buffer.
9. Elute peptides into 100 μL 80% acetonitrile, 0.1% TFA and dry with SpeedVac concentrator.
10. Resuspend samples in 20 μL 3% acetonitrile and 0.1% formic acid and store at –20°C if not directly analyzed by LC–MS/MS. *Note*: It is recommended to add standard peptides for retention time normalization (e.g., iRT kit (Biognosys)), if validation and assay development for targeted proteomics is planned (see Section 4).

3.4 LC–MS/MS Analysis

Pooled peptide fractions from the TAILS and optionally the preTAILS sample are analyzed by tandem mass spectrometry on a hybrid mass spectrometer coupled to a reversed phase (C18) in-line chromatography system. In general, we apply the same settings used for standard iTRAQ quantitative proteomics (peptide labeling) experiments. We have analyzed wound fluid samples processed with the outlined workflow on Thermo Orbitrap XL, Orbitrap Velos, and Orbitrap Q Exactive instruments. All settings have been documented in detail and can be found in our recent publications

(Sabino et al., 2015; Schlage et al., 2014). In particular, we refer to an extensive method chapter by Kockmann et al. that details the most recent and optimized analysis settings for measuring iTRAQ-TAILS peptide samples on a Thermo Orbitrap Q Exactive mass spectrometer (Kockmann, Carte, Melkko, & auf dem Keller, 2015).

3.5 Data Analysis

General differences in interpretation of data from a standard shotgun proteomics experiment and results from an N-terminal enrichment workflow like TAILS are the focus on the peptide level and many inherent protein identifications by single peptides, which require very stringent criteria for spectrum to peptide assignments. These can be achieved, e.g., by combining results from multiple search engines and/or with help of statistical models for secondary validation of assignment scores. Such measures are available in current state-of-the-art commercial and also open access software packages for analysis of MS-based proteomics data. As a standard toolset, we use the Trans-Proteomic Pipeline (TPP) from the Institute of Systems Biology (Seattle, USA) that is freely available and integrates modules for peptide-centric validation as well as iTRAQ-based quantification (Deutsch et al., 2015). The TPP uses vendor-independent data formats and generates output files that can be easily interpreted by custom software. With CLIPPER we have developed an add-on to the TPP that automatically adds positional information for identified N-terminal peptides and allows assignment of proteolytic cleavage events based on statistical interpretation of relative quantification of natural N-termini in protease-exposed and control samples (auf dem Keller & Overall, 2012). CLIPPER is freely available from http://clipserve.clip.ubc.ca/tails and can be installed on top of a working TPP installation on a standard desktop PC. A detailed strategy for interpretation of iTRAQ-TAILS data using the TPP and CLIPPER is available in Kockmann et al. (2015). Example data for processed wound fluids may be downloaded from http://www.proteomexchange.org with identifier PXD001198, and expected outcomes are presented in the accompanying publication by Sabino et al. (2015). Additional resources for the comprehensive analysis of protein N-termini and protease cleavage products are available in form of the extensive public repositories Merops (Rawlings, Barrett, & Finn, 2016) and TopFIND (Fortelny, Yang, Pavlidis, Lange, & Overall, 2015).

4. VALIDATION BY TARGETED PROTEOMICS AND ESTABLISHMENT OF BIOMARKER ASSAYS

Proteome simplification by depletion of high-abundance components together with TAILS N-terminal enrichment is a powerful tool to generate lists of N-termini and protease cleavage events from analyses of complex body fluids. Based on quantitative comparisons of normal and diseased conditions this discovery-based workflow will provide hit candidates that are indicative for the underlying disorder and ideally serve as biomarkers for disease progression. Targeted proteomics approaches, such as selected or parallel reaction monitoring (S/PRM) (Picotti & Aebersold, 2012), can be employed to monitor these N-terminal peptides for validation in a similar but independent experiment as the initial discovery study and to establish assays for repeated analyses of samples of the same type. In many cases, the high sensitivity and specificity of targeted assays replace the need for specific N-terminal enrichment, significantly reducing the effort in sample processing. Often, it might be even possible to measure candidate peptides in tryptic digests of unprocessed body fluid samples, but in our experience depletion of high-abundance proteins might be still required for best results. Moreover, no peptide fractionation is applied, limiting the number of MS runs to one per sample. However, to directly use spectral information from TAILS discovery experiments for assay development, the proteome samples must be labeled at the protein level equally to the original TAILS analysis, and samples from different conditions cannot be mixed and digested together. Separate analysis of the different conditions allows relative label-free quantification of the peptides of interest. Below, we summarize the steps needed to prepare compatible proteomes for monitoring of N-terminal peptides by targeted proteomics that have been selected from iTRAQ-TAILS discovery experiments. For this protocol, it is assumed that samples have been depleted for high-abundance proteins as described earlier.

1. Start with 25 μg total protein (1 μg/μL in TAILS buffer) per condition. For targeted analysis the injection of as little as 1 μg of digested proteome into the mass spectrometer should allow identification and quantification of the peptides of interest.
2. Incubate the sample from each condition at 65°C for 15 min to assure total denaturation. Add 1 m*M* tris(2-carboxy-ethyl)phosphine (TCIP) and incubate at 65°C for 45 min to reduce cysteine residues. Add

5 m*M* iodoacetamide and incubate for 30 min at 65°C in the dark to alkylate the reduced cysteines.

3. Label the proteome of each condition with the same iTRAQ reagent in a 1:4 protein/iTRAQ (w/w) ratio as described in Section 3.2 step 4. Important: The label must have the same chemistry and isobaric mass as the original iTRAQ label. For this, mTRAQ reagent Δ4 is a cost-effective alternative, since it is equivalent to iTRAQ 4plex reagents. If iTRAQ 8plex reagents have been used in the discovery experiment, an 8plex reagent has to be used for whole-proteome labeling.
4. Stop the labeling reaction with 10 m*M* ammonium bicarbonate. Incubate for 30 min at room temperature.
5. Add 6–8 volumes of ice-cold acetone and 1 volume of ice-cold methanol, incubate at least 2 h at −80°C and precipitate by centrifugation at 10,000 × *g* for 20 min at 4°C. Wash the pellet with 1 mL of cold methanol, centrifuge at 10,000 × *g* for 10 min at 4°C, and air dry.
6. Resuspend protein pellet as described in Section 3.2 step 7.
7. Digest overnight with trypsin gold (Promega) at a 1:100 trypsin:protein (w/w) ratio at 37°C. Store at −80°C until further analysis.
8. Desalt samples with C18 resin using OMIX tips or Sep-Pak (Waters) columns (for more than 50 μg of total peptides).
9. Resuspend in a suitable volume of 3% acetonitrile and 0.1% formic acid and store at −20°C prior to MS analysis.

Samples are now ready for analysis by PRM- or SRM-targeted proteomics. Detailed descriptions of assay design, retention time scheduling, and data analysis are beyond the scope of this chapter but can be found in excellent reviews and tutorials (Gallien et al., 2012; Peterson, Russell, Bailey, Westphall, & Coon, 2012; Picotti & Aebersold, 2012; Surinova et al., 2013). The software packages Skyline (open access) from Michael MacCoss' laboratory (MacLean et al., 2010) and SpectroDive™ from Biognosys AG (freely available to academic users) are excellent tools for design and analysis of targeted proteomics experiments and come with extensive documentation and test datasets. An example dataset for SRM analysis of N-terminal peptides in TAILS and preTAILS samples of porcine wound exudates is available as supplementary material with Sabino et al. (2015).

5. CONCLUSIONS

Despite several setbacks in the establishment of biomarkers from body fluids these specimens are still very attractive due to their accessibility and indicative protein composition that changes in response to most disease processes. Monitoring specific PTMs rather than only differential protein abundances might increase the chances of identifying reproducible events with high discriminative power between normal and diseased conditions. Proteomics analysis of protein N-termini does not only record proteolytic cleavages on a global scale but also allows for proteome simplification, and thus vastly extends the number of analytically accessible proteoforms in highly complex body fluid proteomes. With the availability of powerful workflows for depletion of high-abundance proteins and enrichment of protein N-termini, we just begin to explore the proteolytic landscape of body fluid proteomes that will be rapidly extended in the near future and hopefully lead to establishment of new types of proteoform biomarkers in health and disease.

ACKNOWLEDGMENTS

We want to acknowledge S. Werner (ETH Zurich) for her continuous support of our research. A special thanks goes to the proteomics team from the Functional Genomics Center Zurich (FGCZ) for excellent help in mass spectrometry. This research was funded by an unrestricted research grant from HARTMANN AG to U.a.d.K., a predoctoral fellowship from the Portuguese Foundation for Science and Technology (FCT) (SFRH/BD/88564/2012) to F.S., and by funds from the ETH Zurich.

REFERENCES

Agard, N. J., & Wells, J. A. (2009). Methods for the proteomic identification of protease substrates. *Current Opinion in Chemical Biology*, *13*(5–6), 503–509. http://dx.doi.org/10.1016/j.cbpa.2009.07.026.

Ahn, S. M., & Simpson, R. J. (2007). Body fluid proteomics: Prospects for biomarker discovery. *Proteomics. Clinical Applications*, *1*(9), 1004–1015. http://dx.doi.org/10.1002/prca.200700217.

auf dem Keller, U., & Overall, C. M. (2012). CLIPPER: An add-on to the trans-proteomic pipeline for the automated analysis of TAILS N-terminomics data. *Biological Chemistry*, *393*(12), 1477–1483. http://dx.doi.org/10.1515/hsz-2012-0269.

Boschetti, E., & Righetti, P. G. (2008). The ProteoMiner in the proteomic arena: A non-depleting tool for discovering low-abundance species. *Journal of Proteomics*, *71*(3), 255–264. http://dx.doi.org/10.1016/j.jprot.2008.05.002.

Deutsch, E. W., Mendoza, L., Shteynberg, D., Slagel, J., Sun, Z., & Moritz, R. L. (2015). Trans-proteomic pipeline, a standardized data processing pipeline for large-scale reproducible proteomics informatics. *Proteomics. Clinical Applications*, *9*(7–8), 745–754. http://dx.doi.org/10.1002/prca.201400164.

Eckhard, U., Marino, G., Butler, G. S., & Overall, C. M. (2015). Positional proteomics in the era of the human proteome project on the doorstep of precision medicine. *Biochimie*, *122*, 110–118. http://dx.doi.org/10.1016/j.biochi.2015.10.018.

Falk, R., Ramstrom, M., Stahl, S., & Hober, S. (2007). Approaches for systematic proteome exploration. *Biomolecular Engineering*, *24*(2), 155–168. http://dx.doi.org/10.1016/j.bioeng.2007.01.001.

Farrah, T., Deutsch, E. W., Omenn, G. S., Campbell, D. S., Sun, Z., Bletz, J. A., et al. (2011). A high-confidence human plasma proteome reference set with estimated concentrations in PeptideAtlas. *Molecular & Cellular Proteomics*, *10*(9). http://dx.doi.org/10.1074/mcp.M110.006353. M110.006353.

Fortelny, N., Yang, S., Pavlidis, P., Lange, P. F., & Overall, C. M. (2015). Proteome TopFIND 3.0 with TopFINDer and PathFINDer: Database and analysis tools for the association of protein termini to pre- and post-translational events. *Nucleic Acids Research*, *43*(Database issue), D290–D297. http://dx.doi.org/10.1093/nar/gku1012.

Gallien, S., Duriez, E., Crone, C., Kellmann, M., Moehring, T., & Domon, B. (2012). Targeted proteomic quantification on quadrupole-orbitrap mass spectrometer. *Molecular & Cellular Proteomics*, *11*(12), 1709–1723. http://dx.doi.org/10.1074/mcp.O112.019802.

Good, D. M., Thongboonkerd, V., Novak, J., Bascands, J. L., Schanstra, J. P., Coon, J. J., et al. (2007). Body fluid proteomics for biomarker discovery: Lessons from the past hold the key to success in the future. *Journal of Proteome Research*, *6*(12), 4549–4555. http://dx.doi.org/10.1021/pr070529w.

Granger, J., Siddiqui, J., Copeland, S., & Remick, D. (2005). Albumin depletion of human plasma also removes low abundance proteins including the cytokines. *Proteomics*, *5*(18), 4713–4718. http://dx.doi.org/10.1002/pmic.200401331.

Hermes, O., Schlage, P., & auf dem Keller, U. (2011). Wound degradomics—Current status and future perspectives. *Biological Chemistry*, *392*(11), 949–954. http://dx.doi.org/10.1515/BC.2011.092.

Hu, S., Loo, J. A., & Wong, D. T. (2006). Human body fluid proteome analysis. *Proteomics*, *6*(23), 6326–6353. http://dx.doi.org/10.1002/pmic.200600284.

Huesgen, P. F., Lange, P. F., & Overall, C. M. (2014). Ensembles of protein termini and specific proteolytic signatures as candidate biomarkers of disease. *Proteomics Clinical Applications*, *8*(5–6), 338–350. http://dx.doi.org/10.1002/prca.201300104.

International Human Genome Sequencing, Consortium (2004). Finishing the euchromatic sequence of the human genome. *Nature*, *431*(7011), 931–945. http://dx.doi.org/10.1038/nature03001.

Kleifeld, O., Doucet, A., auf dem Keller, U., Prudova, A., Schilling, O., Kainthan, R. K., et al. (2010). Isotopic labeling of terminal amines in complex samples identifies protein N-termini and protease cleavage products. *Nature Biotechnology*, *28*(3), 281–288. http://dx.doi.org/10.1038/nbt.1611.

Kockmann, T., Carte, N., Melkko, S., & auf dem Keller, U. (2015). Identification of protease substrates in complex proteomes by iTRAQ-TAILS on a thermo Q exactive instrument. In J. Grant & H. Li (Eds.), *Analysis of post-translational modifications and proteolysis in neuroscience* (pp. 187–207). New York: Springer Protocols.

Lai, Z. W., Petrera, A., & Schilling, O. (2015). The emerging role of the peptidome in biomarker discovery and degradome profiling. *Biological Chemistry*, *396*(3), 185–192. http://dx.doi.org/10.1515/hsz-2014-0207.

Lange, P. F., & Overall, C. M. (2013). Protein TAILS: When termini tell tales of proteolysis and function. *Current Opinion in Chemical Biology*, *17*(1), 73–82. http://dx.doi.org/10.1016/j.cbpa.2012.11.025.

MacLean, B., Tomazela, D. M., Shulman, N., Chambers, M., Finney, G. L., Frewen, B., et al. (2010). Skyline: An open source document editor for creating and analyzing targeted proteomics experiments. *Bioinformatics*, *26*(7), 966–968. http://dx.doi.org/10.1093/bioinformatics/btq054.

McPherson, J. D., Marra, M., Hillier, L., Waterston, R. H., Chinwalla, A., Wallis, J., et al. (2001). A physical map of the human genome. *Nature*, *409*(6822), 934–941. http://dx.doi.org/10.1038/35057157.

Mebazaa, A., Vanpoucke, G., Thomas, G., Verleysen, K., Cohen-Solal, A., Vanderheyden, M., et al. (2012). Unbiased plasma proteomics for novel diagnostic biomarkers in cardiovascular disease: Identification of quiescin Q6 as a candidate biomarker of acutely decompensated heart failure. *European Heart Journal*, *33*(18), 2317–2324. http://dx.doi.org/10.1093/eurheartj/ehs162.

Nilsson, T., Mann, M., Aebersold, R., Yates, J. R., 3rd, Bairoch, A., & Bergeron, J. J. (2010). Mass spectrometry in high-throughput proteomics: Ready for the big time. *Nature Methods*, 7(9), 681–685. http://dx.doi.org/10.1038/nmeth0910-681.

Pernemalm, M., & Lehtio, J. (2014). Mass spectrometry-based plasma proteomics: State of the art and future outlook. *Expert Review of Proteomics*, *11*(4), 431–448. http://dx.doi.org/10.1586/14789450.2014.901157.

Peterson, A. C., Russell, J. D., Bailey, D. J., Westphall, M. S., & Coon, J. J. (2012). Parallel reaction monitoring for high resolution and high mass accuracy quantitative, targeted proteomics. *Molecular & Cellular Proteomics*, *11*(11), 1475–1488. http://dx.doi.org/10.1074/mcp.O112.020131.

Picotti, P., & Aebersold, R. (2012). Selected reaction monitoring-based proteomics: Workflows, potential, pitfalls and future directions. *Nature Methods*, *9*(6), 555–566. http://dx.doi.org/10.1038/nmeth.2015.

Plasman, K., Van Damme, P., & Gevaert, K. (2013). Contemporary positional proteomics strategies to study protein processing. *Current Opinion in Chemical Biology*, *17*(1), 66–72. http://dx.doi.org/10.1016/j.cbpa.2012.11.026.

Rawlings, N. D., Barrett, A. J., & Finn, R. (2016). Twenty years of the MEROPS database of proteolytic enzymes, their substrates and inhibitors. *Nucleic Acids Research*, *44*(D1), D343–D350. http://dx.doi.org/10.1093/nar/gkv1118.

Rogers, L. D., & Overall, C. M. (2013). Proteolytic post-translational modification of proteins: Proteomic tools and methodology. *Molecular & Cellular Proteomics*, *12*(12), 3532–3542. http://dx.doi.org/10.1074/mcp.M113.031310.

Sabino, F., Hermes, O., Egli, F. E., Kockmann, T., Schlage, P., Croizat, P., et al. (2015). In vivo assessment of protease dynamics in cutaneous wound healing by degradomics analysis of porcine wound exudates. *Molecular & Cellular Proteomics*, *14*(2), 354–370. http://dx.doi.org/10.1074/mcp.M114.043414.

Sandra, K., Verleysen, K., Labeur, C., Vanneste, L., D'Hondt, F., Thomas, G., et al. (2007). Combination of COFRADIC and high temperature-extended column length conventional liquid chromatography: A very efficient way to tackle complex protein samples, such as serum. *Journal of Separation Science*, *30*(5), 658–668.

Schlage, P., Egli, F. E., Nanni, P., Wang, L. W., Kizhakkedathu, J. N., Apte, S. S., et al. (2014). Time-resolved analysis of the matrix metalloproteinase 10 substrate degradome. *Molecular & Cellular Proteomics*, *13*(2), 580–593. http://dx.doi.org/10.1074/mcp.M113.035139.

Shen, Y., Kim, J., Strittmatter, E. F., Jacobs, J. M., Camp, D. G., 2nd, Fang, R., et al. (2005). Characterization of the human blood plasma proteome. *Proteomics*, *5*(15), 4034–4045. http://dx.doi.org/10.1002/pmic.200401246.

Staes, A., Impens, F., Van Damme, P., Ruttens, B., Goethals, M., Demol, H., et al. (2011). Selecting protein N-terminal peptides by combined fractional diagonal chromatography. *Nature Protocols*, *6*(8), 1130–1141. http://dx.doi.org/10.1038/nprot.2011.355.

Surinova, S., Huttenhain, R., Chang, C. Y., Espona, L., Vitek, O., & Aebersold, R. (2013). Automated selected reaction monitoring data analysis workflow for large-scale targeted proteomic studies. *Nature Protocols*, *8*(8), 1602–1619. http://dx.doi.org/10.1038/nprot.2013.091.

Timmer, J. C., Enoksson, M., Wildfang, E., Zhu, W., Igarashi, Y., Denault, J. B., et al. (2007). Profiling constitutive proteolytic events in vivo. *The Biochemical Journal*, *407*(1), 41–48. http://dx.doi.org/10.1042/BJ20070775.

Van Damme, P., Staes, A., Bronsoms, S., Helsens, K., Colaert, N., Timmerman, E., et al. (2010). Complementary positional proteomics for screening substrates of endo- and exoproteases. *Nature Methods*, 7(7), 512–515. http://dx.doi.org/10.1038/nmeth.1469.

Vizovisek, M., Vidmar, R., Fonovic, M., & Turk, B. (2016). Current trends and challenges in proteomic identification of protease substrates. *Biochimie*, *122*, 77–87. http://dx.doi.org/10.1016/j.biochi.2015.10.017.

Wiita, A. P., Hsu, G. W., Lu, C. M., Esensten, J. H., & Wells, J. A. (2014). Circulating proteolytic signatures of chemotherapy-induced cell death in humans discovered by N-terminal labeling. *Proceedings of the National Academy of Sciences of the United States of America*, *111*(21), 7594–7599. http://dx.doi.org/10.1073/pnas.1405987111.

Wildes, D., & Wells, J. A. (2010). Sampling the N-terminal proteome of human blood. *Proceedings of the National Academy of Sciences of the United States of America*, *107*(10), 4561–4566. http://dx.doi.org/10.1073/pnas.0914495107.

Yoshihara, H. A., Mahrus, S., & Wells, J. A. (2008). Tags for labeling protein N-termini with subtiligase for proteomics. *Bioorganic & Medicinal Chemistry Letters*, *18*(22), 6000–6003. http://dx.doi.org/10.1016/j.bmcl.2008.08.044.

CHAPTER TWELVE

Reannotation of Genomes by Means of Proteomics Data

J. Armengaud[1]
CEA, DRF, IBiTec-S, SPI, Li2D, Laboratory "Innovative Technologies for Detection and Diagnostics", Bagnols-sur-Cèze, France
[1]Corresponding author: e-mail address: jean.armengaud@cea.fr

Contents

Abstract

Omics approaches have become popular in biology as powerful discovery tools, and currently gain in interest for diagnostic applications. Establishing the accurate genome sequence of any organism is easy, but the outcome of its annotation by means of automatic pipelines remains imprecise. Some protein-encoding genes may be missed as soon as they are specific and poorly conserved in a given taxon, while important to explain the specific traits of the organism. Translational starts are also poorly predicted in a relatively important number of cases, thus impacting the protein sequence database used in proteomics, comparative genomics, and systems biology. The use of high-throughput proteomics data to improve genome annotation is an attractive option to obtain a more comprehensive molecular picture of a given organism. Here, protocols for reannotating prokaryote genomes are described based on shotgun proteomics and derivatization of protein N-termini with a positively charged reagent coupled to high-resolution tandem mass spectrometry.

Methods in Enzymology, Volume 585
ISSN 0076-6879
http://dx.doi.org/10.1016/bs.mie.2016.09.019

1. INTRODUCTION

Next-generation proteomics allows a deep characterization of any organisms through the identification and quantitation of the most valuable players, i.e., the proteins that are the molecular working horses inside the cell and outside the cell (Armengaud, 2016). This approach relies on proteolysis of proteins into tryptic peptides, use of high-resolution tandem mass spectrometers for characterizing their fragmentation patterns, and protein sequence databases for interpreting the numerous recorded MS/MS spectra. This last point implies that the genome of the organism of interest should be sequenced and annotated. However, this is not always the case, or worse, some errors in terms of annotation are insidiously present because of inherent mistakes during the genome annotation process (Armengaud, Bland, et al., 2011). Dealing with massive numbers of genome sequences currently produced, automatic gene annotation is done based on consensual translation rules and in some cases excessively weight knowledge from sequence conservation. These basic rules may not apply for all organisms. Indeed, in many cases, some genes are considered or not by annotation pipelines in a Russian roulette play. For example, a specific open reading frame may exist in a given organism but homologous sequences may not be found in the closely related species (Fig. 1). On the other way around, an open reading frame may be just missing in the model under scrutiny while the gene is expected because of its hitherto systematic occurrence in a given taxon. Although the genetic code is universal, codons used for translation starts may be different for some specific genes in a given organism or drastically dissimilar in some exotic creatures. For example, *Aeropyrum pernix K1*, an obligatorily thermophilic archaeon isolated from sea sediments samples off the coast of Japan, exhibits a unusual preference for translational initiation codons with UUG being the most used, while AUG and in a lesser extent GUG are ordinarily expected (Yamazaki, Yamazaki, et al., 2006). Noteworthy, numerous errors usually occur during genome annotation when no well-annotated template is available (Armengaud, 2009), and conspicuously, rare taxonomic phyla should be much better considered in terms of sequence and annotation priority (Kumar, Mondal, et al., 2016). As a result of this biological diversity and lack of joint efforts by the scientific community to improve public sequence databases, the information from current genome databases is scrambled both in terms of structural and functional annotation of protein-encoding genes (Pible & Armengaud, 2015).

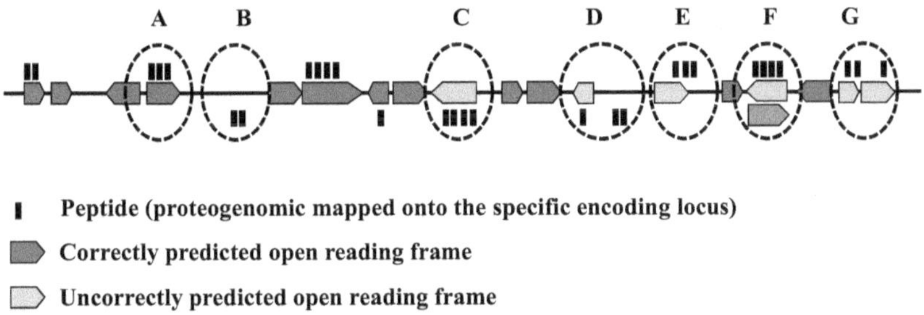

Fig. 1 Examples of genome annotation errors and their detection by proteogenomics. Peptides detected by tandem mass spectrometry are mapped onto the corresponding nucleotide sequence. **Panel (A) shows the validation of a predicted gene through identification of three specific peptide N-terminal peptide sequences corresponding to the encoded polypeptide, panel (B) shows the identification of a new open reading frame that was missed by the annotation software, panel (C) shows the validation of the translational start through detection of the N-terminal peptide, panel (D) shows the identification of an upstream initiation codon, panel (E) shows the identification of a sequence error through the detection of peptides downstream of the predicted open reading frame, panel (F) shows the reversal of the reading sense for a specific locus, and panel (G) shows the validation of a frameshift with the identification of peptides from two adjacent cistrons.

Proteogenomics was first proposed to make use of experimental peptide cartography based on data from proteins identified by shotgun proteomics and directly mapped these peptides on the genome to better define the structural annotation of the corresponding genes (Jaffe, Berg, et al., 2004). As shown in Fig. 1, this strategy allows finding missing genes and confirming the presence of hypothetical protein-encoding genes, as well as identifying introns or specific items such as inteins or frameshifts, or even uncovering genome sequence errors. Furthermore, the systematic search of N-terminal peptides of proteins helps for establishing translational starts of translated polypeptides or posttranslational maturations of secreted or activated proteins, and thus delineating the correct protein sequences (Berry, Steele, et al., 2016; Hartmann & Armengaud, 2014).

1.1 Reannotation of Genomes With Tandem Mass Spectrometry Certified Peptides

Based on a decade of innovative works, numerous reviews have pointed at the power of proteogenomics for refining genome annotation (Ansong, Purvine, et al., 2008; Armengaud, 2013; Nesvizhskii, 2014; Renuse, Chaerkady, et al., 2011). However, a perfect annotation is far to be easily achieved as it requires tremendous investigation efforts for a comprehensive cartography, as well as complementary experimental validations. Most published examples are centered on the reannotation of the genome of a given organism and the most recent studies are relying on next-generation proteomics high-throughput approaches: reannotation of *Methylobacterium extorquens* (Kumar, Mondal, et al., 2014), *Escherichia coli* (Pettersen, Steinsland, et al., 2015), *Bacillus anthracis* (Gao, Wang, et al., 2015), *Candida tropicalis* (Datta, Patil, et al., 2016), *Mycobacterium smegmatis* (Potgieter, Nakedi, et al., 2016), and *Yersinia pestis* (Mao, Yang, et al., 2016) are examples worth citing. Noteworthy, several of these bacteria are among pathogens of great concern (Duriez, Armengaud, et al., 2016). Further, complementation of biochemical evidence with transcriptomics data helps in refining the genome reannotation as recently exemplified with bacteria and eukaryotes (de Groot, Roche, et al., 2014; Krishna, Xia, et al., 2015; Syme, Tan, et al., 2016; Xu, Liu, et al., 2015). RNAseq and ribosome profiling are promising complementary strategies to focus on translated or nascent polypeptides only (Koch, Gawron, et al., 2014; Kumar, Bansal, et al., 2016; Menschaert, Van Criekinge, et al., 2013). Moreover, additional validation by resequencing some genome loci pointed through proteogenomics inconsistencies helps cleaning the genome (Christie-Oleza, Miotello, et al., 2012). Some proteogenomic studies are aimed at a larger vision with reannotation of several related organisms at once. An approach named "comparative proteogenomics" is using multiple proteomes acquired from different members of the same taxon to validate "proteogenomics one-hit wonders" from a given organism (Gupta, Benhamida, et al., 2008). Another approach, namely "ortho-proteogenomics" extends proteogenomics results from one organism to its closely related relatives (Gallien, Perrodou, et al., 2009). The most comprehensive proteogenomics study to date is that proposed by Venter, Smith, et al. (2011) where 46 organisms have been scrutinized with the help of a high-throughput bioinformatics pipeline (Tovchigrechko, Venepally, et al., 2014).

1.2 Exact ORF Delineation With N-Terminomics

High-throughput proteomics data can be searched more specifically for the most N-terminal peptides in order to determine the translational start of proteins or the nature of their maturation if they have been cleft for activation or secretion. Because comprehensive coverage of polypeptide sequences is difficult to obtain for the whole proteome, specific methodologies have been developed along the years in order to enrich for these N-terminal peptides. These approaches have been recently reviewed (Berry et al., 2016; Hartmann & Armengaud, 2014). Among these, labeling of the free amino terminal of the polypeptides with a positively charged reagent such as (*N*-succinimidyloxycarbonylmethyl)tris(2,4,6-trimethoxyphenyl)phosphonium bromide (TMPP) allows after trypsin proteolysis enriching specifically the N-terminal peptides and improving their ionisability (Fig. 2). Such reagent has been successfully used in proteogenomics for reannotating the genome of *M. smegmatis* (Gallien et al., 2009) and *Deinococcus deserti* (Baudet, Ortet, et al., 2010). Combined with a COFRADIC-based strategy such reagent was used to enrich a large panel of N-terminal peptides and identify numerous translational starts from the marine bacterium *Roseobacter denitrificans* (Bland, Hartmann, et al., 2014). In this organism, multiple translational start sites for the same genes were highlighted, a result that challenges

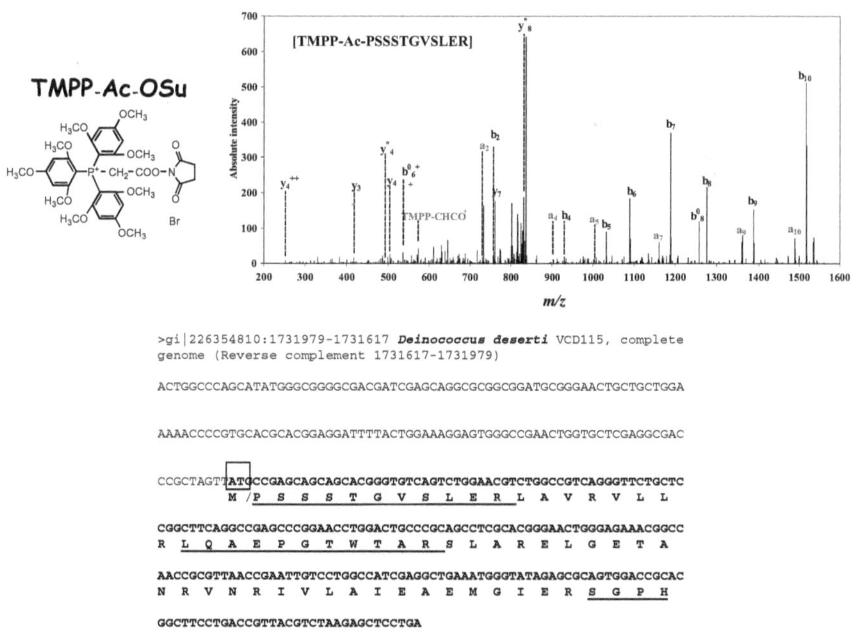

Fig. 2 Discovery of a new protein-encoding gene and delineation of the corresponding translational start by means of TMPP labeling and tandem mass spectrometry. The figure shows the example of Deide14222 from *Deinococcus deserti*, an orphan MS/MS-certified protein. Three peptides have been identified (*underlined*), one being the most N-terminal and TMPP labeled (as shown from its MS/MS spectrum). The corresponding translational start is *boxed*.

the current view of protein translation initiation which is restricted to the annotation of a specific initiator codon only. A more direct enrichment protocol was made possible after the development of antibodies recognizing with high specificity the TMPP moiety and its coupling on magnetic beads (Bland, Bellanger, et al., 2014). In this protocol, enrichment of TMPP-labeled N-terminal peptides is quickly achieved, allowing high-throughput proteogenomics studies. Interestingly, a stable isotope of this reagent may be obtained for further validation or quantitative strategies (Armengaud, 2015; Bertaccini, Vaca, et al., 2013; Shen, An, et al., 2015).

In this chapter, a general experimental procedure for bacterial genome annotation with next-generation proteomics data and proteogenomics-derived applications are presented.

2. BACTERIAL PROTEOGENOMICS: THE METHODS

2.1 General Overview

In order to interpret MS/MS spectra, a proteogenomics protein sequence database should be first constructed. A shortcut that works really well for prokaryotes is a naïve six-frame translation of any nucleotide sequence. The resulting polypeptide sequences can be split into separated open reading frames defined from stop to stop, listed in a FASTA file, and named with a nomenclature that indicates the exact position onto the corresponding nucleic acid sequence. Such direct translation generates many improper sequences, as well as some sequences encompassing the true protein information. Many resulting hypothetical polypeptides are short due to the high frequency of translation stop codons in some nucleic loci. Thus, a first selection of translated sequences with at least 50 residues is a good option to decrease the size of the proteogenomics database. Typically, for *D. deserti* genome which comprises a main chromosome (2.82 Mb) and three mega plasmids (325, 314, and 396 kb), the resulting translated database is rather giant with 65,801 polypeptide sequences, totaling 6,040,642 amino acids while the expected theoretical proteome is 3439 protein-encoding genes for this bacterium (de Groot, Dulermo, et al., 2009). This six-frame proteogenomic database comprises rather short sequences (an average of 92 amino acids per polypeptide). However, it is important to note that because of the inherent problem for accurately predicting translational starts, stop-to-stop sequences introduce systematically extra residues upstream of the correct polypeptide sequences, and in addition, these do not always encompass the good initiator residue. When GUG, a codon usually encoding valine, or UUG a codon coding for leucine, are used as initiation codons, a methionine is beginning the polypeptide sequence. Moreover, this initiator methionine may be further processed by the cell once the polypeptide has been synthesized by the ribosome. Fortunately, a simple search of the MS/MS dataset with valine-to-methionine and leucine-to-methionine replacement modifications can solve this issue.

In order to obtain a comprehensive coverage of the proteome, it is necessary to multiply samples from diverse physiological conditions. As shown previously for generalist bacteria, the most differing experimental proteomes are those from exponential and stationary growth phases (Christie-Oleza, Fernandez, et al., 2012). However, only these two conditions may not be sufficient to get expression of all genes and subsequent transcript translation, and therefore, multiplying conditions such as different culture media is recommended. Furthermore, to get the largest peptide coverage, multiplying proteases such as the use of chymotrypsin, endoGluC, endoLysC, ArgC, beside trypsin, may generate large overlapping but complementary peptidomes (Bland, Bellanger, et al., 2014). Finally, exploring peptidomes with the most recent next-generation proteomics tandem mass spectrometers allows getting almost complete proteomes in a rather limited number of nanoLC-MS/MS runs (Richards, Hebert, et al., 2015). Proteome fractionation either by state-of-the-art chromatography or by gel electrophoresis in denaturing conditions followed by systematic nanoLC-MS/MS analysis may represent a good alternative to uncover the less abundant proteins as previously shown (de Groot et al., 2009).

2.2 Experimental Design

Fig. 3 shows the flow chart of a proteogenomics experiment. Extreme precautions should be taken to handle bacteria and subsequent proteome samples as undesired proteolysis by internal proteases should be avoided. For this, sample handling should remain limited in time and inhibitor of proteases should be introduced in the samples as soon as possible. Cells are usually centrifuged and then disrupted to produce a protein extract. Then, the proteins may be resolved onto a SDS-PAGE gel and after Coomassie Blue staining the lane may be sliced into small polyacrylamide bands, destained, reduced, and treated with iodoacetamide, and then subjected to trypsin proteolysis. The resulting peptides are injected onto a reverse-phase chromatographic column, resolved, and analyzed by tandem mass spectrometry. MS/MS datasets are then assigned against the six-frame

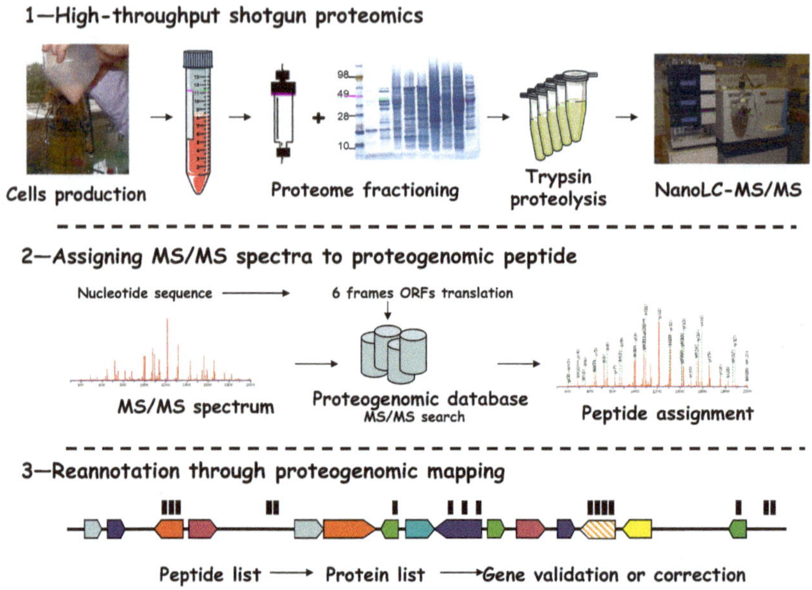

Fig. 3 Flow chart for a bacterial proteogenomics reannotation. Three steps are delineated: a first experimental step is devoted to high-throughput identification of proteins through their peptides identified by tandem mass spectrometry; the second step consists in creating a proteogenomics six-frame translation protein sequence database and interpreting MS/MS spectra; and the third step is the proteogenomics mapping of the peptide sequences onto the nucleotide sequence and reannotation of conflicts.

translation protein sequence database with a search engine such as the MASCOT software (Matrix Science). This query may be done with the following parameters: full-trypsin specificity, maximum of two missed cleavages, and fixed modification of carboxyamidomethylated cysteine. Regarding dynamic modifications, oxidized methionine, deamidation of asparagine and glutamine residues, and N-terminal acetylation, should be considered. Because methionine to isothreonine conversion has been shown to occur frequently and could be a source of false discovery identifications of genetically encoded variants in proteogenomics, this variable modification may also be included (Chernobrovkin, Kopylov, et al., 2015). Mass tolerances of 5 ppm on the parent ion and 0.5/0.02 Da on the MS/MS are usual values depending on the m/z ion analyzer. Note, that if N-terminal peptides arising from GUG or UUG initiator codons are expected, another search with specific modifications as described earlier may be performed.

2.3 Bioinformatics Tools and Data Treatment Strategies

For handling proteogenomics data and visualize the results in the genomic context, use of recently developed bioinformatics tools is recommended. The large number of available tools has been helpfully reviewed (Menschaert & Fenyo, 2016). GenoSuite (Kumar, Yadav, et al., 2013), ProteoAnnotator (Ghali, Krishna, et al., 2014), BPP (Uszkoreit, Plohnke, et al., 2014), PGP (Tovchigrechko et al., 2014), MSProGene (Zickmann & Renard, 2015), PGTools (Nagaraj, Waddell, et al., 2015), and proBAMsuite (Wang, Slebos, et al., 2016) have been shown efficient tools. Furthermore, flexible and accessible Galaxy workflows have been implemented for proteogenomics (Jagtap, Johnson, et al., 2014).

Regarding the strategy to interpret MS/MS data, the unusual inflated size of the six-frame translation protein database compared to a classical genome annotated protein database tends to increase the number of shared peptide sequences between the database and its decoy counterpart (Wright & Choudhary, 2016), underestimate the confidence assigned to peptide-sequence-matches (PSMs), and thus should be taken into account (Blakeley, Overton, et al., 2012; Zhang, Fu, et al., 2015). A two-step database search strategy has been proposed to improve PSM sensitivity, the second database being drastically reduced to the polypeptide hits from the first round of search (Jagtap, Goslinga, et al., 2013). Improvement in MS/MS treatment (Helmy, Sugiyama, et al., 2012) and evaluation of false-positive without decoy database (Gonnelli, Stock, et al., 2015) are stimulating proposals.

3. GENERAL PROTOCOL

3.1 Proteome and Sample Preparation for Shotgun Analysis

1. After cultivation, centrifuge bacterial cells from a 5-mL liquid culture at $6000 \times g$ for 5 min at 4°C (typically the material from a 1-mL liquid culture at OD 600 nm equal to 1.0 is sufficient). Remove the supernatant and centrifuge another round of 2 min at $6000 \times g$ for removing the residual liquid droplets. Weight the tube in order to establish the amount of wet cellular material. Flash frozen the wet pellet and keep at −80°C until use.
2. For protein extraction, add a specific volume of LDS buffer (40% glycerol, 4% LDS, 4% Ficoll-400, 0.8 *M* triethanolamine-Cl, pH 7.6,

0.025% phenol red, 0.025% Coomassie G250, 2 m*M* EDTA disodium, diluted with three volumes of deionized water) to the frozen pellet (60 μL per mg of wet material). Sonicate briefly to dissolve all the cellular particles if necessary.

3. Heat the sample at 99°C for 5 min in a thermomixer.
4. Centrifuge at 6000 × *g* for 2 min.
5. Load 20 μL of the resulting supernatant onto a 4–12% Bis-Tris 10-well gel for a short electrophoresis (5 min) at 200 V in MES 1 × running buffer as previously described (Hartmann, Allain, et al., 2014). Longer electrophoresis (20 min) may be an alternative in order to resolve the proteome along their molecular weight and multiply proteomic samples.
6. After electrophoresis, rinse the gel three times with water. Stain the gel with Coomassie Blue stain for 20 min and then, destain for several hours with ultrapure water.
7. Excise the proteome as polyacrylamide bands with a clean scalpel or razor blade in the most appropriate conditions to avoid dust and keratin contamination.
8. Destain each polyacrylamide band with 200 μL of 50% methanol:50% 50 m*M* ammonium bicarbonate solution in a thermomixer under shaking at 500 rpm for 1 min, and discard the fluid. Repeat this step once.
9. Dehydrate with 200 μL of 50% acetonitrile:50% 50 m*M* ammonium bicarbonate under shaking at 600 rpm for 5 min, and discard the fluid.
10. Dehydrate with 200 μL of pure acetonitrile under shaking at 600 rpm for 1 min, and discard the fluid.
11. Dry the tube(s) in a SpeedVac for 2 min.
12. Rehydrate the gel piece(s) with 100 μL of reduction solution (38.5 mg of dithiothreitol dissolved in 10 mL of 50 m*M* ammonium bicarbonate) and incubate for 20 min at 56°C with shaking at 500 rpm. Discard the fluid.
13. Add 100 μL of alkylation solution (100 mg of iodoacetamide in 10 mL of 50 m*M* ammonium bicarbonate) for 20 min at room temperature in the dark. Note that the alkylation solution once prepared should be kept in the dark. Discard the fluid.
14. Wash with 400 μL of deionized water under shaking at 600 rpm for 1 min. Discard the fluid. Repeat this step once.
15. Dehydrate with 200 μL of 50% acetonitrile:50% 50 m*M* ammonium bicarbonate for 5 min under shaking at 600 rpm. Discard the fluid.

16. Dehydrate with 200 μL of pure acetonitrile for 1 min under shaking at 600 rpm. Discard the fluid.
17. Dry the tube(s) in a SpeedVac for 2 min.
18. Reconstitute lyophilized sequencing-grade trypsin to a final concentration of 0.1 μg/μL in 0.01% trifluoroacetic acid.
19. Rehydrate the polyacrylamide gel piece(s) with 20 μL of trypsin solution. Incubate for 20 min on ice. Remove excess liquid.
20. Add 50 μL of 0.01% ProteaseMax (Promega). Shake briefly. Incubate for 3–4 h at 37°C.
21. Transfer the solution (50 μL) to a clean tube and acidify the sample by addition of 5 μL of 5% trifluoroacetic acid. The peptide mixture can be stored at −80°C prior nanoLC-MS/MS analysis.

3.2 Sample Preparation for TMPP Labeling of Protein N-Termini

Alternatively a protein extract in a buffer compatible for N-terminal labeling with the TMPP reagent may be first produced.

1. Resuspend cells (60 μL per mg of wet material) in a 100 m*M* NaH_2PO_4/Na_2HPO_4 solution buffered at pH 8.2 (at 20°C) complemented with protease inhibitor such as the Mini ethylene diamine tetra acetic acid (EDTA)-free Protease Inhibitor Cocktail solution (Roche) at 25 μL per mL of buffer.
2. Lyse cells in a disruptor or bead-beating device.
3. Pellet the debris by centrifugation for 10 min at 16,000 × *g* at 4°C. Collect the supernatant containing the soluble proteins in a new tube.
4. Desalt the sample by chromatography onto 5 mL HiTrap Desalting columns operated with a chromatographic system at a flow rate of 1 mL/min and equilibrated with 100 m*M* NaH_2PO_4/Na_2HPO_4 solution buffered at pH 8.2 (at 20°C) and containing the protease inhibitor mixture.
5. Perform N-terminal labeling of proteins with TMPP reagent dissolved at 200 m*M* in 20% acetonitrile (ACN)/80% H_2O (v/v) just prior to use. For labeling 1 mL of protein extract at 3 mg/mL, add a volume of 94 μL of 200 m*M* TMPP-AcOSu (approximately, a ratio of 250 TMPP molecules per protein should be reached).
6. Incubate the reaction at room temperature for 60 min under gentle agitation.
7. Stop the reaction by addition of 2.5 mL of 1.5 *M* 2-amino-2-hydroxymethyl-1,3-propanediol/HCl solution buffered at pH 8.5.

8. Incubate the sample at room temperature for 30 min.
9. Desalt the labeled proteins with a 5 mL HiTrap desalting column previously equilibrated with 100 m*M* sodium phosphate solution buffered at pH 7.2 and containing 150 m*M* NaCl.
10. Reduce protein disulfide bonds with tributylphosphine (0.05% final concentration) and incubate for 90 min at room temperature.
11. Alkylate for 30 min free sulfhydryls with 5 m*M* iodoacetamide (Sigma) in the dark at 37°C.
12. Desalt again the sample by chromatography at a flow rate of 1 mL/min with 5 mL HiTrap desalting column and previously equilibrated with 50 m*M* NH_4HCO_3 solution buffered at pH 7.8.
13. Measure the protein concentration of the sample by Bradford assay.
14. Reconstitute lyophilized sequencing-grade trypsin to a final concentration of 0.1 μg/μL in 0.01% trifluoroacetic acid.
15. Subject the proteins to overnight liquid proteolysis in presence of sequencing-grade trypsin at 37°C under gentle mixing.
16. Acidify the peptide sample by addition of 5 μL of 5% trifluoroacetic acid. The peptide mixture can be stored at −80°C prior nanoLC-MS/MS analysis.
17. Resolve the peptides by reverse-phase chromatography in order to remove the main TMPP unbound reagent and fractionate the labeled peptides.
18. Dilute each fraction in 1 mL of 10 m*M* PBS buffered at pH 7.4
19. Coat Dynabead magnetic beads (30 μL suspension at 40 mg magnetic beads per milliliter) with anti-TMPP monoclonal antibodies (40 μg antibody per milligram of magnetic beads). Wash three times with 300 μL of 10 m*M* PBS buffered at pH 7.4. Remove the supernatant after magnetic immobilization of the beads.
20. Add 100 μL of TMPP-treated peptides to the beads and incubate for 2 h at 24°C under vigorous shaking (1000 rpm) to avoid sedimentation of magnetic beads.
21. Remove the supernatant after magnetic immobilization of the beads and wash magnetic particles five times with 300 μL of 10 m*M* PBS buffered at pH 7.4.
22. Elute immunocaptured N-terminal-TMPP-labeled peptides with 200 μL of 50 m*M* HCl.
23. Dry the eluted peptides in a speed vacuum and, then dissolve them in 20 μL of 0.1% TFA.

3.3 Shotgun Mass Spectrometry and Database Search

A wide range of tandem mass spectrometers are currently available for identifying peptides with high accuracy. The use of a recent high-resolution/high-speed tandem mass spectrometer, typically a Q-Exactive HF (Thermo) instrument incorporating an ultra-high field Orbitrap analyzer or any equivalent set-up, is recommended to obtain the most comprehensive coverage of the peptide mixtures. Reverse-phase chromatographic column length and gradient time should be adapted depending on the complexity of each sample.

1. Analyze the peptide mixture(s) by nanoLC-MS/MS with optimized parameters for complex sample.
2. Search the MS/MS data using the search algorithm(s) of choice (e.g., Mascot) against the six-frame translated proteogenomic database on one hand, and the annotated genome database on the other.
3. Compare the list of open reading frames/polypeptides detected in the two database searches.
4. For each discrepant cases, compare the list of detected peptides, i.e., peptides from shotgun analysis and TMPP-peptides signing N-termini of proteins.
5. Analyze the most interesting cases of miss-annotations highlighted by conspicuous coverage discrepancies with the help of a genomic viewer and proteogenomics data integrator.
6. Whenever possible, cross-check conclusive results with additional experiments or data, e.g., protein sequence alignment, nucleic acid (DNA/RNA) sequence alignment, resequencing nucleic acid locus, alternative MS/MS searches including posttranslational modifications, quantitative data.

4. CONCLUSION

Establishing the exact number of proteins encoded in the genome, determining their precise sequences and possible maturations, and monitoring their abundancies, are important to get a comprehensive view of an organism. While genome and transcriptome sequencing is obviously currently easy to perform, their annotation remains a difficult task. Based on protein data acquired by means of next-generation proteomics, the exact number of protein-encoding genes may be established and their annotation in terms of structure may be comprehensively covered. The protocols presented in this chapter are easy to carry out and well adapted to bacteria and archaea. For eukaryotes, due to the presence of numerous introns and alternative splicing events, a more complex approach should be performed. However, the difficulty for correctly predicting splicing and alternative splicing sites can be circumvented with proteogenomics approaches.

REFERENCES

Ansong, C., Purvine, S. O., et al. (2008). Proteogenomics, needs and roles to be filled by proteomics in genome annotation. *Briefings in Functional Genomics & Proteomics*, 7(1), 50–62.

Armengaud, J. (2009). A perfect genome annotation is within reach with the proteomics and genomics alliance. *Current Opinion in Microbiology*, *12*(3), 292–300.

Armengaud, J. (2013). Microbiology and proteomics, getting the best of both worlds! *Environmental Microbiology*, *15*(1), 12–23.

Armengaud, J. (2015). Power of positive thinking in quantitative proteomics. *Proteomics*, *15*(17), 2898–2900.

Armengaud, J. (2016). Next-generation proteomics faces new challenges in environmental biotechnology. *Current Opinion in Biotechnology*, *38*, 174–182.

Armengaud, J., Bland, C., et al. (2011). Microbial proteogenomics, gaining ground with the avalanche of genome sequences. *Journal of Bacteriology & Parasitology*, *S3*, 001.

Baudet, M., Ortet, P., et al. (2010). Proteomics-based refinement of Deinococcus deserti genome annotation reveals an unwonted use of non-canonical translation initiation codons. *Molecular & Cellular Proteomics*, *9*(2), 415–426.

Berry, I. J., Steele, J. R., et al. (2016). The application of terminomics for the identification of protein start sites and proteoforms in bacteria. *Proteomics*, *16*(2), 257–272.

Bertaccini, D., Vaca, S., et al. (2013). An improved stable isotope N-terminal labeling approach with light/heavy TMPP to automate proteogenomics data validation, dN-TOP. *Journal of Proteome Research*, *12*(6), 3063–3070.

Blakeley, P., Overton, I. M., et al. (2012). Addressing statistical biases in nucleotide-derived protein databases for proteogenomic search strategies. *Journal of Proteome Research*, *11*(11), 5221–5234.

Bland, C., Bellanger, L., et al. (2014). Magnetic immunoaffinity enrichment for selective capture and MS/MS analysis of N-terminal-TMPP-labeled peptides. *Journal of Proteome Research*, *13*(2), 668–680.

Bland, C., Hartmann, E. M., et al. (2014). N-terminal-oriented proteogenomics of the marine bacterium Roseobacter denitrificans Och114 using N-succinimidyloxycarbonyl-methyl)tris(2,4,6-trimethoxyphenyl)phosphonium bromide (TMPP) labeling and diagonal chromatography. *Molecular & Cellular Proteomics*, *13*(5), 1369–1381.

Chernobrovkin, A. L., Kopylov, A. T., et al. (2015). Methionine to isothreonine conversion as a source of false discovery identifications of genetically encoded variants in proteogenomics. *Journal of Proteomics*, *120*, 169–178.

Christie-Oleza, J. A., Fernandez, B., et al. (2012). Proteomic insights into the lifestyle of an environmentally relevant marine bacterium. *The ISME Journal*, *6*(1), 124–135.

Christie-Oleza, J. A., Miotello, G., et al. (2012). High-throughput proteogenomics of Ruegeria pomeroyi, seeding a better genomic annotation for the whole marine Roseobacter clade. *BMC Genomics*, *13*, 73.

Datta, K. K., Patil, A. H., et al. (2016). Proteogenomics of Candida tropicalis—An opportunistic pathogen with importance for global health. *OMICS*, *20*(4), 239–247.

de Groot, A., Dulermo, R., et al. (2009). Alliance of proteomics and genomics to unravel the specificities of Sahara bacterium Deinococcus deserti. *PLoS Genetics*, *5*(3), e1000434.

de Groot, A., Roche, D., et al. (2014). RNA sequencing and proteogenomics reveal the importance of leaderless mRNAs in the radiation-tolerant bacterium Deinococcus deserti. *Genome Biology and Evolution*, *6*(4), 932–948.
Duriez, E., Armengaud, J., et al. (2016). Mass spectrometry for the detection of bioterrorism agents, from environmental to clinical applications. *Journal of Mass Spectrometry*, *51*(3), 183–199.
Gallien, S., Perrodou, E., et al. (2009). Ortho-proteogenomics, multiple proteomes investigation through orthology and a new MS-based protocol. *Genome Research*, *19*(1), 128–135.
Gao, Z., Wang, Z., et al. (2015). Experimental validation of Bacillus anthracis A16R proteogenomics. *Scientific Reports*, *5*, 14608.
Ghali, F., Krishna, R., et al. (2014). ProteoAnnotator—Open source proteogenomics annotation software supporting PSI standards. *Proteomics*, *14*(23-24), 2731–2741.
Gonnelli, G., Stock, M., et al. (2015). A decoy-free approach to the identification of peptides. *Journal of Proteome Research*, *14*(4), 1792–1798.
Gupta, N., Benhamida, J., et al. (2008). Comparative proteogenomics, combining mass spectrometry and comparative genomics to analyze multiple genomes. *Genome Research*, *18*(7), 1133–1142.
Hartmann, E. M., Allain, F., et al. (2014). Taking the shortcut for high-throughput shotgun proteomic analysis of bacteria. *Methods in Molecular Biology*, *1197*, 275–285.
Hartmann, E. M., & Armengaud, J. (2014). N-terminomics and proteogenomics, getting off to a good start. *Proteomics*, *14*(23-24), 2637–2646.
Helmy, M., Sugiyama, N., et al. (2012). Mass spectrum sequential subtraction speeds up searching large peptide MS/MS spectra datasets against large nucleotide databases for proteogenomics. *Genes to Cells*, *17*(8), 633–644.
Jaffe, J. D., Berg, H. C., et al. (2004). Proteogenomic mapping as a complementary method to perform genome annotation. *Proteomics*, *4*(1), 59–77.
Jagtap, P., Goslinga, J., et al. (2013). A two-step database search method improves sensitivity in peptide sequence matches for metaproteomics and proteogenomics studies. *Proteomics*, *13*(8), 1352–1357.
Jagtap, P. D., Johnson, J. E., et al. (2014). Flexible and accessible workflows for improved proteogenomic analysis using the Galaxy framework. *Journal of Proteome Research*, *13*(12), 5898–5908.
Koch, A., Gawron, D., et al. (2014). A proteogenomics approach integrating proteomics and ribosome profiling increases the efficiency of protein identification and enables the discovery of alternative translation start sites. *Proteomics*, *14*(23-24), 2688–2698.
Krishna, R., Xia, D., et al. (2015). A large-scale proteogenomics study of apicomplexan pathogens—Toxoplasma gondii and Neospora caninum. *Proteomics*, *15*(15), 2618–2628.
Kumar D., Bansal G., et al. (2016). Integrating transcriptome and proteome profiling, strategies and applications, *Proteomics*, in press.
Kumar, D., Mondal, A. K., et al. (2014). Discovery of rare protein-coding genes in model methylotroph Methylobacterium extorquens AM1. *Proteomics*, *14*(23-24), 2790–2794.
Kumar, D., Mondal, A. K., et al. (2016). Proteogenomics of rare taxonomic phyla, a prospective treasure trove of protein coding genes. *Proteomics*, *16*(2), 226–240.
Kumar, D., Yadav, A. K., et al. (2013). Proteogenomic analysis of Bradyrhizobium japonicum USDA110 using GenoSuite, an automated multi-algorithmic pipeline. *Molecular & Cellular Proteomics*, *12*(11), 3388–3397.
Mao, Y., Yang, X., et al. (2016). Reannotation of Yersinia pestis strain 91001 based on omics data. *The American Journal of Tropical Medicine and Hygiene*, *95*, 562–570.
Menschaert, G., & Fenyo D. (2016). Proteogenomics from a bioinformatics angle: A growing field, *Mass Spectrometry Reviews*, in press.

Menschaert, G., Van Criekinge, W., et al. (2013). Deep proteome coverage based on ribosome profiling aids mass spectrometry-based protein and peptide discovery and provides evidence of alternative translation products and near-cognate translation initiation events. *Molecular & Cellular Proteomics*, *12*(7), 1780–1790.

Nagaraj, S. H., Waddell, N., et al. (2015). PGTools, a software suite for proteogenomic data analysis and visualization. *Journal of Proteome Research*, *14*(5), 2255–2266.

Nesvizhskii, A. I. (2014). Proteogenomics, concepts, applications and computational strategies. *Nature Methods*, *11*(11), 1114–1125.

Pettersen, V. K., Steinsland, H., et al. (2015). Improving genome annotation of enterotoxigenic Escherichia coli TW10598 by a label-free quantitative MS/MS approach. *Proteomics*, *15*(22), 3826–3834.

Pible, O., & Armengaud, J. (2015). Improving the quality of genome, protein sequence, and taxonomy databases, a prerequisite for microbiome metaomics 2.0. *Proteomics*, *15*(20), 3418–3423.

Potgieter, M. G., Nakedi, K. C., et al. (2016). Proteogenomic analysis of Mycobacterium smegmatis using high resolution mass spectrometry. *Frontiers in Microbiology*, *7*, 427.

Renuse, S., Chaerkady, R., et al. (2011). Proteogenomics. *Proteomics*, *11*(4), 620–630.

Richards, A. L., Hebert, A. S., et al. (2015). One-hour proteome analysis in yeast. *Nature Protocols*, *10*(5), 701–714.

Shen, H., An, M., et al. (2015). Evaluation of the accuracy of protein quantification using isotope TMPP-labeled peptides. *Proteomics*, *15*(17), 2903–2909.

Syme, R. A., Tan, K. C., et al. (2016). Comprehensive annotation of the Parastagonospora nodorum reference genome using next-generation genomics, transcriptomics and proteogenomics. *PloS One*, *11*(2), e0147221.

Tovchigrechko, A., Venepally, P., et al. (2014). PGP, parallel prokaryotic proteogenomics pipeline for MPI clusters, high-throughput batch clusters and multicore workstations. *Bioinformatics*, *30*(10), 1469–1470.

Uszkoreit, J., Plohnke, N., et al. (2014). The bacterial proteogenomic pipeline. *BMC Genomics*, *15*(Suppl. 9), S19.

Venter, E., Smith, R. D., et al. (2011). Proteogenomic analysis of bacteria and archaea, a 46 organism case study. *PloS One*, *6*(11), e27587.

Wang, X., Slebos, R. J., et al. (2016). ProBAMsuite, a bioinformatics framework for genome-based representation and analysis of proteomics data. *Molecular & Cellular Proteomics*, *15*(3), 1164–1175.

Wright, J. C., & Choudhary, J. S. (2016). DecoyPyrat, fast non-redundant hybrid decoy sequence generation for large scale proteomics. *Journal of Proteomics & Bioinformatics*, *9*(6), 176–180.

Xu, X., Liu, T., et al. (2015). Proteogenomic analysis of Trichophyton rubrum aided by RNA sequencing. *Journal of Proteome Research*, *14*(5), 2207–2218.

Yamazaki, S., Yamazaki, J., et al. (2006). Proteome analysis of an aerobic hyperthermophilic crenarchaeon, Aeropyrum pernix K1. *Molecular & Cellular Proteomics*, *5*(5), 811–823.

Zhang, K., Fu, Y., et al. (2015). A note on the false discovery rate of novel peptides in proteogenomics. *Bioinformatics*, *31*(20), 3249–3253.

Zickmann, F., & Renard, B. Y. (2015). MSProGene, integrative proteogenomics beyond six-frames and single nucleotide polymorphisms. *Bioinformatics*, *31*(12), i106–i115.

CHAPTER THIRTEEN

Proteogenomics: Recycling Public Data to Improve Genome Annotations

A. McAfee, L.J. Foster[1]
Michael Smith Laboratories, University of British Columbia, Vancouver, BC, Canada
[1]Corresponding author: e-mail address: foster@chibi.ubc.ca

Contents

Methods in Enzymology, Volume 585
ISSN 0076-6879
http://dx.doi.org/10.1016/bs.mie.2016.09.020

Abstract

Massively parallel sequencing is revealing species genomes faster than ever before, but the value of the raw sequence is limited unless the genes can be accurately annotated. This is typically achieved using gene prediction algorithms which, despite continual improvement, still require substantial verification and refinement. For example, in silico methods struggle with annotating splice isoforms accurately and empirical methods are needed to refine and verify the initial bioinformatic gene predictions. RNA-seq is an excellent way to confirm exon–exon boundaries and transcript termini, while mass spectrometry (MS) offers definitive proof that a gene is translated and a secondary means of confirming exon expression, protein termini, and posttranslational modifications. Furthermore, both methods can potentially identify entirely novel genes that were missed by conventional gene predictors. This chapter describes a proteogenomics procedure using information from the proteome, transcriptome, and genome—thus utilizing each component of the central dogma—to annotate genetic elements in eukaryotes. We also discuss gene modeling, integration of RNA-seq and MS data, minimizing false discoveries, proteogenomics software, functional annotation, and sequence validation. We hope that the procedure described here will assist efforts to annotate the genomes of newly sequenced species, as well as sharpen those that have been annotated in the past.

1. INTRODUCTION

Genome sequencing technology has matured substantially over the past 15 years. It took 9 years and about $3 billion to generate the sequence data for the first complete human genome (Venter et al., 2001); now, sequencing and assembly takes a few days at six orders of magnitude lower cost (Miller et al., 2015). NCBI (https://www.ncbi.nlm.nih.gov/genome/) now houses the genomes of thousands of species (Fig. 1). Clearly, generating sequence data is no longer a bottleneck, but annotating the genomes remains cumbersome and experimental validation is often necessary.

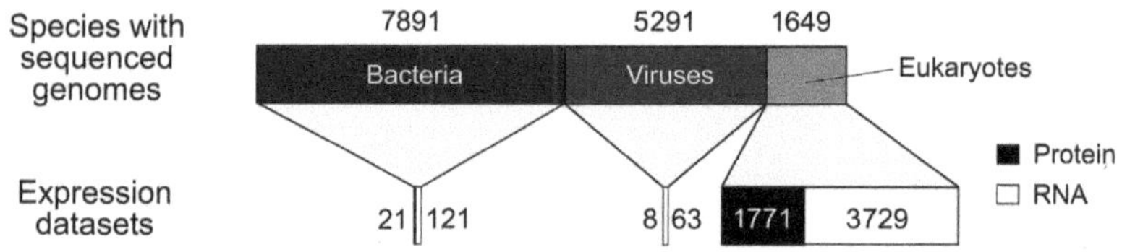

Fig. 1 Summary of publically available data. Genome, transcript, and protein data sources were NCBI genomes (http://www.ncbi.nlm.nih.gov/genome/), the Gene Expression Omnibus (http://www.ncbi.nlm.nih.gov/geo/), and ProteomeXchange (http://www.proteomexchange.org/), respectively, on February 12, 2016.

Initial annotation drafts are almost always produced by gene prediction algorithms (Wilming & Harrow, 2009; Yandell & Ence, 2012; Fig. 2), but this approach has inherent limitations (Brent, 2007). For example, the algorithms are generally not able to accurately predict noncanonical splice junctions, short genes, alternative translational start sites, alternative reading frames (ARFs), or noncoding RNAs (ncRNAs; Korf, 2004; Zickmann & Renard, 2015). Furthermore, these in silico predictions tend to have low sensitivity and specificity due to the frequencies of split genes, merged genes, overzealous predictions, and missing isoforms (Goodswen, Kennedy, & Ellis, 2012). Efforts have been made to combine gene prediction software in order to improve the accuracy of gene predictions (Zickmann & Renard, 2015); however, this approach still propagates problems common across the underlying algorithms.

Many algorithms have been optimized for species-specific genomic properties, such as codon bias, consensus splice sites, and exon length, but perform poorly on "foreign" genomes (see Brent, 2008; Korf, 2004 for more detailed discussions). For example, specialized algorithms are needed to accommodate AT-rich genomes, which tend to have an A or a T in the third nucleotide position more frequently than in GC-rich genomes, creating a skewed distribution of nucleotide composition in protein-coding regions. GC content can vary from 19% to 67% in protists and from 17% to 58%

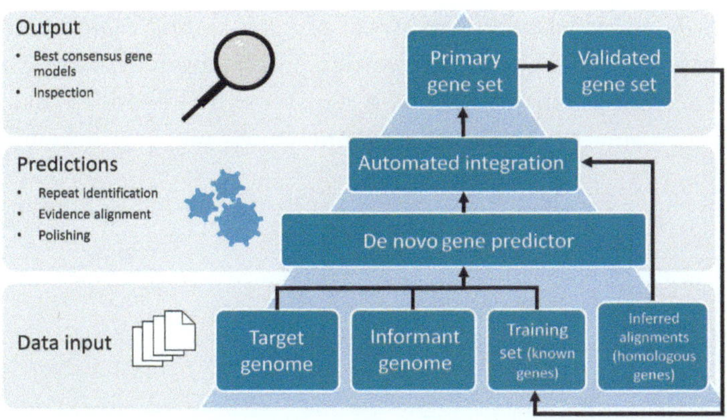

Fig. 2 Schematic of an in silico gene annotation workflow. "Target" refers to the species whose genome is being annotated. "Informant" refers to a species with a well-annotated genome that has similar properties to the target genome. The training gene set consists of known genes of the target or informant species and is used to train the prediction algorithm to identify gene features, whereas inferred alignments are homologous genes supporting the de novo predicted genes. Finally, the validated genes can be used as the training gene set in further annotation iterations.

in animals (NCBI Genomes). Some gene-finding algorithms are specifically trained on each new genome (Korf, 2004), mitigating this problem. However, even this method has its biases: it was developed for compact genomes and even the author states that "the techniques here may not work well in mammalian genomes." To work well, this approach also requires at least one well-annotated genome with similar properties to the target genome, which is not guaranteed with our newfound ability to sequence any species of interest. For example, several efforts have been made to improve the genome annotation of the honey bee (*Apis mellifera*; Elsik et al., 2014; McAfee et al., 2016; Robertson et al., 2007; Weinstock et al., 2006), but this has been challenging in part because the most closely related well-annotated organism is *Drosophila melanogaster*, whose genome is 40% smaller and has 30% higher GC content (Keightley et al., 2009). In fact, it is an ongoing challenge to annotate the honey bee genome because it is unusually AT rich (Elsik et al., 2014). For all these reasons, experimental validation, inspection, and curation of gene sets are always necessary but impractical to do manually, calling for a high-throughput method.

Proteogenomics is a data-driven annotation scheme that utilizes high-throughput data from every level of the central dogma to confirm or edit existing gene models as part of an iterative refinement process (Nesvizhskii, 2014; Fig. 3). An excellent example of proteogenomic refinement of the already highly scrutinized human gene set comes from Wilhelm et al. (2014), who identified 193 novel protein-coding regions, 210 novel protein termini, and 40 cases of exon extensions using mass spectrometry data. Furthermore, Kim et al. (2014) also identified peptides matching to 140 regions formerly annotated as pseudogenes and 9 as noncoding (nc)

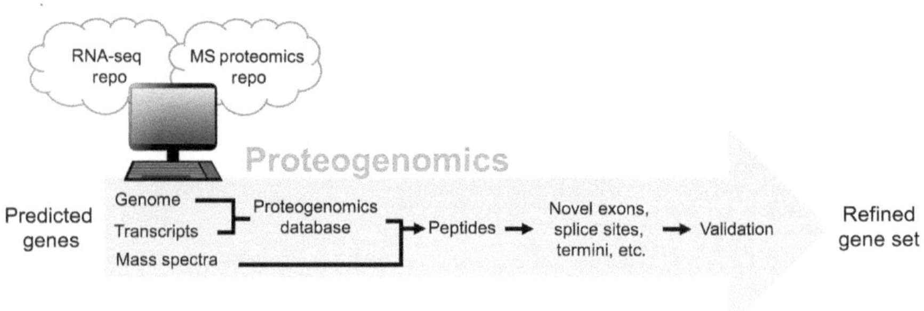

Fig. 3 Schematic of the proteogenomics workflow. Input data may be generated or downloaded from public repositories. Mass spectrometry data is searched against the proteogenomic protein database to identify novel peptides, which indicate unannotated gene features.

RNAs. Proteogenomics has also been used to refine the annotations of several model organisms (Brosch et al., 2011; Brunner et al., 2007; Castellana et al., 2008; McAfee et al., 2016; Merrihew et al., 2008). These projects demonstrate that proteomics and transcriptomics data are complementary and they should be used together to create reliable gene models.

RNA-seq dominates transcriptomic studies because it is more robust, more accurate, and allows deeper coverage than microarrays (Ozsolak & Milos, 2011). However, proteomics is increasingly effective as a complementary means to tackle these issues (for reviews, see Walther & Mann, 2010; Zhang, Fonslow, Shan, Baek, & Yates, 2013). Proteins are at the end point of the central dogma and so detection of a protein is the ultimate validation that the corresponding region of the genome expresses a useful gene product. While it is not yet feasible to detect all proteins in a single proteomics experiment, the sheer amount of proteomic data that is publically available can be used to divulge significant annotation improvements for nonmodel and model organisms alike.

Annotation is an iterative process and public data repositories offer a fantastic opportunity to combine the data from many research groups to upgrade annotations. The push to make raw 'omics data available upon publication has resulted in 3913 GEO datasets (the majority of which are high-throughput sequencing) and 1800 ProteomeXchange projects (MS), each of which can be associated with hundreds of raw data files (Fig. 1). Producing this data requires specialized instrumentation and expertise, but surprisingly, this wealth of data appears to be under-utilized (Lynch & Moore, 2016). While there are potentially endless ways to exploit public data, annotation refinement by proteogenomics is an attractive technique because it benefits the community and can be performed without specialized expertise or spending additional resources on wet lab experiments. Here we provide an outline of how to recycle public data to perform proteogenomics on eukaryotic species.

2. DATA

2.1 Availability

Proteogenomics can be applied to species in all kingdoms of life, whether the genome is already well-annotated or not. We will focus on eukaryotes because that is where the bulk of proteomics and RNA-seq data is directed and eukaryotic genes are typically more challenging to annotate. Data for humans and the most common model organisms make up the majority of online datasets; however, there is an increasing need to annotate newly sequenced and under-studied species. For species that have insufficient data, it may be necessary to generate new proteomics datasets and we direct readers to a review of bottom-up shot-gun proteomics techniques (Zhang et al., 2013). We highly recommend including RNA-seq data in proteogenomic workflows, but if data is not available and not feasible to produce (see Hou et al., 2015; Trapnell et al., 2012 for methods), a large amount of information can still be gained using just proteomics data and the genome sequence (Risk, Spitzer, & Giddings, 2013). Whether downloaded or produced de novo, expression data should ideally have organ- and tissue-level resolution, cross a range of life stages and environmental conditions, and include samples from both sexes. This is important because tissue-specific expression is widespread: in humans, 21% of expressed proteins are tissue-specific (enriched at least fivefold in one tissue or a small group of tissues; Fagerberg et al., 2014) and almost 65% of detectable genes with three or more exons contain a tissue-specific splice isoform (Hestand, Zeng, Coleman, Liu, & MacLeod, 2015). In addition, the genome sequence can impact the annotation process immensely: if the completeness is poor, efforts to improve the reference sequence (see Soon, Hariharan, and Snyder (2013) for a review on genome sequencing technologies) should precede annotation refinement. From here on, we assume that at least a high-quality genome and a rich proteomics data set are available.

2.2 Guide to Data Repositories: Mass Spectrometry

The two major hubs for proteome data are ProteomeXchange (http://www.proteomexchange.org) and Peptide Atlas (http://www.peptideatlas.org), which have some content overlap. The repositories contain data from many different types of mass spectrometry instruments and as a result, the raw data come in a variety of file formats (e.g., mgf, mzXML, mzML, raw, baf, etc.). Depending on the choice of downstream analysis software, it may be necessary to convert data from one file type to another, e.g., using the open source MSconvert tool (ProteoWizard toolkit; Chambers et al., 2012).

2.3 Guide to Data Repositories: RNA-seq

Compared to mass spectrometry, there are far more repositories for RNA-seq data. Some of the largest are the Gene Expression Omnibus (GEO), the Sequence Read Archive (SRA), European Nucleotide Archive (ENA), and the DNA Databank of Japan (DDBJ). The SRA, ENA, and DDBJ are all part of the International Nucleotide Sequence Database Collaboration and their contents are regularly synced, so there is no advantage to obtaining data from more than one of these repositories. The SRA houses raw high-throughput sequencing data, as well as aligned data. However, once data is aligned, sequences that do not match with current gene annotations are often automatically removed, although these may be the most informative of all in a proteogenomics experiment. It is less risky, therefore, to de novo assemble RNA-seq data (see Steijger et al., 2013 for a systematic review of assembly methods) and incorporate this sequence information into the proteogenomics workflow (Fig. 3). Long read RNA-seq data are especially informative for accurate isoform annotation, while short reads can convolute the pattern of exon connectivity (Cho et al., 2014).

3. DATABASES

3.1 The Proteogenomics Database

The underlying theme of proteogenomics is that mass spectrometry-based proteomics data harbors untapped peptide sequence information. In bottom-up proteomics, proteins are digested into peptides with an enzyme (usually trypsin) which are subsequently separated by liquid chromatography and analyzed by the mass spectrometer. Normally, mass spectrometry data is searched against a database of known proteins to identify sequences whose theoretical spectra match the observed spectra. However, this makes it impossible to identify novel sequences. Therefore, a new protein database—referred to as the "proteogenomics database"—must be synthesized in order to identify novel protein sequences. This database contains suspected coding sequences gleaned from one or more of the following sources: a six-frame genome translation, a three-frame transcriptome translation, splice graphs, and occasionally, sequences suspected to have been deleted from previous annotations in error. In the following sections, the assets and limitations of these sequence sources are discussed.

3.2 Producing a Transcript Database

Analyzing large amounts of RNA-seq data to identify novel features is a cumbersome task, but the benefits can be significant (Li et al., 2011). Approximately 25% of tryptic peptides are predicted to span exon–exon junctions (Tanner et al., 2007), so complete knowledge of splice isoforms can substantially increase the proportion of matched spectra. Furthermore, nonsynonymous single-nucleotide polymorphisms (nsSNPs) are usually not included in protein databases, making their polymorphic peptides impossible to match (Schandorff et al., 2007). Krasnov et al. (2015) describe an automated method (PPLine) to achieve one of the major aims in a proteogenomic workflow: pulling out splice isoforms, SNPs, and ARFs from raw RNA-seq data to create a customized transcriptome database. Amazingly, Krasnov et al. uncovered more novel gene features than what was already described by two previous comprehensive human proteome studies (Kim et al., 2014; Wilhelm et al., 2014), including 659 novel SNPs, 475 novel liver proteoforms, and 17 proteins derived from translated ARFs. Zickmann and Renard (2015) offer a similar approach with their software, MSProGene. Both of these are excellent ways to integrate RNA-seq data into the proteogenomic workflow by creating customized transcriptome databases.

3.3 Six-Frame and Three-Frame Translations

Protein databases for proteogenomics are typically larger than those used in conventional proteomic searches because they cast a wide net to include many potentially expressed sequences, rather than only known proteins (basic principles are outlined in Yates, Eng, and McCormack (1995); Fig. 4). For example, the genome translation is meant to catch every potential coding region contained within the genome above a certain cutoff length. The translation algorithm should not require a start codon (so as

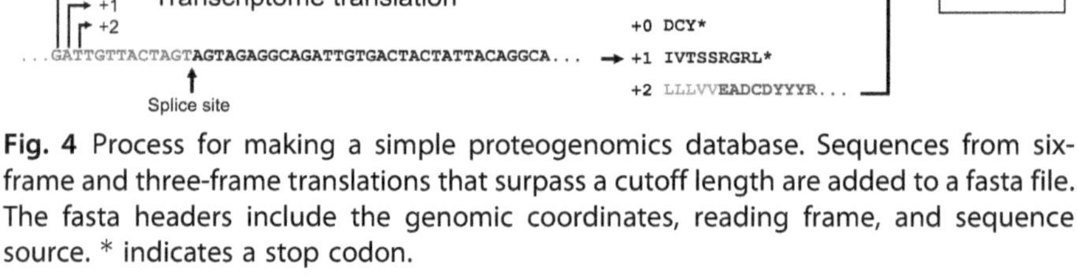

Fig. 4 Process for making a simple proteogenomics database. Sequences from six-frame and three-frame translations that surpass a cutoff length are added to a fasta file. The fasta headers include the genomic coordinates, reading frame, and sequence source. * indicates a stop codon.

to not miss subsequent exons) and just translate each codon based on the genetic code. The size of the resulting database will depend on the minimum sequence length cutoff: a liberal cutoff of 90–120 bases is typically chosen (Chan, Parker, Sun, Deutsch, & Foster, 2011; Wilhelm et al., 2014) because exons shorter than this are rare. The median human exon length is approximately 150 bases (Brent, 2008) and only ~5% of exons are shorter than 129 bases (Gudlaugsdottir, Boswell, Wood, & Ma, 2007), but this will vary somewhat from species to species. Short genes and exons are some of the most often missed during initial gene annotations (in fact, many gene prediction algorithms discard open reading frames that are less than 300 bases; Andrews & Rothnagel, 2014) and may be those best targeted by a proteogenomics experiment. We do not recommend using a cutoff of less than 90 bases because the number of false sequences this generates is high and interferes with the spectral search algorithm's ability to accurately assess the FDR (Krug et al., 2013). Since there will always be some proportion of false sequences incorporated into the search database by this approach, it is always important to confirm novel sequences by at least one secondary means (Section 6).

The three-frame transcriptome translation database (based on the de novo transcript assemblies described in Section 3.2) contains information that overlaps almost entirely with the genome translation but the great benefit to it is that the resulting search space is reduced. It is also more straightforward to identify ARFs, confirm known splice junctions, identify new ones, and find transcripts misannotated as ncRNA. If a genome translation is also being used, it should be combined with the transcriptome translation into a single, nonredundant concatenated database (similar to the approach by Ghali et al., 2014) in order to reduce additive false discoveries incurred by multiple database searches.

3.4 Splice Graphs

Splice graphs are ideal for identifying new exon–exon junctions. A splice graph encodes all possible exon paths to make a transcript, making it possible to use the database to identify peptides formed by previously unobserved exon skipping events (Fig. 5). A tool which is part of the larger proteogenomic-specific software, Enosi, has the capability of producing splice graphs from an input of aligned RNA-seq reads (Woo et al., 2013). The output from this tool is in FASTA format; therefore, this database

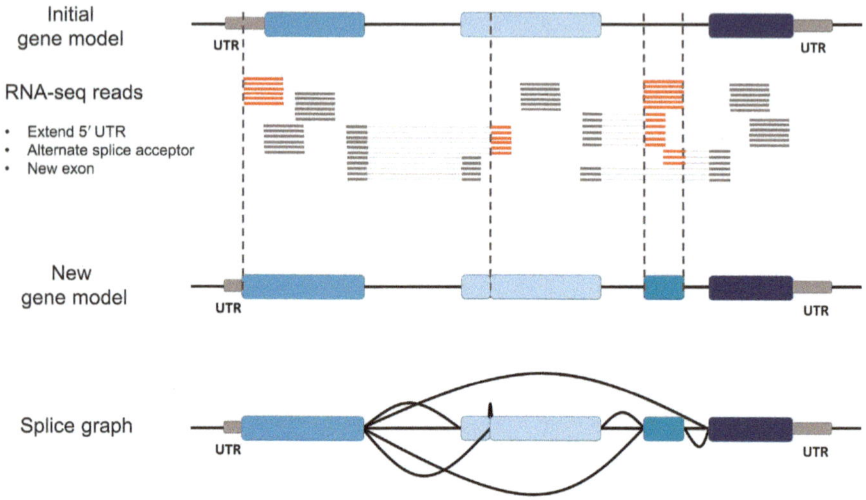

Fig. 5 RNA-seq reads are used to modify gene models. *Red* (*dark gray* in the print version) RNA-seq reads represent reads that indicate transcribed regions not included in the initial gene model. This information is integrated to produce a new gene model, splice graphs of which can further identify unpredicted exon junctions when combined with secondary data (often mass spectrometry).

can be used with any conventional MS/MS search algorithm. However, this particular tool was aimed at identifying only peptides that span the exon–exon junction, not to achieve maximum exon coverage, so a large amount of sequence information was omitted in order to compress the file size. A newer version of the tool (Castellana et al., 2014) includes all sequence information in the search database, but is computationally expensive.

3.5 Incorporating SNPs Into a Protein Database

Knowledge of nonsynonymous SNPs (nsSNPs) is only useful in a proteogenomics workflow if they are included in a protein database but this is not yet standard practice. Including these sequences has the benefit of not only confirming the polymorphism but also accounting for a fraction of the high-quality unmatched spectra. In order to minimize search space expansion, Schandorff et al. (2007) describe a technique that incorporates peptides corresponding to the nsSNP by concatenating them onto the associated protein sequence with a nonamino acid character separating them (Fig. 6). The authors include the associated scripts as supplementary information (http://dx.doi.org/10.1016/bs.mie.2016.09.020).

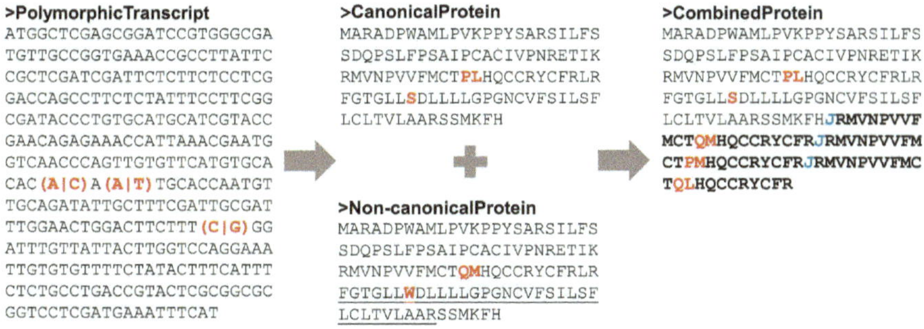

Fig. 6 A minimalist strategy for including nonsynonymous polymorphisms in a protein database. Transcripts containing one or more SNPs are in silico translated, then all possible polymorphic peptides plus flanking peptides are tagged to the canonical protein sequence with a nonamino acid character (e.g., "J"). The enzyme "TrypsinMSIPI" in Mascot is programmed to cleave with trypsin specificity plus C- and N-terminal to J. The *underlined* region represents a peptide that, although polymorphic, would not be included because it is too long to normally be detected by mass spectrometry.

4. PEPTIDE AND PROTEIN IDENTIFICATION

4.1 De Novo Peptide Identification Software

Computer algorithms are necessary for identifying peptides from mass spectra, and there is a multitude of programs to choose from. PEAKS is a popular program for identifying peptides de novo (i.e., without accessing a theoretical protein database) and is particularly useful for species whose genome has not been sequenced (Ma et al., 2003). Some other programs used for the same purpose include PepNovo (Frank & Pevzner, 2005) and AUDENS (Grossmann et al., 2005). However, de novo peptide identification algorithms generally suffer from low sensitivity since very high-quality spectra are necessary for confident peptide sequencing. Another software, ByOnic, uses a hybrid search strategy involving some de novo analysis combined with a conventional database search (Bern, Cai, & Goldberg, 2007). It demonstrates better sensitivity than several classical identification algorithms but it has not become widespread.

4.2 Database-Dependent Peptide Identification Software

Some of the most widely used classical spectra-matching algorithms (i.e., those that rely on protein databases) include MaxQuant (Cox & Mann, 2008; Cox et al., 2009), SEQUEST (Eng, McCormack, & Yates, 1994),

Mascot (Perkins, Pappin, Creasy, & Cottrell, 1999), Comet (Eng, Jahan, & Hoopmann, 2013), and X!tandem (Craig & Beavis, 2004). Each one has its own licensing guidelines, acceptable data files, peak-picking algorithms, and search options; therefore, the user must identify which is the most compatible with the chosen data set(s) and downstream tools. MaxQuant is highly suited to proteogenomics work owing to its highly flexible search options and ability to execute a single search on thousands of data files with different parameters while maintaining a global FDR cutoff. No matter what identification software is used, it is desirable to be able to sort out the unmatched spectra from the matched ones so they can be used in downstream analyses.

4.3 Identifying Posttranslational Modifications

A large fraction of high-quality unmatched spectra often originate from peptides with unanticipated posttranslational modifications (PTMs; Chick et al., 2015). Even if the type of modification is unknown, these spectra can be matched using "error tolerant" or "dependent peptide" functions. Mascot and MaxQuant, e.g., employ these by performing a conventional first search to identify unmodified peptides, then looking for other spectra that might have come from the same peptides with a mass shift corresponding to any of the ~1300 known potential PTMs. When selecting a search engine, a useful feature to consider is the ability to export the unmatched spectra, e.g., de novo peptide identification or searching against a new database. Amino acid substitutions are usually included in the database of potential PTMs; therefore, this approach could be used to confirm nsSNPs that were identified by the Schandorff et al. method (Section 3.5).

5. PROTEOGENOMICS TOOLKITS

5.1 Top-Down Pipelines

Top-down proteomics analyzes intact proteins, which is not commonly performed due to the lack of available tools. At the time of writing, there is only one software program available for top-down proteogenomic analysis: SpectroGene (Kolmogorov, Liu, & Pevzner, 2015). However, since publically available mass spectrometry data is generated almost entirely using the far more established bottom-up approach, SpectroGene will likely have little utility until more top-down proteomics data is produced.

5.2 Bottom-Up Pipelines

Proteogenomics is a growing field and there is an increasing demand for automated tools and pipelines to process the data. Many proteogenomic toolkits are available (see Table 1), and the popular cloud-based data analysis server, Galaxy, has a 140-step workflow for performing the proteogenomic analysis on their platform (Jagtap et al., 2014). There is not one dominant analysis method, though, at least in part because each requires different input data files or formats, and these are not always available for a given species. Sophisticated tools include PGx (Askenazi, Ruggles, & Fenyö, 2016) and PGTools (Nagaraj et al., 2015), which includes a user-friendly graphical interface, an in-line input data format conversion tool and data visualization capabilities. PGTools requires only the mass spectrometry user-input data and includes built-in customized protein databases that are produced using the information available on ENSEMBL (SpliceDB, UTRDB, and SixFrameDB), COSMIC (MutationDB and FusionDB), and NONCODE DB (NonCodeDB). However, these customized databases have only been created for *Homo sapiens*, which is a significant limitation of the software. The authors indicated that they aim to expand the species selection in the future.

Enosi (Castellana et al., 2014; Woo et al., 2013) is more flexible than PGTools and requires only mass spectrometry data and a genome sequence (although it assumes an ab initio predicted gene set is available). Enosi also controls false discoveries using a sophisticated probabilistic scoring system (Castellana et al., 2014). This tool has already undergone two major rounds of development (v0.1 and v1.0) and is probably the most flexible and statistically advanced programs of its kind.

Current proteogenomic workflow tools are very useful for their intended applications but are still somewhat case-specific. In the future, a proteogenomics tool should be created that: (a) employs a user-friendly interface (most biologists are still not adept at a command line interface); (b) accepts a genome sequence, known protein database (.fasta), spectral data files (.raw, .d, .mgf, .mzXML, etc.), and raw RNA-seq data files as inputs; (c) integrates results with genome browser tracks for convenient visualization; (d) feeds newly identified exon sequences directly into a secondary validation tool such as BLAST (Section 6.1) or InterProScan (Section 7.1) and; (e) assigns each peptide and protein sequence a confidence score similar to that implemented in Enosi (Castellana et al., 2014). Until a broadly applicable tool such as this is created, researchers employing proteogenomics will need to piece together different software components in a custom workflow to achieve all aspects of a proteogenomic experiment, especially since none of the current software includes automatic in silico secondary validation.

Table 1 Proteogenomics Software

Software[a]	Input Files	Benefits	Interface	Search Algorithm	Limitations
Peppy[b]	Genome sequence, MS/MS data	Fast, automatic loci mapping, sophisticated PSM scoring	Command line	Custom	No transcript data
Enosi[c]	Genome sequence, mapped RNA-seq reads (.sam), MS/MS data	Automatic novel feature identification, splice-graph compression	Command line	Any (MS-GF+ recommended)	Does not map RNA-seq reads
ProteoAnnotator[d]	Reference transcript coordinates (.gff3), known protein database, MS/MS data	Incorporates proteomics standards initiative	GUI	Omssa, X!tandem	May need to map RNA-seq data independently
PGTools[e]	MS/MS data	Scalable	GUI	MS-GF+, X!tandem, Comet	Only human data
PPLine[f]	Genome sequence, raw RNA-seq data	Raw RNA-seq data processing, incorporates SNP evidence	Command line	N/A	No MS data
SpectroGene[g]	Genome sequence, top-down spectral files	Only top-down proteogenomics tool	Command line	TopPIC	Licensed, no RNA-seq data

NextSearch[h]	Genome sequence, Ensembl reference transcript database (.gtf), MS/MS data	Identifies exon skipping	Command line	MS-GF+	Does not utilize RNA-seq data, only 69 species housed in Ensembl
MSProGene[i]	Genome sequence, raw RNA-Seq data, or customized evidence files, MS/MS data	Reduced search space	Command line	Adaptable (but MS-GF+ is suggested)	Unable to identify ARFs, advanced computational skills required

[a]All are compatible with Unix/Linux.
[b]Risk et al. (2013).
[c]Castellana et al. (2014).
[d]Ghali et al. (2014).
[e]Nagaraj et al. (2015).
[f]Krasnov et al. (2015).
[g]Kolmogorov et al. (2015).
[h]Kim, Park, and Paek (2015).
[i]Zickmann and Renard (2015).

5.3 Custom Workflows

Custom proteogenomics workflows may be required if there are no established pipelines that can accommodate the type of data that is available for the target species, file formats, or computational resources. For example, NextSearch (Table 1) is a sophisticated proteogenomics tool, but it relies on a relatively species-limited Ensembl transcriptome database. Furthermore, we mentioned in Section 4.2 that MaxQuant is a desirable search engine, but its outputs are not directly compatible with the tools discussed thus far. However, as long as the key elements of mapping peptides to genomic loci, identifying false positives, and secondary sequence validation are addressed, a custom workflow is an acceptable alternative to an established pipeline. This approach is aided immensely by basic knowledge of a procedural programming language such as Perl (https://www.perl.org) or Python (https://www.python.org). We have discussed how to obtain public data (Section 2), the elements of a proteogenomics search database (Section 3), as well as software used for matching peptide spectra (Section 4). We will now address some specifics about the proteogenomics search and how the results from such an MS/MS data search can be analyzed.

First, the proteomics data should be searched against the known protein database and matches removed (since they are not novel), while the unmatched spectra are exported for use in subsequent analyses. This will decrease the computational load and eliminate potential false matches of "real" peptide spectra to similar but incorrect sequences in the proteogenomic database. Proteome Discoverer (v1.4 and higher) allows users to set up and run searches with Mascot and SEQUEST, exporting unmatched spectra to produce a new data file that can be used as input for another algorithm (e.g., PEAKS) or reused for a search using a different database. Unfortunately, Proteome Discoverer cannot integrate many of the other common and open source/access search algorithms, including MaxQuant and X!tandem. SpectrumMill software will also export unmatched spectra, but not all search engines have this function, in which case the undepleted dataset must be used.

If matched and unmatched spectra cannot be separated, the workflow must involve a comparison between known protein sequences and the peptides identified in the proteogenomic database to identify the novel matches. Simple scripts can be written that compare the peptide sequences identified in the proteogenomic search to an in silico digestion of the known protein database and output the novel peptides, their parent protein sequences, genomic coordinates (best if encoded in the protein fasta header), and number of peptides per protein. Each search algorithm outputs results in slightly different formats, but in order to illustrate how one might extract genomic loci from a peptide result file, we have provided a sample script that will accomplish this task (Supplementary File 1 (http://dx.doi.org/10.1016/bs.mie.2016.09.020)). The script will accept the "peptides.txt" MaxQuant result files and output the loci in general feature format (GFF). This type of file is compatible with most genome browsers, facilitating manual data inspection. When there are multiple peptides per protein sequence, it can also be useful to trim the protein sequence to span only the region across which peptides are identified since this is really the only region the evidence can support.

One of the most challenging (and most overlooked) tasks when employing a custom workflow is controlling the number of false positives (Fig. 7). A typical proteogenomics database will inevitably contain many sequences that are not really translated, which decreases the accuracy of a conventional target-decoy FDR estimation in the same way that adding a multitude of random sequences to the target protein database would (Fig. 6). An effective way of overcoming this problem is through applying a combination of downstream filtering steps to the "novel" peptides based on (a) some probabilistic scoring method, (b) supporting the novel sequence with one or more lines of orthogonal evidence (Section 6), or (c) both.

Enosi (Castellana et al., 2014) uses a sophisticated but transposable method of statistically pruning its list of novel peptides. It clusters the peptides based on local spatial proximity to one another, then sorts these clusters into different "Events" to which different probabilistic scoring mechanisms were applied. These include "Novel Gene Events," "Distal Events" (gene extensions and novel coding antisense strands), and "Proximal Events" (new exons, overlapping reading frames, and exon junctions). Novel Gene Events, Proximal Events, and Distal Events occur in intergenic regions, near annotated genes, and overlapping with annotated genes, respectively. These diverse categories all fall under the umbrella of novel peptides, but applying the same scoring mechanism to all events would clearly be detrimental to some. One common scoring system is the two-hit model, where sequences with more than one observed peptide are given higher confidence than singleton sequences. This system is effective for Novel Gene Events, where multiple supporting peptides are expected if it is a true-positive result. However, more subtle events like exon extensions may not have more than a

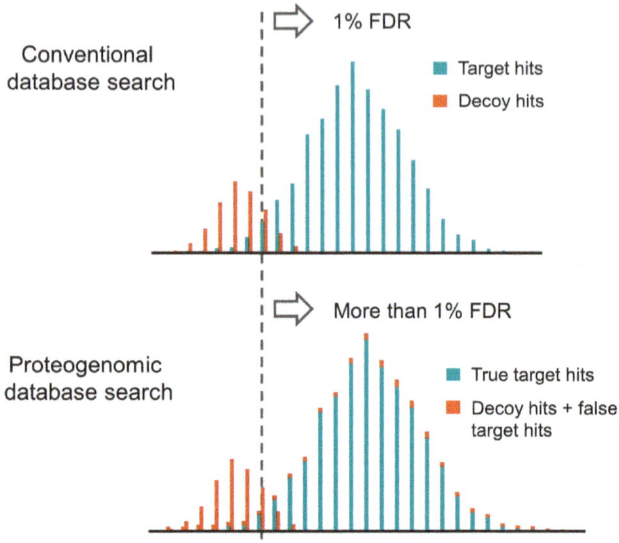

Fig. 7 Illustration depicting FDR inflation. Conventional proteomics searches using target-decoy methods of FDR calculation accurately estimate the true FDR. Proteogenomics searches, where the target database contains many false sequences to begin with, will produce an FDR that is inflated from what is calculated. *Red* (*dark gray* in the print version) *tips* on *blue* (*light gray* in the print version) *bars* represent invisible false positives.

single representative peptide. In these cases, the two-hit approach will render it difficult or impossible to identify events involving short sequences. For this reason, Castellana et al. (2014) introduce the idea of *eventProb*, which also takes into account the quality of the spectral match(es) and show that this method works well for Proximal Events, whereas the two-hit method is more suitable for Distal and Novel Gene Events. It is worth noting that while the two-hit system has lost traction in conventional discovery proteomic analyses (Gupta & Pevzner, 2009), for a proteogenomics analysis where many proteins in the search database are actually incorrect, the rule does in fact reduce false discoveries while maintaining true positives (Castellana et al., 2014). It would be prudent for any custom workflow to employ a similar system of probabilistic scoring in order to assign an informative level of confidence to novel gene features, and this pruning will make subsequent sequence validation and functional annotation more streamline. It should be noted, however, that this approach is only useful with a reasonably complete genome where the average contig size is much larger than an individual gene.

6. SEQUENCE VALIDATION

6.1 BLAST

One of the simplest and easiest ways to validate a novel sequence is to identify homology between the novel peptide or protein sequence to sequences in other organisms. The Basic Local Alignment Search Tool (BLAST; Altschul, Gish, Miller, Myers, & Lipman, 1990) provides an effective means of identifying significant sequence similarity to a very wide range of species. Although BLAST searches can be submitted online to the public BLAST server, we recommend installing BLAST+ (Adzhubei et al., 2010) locally and automating query submissions. The output of a local search is a single concise text file containing all queries that can be easily parsed. One important parameter to consider when executing a search for homology is the type of sequence (i.e., nucleotide vs amino acid). Amino acid comparisons are more accurate than nucleotide comparisons because the scoring system considers chemical similarity rather than just homology. Furthermore, nucleotide searches are also statistically less powerful due to redundancy in the genetic code.

Another critical factor to consider is the choice of database, since larger databases increase the likelihood of matches by chance. An "expect value" is calculated for each query, which represents the predicted number of random matches (based on the database size) with an equal or higher score than the actual reported match. Therefore, the same sequence match will have a higher expect value if it is part of a large database (e.g., the nonredundant sequence list for all species) than a small database (e.g., species-specific sequences). A sufficiently sensitive approach is to include only species within the closest taxonomic grouping as the target species, as this will minimize the database size while retaining the species that are most likely to display some sequence homology.

6.2 Empirical Validation

Although BLAST is an excellent way to provide evidence for genetic elements in silico, the strongest form of validation is experimental. If only a handful of coding regions require validation, RT-PCR is a classical low-throughput, inexpensive method to confirm that the region of DNA in question is also transcribed. This has the added benefit of determining the precise patterns of exon connectivity, which is not achievable with

RNA-seq since reads almost never span more than two exons. Illumina sequencing has a maximum read length of 150 bp, whereas the average exon length for most species is between 150 and 450 bp (McGuire, Pearson, Neafsey, & Galagan, 2008). Primers for RT-PCR should be designed to span the two most distant peptides in the same contiguous element, expanding the range of coverage if required.

Empirical validation of newly discovered protein sequences can also be achieved if new proteomics data contain spectra that can match to these novel proteins. This validation search will be more robust than the initial search using the large proteogenomics database because it will no longer suffer from FDR inflation (Section 5.3). At this point, it is best practice (and required by some journals) to inspect the b- and y-ion annotation of the matched spectra to be sure of the quality of this assignment.

If a gene feature cannot be validated, this does not necessarily mean that it was a false-positive event. The validation dataset, whether it is RNA-seq, mass spectrometry, or RT-PCR, is typically a much smaller sample than the data that was used in the discovery phase. It is quite possible for a novel sequence to be expressed at a specific life stage or tissue and may not be highly abundant. In fact, these are good reasons why a gene feature might be missed in the first place. Therefore, the unvalidated sequences should not be discarded; rather, they should be marked as such and retained so that others may have the opportunity to repeat the validation process with richer data.

6.3 Data Dissemination

After the validation process, ideally, the scientific community can benefit from the newly edited gene structures. One of the simplest ways to make this information broadly accessible is to submit the sequences to a suitable database (e.g., Uniprot (http://www.uniprot.org) or NCBI (http://www.ncbi.nlm.nih.gov)) and by encoding all the relevant information about the gene feature in a supplementary GFF file. Both validated and unvalidated gene features can be included here as long as they are clearly marked. The validated gene features, once added to curated databases, can eventually be used to help better train the gene prediction algorithms (Fig. 2). In fact, as sequencing becomes increasingly feasible, some researchers (Kapheim et al., 2015) are simultaneously sequencing new genomes and generating corresponding RNA-seq data in order to produce a reliable training data set for the initial stages of annotation. By extending this to including proteomic validation of translation, the gene annotation efforts can be made even more accurate. This approach can greatly assist the gene-finding algorithms to optimize pattern recognition for the species in question.

6.4 Functional Annotation

Determining the function of novel genes and their biological relevance is simultaneously the most important and the most difficult aspect of gene annotation. Since these novel sequences tend not to match any conserved domains, as that is often the reason they were missed by traditional gene annotation algorithms in the first place, other homology-based methods are required to assign functions. One of the simplest and most intuitive methods is to use BLAST2GO (Conesa & Gotz, 2008) to identify potential orthologues and transfer the functional description of the orthologue to the novel gene. A common proxy for orthology is the "reciprocal best hit" (RBH) approach (Ward & Moreno-Hagelsieb, 2014), where two sequences are said to be orthologues if the novel gene matches most significantly with a sequence in a related species, and in a reciprocal query the proposed orthologue matches best with the same novel gene. However, orthologues are not always RBH, so some sequences cannot be functionally annotated using this method. A robust alternative to the RBH approach is to use the InterProScan tool (Jones et al., 2014) to identify orthologous proteins. This tool scans a suite of protein databases for orthologues using multiple signature recognition methods rather than relying purely on sequence similarity.

7. LIMITATIONS

7.1 Proteome Coverage

Although MS-based proteomics is becoming increasingly robust, proteome coverage for a given organism is still limited to the subset of peptides that are amenable to ionization, fragmentation, and detection. Trypsin cleaves C-terminal to lysine and arginine residues with high specificity, making it an ideal enzyme for generating ionizable, predictable peptides (Fig. 8). A relatively new and complementary approach is to perform paired-sample digestions with lysargiNase, which cleaves N-terminal to lysine and arginine (Huesgen et al., 2015). Both trypsin and lysargiNase tend to generate peptides with lengths that are highly amenable to detection by mass spectrometry, owing to the natural frequency of arginine and lysine in most proteins.

Fig. 8 Potential protein coverage achievable with different enzymes. *Green* (*light gray* in the print version) *bars* span peptides that are predicted to be amenable to analysis by mass spectrometry based on size. *Dark gray bars* span peptides that are not expected to be observed because they are too short (<7 residues) or too long (>25 residues).

However, some protein regions will inevitably yield peptides that fall outside the ideal size range of 7–25 residues or that fragment too poorly (e.g., internal Asp-Pro) to be analyzed effectively by a mass spectrometer. This means that it is not possible to cover all splice junctions or identify all exons. Alternative enzymes can be used to increase sequence coverage, such as Arg-C, Lys-C, and Glu-C; however, the vast majority of publically available data utilizes trypsin so this approach may require generating entirely new data sets.

7.2 Short Open Reading Frames

The prevalence of short open reading frames (sORFs) in eukaryotes is only just beginning to be appreciated (Andrews & Rothnagel, 2014; Smith et al., 2014). To make matters worse for gene annotation, sORFs are difficult to detect by traditional gene prediction algorithms, as well as by new proteogenomics methods (Andrews & Rothnagel, 2014). The smallest reported sORF contained only six codons (Law, Raney, Heusner, & Morris, 2001), which is too small to be confidently detected by proteomics and would not even be included in a typical proteogenomics database (30 codons is a common minimum length cutoff). Because it is so difficult to distinguish sORFs from false positives within a six-frame or three-frame translation, ribosome profiling (as performed in Smith et al., 2014) offers a more specific means of identifying sequences that might code for short functional peptides. These supported sequences could then be selectively added to the proteogenomics database, avoiding many of the false positives that would accumulate if all short translatable regions were retained.

8. CONCLUDING REMARKS

Proteogenomics has recently emerged as a powerful technique to enhance our ability to annotate protein-coding genes. In the last 3 years, software pipelines have been produced that streamline the proteogenomic workflow; however, future improvements are needed to interface more data types, search algorithms, and validation steps. One major advance has been addressing the problem of FDR inflation through more rigorous statistical methods and best practice guidelines. As the technique matures, it has the potential to become part of the standard practice for annotating the genomes of a wide range of newly sequenced and previously sequenced species alike.

REFERENCES

Adzhubei, I. A., Schmidt, S., Peshkin, L., Ramensky, V. E., Gerasimova, A., Bork, P., et al. (2010). A method and server for predicting damaging missense mutations. *Nature Methods*, 7(4), 248–249.

Altschul, S. F., Gish, W., Miller, W., Myers, E. W., & Lipman, D. J. (1990). Basic local alignment search tool. *Journal of Molecular Biology*, *215*(3), 403–410.

Andrews, S. J., & Rothnagel, J. A. (2014). Emerging evidence for functional peptides encoded by short open reading frames. *Nature Reviews. Genetics*, *15*(3), 193–204.

Askenazi, M., Ruggles, K. V., & Fenyö, D. (2016). PGx: Putting peptides to BED. *Journal of Proteome Research*, *15*(3), 795–799.

Bern, M., Cai, Y., & Goldberg, D. (2007). Lookup peaks: A hybrid of de novo sequencing and database search for protein identification by tandem mass spectrometry. *Analytical Chemistry*, *79*(4), 1393–1400.

Brent, M. R. (2007). How does eukaryotic gene prediction work? *Nature Biotechnology*, *25*(8), 883.

Brent, M. R. (2008). Steady progress and recent breakthroughs in the accuracy of automated genome annotation. *Nature Reviews. Genetics*, *9*(1), 62–73.

Brosch, M., Saunders, G. I., Frankish, A., Collins, M. O., Yu, L., Wright, J., et al. (2011). Shotgun proteomics aids discovery of novel protein-coding genes, alternative splicing, and "resurrected" pseudogenes in the mouse genome. *Genome Research*, *21*(5), 756–767. http://dx.doi.org/10.1101/gr.114272.110.

Brunner, E., Ahrens, C. H., Mohanty, S., Baetschmann, H., Loevenich, S., Potthast, F., et al. (2007). A high-quality catalog of the Drosophila melanogaster proteome. *Nature Biotechnology*, *25*(5), 576–583. http://dx.doi.org/10.1038/nbt1300.

Castellana, N. E., Payne, S. H., Shen, Z., Stanke, M., Bafna, V., & Briggs, S. P. (2008). Discovery and revision of Arabidopsis genes by proteogenomics. *Proceedings of the National Academy of Sciences of the United States of America*, *105*(52), 21034–21038. http://dx.doi.org/10.1073/pnas.0811066106.

Castellana, N. E., Shen, Z., He, Y., Walley, J. W., Cassidy, C. J., Briggs, S. P., et al. (2014). An automated proteogenomic method uses mass spectrometry to reveal novel genes in Zea mays. *Molecular & Cellular Proteomics*, *13*(1), 157–167. http://dx.doi.org/10.1074/mcp.M113.031260.

Chambers, M. C., Maclean, B., Burke, R., Amodei, D., Ruderman, D. L., Neumann, S., et al. (2012). A cross-platform toolkit for mass spectrometry and proteomics. *Nature Biotechnology*, *30*(10), 918–920. http://dx.doi.org/10.1038/nbt.2377.

Chan, Q. W. T., Parker, R., Sun, Z., Deutsch, E. W., & Foster, L. J. (2011). A honey bee (*Apis mellifera* L.) peptide atlas crossing castes and tissues. *BMC Genomics*, *12*(290), 1. http://dx.doi.org/10.1186/1471-2164-12-290.

Chick, J. M., Kolippakkam, D., Nusinow, D. P., Zhai, B., Rad, R., Huttlin, E. L., et al. (2015). A mass-tolerant database search identifies a large proportion of unassigned spectra in shotgun proteomics as modified peptides. *Nature Biotechnology*, *33*(7), 743–749.

Cho, H., Davis, J., Li, X., Smith, K. S., Battle, A., & Montgomery, S. B. (2014). High-resolution transcriptome analysis with long-read RNA sequencing. *PloS One*, *9*(9), e108095.

Conesa, A., & Gotz, S. (2008). Blast2GO: A comprehensive suite for functional analysis in plant genomics. *International Journal of Plant Genomics*, *2008*, 619832. http://dx.doi.org/10.1155/2008/619832.

Cox, J., & Mann, M. (2008). MaxQuant enables high peptide identification rates, individualized ppb-range mass accuracies and proteome-wide protein quantification. *Nature Biotechnology*, *26*(12), 1367–1372.

Cox, J., Matic, I., Hilger, M., Nagaraj, N., Selbach, M., Olsen, J. V., et al. (2009). A practical guide to the MaxQuant computational platform for SILAC-based quantitative proteomics. *Nature Protocols*, *4*(5), 698–705.

Craig, R., & Beavis, R. C. (2004). TANDEM: Matching proteins with tandem mass spectra. *Bioinformatics (Oxford, England)*, *20*(9), 1466–1467. http://dx.doi.org/10.1093/bioinformatics/bth092.

Elsik, C. G., Worley, K. C., Bennett, A. K., Beye, M., Camara, F., Childers, C. P., et al. (2014). Finding the missing honey bee genes: Lessons learned from a genome upgrade. *BMC Genomics*, *15*(1), 1.

Eng, J. K., Jahan, T. A., & Hoopmann, M. R. (2013). Comet: An open-source MS/MS sequence database search tool. *Proteomics*, *13*(1), 22–24.

Eng, J. K., McCormack, A. L., & Yates, J. R. (1994). An approach to correlate tandem mass spectral data of peptides with amino acid sequences in a protein database. *Journal of the American Society for Mass Spectrometry*, *5*(11), 976–989.

Fagerberg, L., Hallstrom, B. M., Oksvold, P., Kampf, C., Djureinovic, D., Odeberg, J., et al. (2014). Analysis of the human tissue-specific expression by genome-wide integration of transcriptomics and antibody-based proteomics. *Molecular & Cellular Proteomics*, *13*(2), 397–406. http://dx.doi.org/10.1074/mcp.M113.035600.

Frank, A., & Pevzner, P. (2005). PepNovo: De novo peptide sequencing via probabilistic network modeling. *Analytical Chemistry*, 77(4), 964–973.

Ghali, F., Krishna, R., Perkins, S., Collins, A., Xia, D., Wastling, J., et al. (2014). ProteoAnnotator–open source proteogenomics annotation software supporting PSI standards. *Proteomics*, *14*(23-24), 2731–2741.

Goodswen, S. J., Kennedy, P. J., & Ellis, J. T. (2012). Evaluating high-throughput ab initio gene finders to discover proteins encoded in eukaryotic pathogen genomes missed by laboratory techniques. *PloS One*, 7(11), e50609.

Grossmann, J., Roos, F. F., Cieliebak, M., Lipták, Z., Mathis, L. K., Müller, M., et al. (2005). AUDENS: A tool for automated peptide de novo sequencing. *Journal of Proteome Research*, *4*(5), 1768–1774.

Gudlaugsdottir, S., Boswell, D. R., Wood, G. R., & Ma, J. (2007). Exon size distribution and the origin of introns. *Genetica*, *131*(3), 299–306.

Gupta, N., & Pevzner, P. A. (2009). False discovery rates of protein identifications: A strike against the two-peptide rule. *Journal of Proteome Research*, *8*(9), 4173–4181.

Hestand, M. S., Zeng, Z., Coleman, S. J., Liu, J., & MacLeod, J. N. (2015). Tissue restricted splice junctions originate not only from tissue-specific gene loci, but gene loci with a broad pattern of expression. *PloS One*, *10*(12), e0144302. http://dx.doi.org/10.1371/journal.pone.0144302.

Hou, Z., Jiang, P., Swanson, S. A., Elwell, A. L., Nguyen, B. K. S., Bolin, J. M., et al. (2015). A cost-effective RNA sequencing protocol for large-scale gene expression studies. *Scientific Reports*. *5*. http://dx.doi.org/10.1038/srep09570.

Huesgen, P. F., Lange, P. F., Rogers, L. D., Solis, N., Eckhard, U., Kleifeld, O., et al. (2015). LysargiNase mirrors trypsin for protein C-terminal and methylation-site identification. *Nature Methods*, *12*(1), 55–58.

Jagtap, P. D., Johnson, J. E., Onsongo, G., Sadler, F. W., Murray, K., Wang, Y., et al. (2014). Flexible and accessible workflows for improved proteogenomic analysis using the galaxy framework. *Journal of Proteome Research*, *13*(12), 5898–5908.

Jones, P., Binns, D., Chang, H. Y., Fraser, M., Li, W., McAnulla, C., et al. (2014). InterProScan 5: Genome-scale protein function classification. *Bioinformatics (Oxford, England)*, *30*(9), 1236–1240. http://dx.doi.org/10.1093/bioinformatics/btu031.

Kapheim, K. M., Pan, H., Li, C., Salzberg, S. L., Puiu, D., Magoc, T., et al. (2015). Social evolution: Genomic signatures of evolutionary transitions from solitary to group living. *Science*, *348*(6239), 1139–1143. http://dx.doi.org/10.1126/science.aaa4788.

Keightley, P. D., Trivedi, U., Thomson, M., Oliver, F., Kumar, S., & Blaxter, M. L. (2009). Analysis of the genome sequences of three Drosophila melanogaster spontaneous mutation accumulation lines. *Genome Research*, *19*(7), 1195–1201. http://dx.doi.org/10.1101/gr.091231.109.

Kim, H., Park, H., & Paek, E. (2015). NextSearch: A search engine for mass spectrometry data against a compact nucleotide exon graph. *Journal of Proteome Research*, *14*(7), 2784–2791.

Kim, M., Pinto, S. M., Getnet, D., Nirujogi, R. S., Manda, S. S., Chaerkady, R., et al. (2014). A draft map of the human proteome. *Nature*, *509*(7502), 575–581.

Kolmogorov, M., Liu, X., & Pevzner, P. A. (2015). SpectroGene: A tool for proteogenomic annotations using top-down spectra. *Journal of Proteome Research*, *15*(1), 144–151.

Korf, I. (2004). Gene finding in novel genomes. *BMC Bioinformatics*, *5*(1), 1.

Krasnov, G. S., Dmitriev, A. A., Kudryavtseva, A. V., Shargunov, A. V., Karpov, D. S., Uroshlev, L. A., et al. (2015). PPLine: An automated pipeline for SNP, SAP, and splice variant detection in the context of proteogenomics. *Journal of Proteome Research*, *14*(9), 3729–3737.

Krug, K., Carpy, A., Behrends, G., Matic, K., Soares, N. C., & Macek, B. (2013). Deep coverage of the escherichia coli proteome enables the assessment of false discovery rates in simple proteogenomic experiments. *Molecular & Cellular Proteomics*, *12*(11), 3420–3430. http://dx.doi.org/10.1074/mcp.M113.029165.

Law, G. L., Raney, A., Heusner, C., & Morris, D. R. (2001). Polyamine regulation of ribosome pausing at the upstream open reading frame of S-adenosylmethionine decarboxylase. *Journal of Biological Chemistry*, *276*(41), 38036–38043.

Li, Z., Zhang, Z., Yan, P., Huang, S., Fei, Z., & Lin, K. (2011). RNA-seq improves annotation of protein-coding genes in the cucumber genome. *BMC Genomics*, *12*(1), 1.

Lynch, S. M., & Moore, J. H. (2016). A call for biological data mining approaches in epidemiology. *BioData Mining*, *9*(1), 1.

Ma, B., Zhang, K., Hendrie, C., Liang, C., Li, M., Doherty-Kirby, A., et al. (2003). PEAKS: Powerful software for peptide de novo sequencing by tandem mass spectrometry. *Rapid Communications in Mass Spectrometry*, *17*(20), 2337–2342.

McAfee, A., Harpur, B. A., Michaud, S., Beavis, R., Kent, C. F., Zayed, A., et al. (2016). Toward an upgraded honey bee (*Apis mellifera* L.) genome annotation using proteogenomics. *Journal of Proteome Research*, *15*(2), 411–421.

McGuire, A. M., Pearson, M. D., Neafsey, D. E., & Galagan, J. E. (2008). Cross-kingdom patterns of alternative splicing and splice recognition. *Genome Biology*, *9*(3), R50.

Merrihew, G. E., Davis, C., Ewing, B., Williams, G., Kall, L., Frewen, B. E., et al. (2008). Use of shotgun proteomics for the identification, confirmation, and correction of C. elegans gene annotations. *Genome Research*, *18*(10), 1660–1669. http://dx.doi.org/10.1101/gr.077644.108.

Miller, N. A., Farrow, E. G., Gibson, M., Willig, L. K., Twist, G., Yoo, B., et al. (2015). A 26-hour system of highly sensitive whole genome sequencing for emergency management of genetic diseases. *Genome Medicine*, 7(100). http://dx.doi.org/10.1038/srep09570.

Nagaraj, S. H., Waddell, N., Madugundu, A. K., Wood, S., Jones, A., Mandyam, R. A., et al. (2015). PGTools: A software suite for proteogenomic data analysis and visualization. *Journal of Proteome Research*, *14*(5), 2255–2266.

Nesvizhskii, A. I. (2014). Proteogenomics: Concepts, applications and computational strategies. *Nature Methods*, *11*(11), 1114–1125.

Ozsolak, F., & Milos, P. M. (2011). RNA sequencing: Advances, challenges and opportunities. *Nature Reviews. Genetics*, *12*(2), 87–98.

Perkins, D. N., Pappin, D. J., Creasy, D. M., & Cottrell, J. S. (1999). Probability-based protein identification by searching sequence databases using mass spectrometry data. *Electrophoresis*, *20*(18), 3551–3567. http://dx.doi.org/10.1002/(SICI)1522-2683(19991201)20:18<3551::AID-ELPS3551>3.0.CO;2-2.

Risk, B. A., Spitzer, W. J., & Giddings, M. C. (2013). Peppy: Proteogenomic search software. *Journal of Proteome Research*, *12*(6), 3019–3025.

Robertson, H. M., Reese, J. T., Milshina, N. V., Agarwala, R., Solignac, M., Walden, K. K., et al. (2007). Manual superscaffolding of honey bee (Apis mellifera) chromosomes 12-16: Implications for the draft genome assembly version 4, gene annotation, and chromosome structure. *Insect Molecular Biology*, *16*(4), 401–410.

Schandorff, S., Olsen, J. V., Bunkenborg, J., Blagoev, B., Zhang, Y., Andersen, J. S., et al. (2007). A mass spectrometry-friendly database for cSNP identification. *Nature Methods*, *4*(6), 465–466.

Smith, J. E., Alvarez-Dominguez, J. R., Kline, N., Huynh, N. J., Geisler, S., Hu, W., et al. (2014). Translation of small open reading frames within unannotated RNA transcripts in saccharomyces cerevisiae. *Cell Reports*, 7(6), 1858–1866.

Soon, W. W., Hariharan, M., & Snyder, M. P. (2013). High-throughput sequencing for biology and medicine. *Molecular Systems Biology*, *9*(1), 640. http://dx.doi.org/10.1038/msb.2012.61.

Steijger, T., Abril, J. F., Engström, P. G., Kokocinski, F., Hubbard, T. J., Guigó, R., et al. (2013). Assessment of transcript reconstruction methods for RNA-seq. *Nature Methods*, *10*(12), 1177–1184.

Tanner, S., Shen, Z., Ng, J., Florea, L., Guigo, R., Briggs, S. P., et al. (2007). Improving gene annotation using peptide mass spectrometry. *Genome Research*, *17*(2), 231–239. http://dx.doi.org/10.1101/gr.5646507.

Trapnell, C., Roberts, A., Goff, L., Pertea, G., Kim, D., Kelley, D. R., et al. (2012). Differential gene and transcript expression analysis of RNA-seq experiments with TopHat and cufflinks. *Nature Protocols*, 7(3), 562–578.

Venter, J. C., Adams, M. D., Myers, E. W., Li, P. W., Mural, R. J., Sutton, G. G., et al. (2001). The sequence of the human genome. *Science (New York, N.Y.)*, *291*(5507), 1304–1351. http://dx.doi.org/10.1126/science.1058040.

Walther, T. C., & Mann, M. (2010). Mass spectrometry-based proteomics in cell biology. *The Journal of Cell Biology*, *190*(4), 491–500. http://dx.doi.org/10.1083/jcb.201004052.

Ward, N., & Moreno-Hagelsieb, G. (2014). Quickly finding orthologs as reciprocal best hits with BLAT, LAST, and UBLAST: How much do we miss? *PloS One*, *9*(7), e101850.

Weinstock, G. M., Robinson, G. E., Gibbs, R. A., Worley, K. C., Evans, J. D., Maleszka, R., et al. (2006). Insights into social insects from the genome of the honeybee Apis mellifera. *Nature*, *443*(7114), 931–949.

Wilhelm, M., Schlegl, J., Hahne, H., Gholami, A. M., Lieberenz, M., Savitski, M. M., et al. (2014). Mass-spectrometry-based draft of the human proteome. *Nature*, *509*(7502), 582–587.

Wilming, L., & Harrow, J. (2009). *Gene annotation methods*. *Bioinformatics*. New York: Springer, pp. 121–136.

Woo, S., Cha, S. W., Merrihew, G., He, Y., Castellana, N., Guest, C., et al. (2013). Proteogenomic database construction driven from large scale RNA-seq data. *Journal of Proteome Research*, *13*(1), 21–28.

Yandell, M., & Ence, D. (2012). A beginner's guide to eukaryotic genome annotation. *Nature Reviews. Genetics*, *13*(5), 329–342. http://dx.doi.org/10.1038/nrg3174.

Yates, J. R., III, Eng, J. K., & McCormack, A. L. (1995). Mining genomes: Correlating tandem mass spectra of modified and unmodified peptides to sequences in nucleotide databases. *Analytical Chemistry*, *67*(18), 3202–3210.

Zhang, Y., Fonslow, B. R., Shan, B., Baek, M., & Yates, J. R., III. (2013). Protein analysis by shotgun/bottom-up proteomics. *Chemical Reviews*, *113*(4), 2343–2394.

Zickmann, F., & Renard, B. Y. (2015). IPred—Integrating ab initio and evidence based gene predictions to improve prediction accuracy. *BMC Genomics*, *16*(1), 1.

CHAPTER FOURTEEN

Targeted In-Depth Quantification of Signaling Using Label-Free Mass Spectrometry

P.R. Cutillas[1]

Barts Cancer Institute, Queen Mary University of London, John Vane Science Centre, London, United Kingdom

[1]Corresponding author: e-mail address: p.cutillas@qmul.ac.uk

Contents

Abstract

Protein phosphorylation encodes information on the activity of kinase-driven signaling pathways that regulate cell biology. This chapter discusses an approach, named TIQUAS (targeted in-depth quantification of signaling), to quantify cell signaling comprehensively and without bias. The workflow—based on mass spectrometry (MS) and computational science—consists of targeting the analysis of phosphopeptides previously identified by shotgun liquid chromatography tandem MS (LC–MS/MS) across the samples that are being compared. TIQUAS therefore takes advantage of concepts derived from both targeted (data-independent) and data-dependent acquisition methods; phosphorylation sites are quantified in all experimental samples regardless of whether or not these phosphopeptides were identified by MS/MS in all runs. As a result, datasets are obtained containing quantitative information on several thousand phosphorylation sites in as many samples and replicates as required in the experimental design, and these rich datasets are devoid of a significant number of missing data points. This chapter discussed the biochemical, analytical, and computational procedures required to apply the approach and for obtaining a biological interpretation of the data in the context of our understanding of cell signaling regulation and kinase–substrate relationships.

Methods in Enzymology, Volume 585
ISSN 0076-6879
http://dx.doi.org/10.1016/bs.mie.2016.09.021

1. INTRODUCTION

Protein phosphorylation is a posttranslational modification (PTM) consisting of the addition of phosphate groups to specific amino acid residues on proteins. This PTM has the potential to alter the stability, subcellular location, and enzymatic activity of proteins with diverse roles in cells. Protein phosphorylation is catalyzed by enzymes known as protein kinases, while the reverse reaction (i.e., removal of phosphates from proteins) is mediated by protein phosphatases (Manning, Whyte, Martinez, Hunter, & Sudarsanam, 2002). In mammals, phosphorylation primarily occurs at serine, threonine, or tyrosine residues, although histidine phosphorylation has also been reported (Steeg, Palmieri, Ouatas, & Salerno, 2003).

Signals emanating from the extracellular space as well as metabolic or stress signals that originate within the cell modulate signaling fluxes are driven by the activity of kinases and phosphatases. Protein phosphorylation is therefore a key mechanism by which cells sense their environment (i.e., the combination of hormones, growth factors, cytokines, and nutrients) and internal metabolic status so that individual cells can orchestrate an appropriate physiological response in the context of the tissue in which they reside (Cohen, 1982).

Processes regulated by protein phosphorylation include bioenergetic metabolism, cell cycle progression, transcription, translation, endocytosis, phagocytosis, and apoptosis. In addition, dysregulation of protein phosphorylation has been implicated in diseases of pressing relevance in the developed countries including diabetes, neurodegeneration, autoimmune disease, and cancer (Forlenza et al., 2011; Janku et al., 2012; Parmenter et al., 2014; Samovski et al., 2015). It is also thought that certain kinase cascades (for example, those downstream of insulin-like growth factor) mediate the process of aging (Harrison et al., 2009; Longo et al., 2015). Consequently, the development of kinase inhibitors as drugs for a variety of diseases is an active area of pharmacological research (Knapp et al., 2013; Workman, Al-Lazikani, & Clarke, 2013). About 30 kinase inhibitors are already approved by the FDA to treat different forms of cancer, and it has been estimated that more than 200 of such compounds are in different states of clinical development (Cohen, Kim, & DeMatteo, in press; Fabbro, 2015; Knapp et al., 2013).

2. PHOSPHOPROTEOMIC ANALYSIS AS A MEANS TO QUANTIFY CELL SIGNALING

Understanding the biological roles of phosphorylation events requires not only cataloging phosphorylated residues in proteins but also measuring how such phosphorylation sites are modulated in cells as a result of experimental conditions, during the process of differentiation, development, and aging or when disease ensues. Therefore, there is a considerable interest in the development of methods to measure the phosphoproteome, which may be defined as the collection of phosphorylation sites present in a given cell population at a particular time (Reinders & Sickmann, 2005).

Although phosphoproteomic methods are developing at a rapid pace, these are not without challenges (Solari, Dell'Aica, Sickmann, & Zahedi, 2015). Several obstacles need to be addressed in order to carry out an efficient analysis of the phosphoproteome. First, methodologies for phosphoproteomics have to be as comprehensive as possible. It has been estimated that >100,000 phosphorylation sites may exists in human cells (Macek, Mann, & Olsen, 2009), and although not all of these sites may be present in a particular cell type under a defined set of conditions, methods to measure the phosphoproteome need to be able to deal with the complex nature of the phosphoproteome. Immunochemical techniques can analyze a few dozen phosphorylation sites simultaneously (Wulfkuhle et al., 2003), but the preferred methodology for global phosphoproteomics is mass spectrometry coupled with liquid chromatography (LC–MS/MS). With this technique, >40,000 phosphorylation sites are detectable in mammalian cells (Sharma et al., 2014).

Second, amino acid residues are often phosphorylated in substoichiometric levels relative to their nonphosphorylated counterparts. A practical consequence of this is that unmodified peptides are detected in preference of phosphorylated peptides in unfractionated cell lysates. Consequently, standard proteomic approaches based on LC–MS/MS fail to detect a significant number of phosphorylation sites. To overcome this problem, chromatographic methods based on immobilized metal affinity chromatography (IMAC) and metal oxide affinity chromatography (MOAC) have been developed over the years (Bodenmiller, Mueller, Mueller, Domon, & Aebersold, 2007; Ficarro et al., 2002; Larsen,

Thingholm, Jensen, Roepstorff, & Jorgensen, 2005; Lee et al., 2007; Pinkse, Uitto, Hilhorst, Ooms, & Heck, 2004). These techniques allow enriching phosphopeptides from digested cell lysates so that these can be efficiently analyzed by LC–MS/MS. Despite these improvements, however, state-of-the-art mass spectrometers still fail to sequence all phosphopeptides present in a given enriched sample. This under-sampling of the phosphoproteome in standard LC–MS/MS runs has repercussions for our ability to reproducibly quantify differences in phosphorylation across samples.

Third, phosphorylation is a reversible modification that is quickly lost by the action of phosphatases, which are constitutively active in cells. From the analytical standpoint, this represents a problem because inconsistencies in sample preparation can significantly alter the nature of the analyzed phosphoproteome. To minimize this problem cells need to be handled and lysed using conditions so that enzymatic activity is inhibited as soon as possible in the experimental workflow.

Fourth, phosphorylation is a relatively labile PTM. Conditions that favor peptide sequencing by MS/MS such as collision-induced dissociation (CID) often fail to pinpoint the precise site of modification within a phosphopeptide in occasions when the peptide has more than one serine, threonine, or tyrosine residue on its sequence. To overcome this potential problem, alternative fragmentation methods based on high-energy collision dissociation (HCD), electron transfer dissociation (ETD), or combinations are being explored as an alternative to CID for phosphopeptide analysis (Frese et al., 2012; Kim & Pandey, 2012).

Fifth, the biological significance of most phosphorylation events remains obscure (Beltrao et al., 2012; Minguez, Letunic, Parca, & Bork, 2013). It is known that some phosphorylation events increase the enzymatic activity of the protein that is receiving the modification; examples include the protein kinases AKT and ERK, whose activity increases when these are phosphorylated on specific residues on their activation loop (Alessi et al., 1996; Boulton et al., 1991). However, other modifications are inhibitory; examples include GSK3β (at Ser9) and BRAF (at Ser365) (Cross, Alessi, Cohen, Andjelkovich, & Hemmings, 1995; Hmitou, Druillennec, Valluet, Peyssonnaux, & Eychene, 2007). While the biochemical consequence of a few modifications is well described, the roles, if any, of the vast majority of them are still unknown (Beltrao et al., 2012; Minguez et al., 2013). It has in fact been argued that most phosphorylation events are not functional (Landry, Levy, & Michnick, 2009; Lienhard, 2008), an assertion that is widely disputed and debated (Tan, Jorgensen, & Linding, 2010). Regardless of their specific biological functions, if any, "phosphorylation" is not synonym with "activation" and although it is now possible to identify and quantify >10,000 phosphorylation sites in "routine" phosphoproteomics experiments, with >40,000–100,000 phosphopeptides quantified in large-scale efforts (de Graaf, Giansanti, Altelaar, & Heck, 2014; Lawrence, Searle, Llovet, & Villen, 2016; Sharma et al., 2014; Wilkes, Terfve, Gribben, Saez-Rodriguez, & Cutillas, 2015), obtaining biological information from such data is not straight forward. To overcome this issue, new computational approaches are being developed with the aim of harness the biological information inherent in phosphoproteomics data.

3. OVERVIEW OF METHODOLOGIES FOR PHOSPHOPROTEOMICS

This chapter describes a phosphoproteomics approach, which we term TIQUAS (targeted in-depth quantification of signaling), designed to address the issues outlined earlier (Fig. 1). TIQUAS uses label-free LC–MS/MS to quantify phosphorylation and derive biological information from

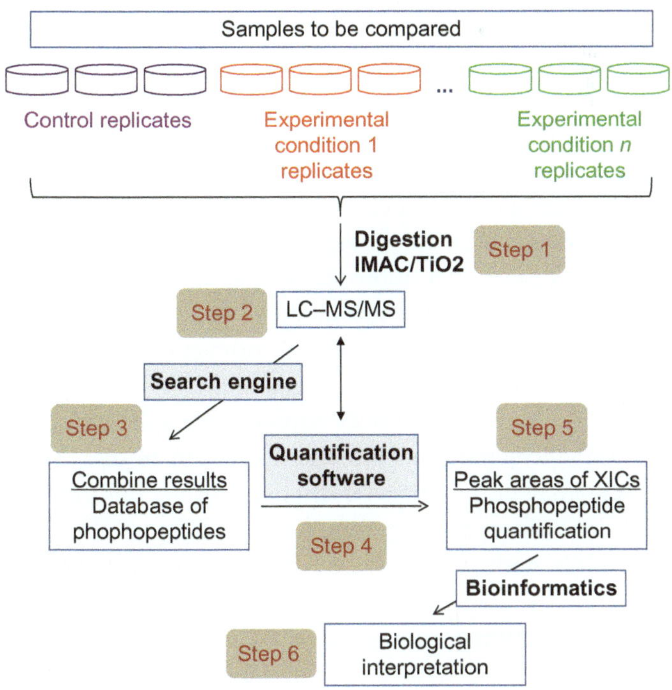

Fig. 1 Workflow for targeted in-depth quantification of signaling (TIQUAS). See text for details.

these data. For detailed descriptions of other methods for phosphoproteomic analysis the reader is directed to comprehensive reviews on this topic (Bensimon, Heck, & Aebersold, 2012; Cutillas & Jorgensen, 2011; Liu, Wang, & Xue, 2013; Nita-Lazar, Saito-Benz, & White, 2008; Riley & Coon, 2016; Thingholm, Jensen, & Larsen, 2009).

Many other phosphoproteomics approaches have been published, and these are suitable for particular applications. Some of the published methods require labeling proteins with isotopes of carbon and nitrogen using either metabolic of chemical labeling, such as SILAC or TMT reagents (Liang, Lahert, Pike, & Ward, 2015; Olsen et al., 2006). After labeling and mixing the samples that are to be compared, proteins (or peptides after proteolytic cleavage) are separated into fraction using complex workflows based on multidimensional chromatography and then analyzed by LC–MS/MS in data-dependent acquisition (DDA) mode. Because of the stochastic nature of DDA, different peptides are identified in consecutive runs and replicate experiments. Thus approaches based on these techniques identify different subsets of the phosphoproteome in different experiments (Alcolea, Kleiner, & Cutillas, 2009).

Alternative approaches to DDA are those based on data-independent analysis (DIA). In contrast to DDA, which involve "sequencing" peptides as these elute from the LC column, DIA allows designing assays to quantify specific peptides by targeting their analysis with predefined instrument settings. These methods, which include selected reaction monitoring (SRM), selected ion monitoring (SIM), and parallel reaction monitoring (PRM), are the gold standard for the quantification of small molecules in sports doping monitoring, forensic, and clinical applications (Ojanpera, Kolmonen, & Pelander, 2012). Newer DIA approaches such as SWATH are promising for the large-scale analysis of unmodified proteins (Collins et al., 2013; Liu, Huttenhain, et al., 2013) but these have not yet been used for phosphoproteomics analysis as the convoluted fragment SWATH data may not offer sufficient specificity to unambiguously sequence phosphopeptides. DIA (by, for example, SRM) overcomes a limitation of DDA in that the same set of phosphopeptides may be quantified in an, in principle, unlimited number of experimental conditions and replicates, thus allowing obtaining datasets from which results with well-powered statistics may be obtained (Parker et al., 2015). This advantage, however, comes with the drawback that there is a limitation to the number of peptides that SRM or PRM can quantify as the number of transitions that can be programmed in current mass spectrometers is limited by their duty cycles (Parker et al., 2015).

In summary, the phosphoproteomics methods discussed earlier pose a tradeoff between the number of phosphopeptides that can be reliable quantified in an experiment and the number of samples that can be compared with reproducibility and reliability. Indeed, methods based on untargeted analysis (namely DDA) have the advantage of producing large datasets but, for these to be reliable, one needs to label proteins or peptides, thus limiting the approach to the analysis of a few conditions and replicates. On the other hand, methods based on targeted analysis (i.e., SRM or PRM) produce reliable data but are restricted to the analysis of few peptides (about 100 at most).

As mentioned earlier, this chapter describes a phosphoproteomics approach, named TIQUAS, that aims to overcome these limitations by using concepts derived from both DDA and DIA so that phosphopeptides can be quantified in a reliable and comprehensive manner and with the ability to compare as many replicates and experimental conditions as required in the experimental design.

4. THE CONCEPTS OF TIQUAS

4.1 Overview

The aim of TIQUAS is to quantify phosphorylation with depth but also in a targeted manner so that datasets are produced without missing data points. The key aspect of the workflow, which is illustrated in Fig. 1, is the creation of a database of phosphopeptides identified (by LC–MS/MS) in at least one of the samples that are being compared. Each phosphopeptide in this database is then quantified across each of sample regardless of whether or not the phosphopeptide was identified by MS/MS in that particular sample. In this, cells or tissues are lysed or homogenized using standard methods. Proteins in these lysates are extracted and digested using a suitable protease and peptides produced as a result are then desalted and subjected to phosphopeptide enrichment using chromatographic methods such as those based on IMAC or MOAC (Fig. 1, step 1). Enriched phosphopeptides are analyzed by LC–MS/MS and identified by database searchers using an appropriate search engine (e.g., Mascot, Fig. 1, step 2). The identified peptides are placed in a database in which key identifiers—including mass to charge ratio (m/z), z, and retention time (tR)—are listed (Fig. 1, step 3). A computer program then performs an extracted ion chromatogram (XIC) for each peptide in the database across all samples that are being compared (Fig. 1, step 4). Peak areas are obtained from these XICs and used as quantitative readouts (Fig. 1, step 5).

After normalization and statistical analysis of the data, bioinformatics methods, such as kinase substrate enrichment analysis (KSEA; Casado, Rodriguez-Prados, et al., 2013), may be used to derive biological information from these phosphoproteomic datasets (Fig. 1, step 6).

Different aspects of the methodology described earlier, and illustrated in Fig. 1, evolved from studies that aimed to quantify two phosphorylation sites on PLCγ2 and for the analysis of unmodified proteins (Cutillas, Geering, Waterfield, & Vanhaesebroeck, 2005; Cutillas & Vanhaesebroeck, 2007). The accuracy and reliability of the approach for the analysis of protein phosphorylation in a global manner was then tested using leukemia cells as a model system (Alcolea, Casado, Rodriguez-Prados, Vanhaesebroeck, & Cutillas, 2012; Casado, Alcolea, et al., 2013; Casado & Cutillas, 2011; Casado, Rodriguez-Prados, et al., 2013) and then applied to the characterization of signal network topology in cancer cells (Terfve, Wilkes, Casado, Cutillas, & Saez-Rodriguez, 2015; Wilkes et al., 2015). More recently, the approach has been used to quantify signaling in leukemia and breast cancer models, to identify PKN2-regulated networks (Quetier et al., 2016) and to discover MAPKAPK-2 as a regulator of mTOR during senescence (Herranz et al., 2015). Some of the concepts behind the TIQUAS approach are similar to the accurate mass tags method developed by Richard Smith to quantify unmodified proteins (Conrads, Anderson, Veenstra, Pasa-Tolic, & Smith, 2000), with the difference that TIQUAS involves constructing a database for each experiment rather than comparing data to databases from historical experiments. Similar approaches have been reported by other groups (de Graaf et al., 2014; Humphrey, Azimifar, & Mann, 2015; Lawrence et al., 2016) and step-by-step procedures to implement the methodology have been published (Montoya, Beltran, Casado, Rodriguez-Prados, & Cutillas, 2011). The following sections describe in more detail the conceptual and practical aspects of TIQUAS.

4.2 Sample Preparation

4.2.1 Lysis

The first step in the workflow involves extracting proteins from cell lysates or tissue homogenates. In principle, any cell lysis method could be used for this purpose. However, it is important to bear in mind that the kinetics of phosphorylation/dephosphorylation reactions means that leaving cells in conditions of stress for any length of time can alter the nature of the phosphoproteome that is being analyzed. Changes in the phosphorylation of proteins in cell lines and tissues are due to a coordinated response to temperature shock and involve the activation of metabolic and growth factor signaling pathways (Casado, Bilanges, Rajeeve, Vanhaesebroeck, & Cutillas, 2014; Mertins et al., 2014). It is thus important to lyse cells as fast as possible using conditions that rapidly denature proteins or inhibit enzymatic activity.

Lysis buffer based on chaotropes such as urea and/or thiourea are normally used to lyse cells (Montoya et al., 2011). These are preferred over conditions based on mild detergents (e.g., triton X-100 or NP40) because chaotropes rapidly denature proteases, phosphatases, and kinases thus avoiding the introduction of artifacts in the phosphoproteome due to sample preparation. More recently, phosphoproteomics workflows using lysis in sodium dodecyl sulfate (SDS)-based buffers were reported (Wisniewski, Zougman, Nagaraj, & Mann, 2009). The advantage of this approach is that SDS is a strong detergent that can solubilize certain protein classes, such as membrane proteins, that are difficult to analyze otherwise (Wu & Yates, 2003). The disadvantage of using SDS or other detergents is that, without further extraction steps, these interfere with downstream MS analysis and can quickly deteriorate chromatographic columns in the LC–MS/MS system thus compromising robustness. For this reason, urea-based lysis buffers are still used in many laboratories as the first step in the phosphoproteomics workflow.

4.2.2 Digestion

After solubilization and denaturation, proteins are normally reduced and alkylated with agents such as DTT and IAM, respectively. Proteins are then proteolyzed, a procedure normally carried out using trypsin as the preferred protease because this enzyme cleaves at the carboxyl terminus of basic amino acid residues (arginine and lysine) in proteins (except when the next residue is a proline). Peptides produced as a result have a basic end that fixes the charge of the peptide, thus producing fragments during CID that retain the charge at the carboxyl end. This feature yields MS/MS spectra enriched in fragments of the y-series, thus aiding in the interpretation and determination of peptide sequences from tandem mass spectra (Steen & Mann, 2004). The efficiency of digestion can be improved by sequential digestion with LysC followed by trypsin (Wisniewski et al., 2009) probably because LyC can function at higher concentrations of urea.

Other studies have evaluated the use of multiple proteases as a means to increase the identification and coverage of the analyzed proteome (Swaney, Wenger, & Coon, 2010). The downside of approaches that use multiple proteases in separate digestions is that the number of analytical runs increases, making it unpractical to perform quantitative studies that require running replicates to obtain statically meaningful data. Therefore, multiple proteases may be used when the aim of the experiment is to identify as many phosphorylation sites in a given cell system as possible and thus obtain a deep coverage of the analyzed proteome or phosphoproteome (Swaney et al., 2010). However, when the aim is to compare signaling events across conditions with statistical rigor, the use of a single protease or sequential use of proteases is recommended as this allows producing reproducible data.

4.2.3 Phosphopeptide Enrichment

Urea and other components of lysis buffers can interfere with further processing steps. Therefore, peptides produced as a result of protein digestion need to be desalted so that they are in a solution compatible with phosphopeptide enrichment (see later). Our laboratory uses reversed-phase solid-phase extraction cartridges to desalt protein-derived peptides and follows the protocol provided by the vendor with the exception that the mobile phase used for elution has the same composition as the first step of the phosphopeptide enrichment protocol.

Several methods to separate phosphopeptides from unmodified peptides exist (for a review, see Beltran & Cutillas, 2012), and these are continuously improving. IMAC has been extensively used to enrich phosphopeptides but methods based on MOAC, such as TiO_2 and ZrO_2, produce superior yields. Techniques based on TiO_2 can routinely produce mixtures in which 80–90% of peptides are phosphopeptides. Recently, hybrid chemistries based on Ti^{4+} IMAC have been described to offer greater performance than standard TiO_2 beads (de Graaf et al., 2014; Han, Ye, & Zou, 2008). Regardless of the type of chromatography used, it is important to consider the ratio between the amounts of protein and chromatographic beads used as this can affect the reproducibility of phosphopeptide extraction and the nature of the phosphoproteome that is enriched. A high ratio of protein to TiO_2 increases the identification of multiply charged phosphopeptides (Li, Ning, Tang, Nie, & Zeng, 2009; Montoya et al., 2011) and limiting the amounts of IMAC or TiO_2 used to extract phosphopeptides can lead to an increase in technical variability (Casado & Cutillas, 2011). Thus quantitative phosphoproteomic analysis using label-free LC–MS/MS requires using an appropriate volume of TiO_2 for the amount of protein being analyzed.

4.3 LC–MS/MS and Database Creation

Once phosphopeptides have been isolated, these are analyzed by LC–MS/MS. The preferred techniques involve nanoflow LC and nano-electrospray (nano-ESI), as these provide superior sensitivity than conventional LC and ESI. The enhanced chromatography is achieved because of the more efficient ionization afforded by low flow rates (Wilm & Mann, 1996), and because narrow bore columns used in nanoflow LC concentrate the analytes in small peak volumes (Cutillas, 2004).

Efficient identification of peptides in complex mixtures requires the use of high-resolution mass spectrometers. Tandem mass analyzers based on quadruple-time-of-flight are suitable but instruments based on orbitrap technology are more popular because of the fast duty cycle and high resolution afforded by these instruments (Scheltema et al., 2014). Several methods for fragmentation have been developed to deduce peptide sequences from tandem mass spectra. CID in ion traps (when hyphenated with orbitrap mass analyzers) or quadrupoles (when the hyphenation is to time-of-flight) is the standard method for MS/MS in shotgun proteomics and phosphoproteomics experiments. Although CID performs well in standard proteomics experiments, a problem of CID in ion traps is that phosphate groups are labile and these readily dissociate from phosphopeptide ions producing a neutral loss ion that dominates the MS/MS spectrum. To fragment this neutral loss ion further, methods based on MS3 were developed but these required larger duty cycles (Gruhler et al., 2005). To harness the principles of MS3 without the need to increase the duty cycle, multistage activation (MSA) methods were developed in which the predicted neutral loss ion is activated and thus fragmented concomitant with the parent ion mass (Schroeder, Shabanowitz, Schwartz, Hunt, & Coon, 2004), thus obviating the need for an additional round of fragmentation. MSA methods give superior identification rates than CID or MS3 (Ulintz et al., 2009) but these still have the problem that it is, on occasion, difficult to precisely pinpoint the site of modification when several modifiable amino acids are present in the peptide that is being sequenced. Methods complementary to CID include HCD and ETD. Combining both fragmentation methods can improve the confidence in site localization within phosphopeptides (Frese et al., 2013).

4.4 Quantification Using Automated Extraction of Mass Chromatograms

Regardless of the methods used to produce MS/MS spectra, a key aspect of TIQUAS is the creation of a database of phosphorylation sites that have been identified at least once in the experiment. Search engines such as Mascot, Sequest, Protein Prospector, or Andromeda can be used to identify peptides and phosphopeptides from MS/MS data. These search engines produce scores that reflect the confidence in peptide sequence determination, and computational methods have been developed to analyze the data further so that additional scores quantify the confidence in site assignment in cases where the phosphopeptide contains more than one serine, threonine, or tyrosine residue (Baker, Medzihradszky, & Chalkley, 2010; Baker, Trinidad, & Chalkley, 2011; Beausoleil, Villen, Gerber, Rush, & Gygi, 2006; Cox et al., 2011; Perkins, Pappin, Creasy, & Cottrell, 1999; Savitski et al., 2011).

The quantification of individual phosphopeptides relies on integrating the areas of peaks in XICs constructed using accurate m/z and retention times for each peptide ion. These XICs are constructed for all peptides in all the LC–MS/MS runs that are compared. Retention time alignment is performed in order to correct for shifts in elution time that may have occurred in consecutive runs. To increase specificity further, XICs may be produced for the first three isotopic peaks, thus allowing obtaining the charge of the peptide. Datasets contain several thousand phosphopeptides which may be quantified in this manner in large number of samples. Thus, in order to make the approach practical, the process should be automated using computer software. Our laboratory uses a computer program named Pescal, for automating the generation of XICs of peptides present in databases of peptides and phosphopeptides (Cutillas & Vanhaesebroeck, 2007). Other computer programs that perform similar functions are available (for an exhaustive review, see Cappadona, Baker, Cutillas, Heck, & van Breukelen, 2012).

The TIQUAS approach assumes that peak areas obtained from XICs are proportional to the amount of peptide present in samples. This assumption may only holds when phosphopeptide extraction procedures produce consistent results, when variability is low and when extraction yields are linearly proportional to phosphopeptide concentrations in samples. However, because the ratio of protein to TiO_2 beads can saturate, phosphopeptide signals in LC–MS/MS runs may not always reflect phosphopeptide concentrations. Therefore, to test the reliability of the workflow as a quantitative method, our laboratory routinely tests the linearity of quantification for all phosphopeptides that may be quantified with the approach using mixing experiments (Casado & Cutillas, 2011). An illustration of these experiments is shown in Fig. 2A. In this, cells treated with pervanadate (a phosphatase inhibitor known to increase phosphorylation) are mixed with untreated cells at different proportions. The mixed cells are then subjected to the TIQUAS workflow as in Fig. 1. There should be a direct relationship between the proportion of treated cells in the mixture and the signals of phosphopeptides that increase as a result of treatment. Fig. 2B shows the XIC of a phosphopeptide that increases as a result of treatment with pV. These types of experiments showed that an appropriate ratio of TiO_2 or IMAC to protein was important to obtain quantitatively linear data (Alcolea et al., 2012; Montoya et al., 2011).

5. OBTAINING BIOLOGICAL INFORMATION FROM PHOSPHOPROTEOMICS DATASETS

Phosphoproteomics experiments generate quantitative information on thousands of phosphorylation sites. Statistical approaches used to identify significant differences in gene expression analysis, such as Limma, may also be used to select significantly modulated phosphopeptides between the comparisons (Wilkes et al., 2015). These statistical analyses produce values of fold difference and significance (*p*-value) for each quantified phosphopeptide.

Once values of relative quantity and significance have been obtained, deriving biological information from such large datasets can be a daunting task. The results of phosphoproteomics experiments are sometimes analyzed using the same bioinformatics tools used for the analysis of gene expression data, namely analysis tools that quantify the enrichment of significant genes in ontologies or pathways (Navarro, Goebel, Feijoo-Carnero, Morrice, & Cantrell, 2011). These pathway and gene ontology tools include Gene Set Enrichment Analysis, Ingenuity Pathway Analysis, Panther, etc. Although these pathway analysis methods can provide useful insights, these are not completely suitable for phosphoproteomics data analysis because phosphoproteins often contain several phosphorylation sites on their sequences, and these show different changes in abundance across experimental conditions. In addition, as outlined earlier, the biological consequence of a phosphorylation site on the activity of the protein that is

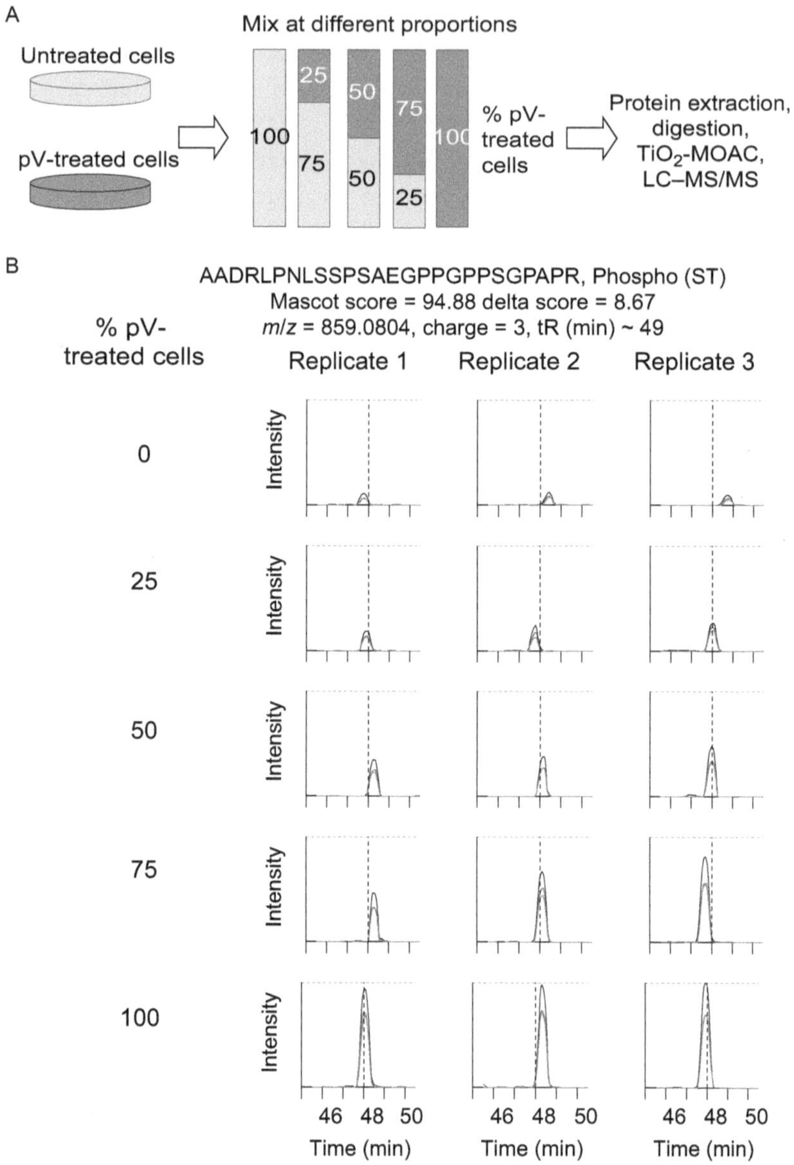

Fig. 2 Experimental design for assessing the quantitative nature of phosphoproteomic data. (A) Cells are either treated with pervanadate (pV) or left untreated and mixed at the proportions shown. Mixtures are then analyzed using the TIQUAS workflow as shown in Fig. 1. (B) Example of extracted ion chromatograms (XICs) of a representative phosphopeptide showing linearity of signal as a function of percentage of pV-treated cells in the mixture. XICs for first, second, and third isotopes of the named phosphopeptide are plotted in different colors. The data can be analyzed using linear regression as in Casado and Cutillas (2011) to assess linearity of quantification for thousands of phosphopeptides simultaneously.

being modified is most cases unknown, thus complicating the interpretation of ontology enrichment analysis.

A way of interpreting phosphoproteomics experiments involves considering the quantitative values of phosphorylation sites that are known markers of pathway activities. Several of these sites are readily detectable by MS-based phosphoproteomics, and their analysis provides a way of quantifying the same pathways as those analyzed by antibody-based approaches (such as reversed-phase protein arrays). Table 1 shows examples of well-known phosphorylation sites that can be detected using the TIQUAS approach, and which can thus be quantified in a single experiment with this workflow. An advantage over immunochemical methods is that, in addition to its specificity and multiplexing nature, MS distinguishes between phosphorylated isoforms, and thus provides an additional level of resolution. For example, ERK1 may be phosphorylated at both T202 and Y204, but the singly phosphorylated peptide at Y204 is also detectable by MS.

Considering sites with known regulatory function in phosphoproteomics datasets, such as those shown in Table 1, provides a means to mine the biological information that can be obtained from such experiments. However, this approach uses a very small proportion of phosphorylation sites present in phosphoproteomics datasets, which contain quantitative information on several thousand phosphorylation events. To address this shortcoming, and to maximize the amount or information that can be obtained from phosphoproteomic experiments, several computational approaches have been developed. One such approach aims to assign biological function to the sites that remain poorly characterized (Beltrao et al., 2012; Minguez et al., 2013). These methodologies are starting to annotate phosphorylation sites with putative functions but experimental confirmation of such computational predictions will take time as experimental confirmation of inferred functions still requires classical biochemical approaches.

An alternative means to derive biological information from phosphoproteomics data consists of using phosphorylation sites as markers of kinase activity. In this approach the function of phosphorylation on the protein receiving the modification is not considered in the analysis; instead phosphorylation is used to infer kinase and hence pathway activities. The basis of these methodologies is that each phosphorylation site is the result of *net* kinase activity (i.e., kinase and phosphatase activities); and therefore, it should in principle be possible to use phosphoproteomics data to infer the activities of all kinases expressed in a given system. A barrier for the successful

Table 1 Examples of Phosphorylation Sites Markers of Pathway Activities That Are Readily Detectable With the TIQUAS Workflow

Gene Name	Synonym	Phosphorylation Sites	Pathway	Effect	References
PRKACA	PKA	T198	cAMP/protein kinase A	Activation	Cheng, Ma, Moore, Hemmings, and Taylor (1998)
PAK2		S141	CDC42/PAK	Activation	Jung and Traugh (2005)
MAP2K1	MEK1	S222	MAPK	Activation	Pham, Arlinghaus, Zheng, Guan, and Singh (1995)
RPS6KA1	P90-S6K	S363	MAPK	Activation	Gavin and Nebreda (1999) and Roux, Richards, and Blenis (2003)
MAPK3	ERK1	T202 + Y204	MAPK	Activation	Boulton and Cobb (1991) and Boulton et al. (1991)
MAPK1	ERK2	Y187	MAPK	Activation	Boulton and Cobb (1991) and Boulton et al. (1991)
MAPK3	ERK1	Y204	MAPK	Activation	Boulton and Cobb (1991) and Boulton et al. (1991)
MAPK1	ERK2	T185 + Y187	MAPK	Activation	Boulton and Cobb (1991) and Boulton et al. (1991)
BRAF		S365	MAPK	Inhibition	Hmitou et al. (2007)
RPS6	Ribosomal S6	S236 + S240	mTORC1	Activation	Flotow and Thomas (1992) and Ruvinsky and Meyuhas (2006)
GSK3B	GSK3b	S9	PI3K/AKT	Inhibition	Cross et al. (1995)
AKT1	PKB	S473	PI3K/AKT	Activation	Alessi et al. (1996)
AKT1S1	PRAS-40	S183	PI3K/AKT/mTOR	Inhibition	Oshiro et al. (2007)
AKT1S1	PRAS-40	T246	PI3K/AKT/mTOR	Inhibition	Vander Haar, Lee, Bandhakavi, Griffin, and Kim (2007)
EIF4EBP2	4EBP2	T46	PI3K/AKT/mTOR	Inhibition	Gingras et al. (1999)
RPS6KB1	P70-S6K	T390 + S394	PI3K/AKT/mTOR	Activation	Han, Pearson, Dennis, and Thomas (1995) and Pearson et al. (1995)
STAT3		S727	STAT	Activation	Gartsbein et al. (2006)
STAT5A		Y694	STAT	Activation	Gouilleux, Wakao, Mundt, and Groner (1994)

application of this concept is that our knowledge of kinase–substrate relationships is limited and some phosphorylation sites can be modified by different kinases. For example, the serine residue at position 25 of the protein stathmin (also known as oncoprotein 18) can be phosphorylated by MAP kinases or CDKs, while serine 16 can be phosphorylate by calmodulin-dependent kinases (Hayashi et al., 2006; Larsson, Melander, Marklund, Osterman, & Gullberg, 1995; le Gouvello, Manceau, & Sobel, 1998; Marklund, Brattsand, Shingler, & Gullberg, 1993). Kinase–substrate relationships may be cell type specific, and such wiring may change upon differentiation or as a consequence of stresses. Thus taking a single phosphorylation site as a marker of a given kinase can lead to erroneous conclusions. To address this problem, algorithms have been developed that quantify the enrichment of kinase substrates in datasets so that the contribution of potential outliers to the final readouts may be diluted. One of such approaches, namely KSEA uses parametric and nonparametric statistics to infer kinase activation from the phosphoproteomics data (Casado, Rodriguez-Prados, et al., 2013). More recently iKAP and CLUE were introduced as machine-learning and clustering based methods, respectively, to estimate kinase activities from these datasets (Mischnik et al., 2015; Yang et al., 2015).

6. CONCLUSIONS

In addition to its uses as a screening tool, TIQUAS allows using phosphoproteomics as a routine assay for systematic and comprehensive quantification of cell signaling. This technology is now reaching maturity as a result of recent developments on biochemical procedures for the enrichment of phosphopeptides, on LC–MS/MS methods and hardware, on software to quantify phosphopeptides using label-free MS, and on computational methods that infer kinase pathway activities from phosphoproteomics datasets.

REFERENCES

Alcolea, M. P., Casado, P., Rodriguez-Prados, J. C., Vanhaesebroeck, B., & Cutillas, P. R. (2012). Phosphoproteomic analysis of leukemia cells under basal and drug-treated conditions identifies markers of kinase pathway activation and mechanisms of resistance. *Molecular & Cellular Proteomics*, *11*(8), 453–466. http://dx.doi.org/10.1074/mcp.M112.017483.

Alcolea, M. P., Kleiner, O., & Cutillas, P. R. (2009). Increased confidence in large-scale phosphoproteomics data by complementary mass spectrometric techniques and matching of phosphopeptide data sets. *Journal of Proteome Research*, *8*(8), 3808–3815. http://dx.doi.org/10.1021/pr800955n.

Alessi, D. R., Andjelkovic, M., Caudwell, B., Cron, P., Morrice, N., Cohen, P., & Hemmings, B. A. (1996). Mechanism of activation of protein kinase B by insulin and IGF-1. *The EMBO Journal*, *15*(23), 6541–6551.

Baker, P. R., Medzihradszky, K. F., & Chalkley, R. J. (2010). Improving software performance for peptide electron transfer dissociation data analysis by implementation of charge state- and sequence-dependent scoring. *Molecular & Cellular Proteomics*, *9*(9), 1795–1803. http://dx.doi.org/10.1074/mcp.M110.000422.

Baker, P. R., Trinidad, J. C., & Chalkley, R. J. (2011). Modification site localization scoring integrated into a search engine. *Molecular & Cellular Proteomics*, *10*(7). http://dx.doi.org/10.1074/mcp.M111.008078. M111 008078.

Beausoleil, S. A., Villen, J., Gerber, S. A., Rush, J., & Gygi, S. P. (2006). A probability-based approach for high-throughput protein phosphorylation analysis and site localization. *Nature Biotechnology*, *24*(10), 1285–1292. http://dx.doi.org/10.1038/nbt1240.

Beltran, L., & Cutillas, P. R. (2012). Advances in phosphopeptide enrichment techniques for phosphoproteomics. *Amino Acids*, *43*(3), 1009–1024. http://dx.doi.org/10.1007/s00726-012-1288-9.

Beltrao, P., Albanese, V., Kenner, L. R., Swaney, D. L., Burlingame, A., Villen, J., … Krogan, N. J. (2012). Systematic functional prioritization of protein posttranslational modifications. *Cell*, *150*(2), 413–425. http://dx.doi.org/10.1016/j.cell.2012.05.036.

Bensimon, A., Heck, A. J., & Aebersold, R. (2012). Mass spectrometry-based proteomics and network biology. *Annual Review of Biochemistry*, *81*, 379–405. http://dx.doi.org/10.1146/annurev-biochem-072909-100424.

Bodenmiller, B., Mueller, L. N., Mueller, M., Domon, B., & Aebersold, R. (2007). Reproducible isolation of distinct, overlapping segments of the phosphoproteome. *Nature Methods*, *4*(3), 231–237. http://dx.doi.org/10.1038/nmeth1005.

Boulton, T. G., & Cobb, M. H. (1991). Identification of multiple extracellular signal-regulated kinases (ERKs) with antipeptide antibodies. *Cell Regulation*, *2*(5), 357–371.

Boulton, T. G., Nye, S. H., Robbins, D. J., Ip, N. Y., Radziejewska, E., Morgenbesser, S. D., … Yancopoulos, G. D. (1991). ERKs: A family of protein-serine/threonine kinases that are activated and tyrosine phosphorylated in response to insulin and NGF. *Cell*, *65*(4), 663–675.

Cappadona, S., Baker, P. R., Cutillas, P. R., Heck, A. J., & van Breukelen, B. (2012). Current challenges in software solutions for mass spectrometry-based quantitative proteomics. *Amino Acids*, *43*(3), 1087–1108. http://dx.doi.org/10.1007/s00726-012-1289-8.

Casado, P., Alcolea, M. P., Iorio, F., Rodriguez-Prados, J. C., Vanhaesebroeck, B., Saez-Rodriguez, J., … Cutillas, P. R. (2013). Phosphoproteomics data classify hematological cancer cell lines according to tumor type and sensitivity to kinase inhibitors. *Genome Biology*, *14*(4), R37. http://dx.doi.org/10.1186/gb-2013-14-4-r37.

Casado, P., Bilanges, B., Rajeeve, V., Vanhaesebroeck, B., & Cutillas, P. R. (2014). Environmental stress affects the activity of metabolic and growth factor signaling networks and induces autophagy markers in MCF7 breast cancer cells. *Molecular & Cellular Proteomics*, *13*(3), 836–848. http://dx.doi.org/10.1074/mcp.M113.034751.

Casado, P., & Cutillas, P. R. (2011). A self-validating quantitative mass spectrometry method for assessing the accuracy of high-content phosphoproteomic experiments. *Molecular & Cellular Proteomics*, *10*(1). http://dx.doi.org/10.1074/mcp.M110.003079. M110 003079.

Casado, P., Rodriguez-Prados, J. C., Cosulich, S. C., Guichard, S., Vanhaesebroeck, B., Joel, S., & Cutillas, P. R. (2013). Kinase-substrate enrichment analysis provides insights into the heterogeneity of signaling pathway activation in leukemia cells. *Science Signaling*, *6*(268), rs6. http://dx.doi.org/10.1126/scisignal.2003573.

Cheng, X., Ma, Y., Moore, M., Hemmings, B. A., & Taylor, S. S. (1998). Phosphorylation and activation of cAMP-dependent protein kinase by phosphoinositide-dependent protein kinase. *Proceedings of the National Academy of Sciences of the United States of America, 95*(17), 9849–9854.

Cohen, P. (1982). The role of protein phosphorylation in neural and hormonal control of cellular activity. *Nature, 296*(5858), 613–620.

Cohen N. A., Kim T. S., & DeMatteo R. P. (2016). Principles of kinase inhibitor therapy for solid tumors, *Annals of Surgery*, in press, http://dx.doi.org/10.1097/SLA.0000000000001740.

Collins, B. C., Gillet, L. C., Rosenberger, G., Rost, H. L., Vichalkovski, A., Gstaiger, M., & Aebersold, R. (2013). Quantifying protein interaction dynamics by SWATH mass spectrometry: Application to the 14-3-3 system. *Nature Methods, 10*(12), 1246–1253. http://dx.doi.org/10.1038/nmeth.2703.

Conrads, T. P., Anderson, G. A., Veenstra, T. D., Pasa-Tolic, L., & Smith, R. D. (2000). Utility of accurate mass tags for proteome-wide protein identification. *Analytical Chemistry, 72*(14), 3349–3354.

Cox, J., Neuhauser, N., Michalski, A., Scheltema, R. A., Olsen, J. V., & Mann, M. (2011). Andromeda: A peptide search engine integrated into the MaxQuant environment. *Journal of Proteome Research, 10*(4), 1794–1805. http://dx.doi.org/10.1021/pr101065j.

Cross, D. A., Alessi, D. R., Cohen, P., Andjelkovich, M., & Hemmings, B. A. (1995). Inhibition of glycogen synthase kinase-3 by insulin mediated by protein kinase B. *Nature, 378*(6559), 785–789. http://dx.doi.org/10.1038/378785a0.

Cutillas, P. (2004). Principles of nanoflow liquid chromatography and applications to proteomics. *Current Nanoscience, 1*(1), 65–71.

Cutillas, P. R., Geering, B., Waterfield, M. D., & Vanhaesebroeck, B. (2005). Quantification of gel-separated proteins and their phosphorylation sites by LC-MS using unlabeled internal standards: Analysis of phosphoprotein dynamics in a B cell lymphoma cell line. *Molecular & Cellular Proteomics, 4*(8), 1038–1051. http://dx.doi.org/10.1074/mcp.M500078-MCP200.

Cutillas, P. R., & Jorgensen, C. (2011). Biological signalling activity measurements using mass spectrometry. *The Biochemical Journal, 434*(2), 189–199. http://dx.doi.org/10.1042/BJ20101974.

Cutillas, P. R., & Vanhaesebroeck, B. (2007). Quantitative profile of five murine core proteomes using label-free functional proteomics. *Molecular & Cellular Proteomics, 6*(9), 1560–1573. http://dx.doi.org/10.1074/mcp.M700037-MCP200.

de Graaf, E. L., Giansanti, P., Altelaar, A. F., & Heck, A. J. (2014). Single-step enrichment by Ti^{4+}-IMAC and label-free quantitation enables in-depth monitoring of phosphorylation dynamics with high reproducibility and temporal resolution. *Molecular & Cellular Proteomics, 13*(9), 2426–2434. http://dx.doi.org/10.1074/mcp.O113.036608.

Fabbro, D. (2015). 25 years of small molecular weight kinase inhibitors: Potentials and limitations. *Molecular Pharmacology, 87*(5), 766–775. http://dx.doi.org/10.1124/mol.114.095489.

Ficarro, S. B., McCleland, M. L., Stukenberg, P. T., Burke, D. J., Ross, M. M., Shabanowitz, J., … White, F. M. (2002). Phosphoproteome analysis by mass spectrometry and its application to *Saccharomyces cerevisiae*. *Nature Biotechnology, 20*(3), 301–305. http://dx.doi.org/10.1038/nbt0302-301.

Flotow, H., & Thomas, G. (1992). Substrate recognition determinants of the mitogen-activated 70 K S6 kinase from rat liver. *The Journal of Biological Chemistry, 267*(5), 3074–3078.

Forlenza, O. V., Torres, C. A., Talib, L. L., de Paula, V. J., Joaquim, H. P., Diniz, B. S., & Gattaz, W. F. (2011). Increased platelet GSK3B activity in patients with mild cognitive impairment and Alzheimer's disease. *Journal of Psychiatric Research, 45*(2), 220–224. http://dx.doi.org/10.1016/j.jpsychires.2010.06.002.

Frese, C. K., Altelaar, A. F., van den Toorn, H., Nolting, D., Griep-Raming, J., Heck, A. J., & Mohammed, S. (2012). Toward full peptide sequence coverage by dual fragmentation combining electron-transfer and higher-energy collision dissociation tandem mass spectrometry. *Analytical Chemistry*, *84*(22), 9668–9673. http://dx.doi.org/10.1021/ac3025366.

Frese, C. K., Zhou, H., Taus, T., Altelaar, A. F., Mechtler, K., Heck, A. J., & Mohammed, S. (2013). Unambiguous phosphosite localization using electron-transfer/higher-energy collision dissociation (EThcD). *Journal of Proteome Research*, *12*(3), 1520–1525. http://dx.doi.org/10.1021/pr301130k.

Gartsbein, M., Alt, A., Hashimoto, K., Nakajima, K., Kuroki, T., & Tennenbaum, T. (2006). The role of protein kinase C delta activation and STAT3 Ser727 phosphorylation in insulin-induced keratinocyte proliferation. *Journal of Cell Science*, *119*(Pt. 3), 470–481. http://dx.doi.org/10.1242/jcs.02744.

Gavin, A. C., & Nebreda, A. R. (1999). A MAP kinase docking site is required for phosphorylation and activation of p90(rsk)/MAPKAP kinase-1. *Current Biology*, *9*(5), 281–284.

Gingras, A. C., Gygi, S. P., Raught, B., Polakiewicz, R. D., Abraham, R. T., Hoekstra, M. F., … Sonenberg, N. (1999). Regulation of 4E-BP1 phosphorylation: A novel two-step mechanism. *Genes & Development*, *13*(11), 1422–1437.

Gouilleux, F., Wakao, H., Mundt, M., & Groner, B. (1994). Prolactin induces phosphorylation of Tyr694 of Stat5 (MGF), a prerequisite for DNA binding and induction of transcription. *The EMBO Journal*, *13*(18), 4361–4369.

Gruhler, A., Olsen, J. V., Mohammed, S., Mortensen, P., Faergeman, N. J., Mann, M., & Jensen, O. N. (2005). Quantitative phosphoproteomics applied to the yeast pheromone signaling pathway. *Molecular & Cellular Proteomics*, *4*(3), 310–327. http://dx.doi.org/10.1074/mcp.M400219-MCP200.

Han, J. W., Pearson, R. B., Dennis, P. B., & Thomas, G. (1995). Rapamycin, wortmannin, and the methylxanthine SQ20006 inactivate p70s6k by inducing dephosphorylation of the same subset of sites. *The Journal of Biological Chemistry*, *270*(36), 21396–21403.

Han, G., Ye, M., & Zou, H. (2008). Development of phosphopeptide enrichment techniques for phosphoproteome analysis. *Analyst*, *133*(9), 1128–1138. http://dx.doi.org/10.1039/b806775a.

Harrison, D. E., Strong, R., Sharp, Z. D., Nelson, J. F., Astle, C. M., Flurkey, K., … Miller, R. A. (2009). Rapamycin fed late in life extends lifespan in genetically heterogeneous mice. *Nature*, *460*(7253), 392–395. http://dx.doi.org/10.1038/nature08221.

Hayashi, K., Pan, Y., Shu, H., Ohshima, T., Kansy, J. W., White, C. L., 3rd, … Bibb, J. A. (2006). Phosphorylation of the tubulin-binding protein, stathmin, by Cdk5 and MAP kinases in the brain. *Journal of Neurochemistry*, *99*(1), 237–250. http://dx.doi.org/10.1111/j.1471-4159.2006.04113.x.

Herranz, N., Gallage, S., Mellone, M., Wuestefeld, T., Klotz, S., Hanley, C. J., … Gil, J. (2015). mTOR regulates MAPKAPK2 translation to control the senescence-associated secretory phenotype. *Nature Cell Biology*, *17*(9), 1205–1217. http://dx.doi.org/10.1038/ncb3225.

Hmitou, I., Druillennec, S., Valluet, A., Peyssonnaux, C., & Eychene, A. (2007). Differential regulation of B-raf isoforms by phosphorylation and autoinhibitory mechanisms. *Molecular and Cellular Biology*, *27*(1), 31–43. http://dx.doi.org/10.1128/MCB.01265-06.

Humphrey, S. J., Azimifar, S. B., & Mann, M. (2015). High-throughput phosphoproteomics reveals in vivo insulin signaling dynamics. *Nature Biotechnology*, *33*(9), 990–995. http://dx.doi.org/10.1038/nbt.3327.

Janku, F., Wheler, J. J., Westin, S. N., Moulder, S. L., Naing, A., Tsimberidou, A. M., … Kurzrock, R. (2012). PI3K/AKT/mTOR inhibitors in patients with breast and gynecologic malignancies harboring PIK3CA mutations. *Journal of Clinical Oncology*, *30*(8), 777–782. http://dx.doi.org/10.1200/JCO.2011.36.1196.

Jung, J. H., & Traugh, J. A. (2005). Regulation of the interaction of Pak2 with Cdc42 via autophosphorylation of serine 141. *The Journal of Biological Chemistry*, *280*(48), 40025–40031. http://dx.doi.org/10.1074/jbc.M509075200.

Kim, M. S., & Pandey, A. (2012). Electron transfer dissociation mass spectrometry in proteomics. *Proteomics*, *12*(4–5), 530–542. http://dx.doi.org/10.1002/pmic.201100517.

Knapp, S., Arruda, P., Blagg, J., Burley, S., Drewry, D. H., Edwards, A., … Zuercher, W. J. (2013). A public-private partnership to unlock the untargeted kinome. *Nature Chemical Biology*, *9*(1), 3–6. http://dx.doi.org/10.1038/nchembio.1113.

Landry, C. R., Levy, E. D., & Michnick, S. W. (2009). Weak functional constraints on phosphoproteomes. *Trends in Genetics*, *25*(5), 193–197. http://dx.doi.org/10.1016/j.tig.2009.03.003.

Larsen, M. R., Thingholm, T. E., Jensen, O. N., Roepstorff, P., & Jorgensen, T. J. (2005). Highly selective enrichment of phosphorylated peptides from peptide mixtures using titanium dioxide microcolumns. *Molecular & Cellular Proteomics*, *4*(7), 873–886. http://dx.doi.org/10.1074/mcp.T500007-MCP200.

Larsson, N., Melander, H., Marklund, U., Osterman, O., & Gullberg, M. (1995). G2/M transition requires multisite phosphorylation of oncoprotein 18 by two distinct protein kinase systems. *The Journal of Biological Chemistry*, *270*(23), 14175–14183.

Lawrence, R. T., Searle, B. C., Llovet, A., & Villen, J. (2016). Plug-and-play analysis of the human phosphoproteome by targeted high-resolution mass spectrometry. *Nature Methods*, *13*(5), 431–434. http://dx.doi.org/10.1038/nmeth.3811.

le Gouvello, S., Manceau, V., & Sobel, A. (1998). Serine 16 of stathmin as a cytosolic target for Ca2+/calmodulin-dependent kinase II after CD2 triggering of human T lymphocytes. *Journal of Immunology*, *161*(3), 1113–1122.

Lee, J., Xu, Y., Chen, Y., Sprung, R., Kim, S. C., Xie, S., & Zhao, Y. (2007). Mitochondrial phosphoproteome revealed by an improved IMAC method and MS/MS/MS. *Molecular & Cellular Proteomics*, *6*(4), 669–676. http://dx.doi.org/10.1074/mcp.M600218-MCP200.

Li, Q. R., Ning, Z. B., Tang, J. S., Nie, S., & Zeng, R. (2009). Effect of peptide-to-TiO2 beads ratio on phosphopeptide enrichment selectivity. *Journal of Proteome Research*, *8*(11), 5375–5381. http://dx.doi.org/10.1021/pr900659n.

Liang, H. C., Lahert, E., Pike, I., & Ward, M. (2015). Quantitation of protein post-translational modifications using isobaric tandem mass tags. *Bioanalysis*, 7(3), 383–400. http://dx.doi.org/10.4155/bio.14.296.

Lienhard, G. E. (2008). Non-functional phosphorylations? *Trends in Biochemical Sciences*, *33*(8), 351–352. http://dx.doi.org/10.1016/j.tibs.2008.05.004.

Liu, Y., Huttenhain, R., Surinova, S., Gillet, L. C., Mouritsen, J., Brunner, R., … Aebersold, R. (2013). Quantitative measurements of N-linked glycoproteins in human plasma by SWATH-MS. *Proteomics*, *13*(8), 1247–1256. http://dx.doi.org/10.1002/pmic.201200417.

Liu, Z., Wang, Y., & Xue, Y. (2013). Phosphoproteomics-based network medicine. *The FEBS Journal*, *280*(22), 5696–5704. http://dx.doi.org/10.1111/febs.12380.

Longo, V. D., Antebi, A., Bartke, A., Barzilai, N., Brown-Borg, H. M., Caruso, C., … Fontana, L. (2015). Interventions to slow aging in humans: Are we ready? *Aging Cell*, *14*(4), 497–510. http://dx.doi.org/10.1111/acel.12338.

Macek, B., Mann, M., & Olsen, J. V. (2009). Global and site-specific quantitative phosphoproteomics: Principles and applications. *Annual Review of Pharmacology and Toxicology*, *49*, 199–221. http://dx.doi.org/10.1146/annurev.pharmtox.011008.145606.

Manning, G., Whyte, D. B., Martinez, R., Hunter, T., & Sudarsanam, S. (2002). The protein kinase complement of the human genome. *Science*, *298*(5600), 1912–1934. http://dx.doi.org/10.1126/science.1075762.

Marklund, U., Brattsand, G., Shingler, V., & Gullberg, M. (1993). Serine 25 of oncoprotein 18 is a major cytosolic target for the mitogen-activated protein kinase. *The Journal of Biological Chemistry*, *268*(20), 15039–15047.

Mertins, P., Yang, F., Liu, T., Mani, D. R., Petyuk, V. A., Gillette, M. A., … Carr, S. A. (2014). Ischemia in tumors induces early and sustained phosphorylation changes in stress kinase pathways but does not affect global protein levels. *Molecular & Cellular Proteomics*, *13*(7), 1690–1704. http://dx.doi.org/10.1074/mcp.M113.036392.

Minguez, P., Letunic, I., Parca, L., & Bork, P. (2013). PTMcode: A database of known and predicted functional associations between post-translational modifications in proteins. *Nucleic Acids Research*, *41*(Database issue), D306–D311. http://dx.doi.org/10.1093/nar/gks1230.

Mischnik, M., Sacco, F., Cox, J., Schneider, H. C., Schafer, M., Hendlich, M., … Klabunde, T. (2015). IKAP: A heuristic framework for inference of kinase activities from phosphoproteomics data. *Bioinformatics*, *32*, 424–431, http://dx.doi.org/10.1093/bioinformatics/btv699.

Montoya, A., Beltran, L., Casado, P., Rodriguez-Prados, J. C., & Cutillas, P. R. (2011). Characterization of a TiO(2) enrichment method for label-free quantitative phosphoproteomics. *Methods*, *54*(4), 370–378. http://dx.doi.org/10.1016/j.ymeth.2011.02.004.

Navarro, M. N., Goebel, J., Feijoo-Carnero, C., Morrice, N., & Cantrell, D. A. (2011). Phosphoproteomic analysis reveals an intrinsic pathway for the regulation of histone deacetylase 7 that controls the function of cytotoxic T lymphocytes. *Nature Immunology*, *12*(4), 352–361. http://dx.doi.org/10.1038/ni.2008.

Nita-Lazar, A., Saito-Benz, H., & White, F. M. (2008). Quantitative phosphoproteomics by mass spectrometry: Past, present, and future. *Proteomics*, *8*(21), 4433–4443. http://dx.doi.org/10.1002/pmic.200800231.

Ojanpera, I., Kolmonen, M., & Pelander, A. (2012). Current use of high-resolution mass spectrometry in drug screening relevant to clinical and forensic toxicology and doping control. *Analytical and Bioanalytical Chemistry*, *403*(5), 1203–1220. http://dx.doi.org/10.1007/s00216-012-5726-z.

Olsen, J. V., Blagoev, B., Gnad, F., Macek, B., Kumar, C., Mortensen, P., & Mann, M. (2006). Global, in vivo, and site-specific phosphorylation dynamics in signaling networks. *Cell*, *127*(3), 635–648. http://dx.doi.org/10.1016/j.cell.2006.09.026.

Oshiro, N., Takahashi, R., Yoshino, K., Tanimura, K., Nakashima, A., Eguchi, S., … Yonezawa, K. (2007). The proline-rich Akt substrate of 40 kDa (PRAS40) is a physiological substrate of mammalian target of rapamycin complex 1. *The Journal of Biological Chemistry*, *282*(28), 20329–20339. http://dx.doi.org/10.1074/jbc.M702636200.

Parker, B. L., Yang, G., Humphrey, S. J., Chaudhuri, R., Ma, X., Peterman, S., & James, D. E. (2015). Targeted phosphoproteomics of insulin signaling using data-independent acquisition mass spectrometry. *Science Signaling*, *8*(380), rs6. http://dx.doi.org/10.1126/scisignal.aaa3139.

Parmenter, T. J., Kleinschmidt, M., Kinross, K. M., Bond, S. T., Li, J., Kaadige, M. R., … McArthur, G. A. (2014). Response of BRAF-mutant melanoma to BRAF inhibition is mediated by a network of transcriptional regulators of glycolysis. *Cancer Discovery*, *4*(4), 423–433. http://dx.doi.org/10.1158/2159-8290.CD-13-0440.

Pearson, R. B., Dennis, P. B., Han, J. W., Williamson, N. A., Kozma, S. C., Wettenhall, R. E., & Thomas, G. (1995). The principal target of rapamycin-induced p70s6k inactivation is a novel phosphorylation site within a conserved hydrophobic domain. *The EMBO Journal*, *14*(21), 5279–5287.

Perkins, D. N., Pappin, D. J., Creasy, D. M., & Cottrell, J. S. (1999). Probability-based protein identification by searching sequence databases using mass spectrometry data. *Electrophoresis*, *20*(18), 3551–3567. http://dx.doi.org/10.1002/(SICI)1522-2683(19991201)20:18<3551::AID-ELPS3551>3.0.CO;2-2.

Pham, C. D., Arlinghaus, R. B., Zheng, C. F., Guan, K. L., & Singh, B. (1995). Characterization of MEK1 phosphorylation by the v-Mos protein. *Oncogene*, *10*(8), 1683–1688.

Pinkse, M. W., Uitto, P. M., Hilhorst, M. J., Ooms, B., & Heck, A. J. (2004). Selective isolation at the femtomole level of phosphopeptides from proteolytic digests using 2D-NanoLC-ESI-MS/MS and titanium oxide precolumns. *Analytical Chemistry*, *76*(14), 3935–3943. http://dx.doi.org/10.1021/ac0498617.

Quetier, I., Marshall, J. J., Spencer-Dene, B., Lachmann, S., Casamassima, A., Franco, C., … Cameron, A. J. (2016). Knockout of the PKN family of Rho effector kinases reveals a non-redundant role for PKN2 in developmental mesoderm expansion. *Cell Reports*, *14*(3), 440–448. http://dx.doi.org/10.1016/j.celrep.2015.12.049.

Reinders, J., & Sickmann, A. (2005). State-of-the-art in phosphoproteomics. *Proteomics*, *5*(16), 4052–4061. http://dx.doi.org/10.1002/pmic.200401289.

Riley, N. M., & Coon, J. J. (2016). Phosphoproteomics in the age of rapid and deep proteome profiling. *Analytical Chemistry*, *88*(1), 74–94. http://dx.doi.org/10.1021/acs.analchem.5b04123.

Roux, P. P., Richards, S. A., & Blenis, J. (2003). Phosphorylation of p90 ribosomal S6 kinase (RSK) regulates extracellular signal-regulated kinase docking and RSK activity. *Molecular and Cellular Biology*, *23*(14), 4796–4804.

Ruvinsky, I., & Meyuhas, O. (2006). Ribosomal protein S6 phosphorylation: From protein synthesis to cell size. *Trends in Biochemical Sciences*, *31*(6), 342–348. http://dx.doi.org/10.1016/j.tibs.2006.04.003.

Samovski, D., Sun, J., Pietka, T., Gross, R. W., Eckel, R. H., Su, X., … Abumrad, N. A. (2015). Regulation of AMPK activation by CD36 links fatty acid uptake to beta-oxidation. *Diabetes*, *64*(2), 353–359. http://dx.doi.org/10.2337/db14-0582.

Savitski, M. M., Lemeer, S., Boesche, M., Lang, M., Mathieson, T., Bantscheff, M., & Kuster, B. (2011). Confident phosphorylation site localization using the Mascot Delta Score. *Molecular & Cellular Proteomics*, *10*(2). http://dx.doi.org/10.1074/mcp.M110.003830. M110 003830.

Scheltema, R. A., Hauschild, J. P., Lange, O., Hornburg, D., Denisov, E., Damoc, E., … Mann, M. (2014). The Q Exactive HF, a benchtop mass spectrometer with a pre-filter, high-performance quadrupole and an ultra-high-field orbitrap analyzer. *Molecular & Cellular Proteomics*, *13*(12), 3698–3708. http://dx.doi.org/10.1074/mcp.M114.043489.

Schroeder, M. J., Shabanowitz, J., Schwartz, J. C., Hunt, D. F., & Coon, J. J. (2004). A neutral loss activation method for improved phosphopeptide sequence analysis by quadrupole ion trap mass spectrometry. *Analytical Chemistry*, *76*(13), 3590–3598. http://dx.doi.org/10.1021/ac0497104.

Sharma, K., D'Souza, R. C., Tyanova, S., Schaab, C., Wisniewski, J. R., Cox, J., & Mann, M. (2014). Ultradeep human phosphoproteome reveals a distinct regulatory nature of Tyr and Ser/Thr-based signaling. *Cell Reports*, *8*(5), 1583–1594. http://dx.doi.org/10.1016/j.celrep.2014.07.036.

Solari, F. A., Dell'Aica, M., Sickmann, A., & Zahedi, R. P. (2015). Why phosphoproteomics is still a challenge. *Molecular BioSystems*, *11*(6), 1487–1493. http://dx.doi.org/10.1039/c5mb00024f.

Steeg, P. S., Palmieri, D., Ouatas, T., & Salerno, M. (2003). Histidine kinases and histidine phosphorylated proteins in mammalian cell biology, signal transduction and cancer. *Cancer Letters*, *190*(1), 1–12.

Steen, H., & Mann, M. (2004). The ABC's (and XYZ's) of peptide sequencing. *Nature Reviews. Molecular Cell Biology*, *5*(9), 699–711. http://dx.doi.org/10.1038/nrm1468.

Swaney, D. L., Wenger, C. D., & Coon, J. J. (2010). Value of using multiple proteases for large-scale mass spectrometry-based proteomics. *Journal of Proteome Research*, *9*(3), 1323–1329. http://dx.doi.org/10.1021/pr900863u.

Tan, C. S., Jorgensen, C., & Linding, R. (2010). Roles of "junk phosphorylation" in modulating biomolecular association of phosphorylated proteins? *Cell Cycle*, *9*(7), 1276–1280.

Terfve, C. D., Wilkes, E. H., Casado, P., Cutillas, P. R., & Saez-Rodriguez, J. (2015). Large-scale models of signal propagation in human cells derived from discovery phosphoproteomic data. *Nature Communications*, *6*, 8033. http://dx.doi.org/10.1038/ncomms9033.

Thingholm, T. E., Jensen, O. N., & Larsen, M. R. (2009). Analytical strategies for phosphoproteomics. *Proteomics*, *9*(6), 1451–1468. http://dx.doi.org/10.1002/pmic.200800454.

Ulintz, P. J., Yocum, A. K., Bodenmiller, B., Aebersold, R., Andrews, P. C., & Nesvizhskii, A. I. (2009). Comparison of MS(2)-only, MSA, and MS(2)/MS(3) methodologies for phosphopeptide identification. *Journal of Proteome Research*, *8*(2), 887–899. http://dx.doi.org/10.1021/pr800535h.

Vander Haar, E., Lee, S. I., Bandhakavi, S., Griffin, T. J., & Kim, D. H. (2007). Insulin signalling to mTOR mediated by the Akt/PKB substrate PRAS40. *Nature Cell Biology*, *9*(3), 316–323. http://dx.doi.org/10.1038/ncb1547.

Wilkes, E. H., Terfve, C., Gribben, J. G., Saez-Rodriguez, J., & Cutillas, P. R. (2015). Empirical inference of circuitry and plasticity in a kinase signaling network. *Proceedings of the National Academy of Sciences of the United States of America*, *112*(25), 7719–7724. http://dx.doi.org/10.1073/pnas.1423344112.

Wilm, M., & Mann, M. (1996). Analytical properties of the nanoelectrospray ion source. *Analytical Chemistry*, *68*(1), 1–8.

Wisniewski, J. R., Zougman, A., Nagaraj, N., & Mann, M. (2009). Universal sample preparation method for proteome analysis. *Nature Methods*, *6*(5), 359–362. http://dx.doi.org/10.1038/nmeth.1322.

Workman, P., Al-Lazikani, B., & Clarke, P. A. (2013). Genome-based cancer therapeutics: Targets, kinase drug resistance and future strategies for precision oncology. *Current Opinion in Pharmacology*, *13*(4), 486–496. http://dx.doi.org/10.1016/j.coph.2013.06.004.

Wu, C. C., & Yates, J. R., 3rd. (2003). The application of mass spectrometry to membrane proteomics. *Nature Biotechnology*, *21*(3), 262–267. http://dx.doi.org/10.1038/nbt0303-262.

Wulfkuhle, J. D., Aquino, J. A., Calvert, V. S., Fishman, D. A., Coukos, G., Liotta, L. A., & Petricoin, E. F., 3rd. (2003). Signal pathway profiling of ovarian cancer from human tissue specimens using reverse-phase protein microarrays. *Proteomics*, *3*(11), 2085–2090. http://dx.doi.org/10.1002/pmic.200300591.

Yang, P., Zheng, X., Jayaswal, V., Hu, G., Yang, J. Y., & Jothi, R. (2015). Knowledge-based analysis for detecting key signaling events from time-series phosphoproteomics data. *PLoS Computational Biology*, *11*(8), e1004403. http://dx.doi.org/10.1371/journal.pcbi.1004403.

CHAPTER FIFTEEN

Biotin Switch Assays for Quantitation of Reversible Cysteine Oxidation

R. Li*,†, J. Kast*,†,‡,1

*The Biomedical Research Centre, University of British Columbia, Vancouver, BC, Canada
†University of British Columbia, Vancouver, BC, Canada
‡Centre for Blood Research, University of British Columbia, Vancouver, BC, Canada
[1]Corresponding author: e-mail address: kast@brc.ubc.ca

Contents

Abstract

Thiol groups in protein cysteine residues can be subjected to different oxidative modifications by reactive oxygen/nitrogen species. Reversible cysteine oxidation, including *S*-nitrosylation, *S*-sulfenylation, *S*-glutathionylation, and disulfide formation, modulate multiple biological functions, such as enzyme catalysis, antioxidant, and other signaling pathways. However, the biological relevance of reversible cysteine oxidation is typically underestimated, in part due to the low abundance and high reactivity of some of these modifications, and the lack of methods to enrich and quantify them. To facilitate future research efforts, this chapter describes detailed procedures to target the different modifications using mass spectrometry-based biotin switch assays. By switching the modification of interest to a biotin moiety, these assays leverage the high affinity between biotin and avidin to enrich the modification. The use of stable isotope labeling and a range of selective reducing agents facilitate the quantitation of individual as well as total reversible cysteine oxidation. The biotin switch assay has been widely applied to the quantitative analysis of *S*-nitrosylation in different disease models and is now also

Methods in Enzymology, Volume 585
ISSN 0076-6879
http://dx.doi.org/10.1016/bs.mie.2016.10.006

emerging as a valuable research tool for other oxidative cysteine modifications, highlighting its relevance as a versatile, robust strategy for carrying out in-depth studies in redox proteomics.

1. INTRODUCTION

Mass spectrometry-based proteomics is the study of proteins in cells or tissues under certain conditions for the purpose of understanding their biological activities (Tyers & Mann, 2003). In order to quantitatively measure a protein/peptide in a mixture, a lot of effort has been put into the field (Bantscheff, Schirle, Sweetman, Rick, & Kuster, 2007; Elliott, Smith, Parker, & Borchers, 2009). Metabolic- and chemical-labeling-based quantitative proteomics are widely applied in human health research to study drug effects, biomarkers of disease, and pathways involved in disease processes. This includes stable isotope labeling with amino acid in cell culture (SILAC) as one of the most accurate quantitative proteomics techniques to date (Mann, 2006).

Posttranslational modifications (PTMs), the modification of proteins after their translation, profoundly affect various aspects of a protein, including protein structure, solubility, activity, location in the cells, and binding to interaction partners. Of the 20 amino acids, cysteine (Cys) is the second least abundant amino acid. It is, however, one of the most susceptible residues for PTMs, including oxidation and disulfide formation in protein folding. All of the PTMs on Cys are due to a change of the oxidation state of the thiol group in the side chain and reflect the redox state of a cell. The branch of proteomics that analyzes oxidative modifications of Cys is named redox proteomics and allows the understanding of the cellular oxidation state as well as redox regulation.

Oxidation of the Cys thiol group by hydrogen peroxide, a reactive oxygen species (ROS), yields sulfenic acid modification (R-SOH), also known as sulfenylation, as a reversible, labile moiety with a mass shift of +16 Da. As the initial PTM generated by oxidation of a thiolate, sulfenic acid can be stabilized within certain protein microenvironments (Phalen et al., 2006), but is generally vulnerable to further irreversible oxidation to sulfinic acid ($R–SO_2H$) and sulfonic acid ($R–SO_3H$) with mass shifts of +32 and +48 Da, respectively. Sulfenic acid can also react with a neighboring amide nitrogen to form the stabilized cyclic 5-membrane ring structure sulfenamide, characterized by a mass shift of 31 Da (Tonks, 2006). Besides ROS, reactive nitrogen species also react with Cys via *S*-nitrosylation to form *S*-nitrosothiol (R–SNO), which is a chemical intermediate. Although found in low abundance, *S*-nitrosylation has been shown to be involved in a variety of dynamic redox-signaling/regulation events, and has been implicated in cardioprotection by several studies (Martínez-Ruiz, Cadenas, & Lamas, 2011; Murphy, Kohr, Sun, Nguyen, & Steenbergen, 2012; Nadtochiy, Burwell, & Brookes, 2007). A leading hypothesis for its protective effect is that reversible *S*-nitrosylation protects critical Cys from subsequent irreversible oxidation (Martínez-Ruiz et al., 2011; Nadtochiy et al., 2007).

Another type of PTM on Cys is the disulfide derivative formed with another protein cysteine (R–S–S–R′, as an intra- or intermolecular interaction), or with a small molecular thiol such as glutathione (GSH) or cysteine. The disulfide bond has initially been regarded as a static structural feature that occurs during protein folding in the oxidizing environment of the endoplasmic reticulum that is preserved throughout the lifetime of the protein. Nowadays, more and more disulfide bonds have been found to be dynamically interchangeable in the cellular redox environment, which alters protein conformation, association, and function (Lindahl, Mata-Cabana, & Kieselbach, 2011). The mixed disulfide bond formed between a protein's Cys thiol and GSH, termed as *S*-glutathionylation, has now been recognized as another potential modulator of redox-sensitive proteins in signal transduction. Similar to *S*-nitrosylation, *S*-glutathionylation has also been proposed to function as a temporary thiol cap to protect critical Cys from potentially irreversible oxidative modification (Hill & Bhatnagar, 2012).

The low abundance of Cys oxidative modifications makes it necessary to include an enrichment procedure in redox proteomic protocols. In order to study *S*-nitrosothiol, Jaffrey et al. introduced the biotin switch method in 2001, which enabled the enrichment and detection of *S*-nitrosylation in complex biological samples (Jaffrey & Snyder, 2001). Since then, several variations of the biotin switch assay have been developed that expanded its range to also include other reversible cysteine modifications. To target sulfenylation, the Eaton group has modified the biotin switch assay by utilizing m-arsenite to selectively reduce the sulfenic acid modification (Saurin, Neubert, Brennan, & Eaton, 2004). Besides arsenite, a widely applied strategy for the identification of sulfenic acid modifications uses dimedone, a specific targeting reagent that eliminates blocking and reducing steps (Furdui & Poole, 2014; Nelson et al., 2010). For *S*-glutathionylation, the modified biotin switch assay employs glutaredoxin (Grx) to enzymatically reduce *S*-GSH-modified Cys (Reynaert et al., 2006). Another biotin switch approach applies a biotin-labeled GSH ester as a cell-permeable

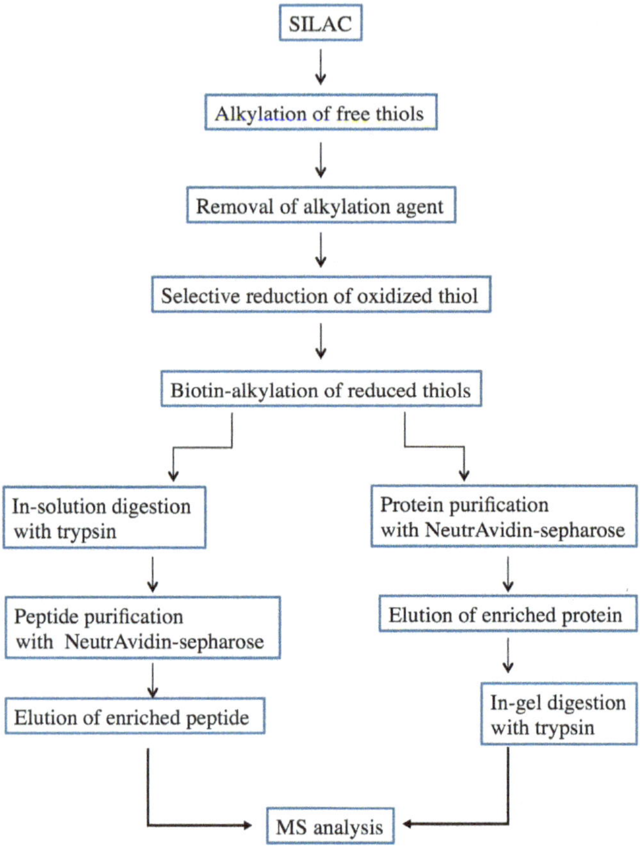

Fig. 1 The general workflow of biotin switch assays for the quantitation of reversible cysteine oxidation. Abbreviations: *SILAC*, stable isotope labeling with amino acid in cell culture; *MS*, mass spectrometry.

reagent to capture *S*-glutathionylation directly (Brennan et al., 2006). Finally, to achieve an unbiased assessment of the entire reversible Cys oxidation proteome including disulfides, the biotin switch assay has been adapted for complete reduction of all reversibly oxidized Cys with tris(2-carboxyethyl)phosphine or dithiothreitol (Li, Huang, & Kast, 2015).

For all modified biotin switch assays, the underlying workflow is the same (Fig. 1), except for a choice between peptide- and protein-level enrichment. In the actual biotin switch part of the workflow, free thiols are first permanently blocked with dedicated cysteine alkylation reagents. Next, the Cys modifications of interest are reduced with specific reducing reagents to recuperate free thiols (Table 1), which are then labeled with a thiol-reactive biotin, or directly labeled with a modification-specific biotin agent (Fig. 2). Once the recuperated modification sites have been labeled with the stable

Table 1 The Available Choices of Selective Reduction Agents and Biotin-Alkylation Agents for Various Types of Reversible Cysteine Oxidations

Cysteine Modification	Selective Reduction Agent	Biotin-Alkylation Agent
S-nitrosylation	Ascorbic acid	Biotin–HPDP, biotin–maleimide
S-sulfenation	Arsenite	Biotin–HPDP, biotin–maleimide, biotin–dimedone[a]
S-glutathionylation	Glutaredoxin	Biotin–HPDP, biotin–maleimide, biotinylated GSH ethyl ester[a]
Reversible cysteine oxidation	TCEP, DTT	Biotin–HPDP, biotin–maleimide

[a]Directly label the modification without reduction.
Abbreviations: *DTT*, DL-dithiothreitol; *TCEP*, tris(2-carboxyethyl)phosphine hydrochloride.

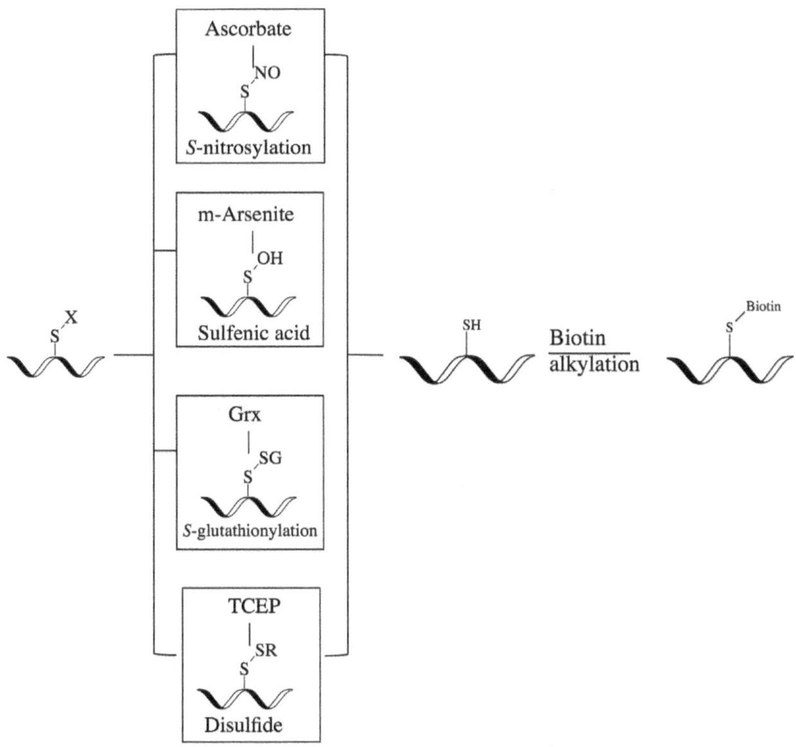

Fig. 2 Schematic diagram of the biotin switch method. Different reversible cysteine oxidative modifications (–S–X) are selectively reduced by their corresponding reducing agents to free thiols (–SH). The newly released free thiols are then alkylated with biotin agent (–S–biotin).

biotin group, they can easily be enriched by avidin beads and analyzed by MS to determine their positions in the corresponding proteins.

2. CHEMICALS

N-ethylmaleimide (NEM), DL-dithiothreitol (DTT), iodoacetamide (IAA), tris(2-carboxyethyl)phosphine hydrochloride (TCEP–HCl), sodium arsenite, and biotin–maleimide can be obtained, e.g., from Sigma (Oakville, ON, Canada). EZ-Link biotin–HPDP and NeutrAvidin resins are available e.g. from Pierce (Rockford, IL, USA).

3. METHOD

3.1 Preparation of Protein Samples

1. The stable isotope labeling with amino acid in cell culture is performed as recommended by the nature protocol (Ong & Mann, 2006).
2. After corresponding treatment, wash SILAC-labeled cells twice with cold-PBS. Pellet the cells and carefully remove PBS.
3. Add 0.5–1 mL lysis buffer to the cell pellet and resuspend the pellet. Transfer the sample into a microcentrifuge tube and incubate on ice for 5 min.

   Lysis buffer typically used: 50 m*M* Tris, 0.1% sodium deoxycholate, 150 m*M* NaCl, 1 m*M* EDTA, 1% NP-40, pH 7.5.

   Note: Add protease inhibitors to lysis buffer before use.

   For the quantitation of sulfenic acid modifications, the pH of lysis buffer has to be maintained at 6.
4. Centrifuge at 16,000 × *g* at 4°C for 15 min. Transfer the supernatant (total lysate) to a new tube.
5. Determine the protein concentration of cell lysate using BCA assay (Adilakshami & Laine, 2002), and dilute the cell lysate with lysis buffer to the final protein concentration to 1 mg/mL.

3.2 Biotin-Alkylation of Proteins

1. Prepare the NEM stock solution of 10 mg/mL in water, and add equal amount of NEM to the cell lysate (1:1, w/w). Incubate the mixture at 37°C for 1 h.
2. Remove excess NEM by buffer exchange using Amicon Centrifugal Filter Units with 10 kDa cutoff.

   Alternative: React extra NEM with excess L-cysteine (fourfold molar ratio to NEM) at 37°C for 1 h.

Note: Cysteine cannot be utilized to quench NEM when quantifying sulfenic acid modifications, because it may react with sulfenic acids to form disulfide bonds that are inaccessible for the selective reducing agent.

3. Biotin-alkylation of targeted cysteine oxidation.

(1) *S*-nitrosylation

Selectively reduce *S*-nitrosothiol with 1 m*M* ascorbic acid (stock solution: 50 m*M* ascorbic acid in deionized water) (Jaffrey & Snyder, 2001), and block the newly released cysteine residues with 1 m*M* biotin–maleimide.

(2) Sulfenylation

Method 1: Harvest the cells from the plate into lysis buffer containing protease inhibitors, 1 m*M* Biotin–dimedone, 200 units/mL catalase, 10 m*M* NEM, and 10 m*M* IAA, to directly label the sulfenic acid-containing proteins at time of lysis. Incubate the mixture on ice for 1 h (Nelson et al., 2010).

Method 2: Selectively reduce sulfenic acids and label the newly released free thiols by adding 200 m*M* arsenite, together with 1 m*M* biotin–maleimide (stock solution: 40 m*M* in acetic acid). Incubate the mixture for 2 h at 37°C.

(3) *S*-glutathionylation

Method 1: Simultaneously add NEM and the stock solution of GSH ethyl ester (BioGEE, 12.5 m*M* in dimethyl sulfoxide, prepared immediately before use) into the cell lysate to a final concentration of 0.25 m*M* (Priora, Coppo, Salzano, Di Simplicio, & Ghezzi, 2010).

Method 2: Selectively reduce *S*-glutathionylated cysteines in a buffer containing 13.5 μg/mL recombinant human Grx, 35 μg/mL GSSG reductase, 1 m*M* GSH, 1 m*M* NADPH, 18 μ*M* EDTA, and 137 m*M* Tris–HCl, pH 8.0 for 30 min (Aesif, Janssen-Heininger, & Reynaert, 2010). Then label the reduced cysteine residues with 1 m*M* biotin–maleimide.

(4) Total reversible cysteine oxidation (including disulfide bond, *S*-nitrosylation, and sulfenic acid modification).

Reduce the reversible cysteine oxidation with TCEP (3 m*M*), and remove extra TCEP by buffer exchange using Amicon Centrifugal Filter Units with 10 kDa cutoff. After that, label the newly released free thiols with biotin–HPDP (dissolve biotin–HPDP in dimethyl sulfoxide and add it to the final concentration of 0.4 m*M*) at 37°C for 2 h.

4. Remove excess biotin agents by buffer exchange using Amicon Centrifugal Filter Units with 10 kDa cutoff.

3.3 NeutrAvidin Enrichment of Biotin-Labeled Peptide/Protein

3.3.1 Peptide Level Enrichment

1. In-solution digest the biotin-labeled proteins (including proteins labeled with biotin–HPDP, biotin–dimedone, or BioGEE) with trypsin overnight at 37°C.

   Digestion buffer: 1% sodium deoxycholate, 50 m*M* NH_4HCO_3, 5 m*M* $CaCl_2$.

   Trypsin stock solution: 1 μg/μL. Add trypsin to a final protease: protein ratio of 1:50 (w/w).
2. After digestion, adjust the pH of solution to 7, and incubate the peptide mixture with sepharose 6B beads for 1 h at room temperature to remove unspecific binding. Collect the supernatant by centrifugation at 500 × *g* for 1 min.
3. Pack resin in a tube by centrifugation at 500 × *g* for 1 min to remove storage solution. Add equal volume of washing buffer on top of the resin bed, centrifuge at 500 × *g* for 1 min to remove buffer.

   Washing buffer: 50 m*M* Tris, 600 m*M* NaCl, 1 m*M* EDTA, 0.5% NP-40, pH 7.5.
4. Mix the supernatant with washed NeutrAvidin agarose resin, and gently rotate for 1.5 h at room temperature. Use approximately 200 μL NeutrAvidin agarose resin per 1 mg protein.
5. After incubation, centrifuge the mixture at 500 × *g* for 10 min, and collect the supernatant as flow-through. Wash the NeutrAvidin agarose resin three times with washing buffer, and three times with cold-PBS.
6. Elute the NeutroAvidin-bound peptides with DTT (10 m*M* in 50 m*M* NH_4HCO_3), and alkylate the newly released cysteines with IAA (55 m*M* in 50 m*M* NH_4HCO_3).
7. Fractionate the purified Cys-containing peptides using an anion exchange tip (Wisniewski, Zougman, & Mann, 2009) into six fractions, and then desalt with a solid phase extraction on C18 stage tips (Rappsilber, Mann, & Ishihama, 2007).

3.3.2 Protein Level Enrichment

1. Pack resin in a tube by centrifugation at 500 × *g* for 1 min to remove storage solution. Add equal volume of washing buffer on top of the resin bed, centrifuge at 500 × *g* for 1 min to remove buffer.

Washing buffer: 50 m*M* Tris, 600 m*M* NaCl, 1 m*M* EDTA, 0.5% NP-40, pH 7.5.

2. Recover the biotin–maleimide-labeled protein mixture from the Amicon centrifugal filter device, dilute the protein mixture with washing buffer to the protein concentration to 1 mg/mL, and incubate the protein mixture with sepharose 6B beads for 1 h at room temperature to remove unspecific binding. After that, collect the supernatant by centrifugation at 500 × *g* for 1 min.
3. Mix the supernatant with washed NeutrAvidin agarose resin, and gently rotate for 1.5 h at room temperature. Use approximately 200 μL NeutrAvidin agarose resin per 1 mg protein.
4. After incubation, centrifuge the mixture at 500 × *g* for 10 min, and collect the supernatant as flow-through. Wash the NeutrAvidin agarose resin three times with washing buffer, and three times with cold-PBS.
5. Elute the enriched proteins by boiling the NeutrAvidin agarose resin in 4 × SDS buffer in the presence of 1 m*M* biotin for 5 min.

   4 × SDS buffer: 400 m*M* Tris, 8% SDS, 40% glycerin, 5 mg/mL Coomassie Brilliant Blue, pH 6.8.
6. Load the eluted proteins onto 10% SDS gel for electrophoretic separation. Stain the gels with Coomassie Brilliant Blue, and cut each lane into five bands.
7. In-gel digestion is performed essentially as described previously (Shevchenko, Tomas, Havlis, Olsen, & Mann, 2006). In brief, reduce the gel pieces with 10 m*M* DTT at 56°C for 30 min, then alkylate with 55 m*M* IAA at room temperature for 45 min. Digest the protein in the gel particles in a buffer containing 50 m*M* NH_4HCO_3, 5 m*M* $CaCl_2$, and 0.1–0.2 μg trypsin per band, and incubate at 37°C for at least 16 h. Extract peptides from the gel pieces by adding extraction buffer (1:2 (v/v) 5% formic acid/acetonitrile), and desalt the peptides via solid phase extraction on C18 stage tips (Rappsilber et al., 2007).

3.4 Mass Spectrometry Analysis

1. Reconstitute dried samples in 6 μL of 0.05% formic acid for analysis by nLC–MS/MS.
2. Perform the MS/MS analysis database searching using Mascot distiller or MaxQuant. Detailed search setup as follows:

   Quantitative method: SILAC

   Database for Mascot searching: Uniprot–Swissprot

   Digestion enzyme: Trypsin/P

A maximum of one missed cleavage
Variable modifications: carbamidomethyl (C), *N*-ethylmaleimide (C), oxidation (M), Sigma-Biotin (C) (biotin–maleimide modification)
1% False discovery rate

3.5 Pyridine-2-Thione Assay to Monitor the Selective Reducing Reaction

To optimize the concentrations and reaction times of the selective reducing agent in different cell lysates, a pyridine-2-thione assay can be performed by utilizing biotin–HPDP. The reaction product of free thiols with biotin–HPDP, pyridine-2-thione, has a molar extinction coefficient of $8.08 \times 10^3\ M^{-1}\ \text{cm}^{-1}$ at 343 nm.

1. After blocking free thiols in cell lysate with NEM and depleting excess NEM, measure and record the absorbance at 343 nm of the protein sample (compared to a blank as the absorbance at time 0 min).
2. Add different concentrations of reducing agents, such as arsenite, ascorbic acid, or Grx, and biotin–HPDP to the protein sample.
3. Measure and record the absorbance at 343 nm of the sample at 15-min intervals for 2 h.
4. For each sample with various concentrations of reducing agent, calculate the absorbance per mg of protein and plot it against time to determine the optimal concentration and reaction time of reducing agent.

3.6 Western Blot of Biotin-Labeled Proteins or Affinity-Enriched Proteins

The global biotin incorporation can be visualized using gel-based techniques, such as antibiotin western blot using a streptavidin-HRP conjugate antibody.

Following step 6 in Section 3.3.2, after loading the gel

1. Transfer protein from SDS-PAGE to preactivated PVDF membrane.
2. Block the PVDF membrane with 5% (w/v) milk powder solved in TBST for 1 h in room temperature.
3. Wash the PVDF membrane three times with TBST for 10 min each time.
4. Incubate membrane with Streptavidin-HRP (at the appropriate dilution) in 10 mL of blocking buffer (1 × TBST with 5% milk powder) with gentle agitation for 1 h at room temperature.
5. Wash three times for 5 min each with 15 mL of TBST.
6. Proceed with detection.

4. APPLICATION

The noncovalent biotin–avidin interaction ($K_d = 10^{-15}\,M$) is employed in a variety of protein detection and purification methods, and the reliance on the biotin–avidin interaction has been driven by the development of biotin switch assays. The original biotin switch assay has been introduced in 2001 to enrich *S*-nitrosothiol (Jaffrey & Snyder, 2001), but has kept evolving since then. The dynamic development of the biotin switch assay includes the incorporation of mass spectrometry analysis to identify *S*-nitrosothiol sites, as well as isotope-labeling methods, e.g., SILAC for quantitation (Zhou et al., 2010). This assay has been used with some success to identify potentially *S*-nitrosylated proteins (for a recent review, see ref. Wojdyla & Rogowska-Wrzesinska, 2015) and has been developed into commercially available kits. For most of the published applications on *S*-nitrosylation, an average of 100 *S*-nitrosothiol sites was identified in various biological systems. However, Lee and colleagues reported the identification of 834 potential *S*-nitrosylated human proteins using a high-density protein microarray chip in conjunction with the biotin switch assay and provided a regulatory mechanism of the ubiquitin proteasome system. Not only did their work define an expanded human *S*-nitrosoproteome, it also provided a high-throughput screening strategy to explore PTMs (Lee et al., 2014). The wide range of applications of the biotin switch assay has identified proteins susceptible to *S*-nitrosothiol in diverse biological models. This includes a model for angiogenesis in endothelial cells, in which 586 unique *S*-nitrosylation sites in 384 proteins were identified, 234 of which were novel (Chen et al., 2010). In this study, the authors discovered several consensus motifs and located nitrosylated cysteines mostly to the cytoplasmic side of membrane proteins, suggesting that tertiary structure and subcellular localization are important factors that control *S*-nitrosylation. The biotin switch assay was also applied to identify *S*-nitrosothiol in a prostate disease model, in which 82 *S*-nitrosothiol sites were determined (Lam et al., 2010). This study further analyzed the sequences of the *S*-nitrosothiol sites and resolved their 3D structure, which indicated that most of the *S*-nitrosothiol modifications resided in hydrophobic pockets or with charged amino acid in their neighborhood. Another study employed the biotin switch assay in a nerve injury model and quantified 30 proteins with increased and 23 proteins with decreased *S*-nitrosothiol (Scheving et al., 2012). The proteins that were identified as being *S*-nitrosylated were involved in mitochondrial function, protein folding and transport, synaptic signaling, and redox control. Besides profiling the *S*-nitrosoproteome, the biotin switch assay was also used to explore *S*-nitrosylation on individual proteins, such as the androgen receptor in prostate cancer (Qin, Dey, Purayil, & Daaka, 2013), X-linked inhibitor of apoptosis in neurodegenerative disorders (Tsang et al., 2009) and calgranulin A in the inflammatory response of neutrophils and macrophages toward lipopolysaccharide and cytokines (Lim et al., 2008; Torta & Bachi, 2012). Together, these studies have significantly expanded our understanding of the (patho)physiological role of *S*-nitrosothiol in health and disease.

Despite the progress in *S*-nitrosylation analysis, most of the studies relied on the exposure of cell lysate to nitrosylating agents, such as *S*-nitrosoglutathione (GSNO), to induce *S*-nitrosothiol, which is an artificial setup that does not represent in vivo conditions. Moreover, the selectivity of ascorbic acid as reducing agent in the assay is not clearly defined. To address these shortcomings, direct labeling methods have recently been developed. For example, a sulfinate-linked biotin agent was used to enrich and detect *S*-nitrosothiol in different mouse tissues (Majmudar et al., 2016). By leveraging the stable thiosulfonate bond that forms between sulfinic acid and *S*-nitrosothiol, this method enriched 175 endogenous *S*-nitrosylated proteins, including metabolic enzymes and proteins with metal coordination sites.

In contrast to the high number of studies carried out on *S*-nitrosylation and their diversity, only a few applications of biotin switch assays on other Cys oxidative modifications have so far been reported. For profiling sulfenic acid modifications, the controversy of the selectivity of the arsenite-based reduction of sulfenic acid modification has seemingly limited its application. One recent study used arsenite in conjunction with two-dimensional electrophoresis to investigate sulfenic acid modifications in the kidney medulla of spontaneously hypertensive rats in comparison to normotensive rats. This study identified 35 sulfenic acid sites, including different peroxiredoxin isoforms, Cu–Zn superoxide dismutase, and ATP synthase (Tyther, Ahmeda, Johns, McDonagh, & Sheehan, 2010). In contrast to the arsenite-based reduction approach, direct labeling with dimedone has been employed more frequently. For example, biotin–dimedone was used to analyze sulfenylated cardiac proteins from rat ventricular myocytes (Charles et al., 2007) and identified potential sulfenylation of various peroxiredoxins, proteins in energy metabolism, and GSH biosynthesis.

For the analysis of *S*-glutathionylation, the biotin switch assay, which involved Grx-catalyzed reduction of *S*-glutathionylation, was employed to in situ evaluate *S*-glutathionylated protein content in mouse tissue from various models of lung injury and fibrosis. The biotin-labeled proteins were visualized through interaction with fluorescent-conjugated streptavidin, which demonstrated the regional changes of *S*-glutathionylation in a variety of diseases (Aesif et al., 2009). Another proteomic study used cell prelabeling with biotin-tagged GSH and identified 61 *S*-glutathionylated proteins in airway epithelial cells, which implied a critical role for DUOX1 in ATP-stimulated *S*-glutathionylation of β-actin, peroxiredoxin 1, the nonreceptor tyrosine kinase Src, and MAPK phosphatase 1 (Hristova et al., 2014).

Our group has adapted the quantitative biotin switch assay to evaluate total reversible Cys oxidation by using TCEP as the reducing agent. The assay was applied to an atherosclerotic model and quantified more than 100 reversible Cys oxidation sites on various proteins, including glycolytic enzymes (glyceraldehyde 3-phosphate dehydrogenase), redox regulation enzymes (thioredoxin and peroxiredoxin), and chaperones (heat shock 70 kDa protein) (Li et al., 2015).

5. CONCLUSION

Since their introduction in 2001, biotin switch assays have evolved into powerful tools that can specifically enrich different types of reversible Cys oxidation in cellular proteins that are produced in response to stimuli or under physiologically relevant conditions. The integration of quantitative workflows now enables the measurement of the degree of modification on individual residues in various biological systems. While extensive studies exist for *S*-nitrosothiol, other types of reversible cysteine oxidation, e.g., sulfenic acid and disulfide, have not been investigated to the same extent. This is partially due to the significance of nitric oxide as a signaling molecule, but also because detailed high-throughput methods for other modifications are lacking. To address this limitation, we provide in this chapter a general protocol for a set of dedicated biotin switch assays, which we have already validated for the quantitation of total reversible cysteine oxidation and sulfenic acid modification in one specific disease model. This should enable researchers interested in less studied oxidative Cys modifications, such as sulfenic acid and *S*-glutathionylation, to directly apply quantitative biotin switch assays to a wide range of biological models. As a result, we expect to see the number of studies involving Cys modifications other than *S*-nitrosylation to increase significantly, and lead to new insights and a greater appreciation of the importance of redox proteomics in general.

REFERENCES

Adilakshami, T., & Laine, R. O. (2002). Ribosomal protein S25 mRNA partners with MTF-1 and La to provide a p53-mediated mechanism for survival or death. *The Journal of Biological Chemistry*, *277*, 4147–4151.

Aesif, S. W., Anathy, V., Havermans, M., Guala, A. S., Ckless, K., Taatjes, D. J., & Janssen-Heininger, Y. M. (2009). In situ analysis of protein S-glutathionylation in lung tissue using glutaredoxin-1-catalyzed cysteine derivatization. *American Journal of Pathology*, *175*(1), 36–45. http://dx.doi.org/10.2353/ajpath.2009.080736.

Aesif, S. W., Janssen-Heininger, Y. M. W., & Reynaert, N. L. (2010). Protocols for the detection of S-glutathionylated and S-nitrosylated proteins in situ. *Methods in Enzymology*, *474*, 289–296. http://dx.doi.org/10.1016/S0076-6879(10)74017-9.

Bantscheff, M., Schirle, M., Sweetman, G., Rick, J., & Kuster, B. (2007). Quantitative mass spectrometry in proteomics: A critical review. *Analytical and Bioanalytical Chemistry*, *389*(4), 1017–1031. http://dx.doi.org/10.1007/s00216-007-1486-6.

Brennan, J. P., Miller, J. I. A., Fuller, W., Wait, R., Begum, S., Dunn, M. J., & Eaton, P. (2006). The utility of N,N-biotinyl glutathione disulfide in the study of protein S-glutathiolation. *Molecular & Cellular Proteomics*, *5*(2), 215–225. http://dx.doi.org/10.1074/mcp.M500212-MCP200.

Charles, R. L., Schröder, E., May, G., Free, P., Gaffney, P. R. J., Wait, R., … Eaton, P. (2007). Protein sulfenation as a redox sensor: Proteomics studies using a novel biotinylated dimedone analogue. *Molecular & Cellular Proteomics*, *6*(9), 1473–1484.

Chen, Y. J., Ku, W. C., Lin, P. Y., Chou, H. C., Khoo, K. H., & Chen, Y. J. (2010). S-alkylating labeling strategy for site-specific identification of the s-nitrosoproteome. *Journal of Proteome Research*, *9*(12), 6417–6439. http://dx.doi.org/10.1021/pr100680a.

Elliott, M. H., Smith, D. S., Parker, C. E., & Borchers, C. (2009). Current trends in quantitative proteomics. *Journal of Mass Spectrometry*, *44*(12), 1637–1660. http://dx.doi.org/10.1002/jms.1692.

Furdui, C. M., & Poole, L. B. (2014). Chemical approaches to detect and analyze protein sulfenic acids. *Mass Spectrometry Reviews*, *33*(2), 126–146. http://dx.doi.org/10.1002/mas.21384.

Hill, B. G., & Bhatnagar, A. (2012). Protein S-glutathiolation: Redox-sensitive regulation of protein function. *Journal of Molecular and Cellular Cardiology*, *52*(3), 559–567. http://dx.doi.org/10.1016/j.yjmcc.2011.07.009.

Hristova, M., Veith, C., Habibovic, A., Lam, Y. W., Deng, B., Geiszt, M., … van der Vliet, A. (2014). Identification of DUOX1-dependent redox signaling through protein S-glutathionylation in airway epithelial cells. *Redox Biology*, *2*, 436–446. http://dx.doi.org/10.1016/j.redox.2013.12.030.

Jaffrey, S. R., & Snyder, S. H. (2001). The biotin switch method for the detection of S-nitrosylated proteins. *Science's STKE*, *2001*(86), pl1. http://dx.doi.org/10.1126/stke.2001.86.pl1.

Lam, Y. W., Yuan, Y., Isaac, J., Babu, C. V. S., Meller, J., & Ho, S.-M. (2010). Comprehensive identification and modified-site mapping of *S*-nitrosylated targets in prostate epithelial cells. *PloS One*, *5*(2). e9075. http://dx.doi.org/10.1371/journal.pone.0009075.

Lee, Y. I., Giovinazzo, D., Kang, H. C., Lee, Y., Jeong, J. S., Doulias, P. T., … Dawson, T. M. (2014). Protein microarray characterization of the S-nitrosoproteome. *Molecular & Cellular Proteomics*, *13*(1), 63–72. http://dx.doi.org/10.1074/mcp.M113.032235.

Li, R., Huang, J., & Kast, J. (2015). Identification of total reversible cysteine oxidation in an atherosclerosis model using a modified biotin switch assay. *Journal of Proteome Research*, *14*(5), 2026–2035. http://dx.doi.org/10.1021/acs.jproteome.5b00133.

Lim, S. Y., Raftery, M., Cai, H., Hsu, K., Yan, W. X., Hseih, H. L., … Geczy, C. L. (2008). S-nitrosylated S100A8: Novel anti-inflammatory properties. *Journal of Immunology*, *181*(8), 5627–5636.

Lindahl, M., Mata-Cabana, A., & Kieselbach, T. (2011). The disulfide proteome and other reactive cysteine proteomes: Analysis and functional significance. *Antioxidants & Redox Signaling*, *14*(12), 2581–2642. http://dx.doi.org/10.1089/ars.2010.3551.

Majmudar, J. D., Konopko, A. M., Labby, K. J., Tom, C. T. M. B., Crellin, J. E., Prakash, A., & Martin, B. R. (2016). Harnessing redox cross-reactivity to profile distinct cysteine modifications. *Journal of the American Chemical Society*, *138*(6), 1852–1859. http://dx.doi.org/10.1021/jacs.5b06806.

Mann, M. (2006). Functional and quantitative proteomics using SILAC. *Nature Reviews. Molecular Cell Biology*, *7*(12), 952–958. http://dx.doi.org/10.1038/nrm2067.

Martínez-Ruiz, A., Cadenas, S., & Lamas, S. (2011). Nitric oxide signaling: Classical, less classical, and nonclassical mechanisms. *Free Radical Biology and Medicine*, *51*(1), 17–29. http://dx.doi.org/10.1016/j.freeradbiomed.2011.04.010.

Murphy, E., Kohr, M., Sun, J., Nguyen, T., & Steenbergen, C. (2012). S-nitrosylation: A radical way to protect the heart. *Journal of Molecular and Cellular Cardiology*, *52*(3), 568–577. http://dx.doi.org/10.1016/j.yjmcc.2011.08.021.

Nadtochiy, S. M., Burwell, L. S., & Brookes, P. S. (2007). Cardioprotection and mitochondrial S-nitrosation: Effects of S-nitroso-2-mercaptopropionyl glycine (SNO-MPG) in cardiac ischemia-reperfusion injury. *Journal of Molecular and Cellular Cardiology*, *42*(4), 812–825. http://dx.doi.org/10.1016/j.yjmcc.2007.01.010.

Nelson, K. J., Klomsiri, C., Codreanu, S. G., Soito, L., Liebler, D. C., Rogers, L. C., … Poole, L. B. (2010). Use of dimedone-based chemical probes for sulfenic acid detection. *Methods in Enzymology*, *473*, 95–115. http://dx.doi.org/10.1016/s0076-6879(10)73004-4.

Ong, S. E., & Mann, M. (2006). A practical recipe for stable isotope labeling by amino acids in cell culture (SILAC). *Nature Protocols*, *1*(6), 2650–2660. http://dx.doi.org/10.1038/nprot.2006.427.

Phalen, T. J., Weirather, K., Deming, P. B., Anathy, V., Howe, A. K., van der Vliet, A., … Heintz, N. H. (2006). Oxidation state governs structural transitions in peroxiredoxin II that correlate with cell cycle arrest and recovery. *The Journal of Cell Biology*, *175*(5), 779–789. http://dx.doi.org/10.1083/jcb.200606005.

Priora, R., Coppo, L., Salzano, S., Di Simplicio, P., & Ghezzi, P. (2010). Measurement of mixed disulfides including glutathionylated proteins. *Methods in Enzymology*, *473*, 149–159. http://dx.doi.org/10.1016/s0076-6879(10)73007-x.

Qin, Y., Dey, A., Purayil, H. T., & Daaka, Y. (2013). Maintenance of androgen receptor inactivation by S-nitrosylation. *Cancer Research*, *73*(22), 6690–6699. http://dx.doi.org/10.1158/0008-5472.can-13-1042.

Rappsilber, J., Mann, M., & Ishihama, Y. (2007). Protocol for micro-purification, enrichment, pre-fractionation and storage of peptides for proteomics using StageTips. *Nature Protocols*, *2*(8), 1896–1906. http://dx.doi.org/10.1038/nprot.2007.261.

Reynaert, N. L., Ckless, K., Guala, A. S., Wouters, E. F. M., van der Vliet, A., & Janssen-Heininger, Y. M. W. (2006). In situ detection of S-glutathionylated proteins following glutaredoxin-1 catalyzed cysteine derivatization. *Biochimica et Biophysica Acta (BBA) - General Subjects*, *1760*(3), 380–387. http://dx.doi.org/10.1016/j.bbagen.2006.01.006.

Saurin, A. T., Neubert, H., Brennan, J. P., & Eaton, P. (2004). Widespread sulfenic acid formation in tissues in response to hydrogen peroxide. *Proceedings of the National Academy of Sciences of the United States of America*, *101*(52), 17982–17987. http://dx.doi.org/10.1073/pnas.0404762101.

Scheving, R., Wittig, I., Heide, H., Albuquerque, B., Steger, M., Brandt, U., & Tegeder, I. (2012). Protein S-nitrosylation and denitrosylation in the mouse spinal cord upon injury of the sciatic nerve. *Journal of Proteomics*, *75*(13), 3987–4004. http://dx.doi.org/10.1016/j.jprot.2012.05.006.

Shevchenko, A., Tomas, H., Havlis, J., Olsen, J. V., & Mann, M. (2006). In-gel digestion for mass spectrometric characterization of proteins and proteomes. *Nature Protocols*, *1*(6), 2856–2860. http://dx.doi.org/10.1038/nprot.2006.468.

Tonks, N. K. (2006). Protein tyrosine phosphatases: From genes, to function, to disease. *Nature Reviews. Molecular Cell Biology*, 7(11), 833–846.

Torta, F., & Bachi, A. (2012). Quantitative analysis of S-nitrosylated proteins. In K. Marcus (Ed.), *Quantitative methods in proteomics* (pp. 405–416). Totowa, NJ: Humana Press.

Tsang, A. H. K., Lee, Y.-I. L., Ko, H. S., Savitt, J. M., Pletnikova, O., Troncoso, J. C., … Chung, K. K. K. (2009). S-nitrosylation of XIAP compromises neuronal survival in Parkinson's disease. *Proceedings of the National Academy of Sciences of the United States of America*, *106*(12), 4900–4905. http://dx.doi.org/10.1073/pnas.0810595106.

Tyers, M., & Mann, M. (2003). From genomics to proteomics. *Nature*, *422*(6928), 193–197.

Tyther, R., Ahmeda, A., Johns, E., McDonagh, B., & Sheehan, D. (2010). Proteomic profiling of perturbed protein sulfenation in renal medulla of the spontaneously hypertensive rat. *Journal of Proteome Research*, *9*(5), 2678–2687. http://dx.doi.org/10.1021/pr1001719.

Wisniewski, J. R., Zougman, A., & Mann, M. (2009). Combination of FASP and StageTip-based fractionation allows in-depth analysis of the hippocampal membrane proteome. *Journal of Proteome Research*, *8*, 5674–5678.

Wojdyla, K., & Rogowska-Wrzesinska, A. (2015). Differential alkylation-based redox proteomics—Lessons learnt. *Redox Biology*, *6*, 240–252. http://dx.doi.org/10.1016/j.redox.2015.08.005.

Zhou, X., Han, P., Li, J., Zhang, X., Huang, B., Ruan, H.-Q., & Chen, C. (2010). ESNOQ, proteomic quantification of endogenous s-nitrosation. *Plos One*, *5*(4), 1–7. http://dx.doi.org/10.1371/10.1371/journal.pone.0010015.g001; 10.1371/journal.pone.0010015.g002.

CHAPTER SIXTEEN

Application of a Thioredoxin-Trapping Mutant for Analysis of the Cellular Nitrosoproteome

M. Benhar[1]

Rappaport Institute for Research in the Medical Sciences, Faculty of Medicine, Technion-Israel Institute of Technology, Haifa, Israel

[1]Corresponding author: e-mail address: benhar@tx.technion.ac.il

Contents

Abstract

Nitric oxide influences a wide range of cellular functions through S-nitrosylation, a redox-dependent posttranslational protein modification that involves attachment of a nitroso moiety to a reactive thiol group. Over the past two decades, S-nitrosylation has emerged as a ubiquitous mechanism for controlling the activity, subcellular localization, and molecular interactions of proteins, thereby influencing many cellular processes. In addition, recent studies have indicated that aberrant S-nitrosylation may lead to cellular dysfunction and damage. Despite significant advances in the field, progress has been hindered by challenges related to the analysis of S-nitrosylation by large-scale proteomic approaches. This chapter describes the application of a thioredoxin-trapping mutant for proteomic analysis of S-nitrosylation. Thioredoxin is a ubiquitous oxidoreductase directly involved in denitrosylation reactions. The presented method relies upon mechanism-based trapping, whereby a recombinant thioredoxin trap mutant captures nitrosylated proteins, which are subsequently isolated and identified by mass spectrometry. This nitrosothiol-trapping procedure can expand upon and complement currently available methods for the analysis of the nitrosoproteome.

Methods in Enzymology, Volume 585
ISSN 0076-6879
http://dx.doi.org/10.1016/bs.mie.2016.09.003

1. INTRODUCTION

Nitric oxide (NO) is a gaseous molecule that plays important functions in diverse organisms. Cellular effects of NO are principally mediated via its interaction with protein metal centers (mainly heme) and through redox modifications of proteins. In particular, S-nitrosylation, the attachment of a nitroso group to a cysteine thiol to form an S-nitrosothiol (SNO), has emerged as a major mechanism by which NO influences a wide range of cellular processes, including metabolism, differentiation, apoptosis, and stress responses (Benhar, 2016; Gould, Doulias, Tenopoulou, Raju, & Ischiropoulos, 2013; Hess, Matsumoto, Kim, Marshall, & Stamler, 2005). Reversible S-nitrosylation modulates protein function in multiple ways, including regulation of enzymatic activity and alteration of protein localization, stability, or interactions with other macromolecules. The nitrosylation state of any protein is determined by the balance between nitrosylation and denitrosylation. SNO formation is largely dependent upon the stimulus-induced activation of one of the three NO synthase (NOS) enzymes, namely nNOS (NOS1), iNOS (NOS2), or eNOS (NOS3) (Anand & Stamler, 2012). SNO reduction can occur by several enzymatic and nonenzymatic mechanisms. Recent studies have indicated that denitrosylation of cellular proteins is largely mediated by glutathione (GSH), thioredoxin (Trx), and their associated redox systems (Benhar, Forrester, & Stamler, 2009). In particular, GSH together with GSNO reductase (GSNOR) and Trx together with Trx reductase (TrxR) are two physiologically relevant denitrosylating systems that control endogenous S-nitrosylation (Benhar, 2015; Benhar et al., 2009). Despite significant progress over the past decade, there is still limited knowledge regarding the components of the proteome that undergo S-nitrosylation across different cell types and conditions; therefore, much is still unknown about the specific roles and functional significance of S-nitrosylation/denitrosylation in cellular function and disease.

Global proteomic approaches hold great potential for broad and unbiased exploration of S-nitrosylation. However, it is currently challenging to comprehensively analyze S-nitrosylation in complex protein samples such as tissue or cell homogenates. The reasons for this difficulty include: (1) the chemical lability of the SNO bond, (2) the transient nature of the modification, and (3) the typically low stoichiometry or occupancy at which it occurs, particularly under physiologically relevant conditions.

Nonetheless, several biochemical approaches have been developed and successfully applied for proteomic analysis of S-nitrosylation (Raju, Doulias, Tenopoulou, Greene, & Ischiropoulos, 2012). Unarguably, the biotin switch technique (BST) continues to be the mainstay in the analysis of S-nitrosylation of endogenous proteins (Jaffrey, Erdjument-Bromage, Ferris, Tempst, & Snyder, 2001). This method has proven to be widely applicable and useful in the functional and proteomic investigation of S-nitrosylation in diverse biological systems. However, various limitations of this indirect method have been noted, as discussed elsewhere (Diers, Keszler, & Hogg, 2014; Forrester, Foster, Benhar, & Stamler, 2009). Of special note, the BST is prone to both false-positive identification of SNO due to incomplete blocking of free thiols and false negatives due to SNO loss during this rather lengthy and harsh procedure. Incomplete thiol reduction or labeling may also compromise assay sensitivity. To overcome some of these limitations, several groups have recently reported alternative approaches for proteomic analysis of endogenous S-nitrosylation. These methods employ different chemical strategies for the direct and selective capture of nitrosylated cysteines, by exploiting the specific reactions between SNO and organic mercury (Doulias et al., 2010), phenylsulfinic acid (Majmudar et al., 2016), or triaryl phosphine (Seneviratne et al., 2016). These approaches have been successfully applied for characterizing the nitrosoproteome of various cell and tissue types, including brain tissue isolated from a mouse model of Alzheimer's disease (Doulias et al., 2010; Majmudar et al., 2016; Seneviratne et al., 2016).

This chapter describes another recently introduced approach for nitrosoproteome analysis involving SNO capture by a Trx-trapping mutant (Ben-Lulu, Ziv, Admon, Weisman-Shomer, & Benhar, 2014). It is based upon recent studies concerning the mechanism of Trx-mediated denitrosylation. These studies have shown that, similar to disulfide reduction, SNO reduction by Trx occurs in a two-step mechanism that involves its Cys-Gly-Pro-Cys active site (Benhar, Forrester, Hess, & Stamler, 2008; Benhar et al., 2009). In the first step of the reaction, the more N-terminal cysteine (Cys32 in human Trx1) attacks the sulfur atom on the SNO moiety of the substrate protein, thereby displacing NO (formally, NO^-) and generating an intermolecular disulfide between Trx and the substrate. In the second step, the disulfide is cleaved via an intramolecular attack by the second active site cysteine (Cys35, known as the "resolving cysteine"), thereby releasing the reduced target protein and the oxidized Trx.

The normally transient disulfide intermediate formed in the first step is stabilized in the case of a Trx mutant that lacks the resolving cysteine. This so-called trap mutant can thus be employed for capturing nitrosylated proteins (Benhar et al., 2008; Ben-Lulu et al., 2014). The Trx trap mutant also harbors a streptavidin-binding peptide (SBP) tag, which facilitates the subsequent enrichment and identification of the Trx–target complexes. The method is schematically illustrated in Fig. 1. The protocol later describes the application of Trx trapping for the analysis of S-nitrosylation of cytosolic proteins obtained from cultured mammalian cells. The method should be applicable to other experimental systems, but adaptations may be required as further discussed later.

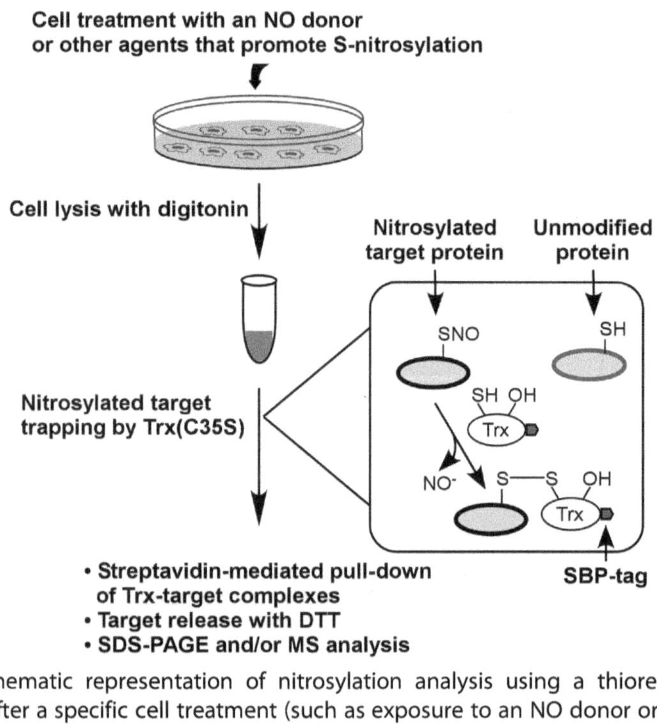

Fig. 1 Schematic representation of nitrosylation analysis using a thioredoxin trap mutant. After a specific cell treatment (such as exposure to an NO donor or cytokines) cell lysates are prepared using digitonin lysis buffer. Protein samples are incubated with a thioredoxin (Trx) trap mutant, Trx(C35S). In the trap mutant, the resolving cysteine is replaced by serine (–OH). The protein also contains a streptavidin-binding peptide (SBP) tag. Trx(C35S) forms mixed disulfide bonds with nitrosylated substrates and the resulting complexes are pulled down using avidin agarose. Captured proteins are released from the complex with dithiothreitol (DTT) and then analyzed by SDS-PAGE and/or liquid chromatography-tandem mass spectrometry (LC-MS/MS).

2. MATERIALS

Recombinant Trx proteins: His-tagged Trx proteins, either wild-type or the C35S trap mutant, are expressed in *Escherichia coli* BL21 cells and purified using nickel affinity chromatography, as described before (Ben-Lulu et al., 2014). These Trx proteins harbor a C-terminally located SBP and hexahistidine tags that facilitate their purification from bacteria or enrichment during the SNO-trapping assay. In these Trx proteins, the three noncatalytic cysteines (residues 62, 69, and 73) are replaced with alanine in order to prevent thiol–disulfide exchange reactions during the trapping procedure. The wild-type Trx protein serves as a negative control to ensure that captured proteins were trapped though the expected mixed disulfide species.

Affinity media: streptavidin agarose beads (Thermo Scientific, Rockford, IL).

Cell lysis buffer: 0.05% digitonin in Tris-buffered saline (TBS), pH 7.5.

Wash buffer 1: 1% Triton X-100 in TBS, pH 7.5.

Wash buffer 2: 0.1% Triton X-100 in TBS, pH 7.5.

Elution buffer: 100 m*M* dithiothreitol (DTT) in TBS, pH 7.5.

3. METHODS

3.1 Cell Treatment

S-nitrosylation can be induced in cultured mammalian cells by applying an exogenous NO/SNO donor or through activation of NOS, with the choice of treatment depending on the experimental system and biological question at hand. Various NO/SNO donating agents can be used such as S-nitrosocysteine, S-nitrosoglutathione, or NONOate compounds. The conditions for application of a specific donor need to be tailored for the cell under investigation. It is usually possible, and highly recommended, that before embarking on large-scale proteomic analyses, small-scale experiments are conducted in order to verify that nitrosylation has occurred. For this purpose, several established methods to assess S-nitrosylation can be employed such as BST or chemical reductive chemiluminescence (Schonhoff & Benhar, 2011).

3.2 Preparation of Cytosolic Fractions

After cell treatment, the cells are washed with phosphate-buffered saline and digitonin cell lysates are prepared by incubating the cells with cell lysis buffer for 10 min at 4°C on a rotating wheel followed by centrifugation (1000 × *g*, 15 min, 4°C) to clear the lysate. Digitonin is used as the cell-permeabilizing agent as it is a weak nonionic detergent, which at low concentrations selectively renders the plasma membrane permeable, causing it to release cytosolic components but to leave other organelles intact. As such, it is optimal for obtaining a cytosolic fraction. Some considerations regarding the preparation of protein samples are as follows. Application of cell lysis methods different from the one used here may be appropriate but need to be verified to be compatible with the trapping method. For example, procedures that disrupts intracellular organelles are likely to produce a lysate that contains many oxidized proteins (for example, various endoplasmic reticulum proteins) resulting in significant protein trapping that is independent of S-nitrosylation. It should also be noted that the present protocol has been optimized for cultured cells. Preparation protein samples from other sources (e.g., animal tissues) will likely require some adaptations. In this regard, one should also consider the risk of protein oxidation that may occur during the initial steps of lysate preparation. If artifactual oxidation presents as a problem, appropriate steps should be taken, such as buffer degassing, or exclusion of metal ions from buffers using Chelex 100 resin.

3.3 Nitrosothiol Capture by Thioredoxin-Trapping Mutant

Proteins samples from the previous step are next subjected to the trapping procedure. Note that the following protocol is suitable for small-scale pilot experiments. The procedure is scaled up for large-scale proteomic analysis (see later). Streptavidin agarose beads (40 μL/sample) are washed three times in TBS. Then, the beads are loaded with recombinant Trx(C35S) protein (50 μg) in the presence of 20 m*M* DTT for 1 h at 4°C and subsequently washed to remove the reductant. The beads are then incubated with supernatants of digitonin-permeabilized cells (3–5 mg protein) for 1 h at 4°C. The trapping reaction is then quenched with *N*-ethylmaleimide (NEM, 100 m*M*) for 15 min at room temperature and thereafter the beads are washed extensively at 4°C as follows: twice with wash buffer 1 containing 10 m*M* NEM and 1 *M* NaCl, once with wash buffer 1 containing 10 m*M* NEM, once with wash buffer 2, and three times with TBS. The pH of all the buffers is adjusted to 7.5. At least 10 mL of each solution is used for each wash.

3.4 Detection of Captured Proteins

Proteins captured by Trx in the preceding step are released by incubating the streptavidin beads with elution buffer for 30 min at room temperature. The samples are then resolved by SDS-PAGE gels, followed by staining with Coomassie blue and destaining using standard procedures. Detection sensitivity can be enhanced by employing suitable fluorescent dyes and imaging devices. After establishing the experimental procedures and obtaining satisfactory results in pilot studies, the next step is to perform large-scale experiments. For this purpose, the earlier-described protocol is scaled up at least fourfold and the eluted material is subjected to mass spectrometric (MS) analysis. The MS analysis is typically performed in a "shotgun" format where proteins are digested to peptides, which are separated and analyzed by liquid chromatography-tandem MS (LC-MS/MS). As discussed elsewhere, absolute or relative protein abundances in the different samples can be measured by metabolic or chemical protein labeling or by label-free methods (Neilson et al., 2011; Nikolov, Schmidt, & Urlaub, 2012). Each approach has its own particular strengths and weaknesses, but in general, labeling approaches provides better accuracy, whereas label-free analysis enables higher proteome coverage and greater dynamic range of quantification (Neilson et al., 2011; Nikolov et al., 2012). In every quantification procedure, it is necessary to set a threshold by which proteins are assigned as candidate nitrosylated targets. This is illustrated in the recent application of the trapping approach to SNO-treated THP-1 monocytes (Ben-Lulu et al., 2014). Based on MS measurements coupled to a label-free analysis using MaxQuant software (Cox & Mann, 2008), a threshold of 2 was chosen, meaning that only those proteins whose abundance was calculated to be higher by a least twofold in the SNO sample vs the control were considered as putative nitrosylated targets (Ben-Lulu et al., 2014). Regardless of the particular analytical approach, technical and biological replicates are important for ensuring the accuracy and biological significance of the MS measurements.

4. CONCLUDING REMARKS

The protocol presented herein represents a simple yet robust method for proteomic analysis of S-nitrosylation. This method has been successfully applied to identify hundreds of novel nitrosylated proteins in SNO-treated human monocytes and in endotoxin and cytokine-stimulated mouse macrophages (Ben-Lulu et al., 2014). The inherent ability of Trx to distinguish nitrosylated protein thiols from the vast excess of reduced thiols and the use of native conditions during the trapping and enrichment procedure are important parameters that contribute to the overall effectiveness of this approach. Indeed, the method was shown to readily identify multiple low abundance SNO targets, which have not been previously identified by BST-based proteomic analyses.

In addition to its strengths, the SNO-trapping method has several limitations and potential pitfalls that need to be considered. First, various SNO proteins may not be recognized and denitrosylated by Trx, and still others may be denitrosylated by Trx via a transnitrosylation reaction that does not involve covalent complex formation (Benhar, 2015; Benhar, Thompson, Moseley, & Stamler, 2010). On the other hand, the Trx trap mutant may form adducts not only with SNO proteins but also with S-oxidized proteins (e.g., disulfides). Therefore, it is important to give consideration to thiol oxidation that may occur in the experimental system under study. In addition, some of the generated SNO groups may rapidly react with available cellular thiols to yield disulfide species (Broniowska & Hogg, 2012). Likewise, intramolecular SNO to disulfide conversions can also occur (Engelman et al., 2013; Engelman, Ziv, Arner, & Benhar, 2016). In such scenarios, Trx may trap the disulfide species rather than the SNO that preceded it. Nevertheless, studies involving SNO-treated monocytes or activated macrophages have shown that trapping by Trx(C35S) was mainly due to S-nitrosylation rather than S-oxidation. This conclusion was based on the finding that protein trapping was largely eliminated by subjecting the lysates to UV light, which efficiently cleaves (photolyzes) SNO bonds (Forrester et al., 2009). Indeed, it is highly recommended to employ such prephotolysis control reactions when designing trapping experiments. Finally, there exists the possibility of nonspecific pull-down of proteins that may associate with trapped targets. However, this issue should be of only minor concern due the stringent wash conditions employed in this protocol.

As in any proteomic analysis, it is important to carry out follow-up validation experiments. The trapping of selected individual proteins (selected from the MS data) should be determined by performing the trapping procedure coupled to Western blot for the protein of interest. In addition, nitrosylation of candidate targets needs to be assessed using an independent approach, such as BST. The trapping approach is expected to enrich SNO proteins that are substrates of Trx. To determine whether

Trx is indeed a *bona fide* denitrosylase of a particular protein requires further experimentation using in vitro and cellular approaches (Ben-Lulu et al., 2014).

In summary, SNO trapping by Trx(C35S) provides a simple yet powerful approach to enrich and identify novel S-nitrosylated proteins. This method can serve as a complementary technique to the existing toolbox available to characterize the dynamic cellular nitrosoproteome.

ACKNOWLEDGMENTS

This work was supported by grants from the Israel Science Foundation, the US-Israel Binational Science Foundation, and the Rappaport Family Institute for Research in the Medical Sciences.

REFERENCES

Anand, P., & Stamler, J. S. (2012). Enzymatic mechanisms regulating protein S-nitrosylation: Implications in health and disease. *Journal of Molecular Medicine (Berlin, Germany)*, *90*, 233–244.

Benhar, M. (2015). Nitric oxide and the thioredoxin system: A complex interplay in redox regulation. *Biochimica et Biophysica Acta*, *1850*, 2476–2484.

Benhar, M. (2016). Emerging roles of protein S-nitrosylation in macrophages and cancer cells. *Current Medicinal Chemistry*, *23*, 2602–2617.

Benhar, M., Forrester, M. T., Hess, D. T., & Stamler, J. S. (2008). Regulated protein denitrosylation by cytosolic and mitochondrial thioredoxins. *Science*, *320*, 1050–1054.

Benhar, M., Forrester, M. T., & Stamler, J. S. (2009). Protein denitrosylation: Enzymatic mechanisms and cellular functions. *Nature Reviews. Molecular Cell Biology*, *10*, 721–732.

Benhar, M., Thompson, J. W., Moseley, M. A., & Stamler, J. S. (2010). Identification of S-nitrosylated targets of thioredoxin using a quantitative proteomic approach. *Biochemistry*, *49*, 6963–6969.

Ben-Lulu, S., Ziv, T., Admon, A., Weisman-Shomer, P., & Benhar, M. (2014). A substrate trapping approach identifies proteins regulated by reversible S-nitrosylation. *Molecular & Cellular Proteomics*, *13*, 2573–2583.

Broniowska, K. A., & Hogg, N. (2012). The chemical biology of S-nitrosothiols. *Antioxidants & Redox Signaling*, *17*, 969–980.

Cox, J., & Mann, M. (2008). MaxQuant enables high peptide identification rates, individualized p.p.b.-range mass accuracies and proteome-wide protein quantification. *Nature Biotechnology*, *26*, 1367–1372.

Diers, A. R., Keszler, A., & Hogg, N. (2014). Detection of S-nitrosothiols. *Biochimica et Biophysica acta*, *1840*, 892–900.

Doulias, P. T., Greene, J. L., Greco, T. M., Tenopoulou, M., Seeholzer, S. H., Dunbrack, R. L., et al. (2010). Structural profiling of endogenous S-nitrosocysteine residues reveals unique features that accommodate diverse mechanisms for protein S-nitrosylation. *Proceedings of the National Academy of Sciences of the United States of America*, *107*, 16958–16963.

Engelman, R., Weisman-Shomer, P., Ziv, T., Xu, J., Arner, E. S., & Benhar, M. (2013). Multilevel regulation of 2-Cys peroxiredoxin reaction cycle by S-nitrosylation. *The Journal of Biological Chemistry*, *288*, 11312–11324.

Engelman, R., Ziv, T., Arner, E. S., & Benhar, M. (2016). Inhibitory nitrosylation of mammalian thioredoxin reductase 1: Molecular characterization and evidence for its functional role in cellular nitroso-redox imbalance. *Free Radical Biology & Medicine*, *97*, 375–385.

Forrester, M. T., Foster, M. W., Benhar, M., & Stamler, J. S. (2009). Detection of protein S-nitrosylation with the biotin-switch technique. *Free Radical Biology & Medicine*, *46*, 119–126.

Gould, N., Doulias, P. T., Tenopoulou, M., Raju, K., & Ischiropoulos, H. (2013). Regulation of protein function and signaling by reversible cysteine S-nitrosylation. *The Journal of Biological Chemistry*, *288*, 26473–26479.

Hess, D. T., Matsumoto, A., Kim, S. O., Marshall, H. E., & Stamler, J. S. (2005). Protein S-nitrosylation: Purview and parameters. *Nature Reviews. Molecular Cell Biology*, *6*, 150–166.

Jaffrey, S. R., Erdjument-Bromage, H., Ferris, C. D., Tempst, P., & Snyder, S. H. (2001). Protein S-nitrosylation: A physiological signal for neuronal nitric oxide. *Nature Cell Biology*, *3*, 193–197.

Majmudar, J. D., Konopko, A. M., Labby, K. J., Tom, C. T., Crellin, J. E., Prakash, A., et al. (2016). Harnessing redox cross-reactivity to profile distinct cysteine modifications. *Journal of the American Chemical Society*, *138*, 1852–1859.

Neilson, K. A., Ali, N. A., Muralidharan, S., Mirzaei, M., Mariani, M., Assadourian, G., et al. (2011). Less label, more free: Approaches in label-free quantitative mass spectrometry. *Proteomics*, *11*, 535–553.

Nikolov, M., Schmidt, C., & Urlaub, H. (2012). Quantitative mass spectrometry-based proteomics: An overview. *Methods in Molecular Biology*, *893*, 85–100.

Raju, K., Doulias, P. T., Tenopoulou, M., Greene, J. L., & Ischiropoulos, H. (2012). Strategies and tools to explore protein S-nitrosylation. *Biochimica et Biophysica Acta*, *1820*, 684–688.

Schonhoff, C. M., & Benhar, M. (2011). Analysis of protein S-nitrosylation. *Current Protocols in Protein Science*, *Chapter 14*. Unit14.6.

Seneviratne, U., Nott, A., Bhat, V. B., Ravindra, K. C., Wishnok, J. S., Tsai, L. H., et al. (2016). S-nitrosation of proteins relevant to Alzheimer's disease during early stages of neurodegeneration. *Proceedings of the National Academy of Sciences of the United States of America*, *113*, 4152–4157.

CHAPTER SEVENTEEN

Combining Click Chemistry-Based Proteomics With Dox-Inducible Gene Expression

J. Gebert*,†, M. Schnölzer‡, U. Warnken‡, J. Kopitz*,†,1
*Institute of Pathology, University Hospital Heidelberg, Heidelberg, Germany
†Cancer Early Detection, German Cancer Research Center (DKFZ), Heidelberg, Germany
‡Functional Proteome Analysis, German Cancer Research Center (DKFZ), Heidelberg, Germany
[1]Corresponding author: e-mail address: juergen.kopitz@med.uni-heidelberg.de

Contents

Abstract

Inactivating mutations in single genes can trigger, prevent, promote, or alleviate diseases. Identifying such disease-related genes is a main pillar of medical research. Since proteins play a crucial role in mediating these effects, their impact on the diseased cells' proteome including posttranslational modifications has to be elucidated for a detailed understanding of the role of these genes in the disease process. In complex disorders, like cancer, several genes contribute to the disease process, thereby hampering the assignment of a proteomic change to the corresponding causative gene. To enable comprehensive screening for the impact of inactivation of a gene, e.g., loss of a tumor suppressor in cancer, on the cellular proteome, we present a strategy based on combination of three technologies that is recombinase-mediated cassette exchange, click chemistry, and mass spectrometry. The methodology is exemplified by the analysis of the proteomic changes induced by the loss of a tumor suppressor gene in colorectal cancer cells. To demonstrate the applicability to screen for posttranslational modification changes, we also describe the analysis of protein glycosylation changes caused by the tumor suppressor inactivation. In principle, this strategy can be applied to analyze

Methods in Enzymology, Volume 585
ISSN 0076-6879
http://dx.doi.org/10.1016/bs.mie.2016.09.022

the effects of any gene of interest on protein expression as well as posttranslational modification by glycosylation. Moreover adaptation of the strategy to an appropriate cell culture model has the potential for application on a broad range of diseases where the disease-promoting mutations have been identified.

1. BACKGROUND

Based on their genetic contribution, human diseases can be classified as monogenic or multifactorial. Monogenic diseases are caused by alterations in a single gene. Examples of monogenic disorders are sickle cell disease, cystic fibrosis, polycystic kidney disease, and Tay–Sachs disease. Monogenic disorders are relatively rare in comparison with human diseases that can be categorized as multifactorial and are also referred to as complex diseases. Examples include cardiovascular disease, cancer, diabetes, and a number of birth defects and psychiatric disorders. Variations in numerous genes were detected in these complex diseases. However, that does not mean, that all the somatic abnormalities present have been involved in the disease process. It is likely that some have made no contribution at all. To embody this concept, the terms "driver" and "passenger" mutation have been coined in tumor biology. A driver mutation is causally implicated in oncogenesis. It has conferred growth advantage on the cancer cell and has been positively selected during tumorigenesis. On the other hand, a passenger mutation has not been selected, has not conferred clonal growth advantage, and has therefore not contributed to cancer development (Stratton, Campbell, & Futreal, 2009). Since disease-promoting mutations, like tumor driver mutations, affect important biological functions, like proliferation and differentiation control in cancer cells, these mutations are reflected by significant changes at the proteome level. Thus, detecting significant proteomic changes that are associated with a certain mutation will help to distinguish disease-promoting mutations from "passenger" mutations. Even more important, characterization of the proteome, representing the cell's "biochemical phenotype," is the key to understand the disease process on a molecular level thereby defining targets for treatment. In the present chapter, we describe a novel experimental design for comprehensive analysis of proteomic changes associated with loss of gene function. To this end we combined three existing technologies: (i) gene reconstitution of a cell line by means of recombinase-mediated cassette exchange (RCME), resulting in a doxycycline-inducible gene expression system. (ii) A click chemistry-based approach enabling the selective extraction of proteins that were synthesized after gene induction. (iii) Read-out by high-resolution mass spectrometry. The experimental approach is exemplified by the reconstitution of the tumor suppressor gene *TGFBR2* in a *TGFBR2*-deficient colorectal cancer cell line. Click chemistry is shown for metabolic protein labeling and for labeling of protein-bound oligosaccharide chains as example for labeling of posttranslational modifications. In principle, this strategy can be adapted to any other cell line and gene. Altogether it is a versatile technique applicable to characterize proteomic changes in cell cultures representing monogenic disorders as well as driver mutations in cell culture models of complex disorders, as outlined later for the combined molecular and proteomic approach for delineating tumor driver-specific effects.

A hallmark of cancer is the loss of normal cell growth and differentiation control due to the acquisition of numerous genetic and epigenetic alterations in the tumor cell genome (Hanahan & Weinberg, 2011). In recent years, large-scale sequencing efforts have exploited the genetic and epigenetic landscape of cancer genomes and provided valuable insights into tumor-relevant (driver) and tumor-irrelevant (passenger) alterations (Kim, Laird, & Park, 2013; The Cancer Genome Atlas Network, 2012; Vogelstein et al., 2013). It turned out that the majority of somatic mutations in most human malignancies represent passenger mutations that occur coincidentally with driver mutations but lack tumor promoting function. In contrast, there appears to exist only a limited set of genes that can act as driver genes for common human malignancies, and it has been proposed that at least three driver gene alterations are required for the development of cancers (Tomasetti, Marchionni, Nowak, Parmigiani, & Vogelstein, 2015). Such driver mutations and their affected proteins are usually members of major signaling pathways that can be part of a network-wide signaling crosstalk and act in a cell context-dependent manner (Massagué, 2012). Several of these pathways are linked to proliferation and survival and play a crucial role in many cancers. This is particularly evident in colorectal tumorigenesis, where progression to malignancy proceeds through clinically and histopathologically well-defined stages which have been associated with several somatically acquired genetic and epigenetic alterations (Fearon & Vogelstein, 1990). In most colorectal tumors, recurrent mutations were discovered in key genes that encode members or regulators of the WNT/β-catenin (*APC*), MAP kinase (*KRAS*), TGFβ (*TGFBR2*, *SMAD2/4*), and DNA-damage-sensing (*TP53*) pathways. On a genomic scale these tumors are classified as chromosomally unstable, because they are usually aneuploid and show large chromosomal deletions, insertions, or rearrangements (Lengauer, Kinzler, & Vogelstein, 1997). An even more severe type of genetic instability, termed microsatellite instability (MSI), is associated with a subset of colorectal tumors (15%) that display distinct molecular and clinico-histopathological features (Boland & Goel, 2010). Due to inactivating genetic or epigenetic alterations in DNA mismatch repair (MMR) genes, these MMR-deficient tumors accumulate numerous insertion/deletion (frameshift) mutations at short coding (and noncoding) repetitive DNA sequences (microsatellites) recognized as the MSI phenotype. In addition, mutations in coding microsatellite sequences (cMS) can abrogate normal function of affected proteins and pathways but only a small but limited number of recurrently mutated genes are assumed to promote tumor growth and thus drive tumorigenesis. The poster child of an MSI tumor driver and prime cMS mutation target (>90%) is the *transforming growth factor β rReceptor type II* (*TGFBR2*) (Markowitz et al., 1995). Since driver mutations play a key role in the development of MSI and other tumors, it is important to understand the mechanisms by which they mediate and regulate the biology of these malignancies. However, MSI and other tumors show significant intratumoral heterogeneity due to different subpopulations of cells. Therefore, model systems are required that reflect the genetic (driver) alterations in the primary tumor and allow to interrogate cellular processes like gene expression, protein synthesis/secretion, or glycosylation in a defined genetic background. However, detecting such alterations still represents a major experimental challenge due to the complexity of the proteome which is further extended by multifaceted posttranslational modifications.

Thus a cell's proteome as well as specific proteomic changes are not exclusively reflected by the expression of different genes. While it is estimated that the human genome consists of approx. 20,000 genes (Pertea & Salzberg, 2010), the number of proteins by far exceeds this number. Posttranscriptional events such as gene splicing and posttranslational changes such as amino acid modifications and proteolytic processing give rise to the high complexity on the protein level. In particular posttranslational modifications (glycosylation, phosphorylation, acetylation, etc.) dramatically increase the number of different protein species in the context of a defined cellular state. In addition, the concentration of expressed proteins in a complex system can vary by several orders of magnitude which makes it extremely challenging to detect and quantify low-abundant proteins. Consequently to study protein changes in more depth and to gain knowledge of the relative abundances it is often necessary to reduce the complexity of the proteome and enrich for the modification of interest. This can be achieved by chemical as well as by affinity-based approaches. Well-established protocols exist for the enrichment of glycoproteins (see later), enrichment of phosphopeptides by a combination of immobilized metal affinity chromatography with titanium dioxide (Thingholm, Jensen, & Larsen, 2009), and for the enrichment of the acetylome using modification-specific antibodies (Choudhary et al., 2009; Svinkina et al., 2015). Another useful approach is the tagging of proteins with biotin and subsequent enrichment with streptavidin. Examples are the biotinylation of the surface proteome by in vivo perfusion (Strassberger, Trüssel, Fugmann, Neri, & Roesli, 2010) and biotin labeling of proteins via click chemistry (Eichelbaum & Krijgsveld, 2014). Mass spectrometry is the method of choice for the read-out using high-end instrumentation like the orbitraps and sophisticated software packages (Cox & Mann, 2008).

Protein glycosylation is the most frequent and most complex posttranslational modification. Protein glycans play crucial roles in various biological processes including cell adhesion, proliferation, cellular signaling, and immune response processes (Christiansen et al., 2014). Consequently changes in protein glycosylation were also recognized as a cause and/or result of malignant processes including tumor development, metastasis, and invasion. Altered glycosylation is an almost universal feature of cancer cells (Stowell, Ju, & Cummings, 2015). Glycosylation changes, arising as a consequence of malignant transformation, influence cell growth as well as the differentiation, adhesiveness, and immunogenicity of cancer cells (Dennis, Granovsky, & Warren, 1999; Gorelik, Galili, & Raz, 2001; Kobata & Amano, 2005). Consequently, tumor glycopeptides represent important tumor-specific targets, and dysregulated protein glycosylation in cancer cells leads to the accumulation of tumor-specific antigens, mainly located on the cells' surface and actively involved in tumor progression and metastasis. Therefore detection of defined carbohydrate and/or glycoprotein tumor antigens and exploitation of abnormal glycosylation pathways in tumors are considered fundamentals for improved diagnostic and therapeutic strategies (Dube & Bertozzi, 2005; Dwek & Brooks, 2004; Vlad & Finn, 2004). Consequently, modern analytical methods of glycomics and glycoproteomics are currently becoming an integral part of cancer research and hold considerable promise to deliver new candidate glycan structures as diagnostic and therapeutic targets (Ahn, Kim, & Yoo, 2015; Frost & Li, 2014; Jensen, Karlsson, Kolarich, & Packer, 2012; Pinho & Reis, 2015).

In addition to the described enrichment techniques the complexity of the target proteome can be reduced by metabolic labeling which can be applied to restrict the detected proteome to proteins that are newly synthesized during a defined experimental period (de novo proteome). Various common strategies to analyze the de novo proteome and glycome by metabolic labeling exist. Traditionally metabolic incorporation of isotope-labeled building blocks of cellular macromolecules has been applied to screen for changes in a cell's biochemical phenotype. These techniques can be applied for in vivo and in vitro metabolic labeling of proteins, glycan structures, lipids, and nucleic acids (Bonifacino, 1998; Diaz & Varki, 2009; Griffiths, 2011). A major advantage of using radioactive precursors for metabolic labeling, e.g., ^{35}S-, ^{14}C-, or ^{3}H-labeled amino acids or monosaccharides, is the high specific radioactivity which enables that even trace labeling can yield a readily measurable incorporation, thereby allowing for relatively short labeling time windows. Although representing the most sensitive method for detection, it often presents difficulties since statutory regulations for the handling of radioactive isotopes are stringent. In addition, radioisotope-based proteomic or glycomic analyses require radiometric instrumentation and elaborate high-resolution prefractionation methods, like 2D electrophoresis in combination with autoradiography or fluorography before mass spectrometric protein identification (Celis & Gromov, 2003). For metabolic labeling with stable isotopes, e.g., ^{2}H-, ^{13}C-, or ^{15}N, no safety regulations have to be considered. Coupling of nano-HPLC to high-resolution mass spectrometers enables the automated analysis of highly complex mixtures of peptides or glycopeptides without additional prefractionation. Thus stable isotope labeling with amino acids in cell culture (SILAC) has become a standard method in comparative proteomics (Beynon & Pratt, 2005). For glycopeptides SILAC can be combined with lectin-based affinity chromatography (Boersema, Geiger, Wisniewski, & Mann, 2013). In contrast to radioactive labeling for effective metabolic labeling by SILAC and related methods, substantial incorporation in the order of minimally 5–10% is required for accurate measurements. When applying the SILAC technology, mass spectrometric detection of labeled proteins has to be conducted in the presence of numerous irrelevant, unlabeled peptides, which hampers the detection of labeled low-abundant proteins (Ballikaya et al., 2014). Similar restrictions apply to metabolic incorporation of stable isotope labels for quantitative glycomics of cultured cells, like the IDAWG technology (isotopic detection of aminosugars with glutamine) (Mechref, Hu, Desantos-Garcia, Hussein, & Tang, 2013;

Orlando et al., 2009). These disadvantages are avoided by restricting the mass spectrometric analyses to the de novo (glyco)proteome via sample extraction. By combination with the molecular approach of focusing on the effects of a single gene this combined strategy allows highly specific in-depth analysis of the effects of a disease-promoting mutation on the cell's biochemical phenotype (Fig. 1).

2. COMBINED EXPERIMENTAL STRATEGY

2.1 Molecular Approach

2.1.1 Principles of Tet-Induced Gene Expression

Assigning accompanying biological changes to a single reconstituted gene is a particular challenging task, especially among a high background of irrelevant genetic alterations as encountered in MSI tumor cells and hence requires a highly specific and temporally controlled system for transgene expression. This has fostered the development of systems for inducible gene expression in mammalian cells involving different inducing agents like metals and hormones (Huang, Marquis, & Gray, 2004; No, Yao, & Evans, 1996). One of the most widely used technologies is the tetracycline controlled gene expression system that has been extensively characterized and improved (Gossen & Bujard, 1992; Hampf & Gossen, 2007; Löw, Heinz, Hampf, Bujard, & Gossen, 2010; Schönig, Bujard, & Gossen, 2010). It is comprised of two essential components. The first component is a transcriptional transactivator (tTA) composed of the DNA-binding domain of the bacterial tetracycline repressor TetR fused to the transactivation domain of the Herpes Simplex Virus VP16 protein. Expression of this component is directed by a promoter that is constitutive and ubiquitous in different cell types (i.e., elongation factor 1 alpha, EF1a or β-actin). The second component consists of seven repeats of the transposon Tn10-derived tet operator sequence fused to a minimal promoter typically derived from the human cytomegalovirus. This Ptet promoter is placed upstream of a gene of interest (GOI). In a modified version, a Tet-inducible bidirectional promoter can concomitantly direct expression of two different genes (Ptet-bi; Baron, Freundlieb, Gossen, & Bujard, 1995). Tetracycline in general and doxycycline (Dox) in particular can bind to the transactivator with high affinity ($\sim 10^{10}\ M^{-1}$) and specificity thereby enabling induction of GOI expression by up to five orders of magnitude (Berens & Hillen, 2003; Gossen & Bujard, 1992). This inducible expression system exists in two different versions depending on the type of transactivator used. In the Tet-Off

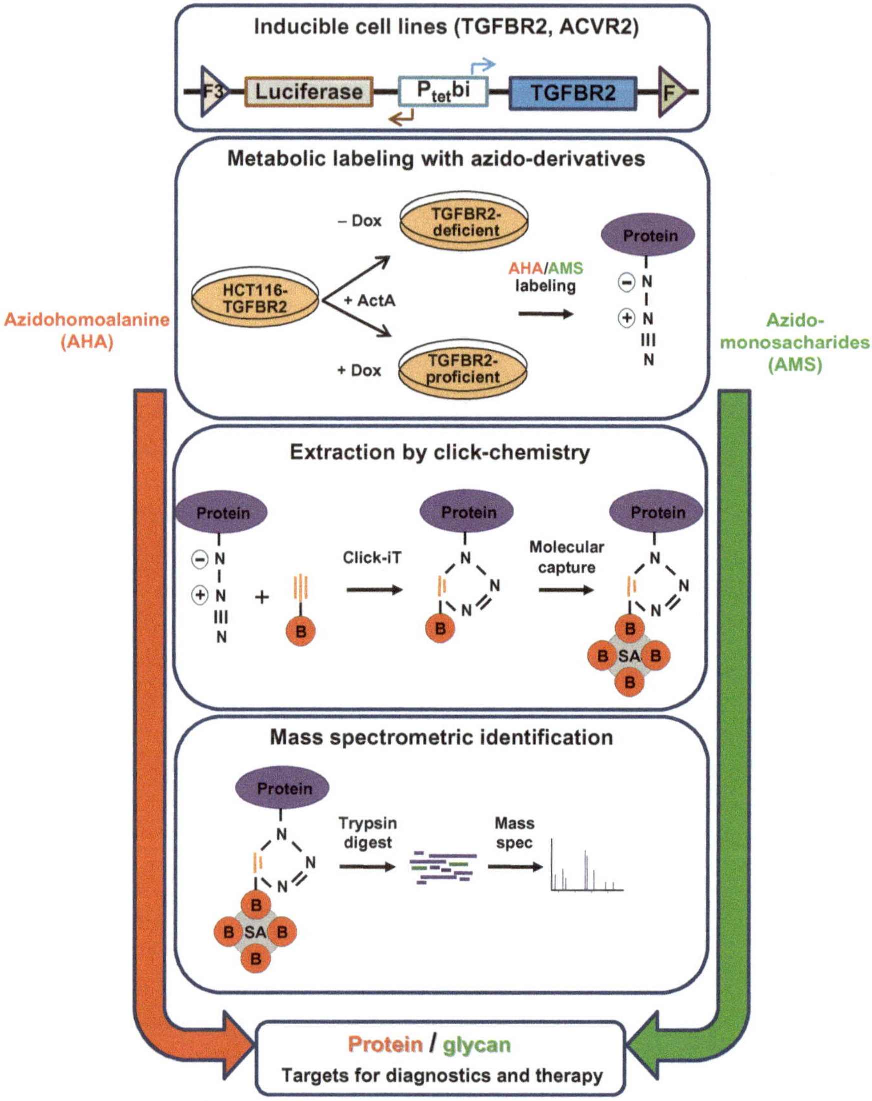

Fig. 1 Recombinase-mediated integration of a bidirectional expression cassette, comprising a driver gene of interest (TGFBR2) and a marker gene (luciferase), into a specific genomic target site of a driver gene-deficient and microsatellite unstable colon cancer cell line (HCT116) allows doxycycline-inducible expression of both genes in a reversible and dose-dependent manner. Newly synthesized proteins or proteoglycans are metabolically labeled with azidohomoalanine (AHA) or azido-monosaccharides (AMS). Click-it-based biotin tagging enables specific extraction of labeled (glyco)proteins by streptavidin-coated beads and upon tryptic on-bead digestion final analysis of captured proteins by mass spectrometry. *B*, biotin; *SA*, streptavidin. (Alternatively, alkyne-derivatized compounds may be applied for metabolic incorporation and clicked to azido-biotin derivatives.)

version, binding of tetracyclines to the tTA suppresses GOI expression whereas in the absence of inducer GOI expression is activated (Gossen & Bujard, 1992; Urlinger et al., 2000; Zhou, Vink, Klave, Berkhout, & Das, 2006). In the complementary Tet-On system, GOI expression is induced upon binding of tetracycline to the reverse transcriptional transactivator (rtTA) but remains suppressed in the absence of the ligand (Gossen et al., 1995). Because of its reversibility, exquisite specificity, and excellent dynamic induction range, this system is particularly apt to study the biology and physiology of specific tumor driver genes at the cellular level and in an otherwise identical genetic background.

2.1.2 Generation of Dox-Inducible Tet-On Cell Lines

Established tumor cell lines are still a valuable model system to study the consequences of specific genetic alterations (Mouradov et al., 2014). Several cMS mutation studies have demonstrated that MSI tumor cell lines and primary tumors carry similar target and driver gene mutations most of them leading to functional inactivation of affected genes (Alhopuro et al., 2012; Duval et al., 2001; Kim et al., 2013; Woerner et al., 2001). Reconstituted GOI expression in a MSI tumor cell line using the Tet technology represents a versatile approach to investigate tumor-specific effects that remain restricted to this gene. For that purpose, both components of the Tet system need to be stably and sequentially integrated into a candidate cell line. For the stepwise protocol outlined below, Tet-controlled (Tet-On) expression of a paradigm MSI driver gene (*TGFBR2*) in a typical MMR- and *TGFBR2*-deficient colorectal cancer cell line (HCT116) has been chosen as an example. In the first genetic modification step (Fig. 2A), the candidate parental cell line is turned into a Tet-On cell line by stably integrating an expression cassette that encompasses the tTA gene linked via a viral internal ribosomal entry site (IRES) to a fluorescence protein encoding marker gene, the latter enabling convenient expression monitoring and screening of both genes (Welman, Barraclough, & Dive, 2006). The tetracycline derivative Dox is used as inducing agent throughout the protocol because it is one of the most efficient effectors that control the Tet-On system (Gossen et al., 1995). Different transactivator constructs are commercially available (Takara, St. Germain-en Laye, France) or can be obtained from academic institutions (Das, Zhou, Metz, Vink, & Berkhout, 2016).

1. Plate 10^7 HCT116 cells on a T175 flask at about 80% confluency and grow them for 24 h in a humidified incubator under standard cell culture conditions (DMEM, 10% FCS, 1% Penicillin/Streptomycin at 37%/5% CO_2).

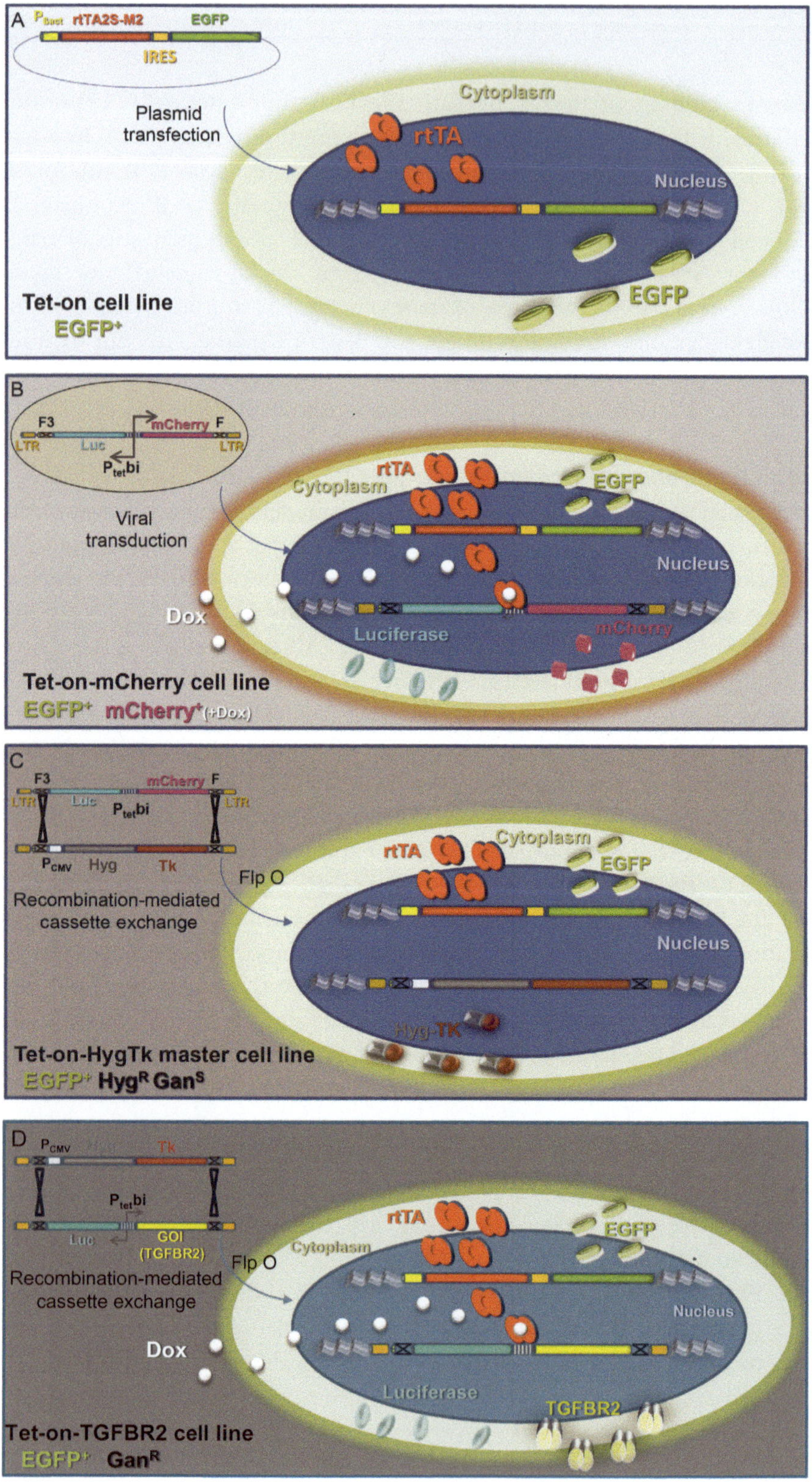

Fig. 2 Experimental strategy for generating cell lines enabling Tet-controlled gene expression.

2. Wash cells twice with 5 mL of 1 × PBS (10 m*M* phosphate-buffered saline, pH 7.2) and detach them with 1 mL prewarmed Trypsin solution for 10 min at 37°C (0.5 mg/mL Trypsin, 0.2 mg/mL EDTA in PBS, and supplemented cell culture media).
3. Stop trypsinization by adding 5 volumes cell culture medium with supplements and harvest cells by centrifugation (7 min, room temperature, 200 × *g*).
4. Wash cells with 2 × volume (10 mL) 1 × PBS.
5. Remove supernatant and resuspend cells in 100 μL electroporation solution.[1] For HCT116 cells the Amaxa Cell Line Nucleofector Kit was used with program D32 of the Nucleofector I device (Lonza, Basel, Switzerland).
6. Mix 10^7 cells with 10–20 μg of linearized transactivator plasmid DNA (pN1pβactinrtTA2S-M2-IRES-EGFP; Welman et al., 2006) carrying the neomycin resistance gene; cells exposed to transfection reagent only (mock) serve as control.
7. Transfer cell/DNA mixture as well as mock-treated cells into certified cuvettes and close cuvettes with caps.
8. Place cuvettes into the electroporator and apply a defined electrical pulse; approximately 10% of cells remain viable after treatment.
9. Quickly add 500 μL of prewarmed supplemented culture medium to the electroporated cells (plasmid and mock) and split each sample onto three 10-cm dishes (selection plates) prepared with supplemented culture medium.
10. Stably transfected clones were selected by growth in culture medium containing a specific antibiotic; the neomycin resistance gene is commonly found on most expression plasmids conferring resistance to the aminoglycoside antibiotic Geneticin (G418, 300–700 μg/mL). Individual G418-resistant cell clones become visible after about 3 weeks when cell death has occurred in all mock-treated cells.
11. Trypsinize and harvest all pooled clones from each of the three selection plates, wash twice, resuspend in 1 × PBS, and dilute to 10^5 cells/mL in 1 × PBS.

[1] Electroporation is recommended for plasmid-based stable transfection of colorectal cancer cell lines, which are usually difficult to transfect. Alternatively, retro-/lentiviral transduction can be used to ensure genomic integration of the transactivator expression construct with high efficiency. Lentiviral vectors efficiently transfer genes to dividing and nondividing cells with long-term gene expression both in vitro and in vivo (Ailles & Naldini, 2002).

12. Screen the clone pools from each selection plate for green fluorescent protein (EGFP marker gene) expression by fluorescence-activated cell sorting (FACS) and sort 96 single EGFP-positive cells from each clone pool (selection plate) into 96-well plates containing 20 μL supplemented culture medium.
13. Select three independent EGFP-positive/G418 resistant clones originating from three different 96-well plates and examine them for rtTA expression via transient supertransfection.
14. Seed 10^6 cells of each clone in quadruplicate onto six-well plates and grow overnight to a cell density of about 80%.
15. Transiently transfect duplicate clones in the presence or absence of Dox (1 μg/mL) by lipofection with a luciferase reporter plasmid that drives luciferase gene expression from a Dox-responsive promoter. A 4:2 ratio (4 μL transfection reagent FuGENE HD (Roche) to 2 μg plasmid DNA pUHC13-3; Gossen & Bujard, 1992) is recommended for HCT116-Tet-On cells. Concomitant transfection of a reference marker gene (for example, a constitutive β-galactosidase or renilla luciferase expression vector) allows normalization of transfection efficiency and inducibility values. Note that the ability of a relatively low number of endogenous rtTA molecules to confer Dox-controlled inducibility to an excess of transfected reporter gene molecules might be limited; and hence, amounts of transfected reporter plasmid DNA have to be carefully titrated.
16. 24 h after transfection, wash cells twice with 1 × PBS and lyse them in 200 μL of luciferase lysis buffer (25 m*M* Tris–PO_4, pH 8.0, 2 m*M* DTT, 2 m*M* CDTA (diaminocyclohexane-tetraacetic acid), 10% glycerol, 1% Triton X-100).
17. Add 10 μL of cell lysate to 250 μL luciferase assay buffer (25 m*M* glycylglycine, 15 m*M* $MgSO_4$, 5 m*M* ATP, 0.2 m*M* D-luciferin) and immediately measure luciferase activity in a luminometer. Note, add luciferin only shortly prior to measurement.
18. Determine β-galactosidase activity by mixing 30 μL of cell lysate with 770 μL Z-buffer (60 m*M* Na_2HPO_4, 40 m*M* NaH_2PO_4, 10 m*M* KCl, 1 m*M* $MgSO_4$, 40 m*M* β-mercaptoethanol: add before use), add 200 μL ONPG (4 mg/mL in 100 m*M* sodium phosphate buffer, pH 7.5), and incubate for 15 min to 2 h at 37°C until the sample turns yellow. Stop reaction with 500 μL 1 *M* Na_2CO_3.
19. Clones that show a significant induction of luciferase expression/activity normalized to β-galactosidase activity in Dox-treated vs

untreated cells and maintain this inducibility throughout continuous passaging in culture represent bona fide Tet-On cell lines amenable to further genetic modification.

2.1.3 Identification of Suitable Genomic Loci for Conditional Gene Expression in Tet-On Cell Lines

Tet-On cell lines have the general capability to act as a genetic switch for GOI expression, if those genes are under control of a Tet-responsive promoter. For analysis of immediate as well as long-term physiological effects of such a Tet-regulated GOI, an appropriate expression cassette has to be integrated into the Tet-On cell genome as described previously (Weidenfeld et al., 2009). Two criteria need to be fulfilled: First, genomic loci must be found that ensure robust, reversible, and dose-dependent Tet/rtTA-regulated GOI expression from the integrated cassette. Second, only a single copy of this expression cassette should be integrated into the Tet-On cell genome to avoid unphysiologically high levels of GOI expression. The former requirement is met by using a bidirectional expression cassette that allows Tet-controlled expression of two marker genes (a fluorescent protein, different from the one being coexpressed with the transactivator and luciferase; Löw, Vigna, Lindemann, Naldini, & Bujard, 2006) that are flanked by two heterologous Flp recombinase target sites (wild type, FRT; mutant, F3; Schlake & Bode, 1994), while the latter requirement is met by employing retroviral transduction of the expression cassette at low multiplicity of infection (MOI). Virus production is performed in 293T cells and based on a transient three plasmid expression system (Soneoka et al., 1995) comprised of a specific proviral plasmid (S2FcLM2CG-FRT; Weidenfeld et al., 2009) and two packaging plasmids (pVPack; pVPack-VSV-G; Stratagene). Transduction of target cells with replication-defective recombinant virus carrying the Tet-responsive bidirectional marker gene expression cassette allows identification of activatable (s/a) genomic loci (Fig. 2B).

1. Seed 10^6 293T cells on 60-mm dishes prior to the transfection day to reach 70–80% confluency.
2. Cotransfect 293T cells by calcium phosphate coprecipitation method with three proviral plasmids (S2FcLM2CG-FRT3, harboring Tet-controlled luciferase and mCherry as marker genes; pVPack-GP; pVPack-VSV-G) each using 5 μg plasmid DNA.
3. After 16–18 h medium should be replaced by 3 mL DMEM supplemented with 5 m*M* Na-butyrate to increase transcription and expression efficiency.

4. Replace medium after 6–8 h by a minimum volume of fresh DMEM, sufficient to cover the cells (<4 mL) in order to obtain high titer virus particles.
5. One day prior to transduction, seed 10^5 HCT116-Tet-On cells onto six-well plates.
6. After 16 h (40 h after the three plasmid transfection), collect virus-containing supernatant from 293T cells and filtrate through 0.45 μm membrane to remove residual cells; viral stocks can be stored for 24 h on ice or for long-term storage at −80°C. Thawing might decrease viral titer. Note, viral titers need to be determined by transduction of recipient cells and counting mCherry-positive cells.
7. Wash HCT116-Tet-On cells (about 30% confluency) once with DMEM; supplement virus-containing supernatant with 5 μg/mL of polybrene to enhance infection ability; and infect at a MOI of 0.01 and 0.05 to ensure single copy virus integration.
8. After 4 h of incubation with virus, wash HCT116-Tet-On cells twice with DMEM to discard the virus-containing supernatant.
9. Induce successfully transduced HCT116-Tet-On cells with 0.2 mg/mL Dox for 16 h and screen for mCherry-positive cells by FACS (On-Sort) into T25 flasks.
10. Select and grow mCherry-positive cells in the absence of Dox for 8 days followed by FACS of mCherry-negative but EGFP-positive cells (Off-Sort) to ensure reversibility of Dox effect.
11. Grow mCherry-negative cells again in the presence of Dox (0.2 mg/mL) for 8 days and subsequently perform FACS screening and sorting of mCherry-positive cells (On-Sort) into 96-well plates. Single mCherry-positive cells are selected for clonal expansion (HCT116-mCherry).
12. Use a luciferase assay for quantification of target gene expression. Suitable genomic loci/clones should confer a rather low basal level of luciferase expression/activity in the absence of Dox (reflecting low locus-dependent chromatin interference) and a large dynamic range of inducibility (>2–3 orders of magnitude) in the presence of Dox.
13. Perform LAM PCR for determination of integration sites and numbers (Lee et al., 2013; Paruzynski et al., 2010).
14. At least two independent HCT116-Tet-On-mCherry clones each marked by integration of a single provirus at a different genomic locus should be used for subsequent studies.

Such genomic loci not only ensure robust and inducible marker gene expression but also facilitate site-specific insertion of any expression construct via Flp recombinase-mediated site-specific recombination. Two rounds of this RMCE procedure (Turan, Zehe, Kuehle, Qiao, & Bode, 2013) are applied using the pTet-inducible cell line to first generate a master cell line containing a Hyg-TK expression unit (HCT116-HygTk cells; Fig. 2C) which in the second round of RMCE is finally exchanged by any Tet-regulated GOI expression cassette flanked by heterologous FRT and F3 sites (HCT116-Tet-On-TGFBR2 cells; Fig. 2D).

1. In the first recombination step (RMCE), cotransfect 5×10^5 HCT116-mCherry cells on six-well plates with 2 μg pE11.F3.HygTK.F plasmid carrying a Hyg-TK expression cassette flanked by two recombination sites (F3/F) and 2 μg pCAGGS-Flpo-IRES-Puro (Raymond & Soriano, 2007).
2. After 12 h, transfer cells onto 10 cm culture dishes, incubate cells with 100 μg/mL of Hygromycin B for at least 10 days, select single mCherry-negative but EGFP-positive cells, and transfer them into 96-well plates after treating cells with Dox.
3. To obtain a sufficient number of cells for the second RMCE, cultivate the cells for another 2 weeks.
4. Seed the resulting HCT116-HygTK master cell clones, which are Hygromycin B resistant and sensitive to Ganciclovir (Hyg^r, Gan^s), at 5×10^5 on six-well plates 1 day prior to the second Flpo-mediated RMCE using a cassette containing Tet-controlled luciferase as a quantitative marker and the GOI, here *TGFBR2*.
5. Cotransfect HCT116-HygTK cells with 2 μg of a plasmid harboring luciferase and TGFBR2 flanked by two recombination sites (F3/F) and 2 μg of pCAGGS-Flpo-IRES-Puro.
6. After 12 h, transfer cells onto 10 cm culture dishes, incubate cells first with 5 μg/mL of puromycin for 36 h, and subsequently change medium supplemented with 50 μg/mL of ganciclovir for at least 12 days until resistant cell clones become visible. Medium that contains ganciclovir has to be changed every 3 days.
7. Pick and transfer individual clones onto 24-well plates for expanded growth in ganciclovir-containing medium.
8. Individual clones are then examined for Dox-inducible target gene expression (GOI/TGFBR2 and luciferase) based on luciferase activity. Successful cassette targeting to the identified genomic locus (see earlier)

is recognized by similar ranges of Dox-inducible luciferase expression in the original mCherry clones (HCT116-Tet-On-mCherry) and the final GOI clones (HCT116-Tet-On-TGFBR2).

9. Apart from confirming luciferase inducibility, Dox-dependent GOI expression and functionality must be examined.

Although the genetic switch built into these Tet-On-GOI cells offers a wide range of comparative analyses solely depending on the presence and absence of the inducing agent (± Dox), it is advisable to always include parental Tet-On cells (± Dox) in order to control for potential unspecific effects, related to the antibiotic itself. This model cell line system becomes even more powerful for biological analyses when this genetic switch is combined with orthogonal (click) labeling approaches for different cellular biomolecules as outlined below.

2.2 Metabolic Labeling and Click Chemistry-Based Extraction

Click chemistry is a term that was introduced by Sharpless to describe reactions that are used to "click" two organic compounds via a heteroatom link (C-X-C) (Kolb, Finn, & Sharpless, 2001). Applications of these easy to perform "spring-loaded" reactions are found in nearly all areas of modern chemistry from drug discovery to materials science (Moses & Moorhouse, 2007). A characteristic example of such reaction is represented by the azide–alkyne cycloaddition (Fig. 3). This reaction also found its way to biological sciences since it provides a versatile tool to click biomolecules, including amino acids, lipids, nucleotides, and monosaccharides with detection molecules, like fluorescent dyes and biotin residues (El-Sagheer & Brown, 2010, 2012; Lim & Lin, 2011; Zhang & Zhang, 2013). Many of the alkyne- or azide-tagged biomolecules are accepted by the cells' translational machinery for metabolic incorporation into cellular macromolecules (Table 1a) (Prescher & Bertozzi, 2005). Therefore these compounds enable specific metabolic labeling of proteins and glycoproteins and their subsequent extraction via the click reaction. To this end the metabolically labeled proteins are clicked to a biotin-derivative (Table 1b) thereby enabling their

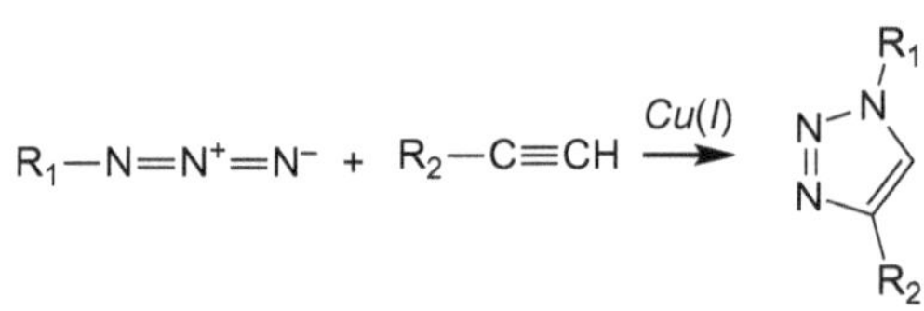

Fig. 3 Principle of the azide–alkyne cycloaddition.

Table 1a Examples of Compounds for Metabolic Labeling of Proteins and Glycoproteins

Application	Compound	Reactive Group	Analog to
Protein labeling	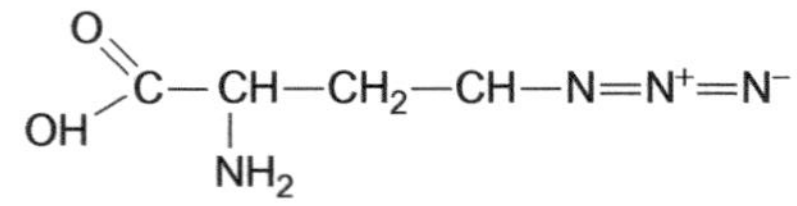 L-Azidohomoalanine	Azide	L-Methionine
Protein labeling	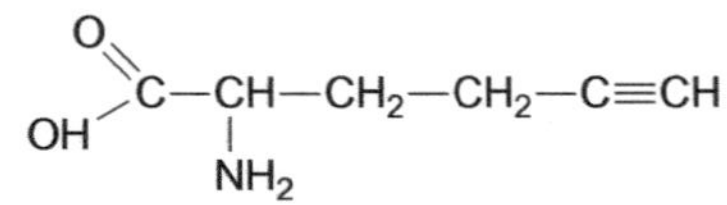 Homopropargylglycine	Alkyne	L-Methionine
Labeling of fucosylated proteins	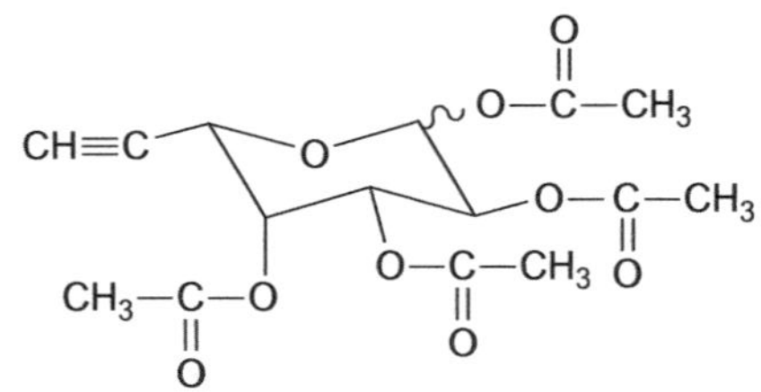 Tetraacetylfucosealkyne	Alkyne	D,L-Fucose
Labeling of O-glycosylated proteins	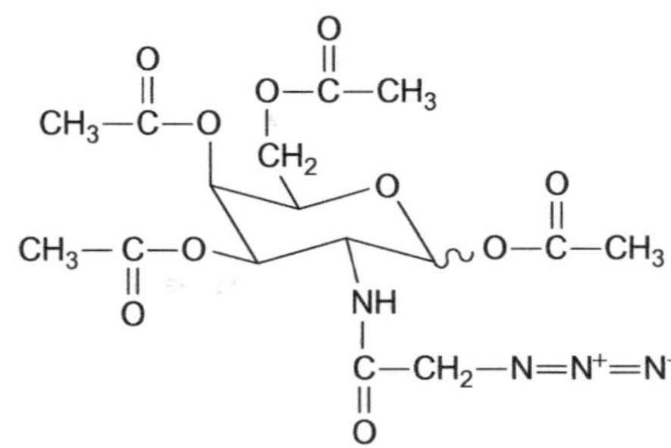 Tetraacetyl-*N*-azidoacetylgalactosamine	Azide	D,L-*N*-Acetylgalactosamine

Continued

Table 1a Examples of Compounds for Metabolic Labeling of Proteins and Glycoproteins—cont'd

Application	Compound	Reactive Group	Analog to
Labeling of sialylated proteins	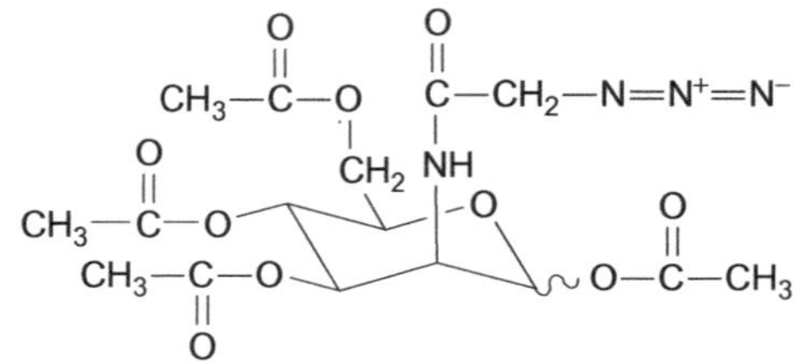 Tetraacetyl-*N*-azidoacetylmannosamine	Azide	D,L-Mannosamine[a]
Labeling of nuclear and cytoplasmatic *O*-GlcNAc-containing glycoproteins	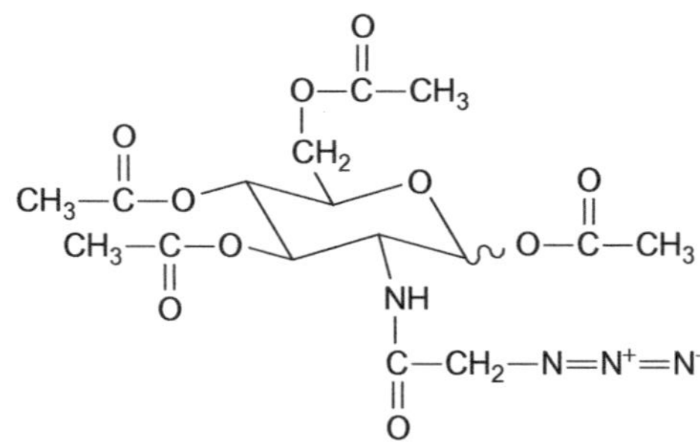 Tetraacetyl-*N*-azidoacetylglucosamine	Azide	D,L-Acetylglucosamine

[a]L-Mannosamine is metabolically converted to *N*-acetylneuraminic acid (sialic acid) and utilized for protein sialylation.

Table 1b Compounds for Biotinylation of Alkyne- or Azide-Labeled Proteins or Glycoproteins

Target	Biotin Derivative
Alkyne	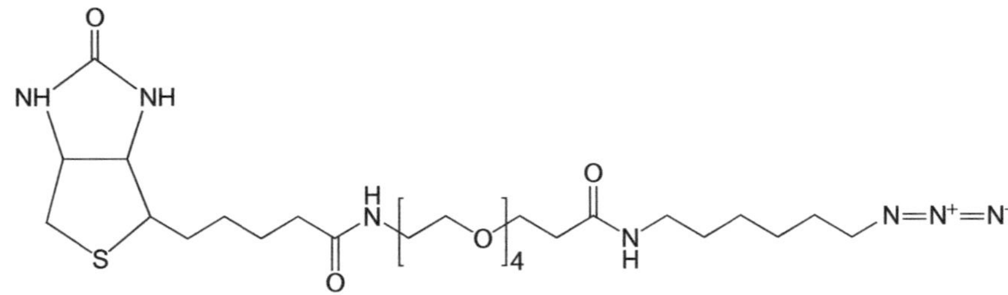 PEG4 carboxamide-6-azidohexanyl-biotin
Azide	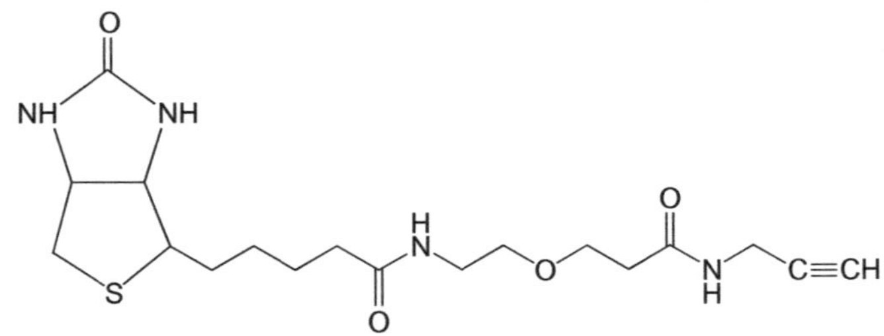 PEG4 carboxamide-propargyl-biotin

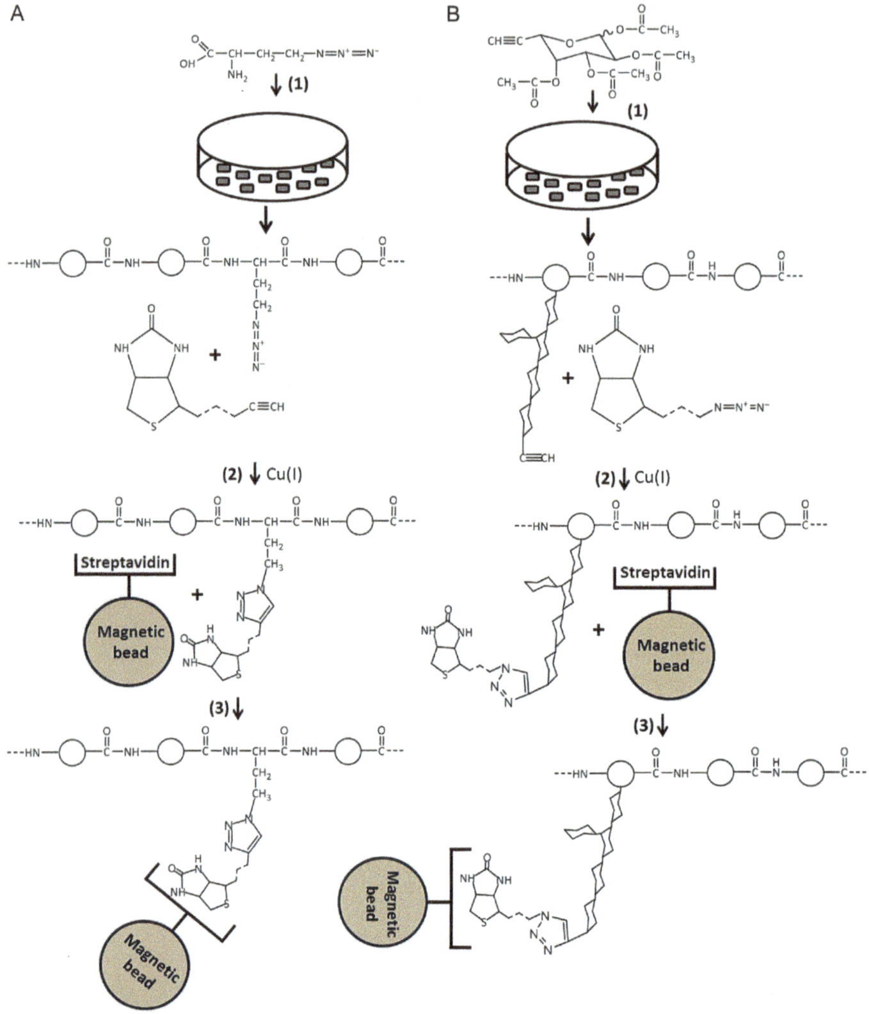

Fig. 4 Schematic presentation of the method for the extraction of the de novo proteome (A) and the glycoproteome (B): (1) Metabolic labeling, (2) click reaction, and (3) streptavidin-based magnetic bead extraction.

extraction with streptavidin beads. A schematic presentation of the method for the extraction of the de novo proteome and the glycoproteome is given in Fig. 4A and B. The detailed protocol is given exemplarily for the azidohomoalanine-labeled proteome and the alkyne-labeled fucosylated glycoproteins.

Protocol:

A. Azidohomoalanine labeling of the de novo proteome

1. Grow cells to approx. 80% confluency in appropriate medium.
2. Change to methionine-free medium to achieve maximum incorporation.
3. Add 40 μ*M* azidohomoalanine in presence or absence of Dox induction as control
4. Incubate for 4–24 h.
5. At the end of the desired labeling time remove medium, wash the cells 3 × with PBS, trypsinize, and harvest by centrifugation.
6. Lyse the cell pellets with 100 μL 1% SDS in 50 m*M* Tris–HCl (pH 8), supplemented with protease inhibitors by sonication for 30 s.
7. After sonication and incubation for 30 min at 4°C on a rotator, centrifuge the cell lysates at 12,000 × *g* for 20 min at 4°C.
8. Determine the protein concentration of the lysate by Bradford assay.
9. Apply 200 μg of protein for the click reaction.
10. Prepare a 4 m*M* PEG4 carboxamide-6-azidohexanyl-biotin stock solution.
11. Apply a commercially available buffer kit (Click-iT® Protein Reaction Buffer Kit; Thermo Scientific) to prepare working solutions for the click reaction.
12. Add the working solutions according to the manufacturer's instructions.
13. Terminate the reaction by methanol/chloroform precipitation (Wessel & Flügge, 1984):
    Add 3 volumes of methanol and mix.
    Add 0.75 volumes of chloroform and mix.
    Add 2 volumes of water and mix.
    Centrifuge for 5 min at 12,000 × *g* for phase separation.
    Carefully remove the upper phase (interphase contains the protein fraction).
    Add 2.25 volumes of methanol and centrifuge for 5 min at 12,000 × *g*.
    Wash the resulting pellet twice with methanol.
    Air dry the pellet.
14. Resolubilize the precipitates in 200 μL RIPA buffer (50 m*M* Tris–HCl, pH 7.5; 150 m*M* NaCl; 1% Triton X-100; 0.5% sodium deoxycholate; 0.1% SDS, 1% DTT, and protease inhibitors).

15. Remove unsoluble material by centrifugation (12,000 × *g*, 20 min).
16. Wash streptavidin-coated magnetic beads (e.g., Dynabeads MyOne Streptavidine T1, Thermo Fischer) thrice with RIPA buffer and prepare a 1:1 slurry.
17. Add 80 μL of the slurry to the solubilized proteins.
18. Incubate for at least 2 h at 4°C on a rotator.
19. Separate the beads on a magnetic stand.
20. Wash 3 × with PBS/Tween to which 2% SDS were added.
21. Conduct final washes with PBS and 40 m*M* ammonium bicarbonate.

B. Labeling of fucosylated proteins with tetraacetylfucosealkyne.

Most steps of the procedure are identical to protocol A with the exception that 40 μ*M* tetraacetylfucosealkyne was applied in step 3, and PEG4 carboxamide-propargyl-biotin was used in steps 10–12.

Various modifications of the described procedure may be considered. In particular, several variations in the composition and length of the spacer arm between the biotin part and the reactive groups in the biotin alkyne/azide are possible. Many of these variants are commercially available. Their suitability for special applications has to be determined empirically. Times for metabolic labeling depend on the doubling and protein synthesis rate of the cells and should be adapted accordingly. As for all metabolic labeling procedures, the labeling time also determines whether the short-lived fraction or the long-lived fraction is preferentially labeled. The magnetic beads can be substituted by streptavidin-linked agarose beads which are separated by centrifugation.

2.3 Analytical Approach

2.3.1 Preparation of Samples for Mass Spectrometry Analysis

Due to the extraordinary stability of the complex between biotin and streptavidin, the release of biotinylated molecules from streptavidin beads remains a major problem (Rösli, Rybak, Neri, & Elia, 2008). Elution requires harsh conditions and is not at all quantitative. Therefore it is advisable to perform an on-beads digestion to obtain tryptic peptides which in turn can be analyzed by mass spectrometry. It is essential to run appropriate controls. These include analysis of streptavidin beads without any further treatment (bead proteome). Commercially available beads are usually capped, e.g., with BSA after covalent attachment of streptavidin.

Protocol for on-beads digestion using streptavidin-coupled dynabeads:

(All experiments should be performed in triplicate.)

Prepare the following buffers and stock solutions:

40 m*M* ammonium bicarbonate (ABC buffer)

10 m*M* dithiothreitol in ABC buffer (DTT)

55 m*M* iodoacetamide in ABC buffer (IAA)

Trypsin 0.5 μg/μL in 1 m*M* HCl (stock solution)

1% and 0.1% trifluoroacetic acid (TFA)

To avoid contamination with keratins all buffers should be prepared prior to use with fresh solvents.

1. Add 100 μL ABC buffer to dynabeads in Eppendorf tubes.
2. Shake gently on a thermomixer (800 rpm), centrifuge shortly.
3. Place tubes in a magnetic rack, remove ABC buffer, and discard.
4. Add 100 μL DTT.
5. Incubate for 1 h at 45°C on a thermomixer at 800 rpm.
6. Centrifuge shortly, place tubes in a magnetic rack, remove supernatant, and discard.
7. Repeat steps 1–3.
8. Add 100 μL IAA.
9. Incubate for 30 min at 25°C on a thermomixer at 800 rpm in the dark.
10. Centrifuge shortly, place tubes in a magnetic rack, remove supernatant, and discard.
11. Repeat steps 1–3.
12. Dilute trypsin stock solution 1:100 with ABC buffer.
13. Add 50 μL of diluted trypsin to dynabeads.
14. Incubate at 37°C overnight on a thermomixer with gentle shaking.
15. Shortly centrifuge.
16. Transfer supernatant containing tryptic peptides into PCR tubes, thereby carefully avoid to carry beads into the tubes.
17. Add 1 μL 1% TFA to peptide mixture.
18. Analyze aliquot by nanoLC-ESI-MS/MS or freeze samples at −20°C until use.

Optional (after step 16)

1. Add 50 μL 0.1% TFA to dynabeads and shake gently.
2. Remove solution from dynabeads and combine supernatants.
3. Evaporate combined supernatants to dryness in a speed vac concentrator.
4. Dissolve tryptic peptides in 5 μL 2.5% hexafluoroisopropanol/0.1% TFA.
5. Analyze by nanoLC-ESI-MS/MS.

2.3.2 Mass Spectrometry Analysis

Depending on the starting amount of cells the entire sample or an aliquot thereof should be injected for mass spectrometry analysis.

Samples can be analyzed by nanoLC-nanoESI MS/MS on various types of instruments. In the following section, MS analysis on an LTQ Orbitrap XL mass spectrometer online coupled with a nanoAcquity UPLC system is described.

1. Dissolve tryptic peptides in 2.5% hexafluoroisopropanol/0.1% TFA.
2. Inject appropriate amount of sample.
3. Trap peptides for desalting and concentration.
   Trapping column: nanoAcquity C18 column (180 μm × 20 mm, particle size 5 μm). Flow: 5 μL/min.
4. Use a 3 h gradient to separate tryptic peptides.
   Separation column: C18 column (BEH 130 C18 100 μm × 100 mm, particle size 1.7 μm)
   Flow: 400 nL/min
   Solvent A: 98.9% water, 1% acetonitrile, 0.1% formic acid
   Solvent B: 99.9% acetonitrile and 0.1% μL formic acid
   Gradient: from 0% to 4% B in 1 min, from 4% to 30% B in 140 min, from 30% to 45% B in 15 min, from 45% to 90% B in 5 min, 10 min at 90% B, from 90% to 0% B in 0.1 min, and 9.9 min at 0% B
5. Use the following parameters for data acquisition on an LTQ Orbitrap XL system:
   Capillary voltage 2400 V; capillary temperature 200°C, normalized collision energy 35 V, activation time 30,000 ms.

The mass spectrometer was operated in data-dependent mode to automatically measure MS1 and MS2. Data are acquired by scan cycles of one FTMS scan with a resolution of 60,000 at *m/z* 400 and a range from 370 to 2000 *m/z* in parallel with six MS/MS scans in the ion trap of the most abundant precursor ions.

2.3.3 Database Search

Numerous pipelines exist for the analysis of mass spectrometry data and for relative quantification of proteins, e.g., MASCOT, commercially available from Matrix Sciences Ltd. (www.matrixsciences.com), Census, which is free of charge (Park, Venable, Xu, & Yates, 2008), and MaxQuant, a software package mainly used for data obtained from Orbitrap instruments (Cox & Mann, 2008). Depending on the MS platform used for data acquisition subsequent database searches and data analysis have to be adapted accordingly.

Below are given some general rules which should be considered independent of the software package which is used for data analysis.

1. Following search parameters should be set:
   Use SwissProt or Uniprot database for searches.
   Taxonomy: human (except the bead proteome which should be searched with taxonomy all to detect streptavidin and nonhuman proteins used for blocking).
   Fixed modification: Carbamidomethyl at cysteine residues.
   Variable modifications: Deamidation of glutamine and asparagine, oxidation of methionine.
   Number of missed cuts: Max 2.
   Enzyme specificity: Trypsin.
   False discovery rate: Set to 1%.
2. Delete well-known contaminants from the list of identified proteins. Common contaminants are human keratins, hornerin, and dermcidin. They mainly result from sample handling and processing and do not reflect any changes of the de novo proteome.
3. Delete proteins found in the "beads proteome." Most of them are bovine proteins and result from blocking of the streptavidin beads.
4. Only proteins identified by at least two significant peptides should be considered.
5. Proteins should be identified in at least two of three biological replicates of Dox-treated (Dox +) and Dox-untreated (Dox −) cells.

The "de novo" proteome is defined by those proteins which show different abundances in Dox-treated vs Dox-untreated cells.

3. APPLICATION AND MODIFICATION

3.1 Advantages

Both parts of this combined strategy—the model system and the orthogonal labeling approach—offer a wide application and modification potential.

With regard to the two-component Tet-On model system, any cell line of interest can be genetically modified to become a transactivator and transgenic cell line only depending on the integration mode and efficiency. Lentiviral systems appear particular useful for this approach because they can infect all cell types including stem cells independent of their proliferation status and also ensure single integration at low MOI. Obviously, these integration events occur randomly and always bear the risk of insertional mutagenesis. Therefore, detailed characterization of these integration sites is obligatory in order to avoid any unpredictable positional effects that potentially could interfere with the physiological consequences exerted by the inducible GOI. Moreover, successful application of this model system is not restricted to the few silent but activatable (s/a) genomic loci (Schönig et al., 2010) identified by us and others (Kovacheva, Zepp, Berger, & Berger, 2014; Lee et al., 2013; Weidenfeld et al., 2009) but should encourage extended screening efforts in search of additional sites. Identifying a larger spectrum of such s/a loci is likely to gain more insights into cell type-specific chromatin structure and properties as well as the abundance and distribution of suitable s/a genomic loci. Once these loci are known, any Tet-regulated expression unit can be precisely integrated at these sites, e.g., by CRISPR/CAS technology thereby overcoming the risk of insertional mutagenesis commonly encountered with plasmid transfections and viral transductions. As a further advantage of such a genetic model system, the biological and biochemical alterations of any GOI can be exploited. Although in MSI tumor cells—as outlined here—reconstituted expression of a mutated (tumor suppressor) gene is a first step to uncover its tumor-specific function, the reverse approach, i.e., suppression of a wild-type gene by inducible siRNA-mediated knock-down, can be easily applied to any other gene and cell type. Finally, this specific targeting to defined s/a genomic loci is not only confined to simple cultured cell lines but also should be applicable to more advanced 3D-culture models like tissue organoids that more faithfully recapitulate, on a cell biological and gene expression level, normal and disease development (Clevers, 2016).

3.2 Improvements

Like with most technical approaches, there is room for modifications and improvements. Evidently, the components and steps required for generating Tet-responsive transgenic cell lines are particularly apt for such modifications. For instance, the use of mutated rtTA transactivator variants that confer increased Tet sensitivity should be advantageous (Zhou et al., 2006). Similarly, different RMCE systems and selection markers associated with these systems are useful improvements (Guan, Magomi, Wang, & Shih, 2013). Also, cutting down on the steps required to generate Tet-controlled transgenic cell lines which involves at least two rounds of stable transfections and/or transductions would reduce time and labor. Hence, combining the Tet-dependent transactivator and Tet-responsive expression cassette into a single unit appears to be a more straightforward approach instead of both components residing on separate vectors. In fact, such retroviral

Tet-regulated all-in-one vectors have been established and successfully applied and shown to confer reduced background expression and improved dynamic range (Heinz et al., 2011).

Also for the metabolic labeling and extraction strategy, several variations can be taken into account. Due to the extremely high affinity of biotin toward streptavidin ($K_D = 10^{-15}$ *M*), the biotinylated molecule/streptavidin interaction is essentially irreversible under physiological conditions. Therefore on-bead digestion was chosen to prepare the sample for mass spectrometric analysis. As an alternative to the on-bead digestion for proteomic analysis of the click chemistry extracted proteins, alkyne- or azide-biotin analogs carrying a cleavable linker may be applied. Representative examples are epoxysuccinate-containing linkers for cleavage with cysteine proteases, vicinal thiols for periodate cleavage (Yang, Hahne, Kuster, & Verhelst, 2013), or dialkoxydiphenylsilane for particularly mild cleavage (Szychowski et al., 2010). As alternative to chemical cleavage also photocleavable linkers are available (Kim, Tallman, Liebler, & Porter, 2009; Olejnik, Sonar, Krzymanska-Olejnik, & Rothschild, 1995; Wang et al., 2010). With these strategies, background protein identifications may be significantly reduced relative to standard on-bead digestion. If specific elution from the streptavidin beads without linker cleavage is desired (desthio)biotin analogs, like PEG4 carboxamide-6-azidohexanyl-(desthio) biotin may be applied. Desthiobiotin binds less tightly to streptavidin ($K_D = 10^{-11}$ *M*), and desthiobiotinylated molecules are therefore more easily eluted from the complex in the presence of excess biotin (Hirsch et al., 2002). The direct conjugation of an alkyne-labeled proteome to a cleavable azide resin utilizing the copper-catalyzed azide–alkyne cycloaddition was suggested to omit the classical streptavidin- and biotin-based affinity enrichment step, thereby representing an operationally simpler, cheaper, and less contaminated alternative for protein extraction (Sibbersen, Lykke, Gregersen, Jorgensen, & Johannsen, 2014).

The copper-catalyzed azide–alkyne cycloaddition is the preferred method for in vitro click reactions. However, use of the copper salt-mediated click reactions for in vivo applications is often highly restricted owing to the toxicity of the copper salt. In studies aimed for in vivo applications such as living cells and biological experiments on animals, a copper-free modification of the click reaction was developed by Bertozzi and coworkers, in which the alkyne is incorporated within a strained cyclooctyne system, which increases its intrinsic reactivity and circumvents the requirement for the copper catalyst (Sletten & Bertozzi, 2011).

Combining click chemistry-based labeling with SILAC enables the quantification of newly synthesized low-abundance proteins in the presence of long-lived high-abundance proteins (Somasekharan et al., 2012). Such approach is a preferable technique for secretome analyses which are frequently hampered due to the limitations in the detection of low-abundance proteins, like cytokines, chemokines, and growth factors, against a background of serum containing media (Eichelbaum, Winter, Berriel-Diaz, Herzig, & Krijgsveld, 2012). The combination of the two labeling techniques allows to study the secretome of cell cultures irrespective of the complexity of the background proteins. Since most secreted proteins are glycosylated also labeling with azido- or alkyne monosaccharide analogs can be applied for secretome analyses (Lai, Nice, & Schilling, 2013).

4. CONCLUSION

The experimental approach outlined here is a suitable tool to investigate and validate the relationship between driver mutations and the biology of tumors thereby facilitating the development of novel diagnostic and therapeutic strategies for the benefit of patients. Moreover the strategy can be adapted to various other diseases where the disease-promoting mutations have been identified.

ACKNOWLEDGMENTS

Work in the authors' labs has been funded by the DFG (Grant GE592/6-2), Deutsche Krebshilfe (Grant number 110121). We thank Kai Schönig (Central Institute of Mental Health, Mannheim, Germany) and Jennifer Lee for critical reading of the manuscript.

REFERENCES

Ahn, Y. H., Kim, J. Y., & Yoo, J. S. (2015). Quantitative mass spectrometric analysis of glycoproteins combined with enrichment methods. *Mass Spectrometry Reviews*, *34*, 148–165.

Ailles, L. E., & Naldini, L. (2002). HIV-1-derived lentiviral vectors. *Current Topics in Microbiology and Immunology*, *261*, 31–52.

Alhopuro, P., Sammalkorpi, H., Niittymäki, I., Biström, M., Raitila, A., Saharinen, J., et al. (2012). Candidate driver genes in microsatellite-unstable colorectal cancer. *International Journal of Cancer*, *130*, 1558–6615.

Ballikaya, S., Lee, J., Warnken, U., Schnölzer, M., Gebert, J., & Kopitz, J. (2014). De Novo proteome analysis of genetically modified tumor cells by a metabolic labeling/azide-alkyne cycloaddition approach. *Molecular and Cellular Proteomics*, *13*, 3446–3456.

Baron, U., Freundlieb, S., Gossen, M., & Bujard, H. (1995). Co-regulation of two gene activities by tetracycline via a bidirectional promoter. *Nucleic Acids Research*, *23*, 3605–3606.

Berens, C., & Hillen, W. (2003). Gene regulation by tetracyclines. Constraints of resistance in bacteria shape TetR for application in eukaryotes. *European Journal of Biochemistry*, *270*, 3109–3121.

Beynon, R. J., & Pratt, J. M. (2005). Metabolic labeling of proteins for proteomics. *Molecular and Cellular Proteomics*, *4*, 857–872.

Boersema, P. J., Geiger, T., Wisniewski, J. R., & Mann, M. (2013). Quantification of the N-glycosylated secretome by super-SILAC during breast cancer progression and in human blood samples. *Molecular and Cellular Proteomics*, *12*, 158–171.

Boland, C. R., & Goel, A. (2010). Microsatellite instability in colorectal cancer. *Gastroenterology*, *138*, 2073–2087.

Bonifacino, J. S. (1998). Metabolic labeling with amino acids. In J. S. Bonifacino (Ed.), *Analysis of proteins* (pp. 1–10). New York, NY: Wiley.

Celis, J. E., & Gromov, P. (2003). Proteomics in translational cancer research: Toward an integrated approach. *Cancer Cell*, *3*, 9–15.

Choudhary, C., Kumar, C., Gnad, F., Nielsen, M. L., Rehman, M., Walther, T. C., et al. (2009). Lysine acetylation targets protein complexes and co-regulates major cellular functions. *Science*, *325*, 834–840.

Christiansen, M. N., Chik, J., Lee, L., Anugraham, M., Abrahams, J. L., & Packer, N. H. (2014). Cell surface protein glycosylation in cancer. *Proteomics*, *14*, 525–546.

Clevers, H. (2016). Modeling development and disease with organoids. *Cell*, *165*, 1586–1597.

Cox, J., & Mann, M. (2008). MaxQuant enables high peptide identification rates, individualized p.p.b.-range mass accuracies and proteome-wide protein quantification. *Nature Biotechnology*, *26*, 1367–1372.

Das, A. T., Zhou, X., Metz, S. W., Vink, M. A., & Berkhout, B. (2016). Selecting the optimal Tet-On system for doxycycline-inducible gene expression in transiently transfected and stably transduced mammalian cells. *Biotechnology Journal*, *11*, 71–79.

Dennis, J. W., Granovsky, M., & Warren, C. E. (1999). Glycoprotein glycosylation and cancer progression. *Biochimica et Biophysica Acta*, *1473*, 21–34.

Diaz, S., & Varki, A. (2009). Metabolic radiolabeling of animal cell glycoconjugates. *Current Protocols in Protein Science*, *57*(12.2), 1–55.

Dube, D. H., & Bertozzi, C. R. (2005). Glycans in cancer and inflammation—Potential for therapeutics and diagnostics. *Nature Reviews Drug Discovery*, *4*, 477–488.

Duval, A., Rolland, S., Compoint, A., Tubacher, E., Iacopetta, B., Thomas, G., et al. (2001). Evolution of instability at coding and non-coding repeat sequences in human MSI-H colorectal cancers. *Human Molecular Genetics*, *10*, 513–518.

Dwek, M. V., & Brooks, S. A. (2004). Harnessing changes in cellular glycosylation in new cancer treatment strategies. *Current Cancer Drug Targets*, *4*, 425–442.

Eichelbaum, K., & Krijgsveld, J. (2014). Combining pulsed SILAC labeling and click-chemistry for quantitative secretome analysis. *Methods in Molecular Biology*, *1174*, 101–114.

Eichelbaum, K., Winter, M., Berriel-Diaz, M., Herzig, S., & Krijgsveld, J. (2012). Selective enrichment of newly synthesized proteins for quantitative secretome analysis. *Nature Biotechnology*, *30*, 984–990.

El-Sagheer, A. H., & Brown, T. (2010). Click chemistry with DNA. *Chemical Society Reviews*, *39*, 1388–1405.

El-Sagheer, A. H., & Brown, T. (2012). Click nucleic acid ligation: Applications in biology and nanotechnology. *Accounts of Chemical Research*, *45*, 1258–1267.

Fearon, E. R., & Vogelstein, B. (1990). A genetic model for colorectal tumorigenesis. *Cell*, *61*, 759–767.

Frost, D. C., & Li, L. (2014). Recent advances in mass spectrometry-based glycoproteomics. *Advances in Protein Chemistry and Structural Biology*, *95*, 71–123.

Gorelik, E., Galili, U., & Raz, A. (2001). On the role of cell surface carbohydrates and their binding proteins (lectins) in tumor metastasis. *Cancer and Metastasis Reviews*, *20*, 245–277.

Gossen, M., & Bujard, H. (1992). Tight control of gene expression in mammalian cells by tetracycline-responsive promoters. *Proceedings of the National Academy of Sciences of the United States of America*, *89*, 5547–5551.

Gossen, M., Freundlieb, S., Bender, G., Müller, G., Hillen, W., & Bujard, H. (1995). Transcriptional activation by tetracyclines in mammalian cells. *Science*, *268*, 1766–1769.

Griffiths, R. (2011). In vivo labeling. In R. J. Slater (Ed.), *Radioisotopes in biology: A practical approach* (pp. 109–126). Oxford, UK: University Press.

Guan, B., Magomi, T., Wang, T. L., & Shih, I. M. (2013). Establishing isogenic inducible cell lines using founder reporter lines and recombinase-mediated cassette exchange. *Biotechniques*, *55*, 233–242.

Hampf, M., & Gossen, M. (2007). Promoter crosstalk effects on gene expression. *Journal of Molecular Biology*, *365*, 911–920.

Hanahan, D., & Weinberg, R. A. (2011). Hallmarks of cancer: The next generation. *Cell*, *144*, 646–674.

Heinz, N., Schambach, A., Galla, M., Maetzig, T., Baum, C., Löw, R., et al. (2011). Retroviral and transposon-based tet-regulated all-in-one vectors with reduced background expression and improved dynamic range. *Human Gene Therapy*, *22*, 166–176.

Hirsch, J. D., Eslamizar, L., Filanoski, B. J., Malekzadeh, N., Haugland, R. P., Beechem, J. M., et al. (2002). Easily reversible desthiobiotin binding to streptavidin, avidin, and other biotin-binding proteins: Uses for protein labeling, detection, and isolation. *Analytical Biochemistry*, *308*, 343–357.

Huang, E. P., Marquis, C. P., & Gray, P. P. (2004). Process development for a recombinant Chinese hamster ovary (CHO) cell line utilizing a metal induced and amplified metallothionein expression system. *Biotechnology and Bioengineering*, *88*, 437–450.

Jensen, P. H., Karlsson, N. G., Kolarich, D., & Packer, N. H. (2012). Structural analysis of N- and O-glycans released from glycoproteins. *Nature Protocols*, *7*, 1299–1310.

Kim, T.-M., Laird, P. W., & Park, P. J. (2013). The landscape of microsatellite instability in colorectal and endometrial cancer genomes. *Cell*, *155*, 858–868.

Kim, H. Y., Tallman, K. A., Liebler, D. C., & Porter, N. A. (2009). An azido-biotin reagent for use in the isolation of protein adducts of lipid-derived electrophiles by streptavidin catch and photorelease. *Molecular and Cellular Proteomics*, *8*, 2080–2089.

Kobata, A., & Amano, J. (2005). Altered glycosylation of proteins produced by malignant cells, and application for the diagnosis and immunotherapy of tumours. *Immunology and Cell Biology*, *83*, 429–439.

Kolb, H. C., Finn, M. G., & Sharpless, K. B. (2001). Click chemistry: Diverse chemical function from a few good reactions. *Angewandte Chemie (International Ed. in English)*, *40*, 2004–2021.

Kovacheva, M., Zepp, M., Berger, S. M., & Berger, M. R. (2014). Sustained conditional knockdown reveals intracellular bone sialoprotein as essential for breast cancer skeletal metastasis. *Oncotarget*, *5*, 5510–5522.

Lai, Z. W., Nice, E. C., & Schilling, O. (2013). Glycocapture-based proteomics for secretome analysis. *Proteomics*, *13*, 512–525.

Lee, J., Ballikaya, S., Schönig, K., Ball, C. R., Glimm, H., Kopitz, J., et al. (2013). Transforming growth factor beta receptor 2 (TGFBR2) changes sialylation in the microsatellite unstable (MSI) Colorectal cancer cell line HCT116. *PLoS One*, *8*, e57074.

Lengauer, C., Kinzler, K. W., & Vogelstein, B. (1997). Genetic instability in colorectal cancers. *Nature*, *386*, 623–627.

Lim, R. K., & Lin, Q. (2011). Photoinducible bioorthogonal chemistry: A spatiotemporally controllable tool to visualize and perturb proteins in live cells. *Accounts of Chemical Research*, *44*, 828–839.
Löw, R., Heinz, N., Hampf, M., Bujard, H., & Gossen, M. (2010). Improved Tet-responsive promoters with minimized background expression. *BMC Biotechnology*, *10*, 81.
Löw, R., Vigna, E., Lindemann, D., Naldini, L., & Bujard, H. (2006). Retroviral vectors containing Tet-controlled bidirectional transcription units for simultaneous regulation of two gene activities. *Journal of Molecular and Genetic Medicine*, *2*, 107–118.
Markowitz, S., Wang, J., Myeroff, L., Parsons, R., Sun, L., Lutterbaugh, J., et al. (1995). Inactivation of the type II TGF-beta receptor in colon cancer cells with microsatellite instability. *Science*, *268*, 1336–1338.
Massagué, J. (2012). TGFβ signaling in context. *Nature Reviews Molecular Cell Biology*, *13*, 616–630.
Mechref, Y., Hu, Y., Desantos-Garcia, J. L., Hussein, A., & Tang, H. (2013). Quantitative glycomics strategies. *Molecular and Cellular Proteomics*, *12*, 874–884.
Moses, J. E., & Moorhouse, A. D. (2007). The growing applications of click chemistry. *Chemical Society Reviews*, *36*, 1249–1262.
Mouradov, D., Sloggett, C., Jorissen, R. N., Love, C. G., Li, S., Burgess, A. W., et al. (2014). Colorectal cancer cell lines are representative models of the main molecular subtypes of primary cancer. *Cancer Research*, *74*, 3238–3247.
No, D., Yao, T. P., & Evans, R. M. (1996). Ecdysone-inducible gene expression in mammalian cells and transgenic mice. *Proceedings of the National Academy of Sciences of the United States of America*, *93*, 3346–3351.
Olejnik, J., Sonar, S., Krzymanska-Olejnik, E., & Rothschild, K. J. (1995). Photocleavable biotin derivatives: A versatile approach for the isolation of biomolecules. *Proceedings of the National Academy of Sciences of the United States of America*, *92*, 7590–7594.
Orlando, R., Lim, J. M., Atwood, J. A., 3rd, Angel, P. M., Fang, M., Aoki, K., et al. (2009). IDAWG: Metabolic incorporation of stable isotope labels for quantitative glycomics of cultured cells. *Journal of Proteome Research*, *8*, 3816–3823.
Park, S. K., Venable, J. D., Xu, T., & Yates, J. R., 3rd. (2008). A quantitative analysis software tool for mass spectrometry-based proteomics. *Nature Methods*, *5*, 319–322.
Paruzynski, A., Arens, A., Gabriel, R., Bartholomae, C. C., Scholz, S., Wang, W., et al. (2010). Genome-wide high-throughput integrome analyses by nrLAM-PCR and next-generation sequencing. *Nature Protocols*, *5*, 1379–1395.
Pertea, M., & Salzberg, S. L. (2010). Between a chicken and a grape: Estimating the number of human genes. *Genome Biology*, *11*, 206.
Pinho, S. S., & Reis, C. A. (2015). Glycosylation in cancer: Mechanisms and clinical implications. *Nature Reviews Cancer*, *15*, 540–555.
Prescher, J. A., & Bertozzi, C. R. (2005). Chemistry in living systems. *Nature Chemical Biology*, *1*, 13–21.
Raymond, C. S., & Soriano, P. (2007). High-efficiency FLP and PhiC31 site-specific recombination in mammalian cells. *PLoS One*, *2*, e162.
Rösli, C., Rybak, J. N., Neri, D., & Elia, G. (2008). Quantitative recovery of biotinylated proteins from streptavidin-based affinity chromatography resins. *Methods in Molecular Biology*, *418*, 89–100.
Schlake, T., & Bode, J. (1994). Use of mutated FLP recognition target (FRT) sites for the exchange of expression cassettes at defined chromosomal loci. *Biochemistry*, *33*, 12746–12751.
Schönig, K., Bujard, H., & Gossen, M. (2010). The power of reversibility regulating gene activities via tetracycline-controlled transcription. *Methods in Enzymology*, *477*, 429–453.

Sibbersen, C., Lykke, L., Gregersen, N., Jorgensen, K. A., & Johannsen, M. (2014). A cleavable azide resin for direct click chemistry mediated enrichment of alkyne-labeled proteins. *Chemical Communications*, *50*, 12098–12100.

Sletten, E., & Bertozzi, C. R. (2011). From mechanism to mouse: A tale of Two bioorthogonal reactions. *Accounts of Chemical Research*, *44*, 666–676.

Somasekharan, S. P., Stoynov, N., Rotblat, B., Leprivier, G., Galpin, J. D., Ahern, C. A., et al. (2012). Identification and quantification of newly synthesized proteins translationally regulated by YB-1 using a novel Click-SILAC approach. *Journal of Proteomics*, *77*, e1–e10.

Soneoka, Y., Cannon, P. M., Ramsdale, E. E., Griffiths, J. C., Romano, G., Kingsman, S. M., et al. (1995). A transient three-plasmid expression system for the production of high titer retroviral vectors. *Nucleic Acids Research*, *23*, 628–633.

Stowell, R. S., Ju, T., & Cummings, R. D. (2015). Protein glycosylation in cancer. *Annual Review of Pathology*, *10*, 473–510.

Strassberger, V., Trüssel, S., Fugmann, T., Neri, D., & Roesli, C. (2010). A novel reactive ester derivative of biotin with reduced membrane permeability for in vivo biotinylation experiments. *Proteomics*, *10*, 3544–3548.

Stratton, M. R., Campbell, P. J., & Futreal, A. P. (2009). The cancer genome. *Nature*, *458*, 719–724.

Svinkina, T., Gu, H., Silva, J. C., Mertins, P., Qiao, J., Fereshetian, S., et al. (2015). Deep, quantitative coverage of the lysine acetylome using novel anti-acetyl-lysine antibodies and an optimized proteomic workflow. *Molecular and Cellular Proteomics*, *14*, 2429–2440.

Szychowski, J., Mahdavi, A., Hodas, J. J., Bagert, J. D., Ngo, J. T., Landgraf, P., et al. (2010). Cleavable biotin probes for labeling of biomolecules via azide-alkyne cycloaddition. *Journal of the American Chemical Society*, *132*, 18351–18360.

The Cancer Genome Atlas Network (2012). Comprehensive molecular characterization of human colon and rectal cancer. *Nature*, *487*, 330–337.

Thingholm, T. E., Jensen, O. N., & Larsen, M. R. (2009). Analytical strategies for phosphoproteomics. *Proteomics*, *9*, 1451–1468.

Tomasetti, C., Marchionni, L., Nowak, M. A., Parmigiani, G., & Vogelstein, B. (2015). Only three driver gene mutations are required for the development of lung and colorectal cancers. *Proceedings of the National Academy of Sciences of the United States of America*, *112*, 118–123.

Turan, S., Zehe, C., Kuehle, J., Qiao, J., & Bode, J. (2013). Recombinase-mediated cassette exchange (RMCE)—A rapidly-expanding toolbox for targeted genomic modifications. *Gene*, *515*, 1–27.

Urlinger, S., Baron, U., Thellmann, M., Hasan, M. T., Bujard, H., & Hillen, W. (2000). Exploring the sequence space for tetracycline-dependent transcriptional activators: Novel mutations yield expanded range and sensitivity. *Proceedings of the National Academy of Sciences of the United States of America*, *97*, 7963–7968.

Vlad, A. M., & Finn, O. J. (2004). Glycoprotein tumor antigens for immunotherapy of breast cancer. *Breast Disease*, *20*, 73–79.

Vogelstein, B., Papadopoulos, N., Velculescu, V. E., Zhou, S., Diaz, L. A., Jr., & Kinzler, K. W. (2013). Cancer genome landscapes. *Science*, *339*, 1546–1558.

Wang, Z., Udeshi, N. D., O'Malley, M., Shabanowitz, J., Hunt, D. F., & Hart, G. W. (2010). Enrichment and site mapping of O-linked N-acetylglucosamine by a combination of chemical/enzymatic tagging, photochemical cleavage, and electron transfer dissociation mass spectrometry. *Molecular and Cellular Proteomics*, *9*, 153–160.

Weidenfeld, I., Gossen, M., Low, R., Kentner, D., Berger, S., Gorlich, D., et al. (2009). Inducible expression of coding and inhibitory RNAs from retargetable genomic loci. *Nucleic Acids Research*, *37*, e50.

Welman, A., Barraclough, J., & Dive, C. (2006). Generation of cells expressing improved doxycycline-regulated reverse transcriptional transactivator rtTA2S-M2. *Nature Protocols*, *1*, 803–811.

Wessel, D., & Flügge, U. I. (1984). A method for the quantitative recovery of protein in dilute solution in the presence of detergents and lipids. *Analytical Biochemistry*, *138*, 141–143.

Woerner, S. M., Gebert, J., Yuan, Y. P., Sutter, C., Ridder, R., Bork, P., et al. (2001). Systematic identification of genes with coding microsatellites mutated in DNA mismatch repair-deficient cancer cells. *International Journal of Cancer*, *93*, 12–19.

Yang, Y., Hahne, H., Kuster, B., & Verhelst, S. H. (2013). A simple and effective cleavable linker for chemical proteomics applications. *Molecular and Cellular Proteomics*, *12*, 237–244.

Zhang, X., & Zhang, Y. (2013). Applications of azide-based bioorthogonal click chemistry in glycobiology. *Molecules*, *18*, 7145–7159.

Zhou, X., Vink, M., Klave, B., Berkhout, B., & Das, A. T. (2006). Optimization of the Tet-On system for regulated gene expression through viral evolution. *Gene Therapy*, *13*, 1382–1390.

CHAPTER EIGHTEEN

Analysis of the Proteome of Hair-Cell Stereocilia by Mass Spectrometry

J.F. Krey*,†, P.A. Wilmarth‡, L.L. David‡, P.G. Barr-Gillespie*,†,‡,1

*Oregon Hearing Research Center, Oregon Health & Science University, Portland, OR, United States
†Vollum Institute, Oregon Health & Science University, Portland, OR, United States
‡Oregon Health & Science University, Portland, OR, United States
[1]Corresponding author: e-mail address: gillespp@ohsu.edu

Contents

Abstract

Characterization of proteins that mediate mechanotransduction by hair cells, the sensory cells of the inner ear, is hampered by the scarcity of these cells and their sensory organelle, the hair bundle. Mass spectrometry, with its high sensitivity and identification precision, is the ideal method for determining which proteins are present in bundles

Methods in Enzymology, Volume 585
ISSN 0076-6879
http://dx.doi.org/10.1016/bs.mie.2016.09.023

and what proteins they interact with. We describe here the isolation of mouse hair bundles, as well as preparation of bundle protein samples for mass spectrometry. We also describe protocols for data-dependent (shotgun) and parallel reaction monitoring (targeted) mass spectrometry that allow us to identify and quantify proteins of the hair bundle. These sensitive methods are particularly useful for comparing proteomes of wild-type mice and mice with deafness mutations affecting hair-bundle proteins.

1. INTRODUCTION

The inner ear, with its auditory and vestibular divisions, measures sound and detects head movements, transmitting this information to the VIIIth cranial nerve and, eventually, the central nervous system. Auditory and vestibular hair cells, the sensory cells, mediate the mechanical-to-electrical transduction at the heart of this process (Fettiplace & Kim, 2014). To carry out mechanotransduction, hair cells use an apical cluster of actin-rich stereocilia, which are coupled together by various linkages to form a hair bundle (Fig. 1). The stereocilia in the bundle are arranged in staggered rows, giving the structure a beveled appearance. Deflection of bundles by external stimuli, which are transmitted to the hair cells through structures in the outer, middle, and inner ears, leads to opening and closing of cation-conducting transduction channels. Special links within the bundle, the tip links, run up the bevel and control the transduction channels; deflection of bundles toward the tallest stereocilia opens channels and depolarizes hair cells. Hair cell depolarization, in turn, triggers neurotransmitter release and the ensuing signal transmission to the central nervous system.

Knowledge of the identity, concentrations, and interactions of the proteins that make up hair bundles and control transduction is critical for understanding how these structures operate to carry out transduction with impressive sensitivity (Barr-Gillespie, 2015). Unfortunately, there are few hair cells in the inner ear; for example, each ear of the mouse has $\sim$10,000 hair cells combined in its cochlea (auditory system) and its five vestibular organs. Moreover, hair bundles account for $<1\%$ of the protein in a hair cell, and thus are scarce even within the inner ear. Even more daunting, transduction molecules like the tip link and channel may be present at <1 part in 10^5 in the bundle. Finally, hair cells are housed in the periotic bone, one of the hardest in the body, and these cells' sensitivity to minute stimuli renders them extremely delicate, readily damaged in

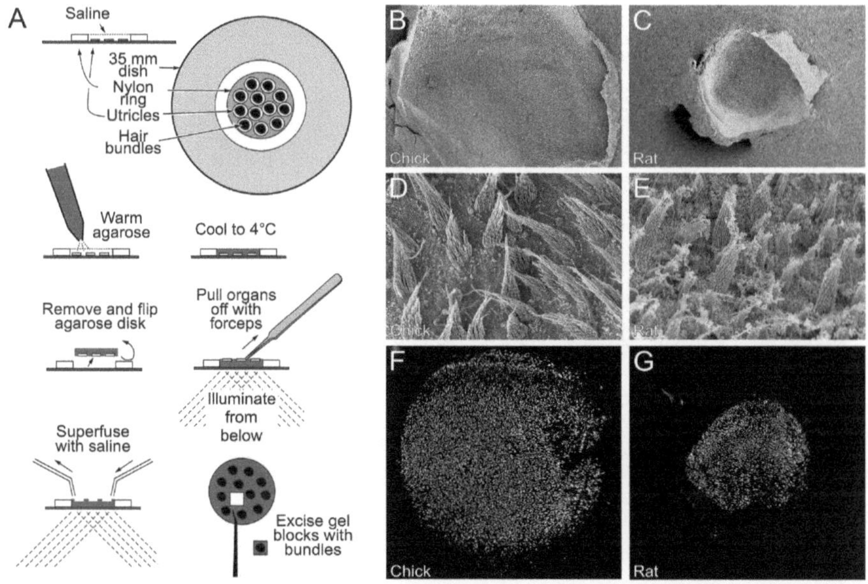

Fig. 1 Isolation of utricle hair bundles. (A) Illustration of twist-off method. Utricles are adhered to the bottoms of plastic culture dishes and are surrounded by a nylon ring. Warm agarose is applied to the utricles, which is decanted to leave agarose only in the volume bounded by the ring. After cooling to gel the agarose, the disk is removed and flipped; the utricles remain in the agarose and their back surfaces are exposed. A quick perpendicular pull on the utricle using forceps leaves the bundles in agarose. Bundles are superfused with saline and cleaned up under a dissecting microscope; small blocks of agarose containing bundles from each ear are excised and frozen. (B and C), Low-power scanning electron micrograph showing chick (B) and rat (C) utricles. (D and E), high-power scanning electron micrograph showing chick (D) and rat (E) hair bundles. (F and G), Hair bundles in agarose, isolated from chick (F) or rat (G) utricles, labeled with phalloidin. Panel full widths: (B and C) and (F and G) 1170 μm; (D and E) 41 μm.

attempts to extricate them. Taken together, these difficulties conspire to slow molecular characterization of this system as compared to others.

Nevertheless, methods have been developed to allow isolation, identification, and quantitation of proteins critical for function of the hair bundle. In particular, the "twist-off" method allows for a clean isolation of bundles from vestibular epithelia, including those of frog, mouse, rat, and chicken. Bundle proteins can be analyzed by shotgun mass spectrometry, which allows determination of the diversity of proteins present and estimation of their concentrations, and by targeted mass spectrometry, which allows accurate measurement of protein abundance. Together these biochemical techniques complement other methods used for characterizing bundle function, including electrophysiology, imaging, and genetics.

2. ISOLATION OF HAIR BUNDLES

Hair bundles are isolated relatively easily because of the unique structure of their constituent stereocilia (Fig. 1). While stereocilia are 200–500 nm in diameter throughout most of their length, as they approach the apical surface, they taper down to few tens of nanometers wide. Because of this tapering, the stereocilia both flex easily—as part of their physiological role—and are mechanically weak, which means that they can be readily sheared off of the cell. The twist-off method for bundle isolation, originally developed for bullfrog saccule hair bundles (Gillespie & Hudspeth, 1991) and later adapted for hair bundles of mouse (Dumont, Zhao, Holt, Bahler, & Gillespie, 2002; Krey, Sherman, Jeffery, Choi, & Barr-Gillespie, 2015), rat (Shin, Pagana, & Gillespie, 2009; Wilmarth et al., 2015), and chick (Shin et al., 2007, 2013) utricles, isolates bundles by first trapping them in an agarose matrix, then mechanically displaces them from the utricular epithelium (Fig. 1A).

We describe here isolation of hair bundles from the mouse utricle, a vestibular organ. We have used both shotgun (Ebrahim et al., 2016; Francis et al., 2015; Krey et al., 2015, 2016; Wilmarth et al., 2015) and targeted mass spectrometry (Ebrahim et al., 2016; Krey et al., 2016) to characterize mouse utricle proteins; similar methods are used with chick and rat hair bundles (Fig. 1B–G).

2.1 Mouse Utricle Dissection and Twist-Off Bundle Isolation

Originally named for the twisting motion applied to the agarose relative to the fixed utricular epithelia, in the twist-off procedure, the epithelia are adhered to the bottom of a chamber that is formed with a small nylon washer and coverslip, and the buffer solution in this chamber is displaced with molten low-melting point agarose. After the agarose infiltrates the bundles, it is cooled to allow it to gel firmly. The epithelia are then tugged off, parallel to the agarose, and leaving the trapped bundles in the agarose. With the right illumination angle, bundles (and cellular debris) can be visualized on a dissecting microscope, allowing cleaning of the embedded bundles prior to excision.

2.1.1 Dissecting Mouse Utricles

1. Prepare dissection solution by adding 1 *M* HEPES at pH 7.4 to Leibovitz's L-15 Medium (without phenol red) to a final concentration

of 5 m*M*. Hank's balanced salt solution (HBSS) can also be used. Oxygenate for 5–15 min.

2. Euthanize mice using CO_2 asphyxiation; use decapitation for animals younger than postnatal day 8 (P8). Utricles from 20 to 25 mice can be processed at one time.
3. Decapitate mice, cut the skull sagittally along the midline, remove brain, and cut out periotic bone containing ear. Place the temporal bones in dissection solution in petri dish on ice. Repeat for all mice.
4. Transfer each periotic bone to new petri dish with cold dissection solution. For mice older than P8, use dissection scissors to cut the bones encasing the semicircular canals, then use forceps to chip open the inner-ear capsule and expose the utricle. This procedure can be done entirely with forceps for younger animals. Carefully remove the utricle, using forceps to cut the nerve and semicircular canals. Transfer utricles to new dish; a cut 200 μL plastic pipette tip that has been passivated with a 10% BSA solution works well as a transfer pipette for mouse utricles.
5. Use forceps to complete removal of the nerve from back of utricle and remove the endolymphatic sac. Carefully remove the otolithic membrane using a fine eyelash. Once all utricles have been dissected, transfer them to new petri dish with fresh dissection solution. Wash 1–2 times with dissection solution.

2.1.2 Embedding Utricles in Agarose

1. Prepare a 4.5% agarose solution by adding 10 mL dissection medium to 0.45 g low-melting point agarose, swirling to mix. Microwave until agarose is dissolved (~30 s), stopping to swirl the tube every 5 s. Filter through a 5-μm syringe filter into a new 50-mL conical tube and allow to cool to 40–42°C in a water bath.
2. While cooling, fill a new uncoated 35-mm bacteriological petri dish (BD Falcon; cat #351008) with dissection solution. Carefully transfer utricles to the dish with the transfer pipette. Under a dissection microscope, use forceps to stick down the organs (nerve side down) to the bottom of the dish. Use leftover tissue flaps from the endolymphatic sac to help flatten the utricles as much as possible without touching the sensory epithelium. To give a guide to utricles placement, trace the inside of the washer on the bottom side of the dish with a marker prior to adding dissection solution.
3. Once all utricles are adhered to the dish, carefully place a nylon washer ring into the dish (angling it into the surface of the liquid) to surround the

utricles. Using a disposable plastic transfer pipette, slowly and carefully pipette the agarose solution onto the utricles and fill the inside of the washer ring (the agarose should be level with or higher than the top of the washer ring). It is helpful to do this under a dissection microscope to ensure even dispersal of the agarose.

4. Leave the dish untouched at room temperature (RT) until the agarose has turned opaque, then remove the remaining saline. Place the covered dish at 4°C for at least 15 min to fully solidify the agarose.

2.1.3 Twist-Off

1. Set up the isolation apparatus in a laminar flow hood. The apparatus includes a dissecting microscope and stage (30 cm wide × 13 cm deep × 7.5 cm tall) that has a 7.5 × 3.8 cm rectangular opening that allows placement of a glass slide, on which a nylon ring is permanently attached. A fiber optic light is placed underneath for visualization of the bundles within the agarose. Superfusion and suction tubes are attached to the stage and a peristaltic pump is used to pump cold dissection solution across the agarose during the procedure.
2. Use a clean scalpel blade to cut away excess agarose from around the washer. Turn washer on its side and slice away any excess agarose from the top so that only an even disk of agarose inside of the washer remains. The organs should remain embedded in the agarose on one side. Remove the agarose disk from the washer and place it organ side up inside the washer affixed to the glass slide on the stage. Illuminate from below with the fiber optic light and start superfusion with cold dissection media.
3. Using forceps, grab the peripheral edge of each organ and quickly pull from the agarose, parallel to the surface of the agarose disk. Discard the organs. Adjust the position of the light to allow clear visualization of the bundles; they act as light pipes, so oblique illumination increases their contrast with the surrounding agarose. Use a fine scalpel blade or tungsten needle to cut away any agarose surrounding the bundles that contains cellular debris. It is easiest to carve around the regions containing bundles, leaving them on an agarose pedestal. Use a fine tungsten needle to slice underneath the bundles and then pierce the chunk of agarose containing the bundles and transfer it to a low protein-binding microfuge tube. Store the bundles at −80°C until processing.

Purified hair bundles always contain some contaminants from the whole epithelium. Purity of bundles can be assessed in several ways. First, bundles and contaminants can be visualized by fluorescence microscopy directly in the agarose; stereocilia actin can be stained with fluorescent phalloidin, nuclei with DAPI or any other nuclear dye, and contaminants with selective antibodies (Krey et al., 2015; Shin et al., 2007, 2013). While not quantitative, fluorescence microscopy can reveal whether certain contaminants (e.g., cuticular plates) are present along with bundles. Mass spectrometry can also be used to assess purity (see later).

2.2 Biochemical Manipulation of Hair Bundles in Agarose

Hair bundles embedded in agarose can be subjected to biochemical manipulation, including selective extraction to determine properties of individual proteins (Gillespie & Hudspeth, 1991; Walker, Hudspeth, & Gillespie, 1993) or manipulation of nucleotide state (Gillespie, Wagner, & Hudspeth, 1993; Yamoah & Gillespie, 1996). Although purified bundles are quite dilute—bundles from one ear, ~5 ng of total protein can be recovered in about 1 μL of agarose—the agarose prevents the bundles from being lost to tube surfaces and allows easy sedimentation. Extraction of bundle proteins, therefore, requires addition of a carrier, such as a purified protein, that prevents nonspecific binding but does not interfere with analysis; in different experiments, we have successfully used lysozyme, myoglobin, hemoglobin, bovine serum albumin, and maltose-binding protein (Gillespie & Hudspeth, 1991; Gillespie et al., 1993; Walker et al., 1993). We do not add carrier proteins, however, when carrying out mass spectrometry analyses.

2.3 Sample Preparation for Mass Spectrometry Using In-Gel Digestion

Prior to mass spectrometry, hair-bundle proteins must be reduced, alkylated, and digested by trypsin to peptides. Because of the small amounts of protein, efficiency of these steps is of utmost importance. In-gel digestion of proteins (Shevchenko, Tomas, Havlis, Olsen, & Mann, 2006) relies on SDS-PAGE to eliminate contaminants that otherwise might interfere with mass spectrometry; moreover, little handling is required, and proteins remain in high concentrations of SDS, minimizing protein loss. In-gel digestion has limitations; however, some proteins aggregate strongly and never enter the gel matrix, preventing their analysis; in addition, small proteins can diffuse out of the gel matrix and are lost during initial steps. Finally, unless handled with extreme precautions, gels attract keratins during processing and thus usually have higher levels of contamination than in-solution digests. Recovery from gels can be improved by addition of a degradable surfactant (ProteaseMAX or RapiGest) to digestion solution. In-gel digestion also allows for fractionation of the sample by cutting the gel lane into multiple pieces corresponding to different protein masses. While fractionation is associated with some additional sample loss, the multiple shotgun runs performed on each sample can increase proteome coverage.

2.3.1 Sample Preparation

To avoid keratin contamination, all steps are performed in a laminar flow hood, with lab coat and nitrile gloves, taking care to rinse gloves often. All solutions are made fresh on the day of the procedure. Sample preparation, including SDS-PAGE and reduction and alkylation, has been described in detail elsewhere (Krey et al., 2015; Shin et al., 2010, 2013; Wilmarth et al., 2015). Isolated hair-bundle samples are thawed on ice, then run on a 4–12% Bis–Tris 1.5 mm 10-well gradient gel (Invitrogen) at 150 V for 10 min until top of dye front is ~1 cm from bottom of wells. The gel is washed with water, stained for 5 h with Imperial Stain (Thermo Fisher Scientific, Waltham, MA) at RT, and washed with water overnight. Keep gel covered at all times. Each lane of the gel is then cut into six 1–2 mm-wide slices, each of which is cut into four pieces.

Reduction is carried out with 25 m*M* DTT in 50 m*M* ammonium bicarbonate pH 8 for 20 min at 56°C, then the sample is alkylated with 55 m*M* iodoacetamide in 50 m*M* ammonium bicarbonate pH 8 in the dark for 20 min at RT (Krey et al., 2015; Shin et al., 2010, 2013; Wilmarth et al., 2015). The gel slices are then washed, stepped into 100% acetonitrile, and are dried by vacuum concentration or air-drying for 5 min in a laminar flow hood.

2.3.2 In-Gel Digestion

We have had the greatest success using the procedure outlined in the ProteaseMAX manual (Promega, Madison, WI) for low-abundance protein samples. Gel pieces are rehydrated with 30 μL of 6 ng/μL trypsin, 0.01% ProteaseMAX, 50 m*M* ammonium bicarbonate pH 8 for 30 min at 4°C. The samples are overlaid with 20 μL of 0.01% ProteaseMAX and digested for 3 h at 37°C. The digest solution is removed after centrifugation, then

30 μL of 2.5% trifluoroacetic acid (in HPLC-grade H_2O) is added to gel pieces and vortexed for 15 min. The combined supernatants are transferred to 0.45 μm-filter tubes (Ultrafree centrifugal filters, #UFC0HV00; EMD Millipore, Billerica, MA), and are spun 5 min at 4000 rpm. The filtrated is dried in a vacuum concentrator until all solution is evaporated; the tubes with dried peptides are stored at −80°C until analysis.

2.4 Sample Preparation for Mass Spectrometry Using Enhanced Filter-Assisted Sample Preparation

In-solution digestion is the least labor-intensive and allows for optimal peptide recovery as samples are not transferred from tube to tube. We have achieved the best results with bundles using the enhanced filter-assisted sample preparation (eFASP) method (Erde, Loo, & Loo, 2014). The eFASP method carries out each of the preparation steps in an ultrafiltration unit, which allows solution exchange and prevents losses from tube-to-tube transfers. Since solution exchange occurs, the eFASP method is compatible with samples solubilized in SDS, and subsequently uses the surfactant deoxycholate to enhance trypsinization. The methodology also removes traces of agarose that interfere with mass spectrometry and is compatible with multidimension chromatography separations.

To carry out eFASP-mediated reduction, alkylation, and trypsin digestion, we follow the "eFASP with Passivated Ultrafiltration Unit" procedure described by Erde et al. (2014). Proteins are reduced with TCEP, alkylated with iodoacetamide, and digested with 200 ng sequencing-grade modified trypsin (Promega) in the filter unit; a total volume of 100 μL digestion buffer is used and the reaction is carried out at 37°C for 12–16 h. After isolating peptides by centrifugation, they are extracted with ethyl acetate to remove remaining deoxycholic acid.

3. SHOTGUN MASS SPECTROMETRY

Shotgun mass spectrometry, which uses the mass spectrometer's data-dependent acquisition (DDA) modality, is the method of choice when the proteins present in a sample are unknown (Figs. 2 and 3). Peptides eluting from the LC system are first subjected to mass analysis (the MS1 spectrum), and the instrument then chooses the most abundant 5–10 precursor ions for fragmentation. While biased for more abundant peptides, some breadth of coverage is obtained by the use of a dynamic exclusion list where recently fragmented ions are ignored for a set period of time (e.g., 30 s), increasing

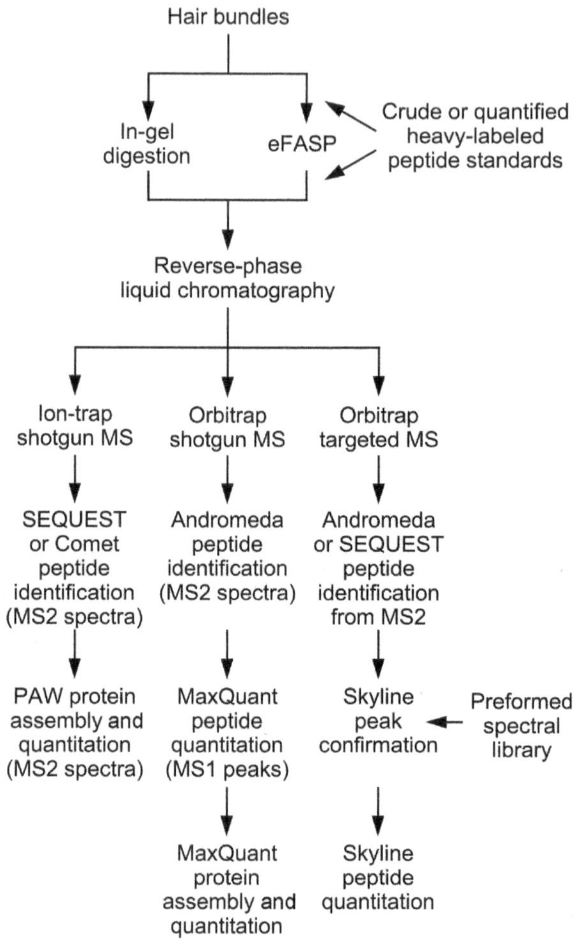

Fig. 2 Flow chart illustrating sample preparation and analysis by shotgun and targeted mass spectrometry. Hair-bundle proteins can be prepared for mass spectrometry using either in-gel or eFASP digestion; if used, standards (e.g., heavy isotope-labeled peptides) are added after digestion. Peptides are separated by reverse-phase liquid chromatography, introduced into the mass spectrometer by electrospray ionization, and are subjected to tandem mass spectrometry. If data are acquired on an ion-trap mass spectrometer, MS2 spectra are analyzed by Comet peptide searching and PAW protein quantitation. Shotgun data acquired on an Orbitrap (high-resolution) mass spectrometer are analyzed by Andromeda peptide searching of MS2 data and MaxQuant protein quantitation using MS1 peak intensities, while targeted data are analyzed by Skyline after Andromeda or Protein Discoverer peptide searching.

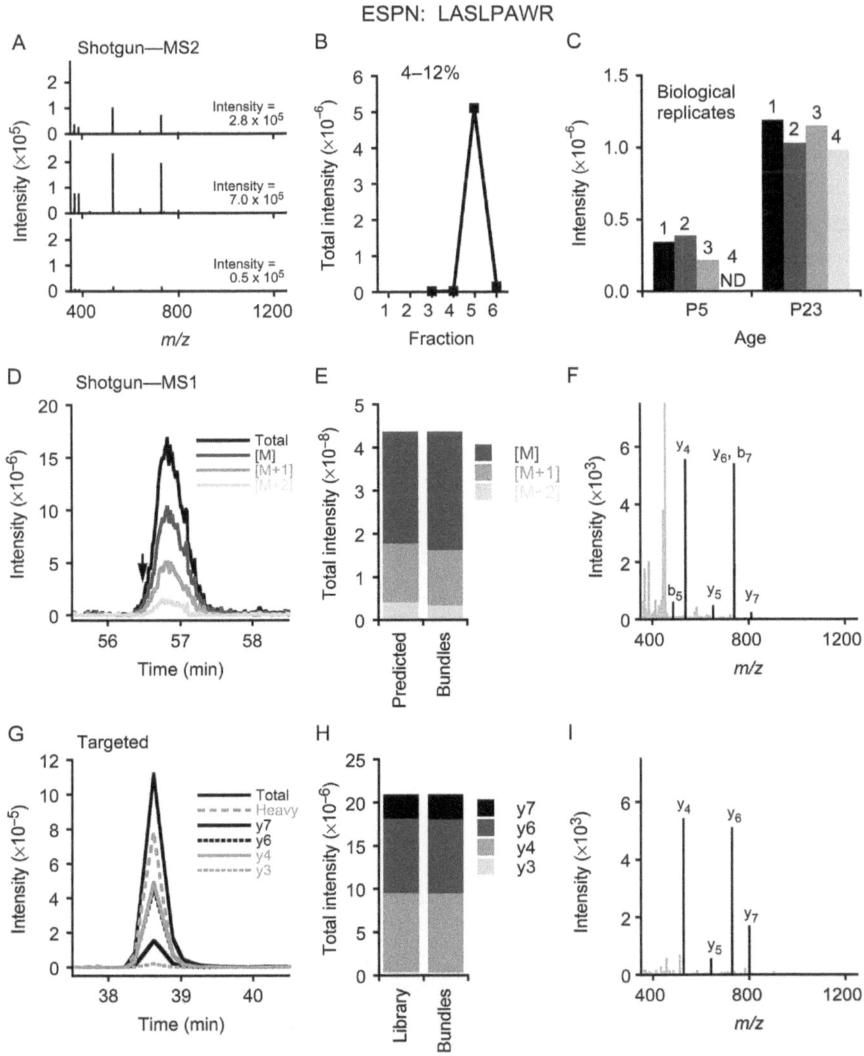

Fig. 3 Shotgun and targeted mass spectrometry. Comparison of ESPN peptide LASLPAWR using shotgun MS2 (A–C), shotgun MS1 (D–F), and targeted (G–I) analyses. Data for A–C are from ProteomeXchange PXD002167, while data for (G–I) are from Ebrahim et al. (2016). (A) Examples of three matched LASLPAWR spectra, all from a P23 mouse hair-bundle replicate, and the summed intensities of the matched ions. (B) All matched intensities for four P23 bundle samples, separated by gel slice number (high to low mass). Note that almost all LASLPAWR intensity is in the fifth fraction, appropriate for the small ESPN isoform molecular masses of 28–34 kDa. (C) Reproducibility of LASLPAWR detection among mouse bundle biological replicates. ND, not detected. (D) Chromatographic profile for LASLPAWR peptide; the single MS2 identification is indicated with *arrow*. Isotopic distribution is shown; [M] corresponds to the monoisotopic *(Continued)*

the chance that lower-abundance ions can be analyzed. Ions chosen are fragmented by collision-induced or higher-energy-induced dissociation and the *m/z* patterns of fragment ions are acquired. All of the *m/z* patterns (the MS2 spectra) are searched against theoretical MS2 spectra, derived from a protein database. Despite some biases against small proteins, proteins with unusual amino acid compositions, and membrane proteins, as long as a protein is present in the database, theoretically it can be detected. Given the poorly characterized proteome of the hair bundle, shotgun mass spectrometry has been the logical first step in defining which proteins are present.

3.1 Instrumentation

We have used several instrument configurations for shotgun mass spectrometry of hair-bundle samples. These instruments include two Orbitrap (high-resolution) instruments: (1) a Thermo Fisher Scientific Electron Orbitrap Velos mass spectrometer and a Protana (Toronto, Ontario, Canada) nanospray ion source, interfaced to a reversed-phase capillary column of 8 cm length × 75 μm internal diameter, self-packed with Phenomenex Jupiter C18 particles of 10 μm size; and (2) a Thermo Fisher Scientific Orbitrap Fusion Tribrid mass spectrometer coupled to a Thermo Fisher Scientific/ Dionex Ultimate 3000 Rapid Separation UPLC system and EasySpray source, using an Acclaim PepMap C18 trap and a 25 cm × 75 μm EasySpray PepMap RSLC C18 column, with 2 μm particle size.

Ion-trap configurations were: (3) a Thermo Fisher Scientific LTQ linear ion-trap mass spectrometer and a Thermo Fisher Scientific Ion Max source,

---

Fig. 3—Cont'd peak (parent), while [M+1] and [M+2] are the ions with 1 or 2 additional mass units due to natural heavy isotopes. Together the isotopic peaks have an intensity of 4.4×10^8 units; to quantify ESPN, iBAQ values for all peptides matched to ESPN are summed and divided by the number of peptides theoretically observable (i.e., between 6 and 30 amino acids). (E) Isotopic peak distribution predicted based on peptide sequence and measured from bundle ESPN peptide. (F) MS2 spectrum identified from peptide peak region. Matched y- and b-ions are labeled and are *black*; unmatched peaks are *gray*. (G) Chromatographic profile for LASLPAWR peptide; there were 12 MS2 spectra acquired across the peak (not shown). A heavy-atom peptide standard was coinjected, and it coelutes with the quantified bundle peptide peak. The four monitored daughter ions are indicated. To quantify ESPN, the areas for each daughter ion were summed, then the amount was determined by comparison to the heavy peptide area and its starting concentration. Note that LASLPAWR elutes earlier in (G) because of the shorter gradient than that in (D). (H) Intensities for each daughter ion, as compared to those in the spectral library. The dot product of 0.99 indicates high correspondence. (I) Example MS2 spectrum identified from peptide peak region.

with a Michrom trap cartridge (Bruker-Michrom, Auburn, CA) and an Agilent 25 cm × 500 μm Zorbax SB-C18 column with 5 μm particles controlled with an Agilent 1100 series capillary LC system (Agilent, Santa Clara, CA); (4) a Thermo Fisher Scientific LTQ Velos linear ion-trap mass spectrometer and an Ion Max source, also with Michrom trap and Zorbax column controlled by an Agilent 1100 series LC; (5) a Thermo Fisher Scientific LTQ Velos linear ion-trap mass spectrometer and a Microm CaptiveSpray source, with a Symmetry C18 trap cartridge and a Waters (Milford, MA) 25 cm × 75 μm nanoACQUITY BEH130C18 column with 1.7 μm particles controlled with a Waters nanoACQUITY LC system.

Prior to mass spectrometry, peptide samples are dissolved in 5% formic acid and transferred to autosampler vials for injection. Details of sample loading, chromatography, electrospray ionization, and mass spectrometry are described elsewhere (Ebrahim et al., 2016; Francis et al., 2015; Krey et al., 2015, 2014, 2016; Shin et al., 2013; Wilmarth et al., 2015). Data from these ThermoFisher instruments are saved as RAW files, which are then analyzed using the PAW or MaxQuant pipelines described later.

3.2 Data Analysis Using the PAW Pipeline

We use two different methods for analysis of shotgun mass spectrometry data. For all ion-trap data, and for some data collected on Orbitrap instruments, we used analysis that includes peptide identification with SEQUEST or Comet, followed by peptide validation and protein assembly using the PAW pipeline developed at OHSU (Wilmarth, Riviere, & David, 2009). The PAW pipeline employs a "best practices" philosophy for proteomics data analysis. Linear discriminant score transformations are used to increase sensitivity (Keller, Nesvizhskii, Kolker, & Aebersold, 2002), target/decoy database strategies are used to control identification error rates (Elias & Gygi, 2007), and false discovery rates are controlled independently on different subclasses of peptides (Ma et al., 2009). Support for accurate mass on high-resolution instruments has been added; narrow tolerances on mass differences between measured and calculated peptide masses are used to generate conditional score distributions, which are used in conjunction with wider tolerance searches to maximize sensitivity. This approach to Orbitrap data was developed several years ago (Hsieh, Hoopmann, MacLean, & MacCoss, 2010) but has been largely ignored in proteomics data analysis.

Typical analyses start with conversions of the instrument RAW files into human-readable MS2-formatted files (Tabb, McDonald, & Yates, 2002) using MSConvert from the Proteowizard toolkit (Chambers et al., 2012). Protein sequence databases are prepared using utilities available at http://www.ProteomicAnalysisWorkbench.com. Database searches are performed using Comet (Eng, Jahan, & Hoopmann, 2013), an open source implementation of the original SEQUEST algorithm (Eng, McCormack, & Yates, 1994). Some Comet parameter settings depend on whether the instrument is a low-resolution ion trap or an Orbitrap; other parameters are independent of instrument. Common independent search parameters are protein database, tryptic enzyme specificity, missed enzymatic cleavages (maximum of two), static modification on cysteine residues (alkylation +57.0215 Da), variable oxidation of methionine modification (+15.9949 Da), and monoisotopic fragment-ion mass tolerance of 1.0005. For low-resolution instruments, the parent-ion mass tolerance is an average mass window of 2.5 Da. For Orbitraps, we use a monoisotopic parent-ion mass tolerance of 1.25 Da. The delta mass values in Da (difference between measured peptide mass and assigned sequence calculated mass) for all identifications are histogrammed in small bins separately for target identifications and for decoy identifications. Narrow windows are set around the 0 Da peak, and the 0.984 Da/1.003 Da doublet (deamidation and first isotopic peaks, respectively) to create sets of score histograms conditioned on accurate delta mass. Score histograms for all delta masses not located inside of these two narrow regions are also generated. Orbitrap instruments frequently cannot determine accurate monoisotopic masses or charge states for low-abundance peptide signals, such as those from hair-bundle samples. Many correct identifications do not have precisely measured parent-ion masses, and these identifications can be lost in searches with narrow parent-ion mass tolerances.

Postprocessing of search results to filter peptides to desired false discovery rates, as well as protein inference using standard and extended parsimony principles, are done with a suite of Python programs. Spectral counting (Liu, Sadygov, & Yates, 2004) and MS2 intensity-weighted spectral counting (Krey et al., 2014; Spinelli et al., 2012) are supported.

When using PAW to analyze ion-trap mass spectrometry data from purified hair bundles (Krey et al., 2015; Shin et al., 2013; Wilmarth et al., 2015), we exclusively use MS2 intensities for quantitation (Fig. 3A and B). All intensities from matched MS2 spectra for a given peptide, for example, the ESPN (espin) peptide LASLPAWR (Fig. 3A and C), are summed for a given biological replicate. Technical replicates are not used for purified-bundle experiments. Because a single sample is fractionated into six gel slices, each of which is processed and run separately, the peptide's identification can occur in different fractions (Fig. 3B). Detection of a single peptide is reasonably reproducible when it is present at moderate to high levels, but becomes more stochastic as its levels drop (Fig. 3C). To generate the intensity for a single protein in a single replicate, intensities from all identified peptides for that protein are summed. Further processing for quantitation is described in Section 3.5.

3.3 Data Analysis Using MaxQuant

We also use MaxQuant (Cox & Mann, 2008) to analyze data collected on Orbitrap mass spectrometers. MaxQuant identifies peptides using the search engine Andromeda (Cox et al., 2011) and quantifies them by measuring MS1 peak intensity (Fig. 3D–F); the program carries out intensity-based absolute quantitation (iBAQ), which we use for quantitation (Krey et al., 2014; Shin et al., 2013).

MaxQuant (available at http://www.coxdocs.org/doku.php?id=maxquant:start) is installed as described. With the exception of iBAQ quantitation and "Match Between Runs" (occasionally used), all settings are the default ones. The principal output (proteinGroups.txt) is processed further using a custom Mathematica program and using Excel spreadsheet manipulation. The Mathematica program (a) rearranges the table to reduce the number of columns; (b) replaces the description and symbol for the "best protein"; (c) removes entries flagged by MaxQuant as contaminants (reversed entries are maintained to allow calculation of the false discovery rate); (d) finds symbols for all matched identifiers; (e) finds all the MaxQuant groups that share a defined fraction of their peptides, which is defined by the user (we usually use $\geq$20%); (f) consolidates those groups with shared peptides together into a "super-group," making a list with all contributing protein symbols; and (g) sums the number of unique peptides and the iBAQ values for the super-group; and writes an output file. Further processing for quantitation is described in Section 3.5.

3.4 Protein Databases

We generally use Ensembl databases for searching shotgun data, acquired using the BioMart tool (http://uswest.ensembl.org/biomart/martview/). Because they include predicted and verified splice products, Ensembl databases are relatively large and have greater tryptic peptide redundancy. This increases processing times and increases peptide-level quantification ambiguity (which we address through protein grouping). Advantages of Ensembl databases are completeness of gene products and rich gene-level annotation.

3.5 Quantitation of Proteins in Bundles

For samples analyzed by Comet and PAW, the summed intensities (i) for each protein or protein group in each separate replicate are divided by the protein's molecular mass; each i/MW value is then divided by the sum of all nonreversed, noncontaminant i/MW values for that run, giving the relative molar intensity for each protein (i_m). i_m values are averaged across all replicate runs, giving an average *im*. Stochastic sampling of MS2 spectra during shotgun analysis leads to considerable variability, and at least four biological replicates are favored for each condition.

For samples analyzed by MaxQuant, the Mathematica output file is opened in Excel and iBAQ values for all nonreversed, noncontaminant entries are summed for each replicate; the iBAQ for a given protein from that replicate is then divided by the summed iBAQ, yielding riBAQ. For each protein, riBAQ values from each replicate are averaged together to give the mean riBAQ for that protein under that condition.

Analysis using known standards diluted in a complex protein background showed that i_m and riBAQ methods for protein quantitation are on average accurate (Krey et al., 2014), with somewhat better performance by riBAQ. While this correspondence does hold on average, it is not necessarily true for an individual protein. Thus the estimates of the amounts of each protein present in a proteome—as determined with shotgun mass spectrometry—must be used with caution.

Bundle-to-epithelium enrichment, which helps identify which proteins are concentrated in hair bundles and which are present as contaminants, is calculated by dividing each protein's mean i_m or riBAQ value for bundles by the mean *im* or riBAQ value for the epithelium. In addition, we calculate molecules per stereocilium for each bundle protein; the i_m or riBAQ value for each protein is divided by the value for actin, then multiplied by 400,000 actin monomers per stereocilium. While this value for actin varies from species to species and certainly within a bundle (stereocilia in a single chick or rodent utricle bundle range from 2 to 15 μm in length), it is useful to use a single value to allow comparison between species.

We assess the degree of contamination by the epithelium in a hair-bundle sample by quantifying its histone content. Relative molar abundance for each contaminant is calculated as discussed later; as no histones should be present in the bundle samples, the bundle-to-epithelium ratio for each histone directly indicates the degree of contamination. Histone analysis indicates that chick or mouse bundle preparations have 2–20% contamination, depending on the particular experiment. Other contaminants may be present at higher levels; however, if they are more readily captured by the agarose than is whole epithelium debris. For example, small amounts of apical microvilli, cuticular plates, and mitochondria all can be visualized in the bundle preparation using transmission electron microscopy (Gillespie & Hudspeth, 1991). If proteins making up these components are known, their enrichment can be calculated separately from the shotgun proteomics data; their enrichment can be compared to histone enrichment to determine if they are selectively isolated by the twist-off method.

4. TARGETED MASS SPECTROMETRY

While shotgun mass spectrometry allows determination of the spectrum of proteins present in a given sample, accuracy and precision of protein abundance in these experiments suffers from protein-to-protein variability inherent with DDA. In addition, shotgun proteomics tends to be biased toward detection of higher abundance peptides, often failing to detect very low-abundance proteins. Targeted proteomics offers a different strategy for protein detection and quantitation, and is well suited for comparison of proteins to standards and thus absolute quantitation (Liebler & Zimmerman, 2013). In targeted assays, specific peptides from each protein of interest are isolated and fragmented within the mass spectrometer, and the intensity of specific fragment ions from each peptide are monitored over the entire chromatographic elution (Fig. 3G–I). Targeted proteomics can be performed using selected reaction monitoring (SRM) or parallel reaction monitoring (PRM), depending on the instrument. We use PRM (Gallien & Domon, 2015), which takes advantage of a high-resolution hybrid mass spectrometer (quadrupole-Orbitrap), the same instrument we now also use for shotgun analysis of bundle samples. While both SRM and PRM have comparable sensitivity, reproducibility, and precision, PRM requires less effort in building assay methods. PRM acquisition produces high-resolution fragmentation (MS2) spectra across a wide m/z range that can also be used to confirm the sequence of each peptide, and chromatographic peaks from each fragment ion that can be quantified to accurately measure the amount of the peptide in the sample. The intensities of multiple peptides are then used to infer the abundance of the corresponding protein. Targeted assays have been shown to quantify proteins in the attomole range and thus are well suited toward detection of scarce proteins in low-protein amount samples like purified hair bundles.

4.1 Designing a Targeted Mass Spectrometry Method

Targeted proteomic assays for hair-bundle proteins require optimal proteotypic peptides for each protein. Proteotypic peptides are unique to the protein of interest and, when analyzed by mass spectrometry, have favorable fragmentation characteristics and high-intensity per mole (Mallick et al., 2007). Response characteristics of peptides derived from any protein vary enormously due to differences in sequence specific physicochemical properties; for this reason, choosing the highest responding peptides for each protein can greatly boost sensitivity. Proteotypic peptides can be identified from shotgun experiments, retrieved from public spectral databases (http://www.peptideatlas.org, http://www.thegpm.org), or predicted using online tools (e.g., Prego, CONSeQuence, ESP). Proteotypic peptides must be unique to the protein of interest and typically are 7–26 amino acids in length, have doubly or triply charged m/z values within the mass range of the instrument, are efficiently produced during enzymatic digestion, and have no known modification sites or amino acids prone to variable modification during sample processing (e.g., methionine oxidation).

We have successfully used peptides that were routinely identified in shotgun runs with relatively high MS1 peak intensities. The shotgun runs also provide useful instrument-specific information for these peptides, including retention time, precursor charge state, fragmentation pattern, and relative fragment-ion intensities. When targeting very low-abundance proteins with minimal shotgun data, databases or prediction algorithms can be used, although peptides predicted by these approaches are not always detected. Digestion methods used and the modification state of the endogenous protein both can influence whether theoretical tryptic peptides actually get cleaved or detected. Empirical data acquired by analyzing all possible tryptic peptides using a recombinant protein are perhaps the ideal approach for identifying the highest responding peptides for a protein, accounting for variables associated with chosen digestion and LC–MS/MS methods (Bollinger, Stergachis, Johnson, Egertson, & MacCoss, 2016).

Once optimal peptides have been identified for each protein, multiplexed targeted experiments are designed to measure the best 3–5 peptides for each of the group of proteins to be assayed simultaneously. With PRM, because a full MS2 spectrum is collected for each peptide, only the precursor m/z (and charge state) of the targeted peptide is entered into the isolation list. Precursor isolation lists for all peptides are exported from Skyline (http://proteome.gs.washington.edu/software/skyline/), a freely available software package (MacLean et al., 2010), into Orbitrap control software.

4.2 Internal Standards for Targeted Quantitation Methods

Quantitation with targeted proteomics can use either label-free or label-based methods. In label-based quantitation methods, heavy isotope-labeled peptides or proteins are spiked into the sample during processing and are used to calibrate signals of the endogenous peptides. If the concentration of the heavy peptide or protein is known, "absolute" quantitation of the endogenous peptide can be carried out. If heavy peptides are not used, label-free quantitation methods can be used to measure the relative amounts of each peptide in different samples. While heavy peptides or proteins can both be used as internal standards, proteins offer some advantages as they are spiked into the sample at the beginning of processing, minimizing variability introduced during sample processing. Indeed, the use of heavy peptides assumes complete efficiency of trypsin digestion for the endogenous peptide, which is not always the case. Full-length heavy protein standards are very expensive and labor-intensive to make; however, so the use of heavy peptides is advantageous when seeking to quantify multiple proteins within a sample. We will focus on the use of heavy peptides as internal standards, the approach we have taken with targeted quantitation of hair-bundle proteins.

4.2.1 Label-Based Quantitation

The most commonly used internal standards for label-based quantitation are heavy isotope-labeled peptides with ^{13}C and ^{15}N atoms in the C-terminal arginine or lysine residue. These peptides are ordered to be identical in sequence (with any modifications) to chosen endogenous proteotypic peptides. While the precursor mass for the heavy peptide will be higher (and thus distinguishable from the endogenous peptide in the mass spectrometer), properties such as the retention time and fragmentation pattern will be identical to the endogenous peptide. The peptides are spiked into the samples following trypsin digestion (or along with trypsin if eFASP processing or SpikeTides peptides are used, see later). By including them in the sample processing, one can detect and correct for any technical variability between runs and thereby achieve more reproducible quantitation. Both heavy and light peptide precursor masses are included in the isolation list used during the LC–MS/MS run, and the heavy/light ratio of the coeluted fragment-ion peak areas is calculated to determine the endogenous peptide amount (Fig. 2D). For absolute quantitation, the heavy peptide must be highly purified (with high isotopic purity) so that accurate quantitation of the synthesized peptide can be performed. Purified heavy peptides can be very expensive but we have had success with the relatively inexpensive SpikeTides_TQL from JPT Technologies (http://www.jpt.com). To avoid artifacts, the spiked-in heavy peptide should produce a peak area that is similar to that of the endogenous peak. A calibration curve for each peptide should be produced to ensure that it performs linearly around the measurement range and that the measurements are not being performed outside the linear range of quantitation. Heavy-labeled peptides can also be used in relative quantitation experiments (e.g., bundles collected at different developmental time points), spiking the same amount into each sample. In this case, relatively inexpensive crude heavy peptides, which are supplied with an imprecise concentration, can be used. In this case, the ratio between heavy and light peptide peaks is calculated and compared between samples to determine the relative concentration of each protein.

4.2.2 Label-Free Quantitation

Label-free quantitation, a much simpler and more cost-effective approach, is ideal for experiments in which only semiquantitative measurements need to be made or in which large changes in protein abundance are expected between the samples (e.g., when comparing hair bundles collected from wild-type and knockout mice). In a label-free approach, equal amounts of bundle samples are digested and run without including any heavy peptides. Fragment-ion peak areas are normalized to the peak areas for actin peptides, which are also spiked into each sample to control for differences in bundle recovery and concentration within each sample. While truly label-free targeted proteomics can be carried out, use of crude heavy-labeled peptides (e.g., SpikeTides_L from JPT) is highly recommended to demonstrate that the monitored peptide is correctly identified by having the same elution time, parent-ion isotopic distribution, and fragmentation pattern as the heavy peptide.

4.3 Peptide Analysis Using PRM

We perform our targeted assays using an Orbitrap Tribrid Fusion quadrupole-Orbitrap hybrid mass spectrometer coupled to a Thermo/Dionex

Ultimate 3000 Rapid Separation UPLC system and EasySpray nanosource (Thermo Fisher Scientific). Samples are loaded onto an Acclaim PepMap C18, 5 μm particle, 100 μm × 2 cm trap using a 5 μL/min flow rate and then separated on a EasySpray PepMap RSLC, C18, 2 μm particle, and 75 μm × 25 cm column (Thermo Fisher Scientific) at a 300 nL/min flow rate. Solvent A is water and solvent B is acetonitrile, each containing 0.1% (v/v) formic acid. After loading at 2% B for 5 min, peptides are separated using a 55 min gradient from 7.5% to 30% B, 10 min gradient from 30% to 90% B, 6 min at 90% B, followed by a 19 min reequilibration at 2% B. This chromatographic gradient produces sharp peaks with an average width of ~30 s. Increasing the length of the gradient is not advisable for targeted runs as it will decrease the peak sharpness. In the mass spectrometer, precursor ions for each peptide are isolated within the quadrupole, then full high-resolution MS2 scans are performed on each precursor with the Orbitrap detector at 30,000 resolution following higher-energy collisional dissociation. Targeted MS2 parameters for our bundle experiments typically include an isolation width of 2 *m/z* for each precursor of interest, collision energy of 30%, automatic gain control (AGC) target of 5×10^4, maximum ion injection time of 100 ms, spray voltage of 2400 V, and ion transfer temperature of 275°C. These parameters should be optimized, however, for the particular sample, instrument, and chromatographic gradient used. For reliable quantification, at least 6–8 scans are acquired across the chromatographic peak for each peptide. The cycle time, or time it takes the instrument to cycle through the entire list of peptides, should be adjusted based on the average peak width to ensure adequate sampling across the peak. The cycle time is determined by the number of peptides in the isolation list, the injection time, and the resolution used to collect the MS2 scan. With our instrument, we have found empirically that no more than 75 precursors can be targeted in each run. Because hair-bundle samples have very small amounts of protein, we use a relatively long injection time, which allows collection of as many ions as possible and increases the sensitivity for low-abundance proteins. To shorten cycle time and increase peak coverage without changing injection times, the number of peptides analyzed in each run can be reduced, the resolution can be decreased, or peptide acquisition can be scheduled for a specific time during the gradient. Higher abundance peptides are more likely to reach the AGC target before the maximum injection time is reached and thus cycle time will be shorter for these.

4.4 Data Analysis Using Skyline

Skyline software is optimized for the analysis of targeted MS/MS data (MacLean et al., 2010). Tutorials, webinars, and other resources are available at the Skyline website and are extremely useful for training of new users. Chromatographic and spectral data from the RAW files are loaded into Skyline and manually analyzed to confirm fragment-ion peaks corresponding to each peptide, most often using spectral libraries generated from shotgun results. RAW files are also processed using Proteome Discoverer (ThermoFisher) software in order to match MS2 spectra to an Ensembl protein database using SEQUEST HT. Once the data have been imported, the different precursor masses targeted for each peptide can be scanned and the extracted ion chromatograms of all detected fragment ions, retention times, and quality of match to MS2 spectra in a spectral library can be visualized. The mass error (difference between the expected and measured mass) of the fragment ions can also be examined, helping to identify background peaks within the chromatogram.

Skyline automatically identifies the boundaries of each chosen peak, but manual inspection and correction of the boundaries should be performed. The peak boundaries of the light and heavy peptides should be identical. Fragment-ion peaks for each peptide are chosen according to the following criteria: (1) three or more coeluting fragment ions contribute to the peak signal; (2) six or more data points collected across the peak; and (3) one or more MS2 spectra within the peak match to the correct peptide sequence within the spectral database. Any fragment-ion peaks that have coeluting, interfering background ions can be removed. When MS2 spectra are not directly identified by SEQUEST, peaks are integrated provided (a) the retention time of the chosen peak is within 2 min of the retention time of an identified peak for that peptide in another sample and (b) the type of daughter ions contributing to the peak match the identified peptide peak from another sample. If no peaks matching these criteria are found in a particular sample, the peak area is counted as zero. Chromatographic peak areas from all detected fragment ions for each peptide are integrated and summed to give a final peptide peak area. The peptide peak areas for each protein of interest are averaged for each sample. For calibration curve samples, a linear regression of the heavy peptide peak area in each of four calibration samples is performed and tested for linearity in the measurement range. Peptides that do not perform linearly ($R^2 > 0.98$) are excluded from analysis.

4.5 Quantitation Using Targeted Mass Spectrometry

For samples including spiked-in peptides, chromatographic peak areas from all detected fragment ions for the light and heavy version of each peptide are integrated and summed, and then the peak area ratio between the light and heavy peptides is calculated. This ratio is multiplied by the amount of spiked heavy peptide to give the molar amount of each light (endogenous) peptide in the sample. With purified hair bundles, because the amount of bundle protein varies from replicate to replicate, we reference each peptide's signal to that of actin in that same replicate. As with shotgun data, we then use a constant for actin (400,000 molecules per typical stereocilium) to calculate the protein's abundance. The normalized amounts for different peptides from a protein are averaged for each replicate, then the protein amounts are averaged for (typically) four biological replicates of each condition to give an average abundance for each protein.

In the absence of calibrated heavy standards, relative protein abundance can be estimated by comparing two or more conditions, for example, hair bundles from wild-type and knockout mice (Krey et al., 2016). Again, results are improved by normalizing the signal in each sample to actin or another protein that is not expected to change between the samples.

5. SUMMARY

Mass spectrometry has had a profound impact on the biochemical characterization of hair bundles and bundle function. While antibody-based methods can be sufficiently sensitive to characterize the negligible amounts of protein present in purified bundles, the performance of antibodies varies dramatically, necessarily focusing attention only on those proteins for which antibodies work. By contrast, proteins are much more democratically detected with mass spectrometry, with little bias between larger and reasonably abundant proteins where many distinct peptides can be detected. While shotgun mass spectrometry is outstanding for detecting the breadth of proteins present in a sample such as purified bundles, it is less favorable for quantitation. Instead, targeted proteomics, particularly PRM, offers much more precise and accurate quantitation of bundle proteins.

Shotgun and targeted mass spectrometry estimates of protein concentration are similar (Fig. 4), but targeted analysis requires considerably less protein and is usually more precise. Other factors can influence quantitation, such as the loss of small proteins during the long processing steps for in-gel

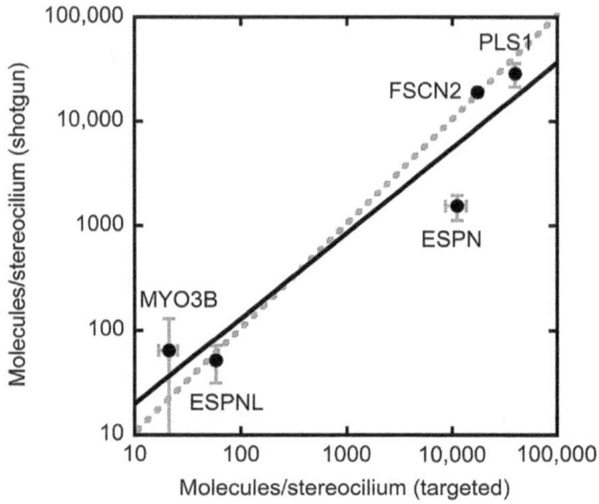

Fig. 4 Comparison of shotgun and targeted quantitation. Estimated molecules per stereocilum (±SEM) for five proteins with both shotgun and targeted mass spectrometry data. Linear regression (log–log plot) gives $y = 3.01 \times 0.82$, with $R^2 = 0.87$. *Dashed line*, unity line. Note that ESPN is an outlier; its small size (several isoforms are ~30 kDa) allow it to diffuse out of SDS gels used in in-gel sample preparation for shotgun mass spectrometry; by contrast, ESPN is not lost in the eFASP procedure, used for targeted mass spectrometry.

digestion (Fig. 4; see ESPN). Care therefore must be taken at every step to ensure that the values obtained are accurate.

In addition to characterizing the identities and concentrations of components of the hair-bundle proteome, both mass spectrometry methods are also valuable for other biochemical characterization experiments with bundles. For example, we are using both shotgun and targeted proteomics to analyze experiments immunoprecipitating key bundle proteins; quantitation with either method, especially when comparing a protein's concentration in the starting material with that in the immunoprecipitate, allows us to identify proteins that specifically coprecipitate with the protein of interest. Using these experiments, we will be able to build on our quantitation information for bundle proteins, characterizing their interactions as well. This information will be invaluable when describing how hair bundles operate at a biochemical level.

ACKNOWLEDGMENTS

P.G.B.G. was supported by NIH grants R01DC002368, R01DC011034, and P30DC005983; J.F.K. by R03DC014544; and P.A.W. and L.L.D. by R01EY007755 and P30EY010572. We obtained mass spectrometry support from the University of Virginia W.M. Keck Biomedical Mass Spectrometry Lab and from the OHSU Proteomics Shared Resource (partial support from NIH core grants P30 EY010572 and P30 CA069533; Orbitrap Fusion S10 OD012246).

REFERENCES

Barr-Gillespie, P. G. (2015). Assembly of hair bundles, an amazing problem for cell biology. *Molecular Biology of the Cell*, *26*, 2727–2732.

Bollinger, J. G., Stergachis, A. B., Johnson, R. S., Egertson, J. D., & MacCoss, M. J. (2016). Selecting optimal peptides for targeted proteomic experiments in human plasma using in vitro synthesized proteins as analytical standards. *Methods in Molecular Biology*, *1410*, 207–221.

Chambers, M. C., et al. (2012). A cross-platform toolkit for mass spectrometry and proteomics. *Nature Biotechnology*, *30*, 918–920.

Cox, J., & Mann, M. (2008). MaxQuant enables high peptide identification rates, individualized p.p.b.-range mass accuracies and proteome-wide protein quantification. *Nature Biotechnology*, *26*, 1367–1372.

Cox, J., Neuhauser, N., Michalski, A., Scheltema, R. A., Olsen, J. V., & Mann, M. (2011). Andromeda: A peptide search engine integrated into the MaxQuant environment. *Journal of Proteome Research*, *10*, 1794–1805.

Dumont, R. A., Zhao, Y. D., Holt, J. R., Bahler, M., & Gillespie, P. G. (2002). Myosin-I isozymes in neonatal rodent auditory and vestibular epithelia. *Journal of the Association for Research in Otolaryngology*, *3*, 375–389.

Ebrahim, S., et al. (2016). Stereocilia-staircase spacing is influenced by myosin III motors and their cargos espin-1 and espin-like. *Nature Communications*, 7, 10833.

Elias, J. E., & Gygi, S. P. (2007). Target-decoy search strategy for increased confidence in large-scale protein identifications by mass spectrometry. *Nature Methods*, *4*, 207–214.

Eng, J. K., Jahan, T. A., & Hoopmann, M. R. (2013). Comet: An open-source MS/MS sequence database search tool. *Proteomics*, *13*, 22–24.

Eng, J. K., McCormack, A. L., & Yates, J. R. (1994). An approach to correlate tandem mass spectral data of peptides with amino acid sequences in a protein database. *Journal of the American Society for Mass Spectrometry*, *5*, 976–989.

Erde, J., Loo, R. R., & Loo, J. A. (2014). Enhanced FASP (eFASP) to increase proteome coverage and sample recovery for quantitative proteomic experiments. *Journal of Proteome Research*, *13*, 1885–1895.

Fettiplace, R., & Kim, K. X. (2014). The physiology of mechanoelectrical transduction channels in hearing. *Physiological Reviews*, *94*, 951–986.

Francis, S. P., et al. (2015). A short splice form of Xin-actin binding repeat containing 2 (XIRP2) lacking the Xin repeats is required for maintenance of stereocilia morphology and hearing function. *Journal of Neuroscience*, *35*, 1999–2014.

Gallien, S., & Domon, B. (2015). Advances in high-resolution quantitative proteomics: Implications for clinical applications. *Expert Review of Proteomics*, *12*, 1–10.

Gillespie, P. G., & Hudspeth, A. J. (1991). High-purity isolation of bullfrog hair bundles and subcellular and topological localization of constituent proteins. *Journal of Cell Biology*, *112*, 625–640.

Gillespie, P. G., Wagner, M. C., & Hudspeth, A. J. (1993). Identification of a 120 kd hair-bundle myosin located near stereociliary tips. *Neuron*, *11*, 581–594.

Hsieh, E. J., Hoopmann, M. R., MacLean, B., & MacCoss, M. J. (2010). Comparison of database search strategies for high precursor mass accuracy MS/MS data. *Journal of Proteome Research*, *9*, 1138–1143.

Keller, A., Nesvizhskii, A. I., Kolker, E., & Aebersold, R. (2002). Empirical statistical model to estimate the accuracy of peptide identifications made by MS/MS and database search. *Analytical Chemistry*, *74*, 5383–5392.

Krey, J. F., Sherman, N. E., Jeffery, E. D., Choi, D., & Barr-Gillespie, P. G. (2015). The proteome of mouse vestibular hair bundles over development. *Scientific Data*, *2*, 150047.

Krey, J. F., et al. (2014). Accurate label-free protein quantitation with high- and low-resolution mass spectrometers. *Journal of Proteome Research*, *13*, 1034–1044.

Krey, J. F., et al. (2016). Annexin A5 is the most abundant membrane-associated protein in stereocilia but is dispensable for hair-bundle development and function. *Scientific Reports*, *6*, 27221.

Liebler, D. C., & Zimmerman, L. J. (2013). Targeted quantitation of proteins by mass spectrometry. *Biochemistry*, *52*, 3797–3806.

Liu, H., Sadygov, R. G., & Yates, J. R. (2004). A model for random sampling and estimation of relative protein abundance in shotgun proteomics. *Analytical Chemistry*, *76*, 4193–4201.

Ma, Z. Q., et al. (2009). IDPicker 2.0: Improved protein assembly with high discrimination peptide identification filtering. *Journal of Proteome Research*, *8*, 3872–3881.

MacLean, B., et al. (2010). Skyline: An open source document editor for creating and analyzing targeted proteomics experiments. *Bioinformatics*, *26*, 966–968.

Mallick, P., et al. (2007). Computational prediction of proteotypic peptides for quantitative proteomics. *Nature Biotechnology*, *25*, 125–131.

Shevchenko, A., Tomas, H., Havlis, J., Olsen, J. V., & Mann, M. (2006). In-gel digestion for mass spectrometric characterization of proteins and proteomes. *Nature Protocols*, *1*, 2856–2860.

Shin, J. B., Pagana, J., & Gillespie, P. G. (2009). Twist-off purification of hair bundles. *Methods in Molecular Biology*, *493*, 241–255.

Shin, J. B., et al. (2007). Hair bundles are specialized for ATP delivery via creatine kinase. *Neuron*, *53*, 371–386.

Shin, J. B., et al. (2010). The R109H variant of fascin-2, a developmentally regulated actin crosslinker in hair-cell stereocilia, underlies early-onset hearing loss of DBA/2 J mice. *Journal of Neuroscience*, *30*, 9683–9694.

Shin, J. B., et al. (2013). Molecular architecture of the chick vestibular hair bundle. *Nature Neuroscience*, *16*, 365–374.

Spinelli, K. J., et al. (2012). Distinct energy metabolism of auditory and vestibular sensory epithelia revealed by quantitative mass spectrometry using MS2 intensity. *Proceedings of the National Academy of Sciences of the United States of America*, *109*, E268–E277.

Tabb, D. L., McDonald, W. H., & Yates, J. R. (2002). DTASelect and contrast: Tools for assembling and comparing protein identifications from shotgun proteomics. *Journal of Proteome Research*, *1*, 21–26.

Walker, R. G., Hudspeth, A. J., & Gillespie, P. G. (1993). Calmodulin and calmodulin-binding proteins in hair bundles. *Proceedings of the National Academy of Sciences of the United States of America*, *90*, 2807–2811.

Wilmarth, P. A., Krey, J. F., Shin, J. B., Choi, D., David, L. L., & Barr-Gillespie, P. G. (2015). Hair-bundle proteomes of avian and mammalian inner-ear utricles. *Scientific Data*, *2*, 150074.

Wilmarth, P. A., Riviere, M. A., & David, L. L. (2009). Techniques for accurate protein identification in shotgun proteomic studies of human, mouse, bovine, and chicken lenses. *Journal of Ocular Biology, Diseases, and Informatics*, *2*, 223–234.

Yamoah, E. N., & Gillespie, P. G. (1996). Phosphate analogs block adaptation in hair cells by inhibiting adaptation-motor force production. *Neuron*, *17*, 523–533.

CHAPTER NINETEEN

Proteomic Analysis of Metabolic Responses to Biofuels and Chemicals in Photosynthetic Cyanobacteria

T. Sun[*,†,‡], L. Chen[*,†,‡], W. Zhang[*,†,‡,1]

[*]Laboratory of Synthetic Microbiology, School of Chemical Engineering & Technology, Tianjin University, Tianjin, PR China
[†]Key Laboratory of Systems Bioengineering (Ministry of Education), Tianjin University, Tianjin, PR China
[‡]SynBio Research Platform, Collaborative Innovation Center of Chemical Science and Engineering (Tianjin), Tianjin, PR China
[1]Corresponding author: e-mail address: wwzhang8@tju.edu.cn

Contents

Abstract

Recent progresses in various "omics" technologies have enabled quantitative measurements of biological molecules in a high-throughput manner. Among them, high-throughput proteomics is a rapidly advancing field that offers a new means to quantify metabolic changes at protein level, which has significantly facilitated our understanding of cellular process, such as protein synthesis, posttranslational modifications, and degradation in responding to environmental perturbations. Cyanobacteria are autotrophic prokaryotes that can perform oxygenic photosynthesis and have recently attracted significant attentions as one promising alternative to traditionally biomass-based

Methods in Enzymology, Volume 585
ISSN 0076-6879
http://dx.doi.org/10.1016/bs.mie.2016.09.024

"microbial cell factories" to produce green fuels and chemicals. However, early studies have shown that the low tolerance to toxic biofuels and chemicals represented one major hurdle for further improving productivity of the cyanobacterial production systems. To address the issue, metabolic responses and their regulation of cyanobacterial cells to toxic end-products need to be defined. In this chapter, we discuss recent progresses in interpreting cyanobacterial responses to biofuels and chemicals using high-throughput proteomics approach, aiming to provide insights and guidelines on how to enhance tolerance and productivity of biofuels or chemicals in the renewable cyanobacteria systems in the future.

1. INTRODUCTION

Several popular "omics" strategies including transcriptomics, proteomics, metabolomics, interactomics, and fluxomics have been developed and applied to quantitative measurements of the various biological molecules in a high-throughput manner, thus allowing a better determination of their variation between different biological states on a genomic scale (Zhang, Li, & Nie, 2010). Among them, high-throughput proteomics is a rapidly advancing field targeting identification and quantitation of proteins. Proteins are major components for building cellular structure and serve as catalytic enzymes in metabolic pathways as well as signal transduction in regulatory pathways (Graham, Graham, & McMullan, 2007). With the help of high-resolution liquid chromatography (LC), advanced tandem mass spectrometers (MS), as well as the stable metabolic labeling (isotope labeling) or chemical protein and peptide labeling (iTRAQ) techniques, current proteomics approaches are able to provide information related to proteins expressed in any given system, such as abundance, activity, structure, posttranslational regulation, or the interaction between these proteins (Wright, Noirel, Ow, & Fazeli, 2012).

As organisms are living in environments with various biotic and abiotic challenges, there is always a great interest in understanding cellular responses and their regulation to changing conditions, such as changes in temperature, osmolality, pH, and nutrition limitation (Tomanek, 2011). Meanwhile, since metabolic responses typically involve multiple pathways, metabolic modules, or even networks, global-based analytical tools, such as proteomics, were believed to play better roles in defining the responses than the traditional approaches that focus on a limited number of biomolecules (Zhang, Wu, Stenoien, & Pasa-Tolic, 2014). In the past decades, increasing studies have used proteomics to study metabolic responses to environmental perturbations in animals, plants, and microorganisms (Tomanek, 2011). For example, Magdeldin, Blaser, Yamamoto, and Yates (2015) employed quantitative tandem mass tag (TMT) isobaric labeling workflow to detect altered protein candidates in zebrafish in responding to electric shock exposure and demonstrated a wide range of metabolic responses, including proteins contributing crucially in catalytic activities, stress response, cation transport, and motor activities (Magdeldin et al., 2015). In plants, Neilson, Mariani, and Haynes (2011) investigated the proteomic responses of rice seedlings to cold stress using both label-free and iTRAQ approaches and revealed differential expression of proteins involving transport, photosynthesis, generation of precursor metabolites, and energy (Neilson et al., 2011). For microorganisms, Sivagnanam et al. (2011) identified 894 proteins in *Clostridium acetobutylicum* from the acetone–butanol–ethanol fermentation process by two-dimensional LC–tandem mass spectrometry method and found 22 proteins were significantly differentially expressed between glucose and xylose substrates (Sivagnanam et al., 2011). All these studies demonstrated the wide application of proteomics techniques in interpreting biotic and abiotic responses, and its application has contributed significantly to the better understanding of the global stress responses of host cells.

Cyanobacteria are autotrophic prokaryotes that can perform oxygenic photosynthesis (Stanier & Cohen-Bazire, 1977). Recently, photosynthetic cyanobacteria have attracted significant attentions as one promising alternative to the traditionally biomass-based "microbial cell factories" for producing green fuels and chemicals, due to their capabilities to absorb solar irradiation and carbon dioxide (CO_2) as the sole energy and carbon sources, respectively (Gao, Sun, Pei, Chen, & Zhang, 2016; Zhou, Zhu, Cai, & Li, 2016). It is thus expected that cyanobacteria-based biosynthetic processes would have the advantages of using clean solar energy and low-cost CO_2 carbon source for value-added products. The first synthetic pathway for biofuel production in cyanobacteria was reported by Deng and Coleman (1999) who introduced pyruvate decarboxylase (*pdc*) and alcohol dehydrogenase II (*adh*)-encoding genes from *Zymomonas mobilis* into *Synechococcus* sp. PCC 7942 (hereafter *Synechococcus* 7942), achieving a production of 1.4–1.7 m*M* ethanol after 3 weeks' cultivation (Deng & Coleman, 1999). Since then progresses have been made to redirect photosynthetically fixed carbon from several key intracellular metabolites (i.e., pyruvate, acetyl-CoA) toward commercially important chemicals by synthesizing novel pathways in cyanobacteria. So far more than several dozens of compounds have been successfully synthesized in cyanobacterial (Gao et al., 2016). However, current productivity for the biofuels or chemicals in cyanobacterial chassis is still several orders of magnitude lower than sugar-based platforms such as *Escherichia coli* and *Saccharomyces cerevisiae* (Savakis & Hellingwerf, 2014), far away from the level required for possible commercial application (Gao et al., 2016). The low tolerance to toxic end-products has thus been considered as one of the major hurdles for further improving productivity of the cyanobacterial systems (Dunlop, 2011; Jin, Chen, Wang, & Zhang, 2014; Nicolaou, Gaida, & Papoutsakis, 2010; Wang et al., 2014). Toxicity of several potential biofuels and chemicals to model cyanobacterium *Synechocystis* sp. PCC 6803 (hereafter *Synechocystis*) has been studied previously and was illustrated in Fig. 1 (Gao et al., 2016). To fully understand toxic effects of biofuels and chemicals on the cyanobacterial cells so that a rational construction of more robust producing cyanobacterial hosts could be conducted, several recent studies have employed a quantitative proteomics to reveal global metabolic responses of *Synechocystis* to exogenously added or internally produced biofuels or chemicals (Borirak et al., 2015; Qiao et al., 2012; Ren, Shi, Chen,

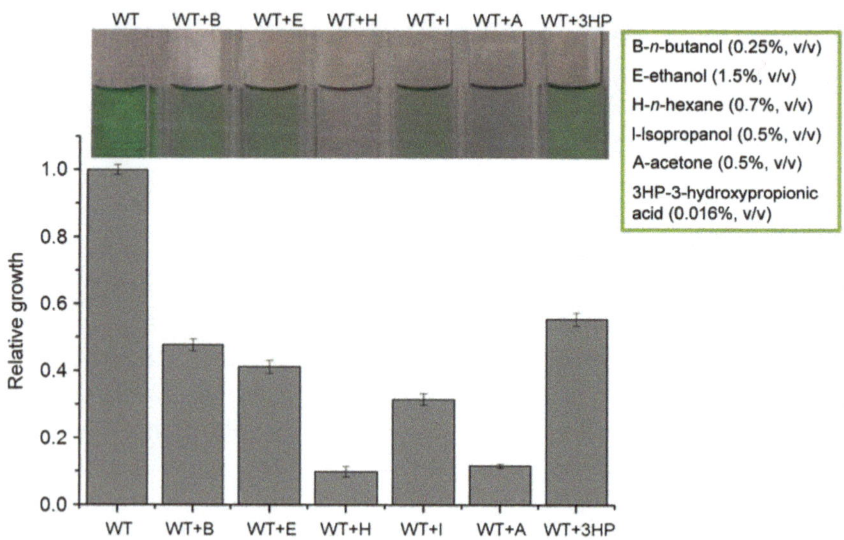

Fig. 1 A random toxicity test of different fuels and chemicals to *Synechocystis*. OD_{730} of wild-type *Synechocystis* control as well as *Synechocystis* adding a series of fuels and chemicals (i.e., *n*-butanol, ethanol, *n*-hexane, isopropanol, acetone, and 3-hydroxypropionic acid) was measured after 48 h cultivation. The results were shown as relative growth to wild-type control.

Wang, & Zhang, 2014; Tian, Chen, Wang, Qiao, & Zhang, 2013), and the results showed that *Synechocystis* cells may employ a combination of multiple mechanisms like membrane modifications, two-component systems (TCSs), and transporters against biofuel toxicity. In this chapter, we discuss recent progresses in deciphering cyanobacterial responses to biofuels and chemicals utilizing proteomics technology, aiming to provide valuable knowledge base for engineering tolerance of cyanobacteria chassis for producing biofuels or chemicals. In addition, several popular high-throughput proteomics methods utilized in analyzing metabolic responses to biofuels and chemicals are also briefly introduced.

2. CURRENT PROTEOMICS METHODS USED IN ANALYZING METABOLIC RESPONSES

Many strategies of proteomic analysis have been established in the past decades (Abdallah, Dumas-Gaudot, Renaut, & Sergeant, 2012). Here we focus on two stable labeling methods, e.g., metabolic labeling and chemical labeling. For other proteomics strategies such as two-dimensional gel electrophoresis and label-free quantification methods, interested readers may refer to several excellent reviews published recently (Bantscheff, Lemeer, Savitski, & Kuster, 2012; Zhang et al., 2014).

2.1 Metabolic Labeling

In metabolic labeling strategies, isotope labeling is introduced into every protein molecule during cell growth process, generating a labeled standard for each protein in a sample of interest. One of the most popular metabolic labeling methods is stable isotope labeling with amino acids in cell cultures (SILAC) (Ong et al., 2002). Typically, in the traditional SILAC experiment, isotopically labeled arginine and lysine (^{13}C, ^{15}N) are added to cultivation medium, ensuring that most of the peptides following tryptic digestion contain at least one labeled amino acid. Relative quantification is achieved by comparing intensities of the isotope clusters of the labeled and unlabeled peptides in peptide ion mass spectra. In recent years, several advanced quantification strategies based on SILAC were also developed: (i) the pulsed SILAC approach, in which cells or model organisms are pulse labeled with heavy isotopes for a certain period of time, and subsequently protein turnover can be estimated by determining ratios of heavy and light peptides. The ratio of heavy to low signals in such experiments is a good measure of protein synthesis and degradation; (ii) the super SILAC strategy, in which unlabeled samples are compared with a representative mixture of heavy SILAC-labeled cells that act as a common internal standard. In contrast to ordinary SILAC, the super SILAC method is a two-step procedure in which, first, the proteins from each (unlabeled) sample are quantified against the (labeled) reference standard, and second, the differences between the samples are quantified by calculating the ratio of the two SILAC ratios obtained in the first step (Bantscheff et al., 2012).

2.2 Chemical Protein and Peptide Labeling

Labeling of proteins and peptides via chemical methods after biosynthesis has been widely used in proteomics in recent years (Chandramouli & Qian, 2009). In chemical protein and peptide labeling strategies, most current techniques target N-terminus of peptides/proteins and the ε-amino group of lysine. Two most popular chemical labeling methods nowadays are TMTs and isobaric tags for absolute and relative quantification (iTRAQ). The iTRAQ reagent structure and the principle of iTRAQ quantitative proteomics are shown in Fig. 2A and B. Both methods target primary amines or so-called isobaric tags. In contrast to most of other labeling methods, isobaric labeling is well suited for the multiplexed analysis of different biological samples because: (i) reagents are available in up to eight different versions; (ii) complexity of LC separations is not increased because the labeled peptides are precisely coeluted; (iii) complexity of peptide mass spectra is not increased because the differentially labeled peptides are isobaric (Bantscheff et al., 2012).

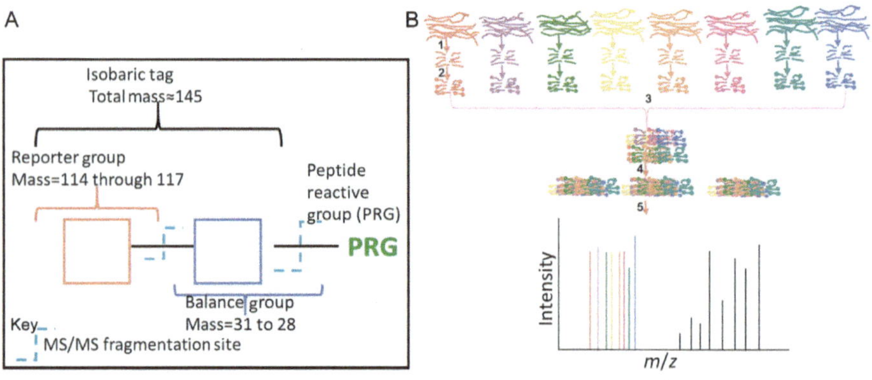

Fig. 2 The iTRAQ reagent structure (A) and the principle (B) of iTRAQ quantitative proteomics.

3. METABOLIC RESPONSES OF CYANOBACTERIA TO BIOFUELS REVEALED BY PROTEOMICS

3.1 Bioethanol

Bioethanol is the most commonly used biofuel, mainly as a biofuel additive for gasoline. Nowadays, bioethanol has been widely used in Brazil and the United States, whose production accounts for 87.1% of the total bioethanol production in the world in 2011 (Liu, 2014). Since 1999, studies have been conducted to utilize cyanobacteria as a host for bioethanol production (Deng & Coleman, 1999; Dexter & Fu, 2009) and the final production has been improved to 5.5 g/L in *Synechocystis* (Gao, Zhao, Li, Tan, & Lu, 2012). Bioethanol is well known to be toxic to cyanobacterial cells (Fig. 1); meanwhile, their metabolic responses to toxic bioethanol are still poorly understood.

To address the issue, Qiao et al. (2012) applied iTRAQ LC–MS/MS proteomic analysis for both control and 1.5% ethanol-treated *Synechocystis* collected at 24 h (middle-exponential phase) and 48 h (exponential phase), each sample with two replicates (Qiao et al., 2012). After data filtering, 1509 unique proteins were identified, representing approximately 42% of the 3569 predicted proteins in the *Synechocystis* (Kaneko et al., 2003). Using a cutoff of $\geq$1.5-fold change and a p-value $<$ 0.05, a total of 135 and 293 unique proteins were differentially regulated after ethanol treatment compared to control at 24 and 48 h, respectively.

Heat-shock proteins are the most common protective proteins in responding to stress conditions in cells (Ding et al., 2009). The proteomic analysis showed that two heat-shock proteins GroES (Slr2075) and GrpE (Sll0057), and a universal stress protein (Slr1101) were upregulated by ethanol. GroES is an oligomeric molecular chaperone functioning in protein folding and possibly in intercellular signaling. Consistent with this study, GroES in *Z. mobilis* was also found with an induced expression in responding to bioethanol (Barbosa, Yomano, & Ingram, 1994). In addition, GrpE was previously found involved in cellular behavior in adaptation to *n*-butanol tolerance in *C. acetobutylicum* (Janssen, Grimmler, Ehrenreich, Bahl, & Fischer, 2012).

Oxidative stress responses have been reported for cells under organic solvent stress as they induce production of highly reactive oxygen species (ROS) (Du & Takagi, 2007). The proteomic analysis showed that glutathione peroxidase (Slr1992) was upregulated at both 24 and 48 h after ethanol exposure. This enzyme catalyzed lipid hydroperoxides to their corresponding alcohols, and its ability against oxidative stress has been reported in various microbes (Arenas et al., 2010; Du & Takagi, 2007). In addition, two methionine sulfoxide reductases (MSRs), i.e., Sll1394 and Sll1680, were found upregulated. Methionine sulfoxides can be reduced back to methionine by MSR, providing a mechanism to repair proteins damaged by ROS (Zhang, Culley, Hogan, Vitiritti, & Brockman, 2006). The upregulation of Sll1394 and Sll1680 suggested that antioxidative responses could be important in combating stress of biofuels and chemicals in photosynthetic *Synechocystis*.

Cross-membrane transporters for small molecules have been suggested as one important mechanism against ethanol toxicity in early studies with yeasts (Ding et al., 2009). In cyanobacteria, transporters were also previously found involved in many different types of stress responses, such as arsenate resistance in *Anabaena variabilis*, Cu^{2+} resistance in *Nostoc calcicola*, salinity stress in *Synechococcus* 7942, and *Synechocystis* (Mikkat, Hagemann, & Schoor, 1996; Nomura, Ishitani, Takabe, Rai, & Takabe, 1995; Thiel, 1988; Verma & Singh, 1991). The quantitative proteomic analysis identified five putative transporters with induced expression levels: Sll0064 functioning as a putative polar amino acid transport system substrate-binding protein, Slr0040 and Slr0043 functioning as bicarbonate transporters, a Na^+/H^+ antiporter for Sll0689 and Slr1295 serving as an iron transport protein. The differentially regulated transporters with a wide range of substrate specificity suggested that *Synechocystis* may employ several different resistance mechanisms simultaneously against ethanol stress, while hexose and amino acid transporters seemed predominant in yeasts (Stanley, Bandara, Fraser, Chambers, & Stanley, 2010).

Recently, several genome-wide analyses showed that genes involved in cell mobility and chemotaxis were upregulated by ethanol or *n*-butanol in *C. acetobutylicum* (Alsaker, Spitzer, & Papoutsakis, 2004; Tomas, Welker, & Papoutsakis, 2003). The proteomic analysis showed that three CheY family proteins (i.e., Sll0039, Sll1292, Slr1982) involved in chemotaxis were upregulated at 48 h after bioethanol treatment in *Synechocystis*. Three regulatory proteins including two response regulators of the NarL subfamily TCSs (Slr1783, Slr1909) and an uncharacterized HTH-type transcriptional regulator (Sll1512) were also found with increased expression levels. Slr1783 was also annotated as Rre1, which has been previously found involved in salt stress (Shoumskaya et al., 2005). In addition, a LysR transcriptional regulator (Sll0998) was downregulated. The *sll0998* gene homologous to proteobacterial Calvin cycle regulators (CbbR) has been found crucial to Calvin cycle and cell viability under autotrophic conditions (Figge, Cassier-Chauvat, Chauvat, & Cerff, 2001). Finally, differentially regulated proteins related to photosynthesis systems and circadian rhythms like Slr0172, Ssr2831, Sll0629, and Sll1328 as well as central metabolic process proteins like Sll1028, Sll1029, Slr1839, Slr1838, Slr0436, Sll1030, and Sll1031 were also found.

In a more recent study, Borirak et al. (2015) applied Fourier transform mass spectrometry (LC–FT-MS/MS) quantitative proteomics analyses of ^{14}N-labeled *Synechocystis* mutant SAA012 with a bioethanol biosynthetic pathway (carrying *pdc* and alcohol dehydrogenase, *adhII*, from *Z. mobilis*) as well as ^{15}N-labeled wild type (WT) to investigate physiological responses to bioethanol (Borirak et al., 2015). The analysis identified a total of 761 and 826 proteins in WT and SAA012, respectively, among which 168 proteins showed a significantly altered abundance in the mutant SAA012 using boundary conditions (Significance $\alpha = 0.01$ if p-value $< 1.214 \times 10^{-5}$). The following KEGG pathway enrichment analyses found that in SAA012, the upregulated pathways were determined as syn00710—carbon fixation in photosynthetic organisms, and syn00480—glutathione metabolism, while the downregulated pathways identified were syn03010—ribosome, and syn00196—photosynthesis-antenna proteins ($p < 0.05$).

Given the ethanol-producing strain showed a higher total carbon fixation rate (q_{CO_2}) than WT, one question is whether the higher carbon fixation rate is due to an increase in the activity of ribulose-1,5-bisphosphate carboxylase (RuBisCO) by release of product inhibition or that more RuBisCO enzyme is increasingly expressed in responding to the ethanol biosynthetic pathway (i.e., the carbon sink). The proteomic analysis suggested that the higher q_{CO_2} was not only a result of the upregulation of the RuBisCO enzyme (which is ~1.4-fold) but also some of the downstream enzymes involved in the Calvin cycle, including phosphoglycerate kinase (Pgk), fructose-bisphosphate aldolase class 1 (Fda), fructose-1,6-bisphosphatase class 1 (Slr0952), and glucose-6-phosphate 1-dehydrogenase (Zwf) functioning in pentose phosphate pathway. In addition, carbonic anhydrase (Slr0051) and proteins involved in the CO_2 concentrating mechanism (CcmM and CcmK1) assisting to increase intracellular CO_2 level were also considerably increased. Together the results led to a conclusion that a high rate of ethanol production increases the overall rate of carbon fixation of the engineered bioethanol-producing cyanobacterial cells.

In the bioethanol-producing strain SAA012, considerable ribosomal proteins and chaperonins were downregulated, which may be due to growth retardation caused by the rechanneling of intermediary metabolites in SAA012. Besides the decreased protein synthesis machinery, the components of the phycobilisome light-harvesting complex, like CpcA, CpcB, CpcC1, CpcC2, CpcG2, PbsA1 (heme oxygenase 1), HemE, and HemF, were also downregulated, while the key components of the two photosystems were found unchanged in the bioethanol-producing strain. The high-level ethanol-producing *Synechocystis* cells did not show any increase in their general stress responses, while moderate induction of the oxidative stress response was observed as the levels of glutathione synthetase (GshB) and hydroperoxy fatty acid reductase (Slr1992) increased. This led to a conclusion that intracellular ethanol production reduced expression of the protein synthesis machinery and the phycobilisomes but activated an oxidative stress response.

Although it was suggested that biofuels produced intracellularly may be more toxic at lower concentration than the exogenously added biofuels, and the metabolic responses revealed from the engineering strain carrying internal biosynthetic pathways may be more close to the real situation, these two studies that examined the response mechanisms to bioethanol from two different aspects (i.e., externally added ethanol and internally biosynthesized ethanol) shared some similar findings such as upregulation of oxidative stress response protein Slr1992, which is consistent with previous study that general metabolic responses to externally added *n*-hexane were similar to that caused by *n*-hexane generated intracellularly (Liu, Chen, Wang, Qiao, & Zhang, 2012).

3.2 *n*-Butanol

n-Butanol has been considered as a promising alternative to traditional gasoline fuels due to its better chemical properties and higher energy content (Durre, 2007). In a recent study, by introducing five heterologous genes into cyanobacterium *Synechococcus* 7942 to convert acetyl-CoA to 1-butanol, Lan and Liao (2011) achieved a production of *n*-butanol from CO_2 at a level of 14.5 mg/L (Lan & Liao, 2011). The productivity of *n*-butanol was further improved to 404 mg/L by modifying ATP and cofactor driving force (Lan & Liao, 2012; Lan, Ro, & Liao, 2013). More recently, Anfelt et al. (2015) constructed the *n*-butanol biosynthetic pathway in a *Synechocystis* mutant deficient in polyhydroxybutyrate synthesis and achieved a production at

37 mg/L by nitrogen starvation and acetyl-CoA enhancement (Anfelt et al., 2015). Nevertheless, the low tolerance to *n*-butanol (Fig. 1) has raised urgent needs to acquire a better understanding of the molecular mechanisms involved in *n*-butanol tolerance and to further construct more robust butanol-producing cyanobacterial hosts by gene manipulations.

To reveal global metabolic responses of *Synechocystis* to external *n*-butanol stress, Tian et al. (2013) cultivated and collected both control and 0.2% *n*-butanol (v/v)-treated *Synechocystis* samples of 24 and 48 h, each with two replicates, for iTRAQ LC–MS/MS proteomics analysis (Tian et al., 2013). Using a cutoff of 1.5-fold change and a p-value < 0.05, 126 and 177 unique proteins were found differentially regulated after *n*-butanol treatment at 24 and 48 h, respectively. Consistent with the lower growth rate under *n*-butanol, the primary metabolism of *Synechocystis* was downregulated significantly in *n*-butanol-treated cells, as reflected by the decreased expression of ribosomal proteins, translation elongation factors, translation initiation factors, translation trigger factors, and tRNA synthetase.

The quantitative proteomics identified 10 heat-shock proteins, among which only DnaJ1 (Slr0093) met the cutoff, and was significantly induced at 48 h after *n*-butanol treatment. However, detailed analysis of other identified heat-shock proteins including GrpE (Sll0057) and Hsp90 (Sll0430) was also slightly induced (mostly around 1.2- to 1.3-fold), suggesting they could also be involved in *n*-butanol resistance. GrpE was also found upregulated under ethanol stress (discussed earlier in Section 3.1; Qiao et al., 2012), indicating it may be involved in tolerance to multiple biofuels, although further confirmation is still needed.

The proteomic analysis identified eight putative transporters with induced expression levels. An NrtD-like transporter Slr0042 and an NrtC-like transporter protein Slr0043 were found upregulated significantly at both 24 and 48 h. In *Synechocystis*, it has been suggested that the *slr0040*, *slr0041*, *slr0043*, and *slr0044* genes form an operon with a putative porin gene *slr0042*, encoding a high-affinity bicarbonate transporter transcribing only under CO_2-limited conditions (Omata, Gohta, Takahashi, Harano, & Maeda, 2001). In addition, NrtC protein has been previously found involved in high-salt resistance, redox-response to redox change, and light quality change in *Synechocystis* (Hihara, Sonoike, Kanehisa, & Ikeuchi, 2003; Liang et al., 2011; Singh et al., 2009), suggesting that these types of proteins have a wide range of functions against stress conditions. A Na^+/H^+ antiporter (Sll0689) was upregulated at 48 h after butanol treatment, and its upregulation was also found in ethanol stress (Qiao et al., 2012). Currently no evidence is available whether these responsive transporters are involved in translocating *n*-butanol directly or indirectly as secondary responses to deal with the cellular damages caused by *n*-butanol.

Early studies have found that solvent-tolerant microbes enhanced tolerance by modifying fatty acid composition or other accessory molecules of their cell membrane to block entry of solvents (Dunlop, 2011). The proteomic analysis identified six regulated proteins involved in membrane modification and lipid synthesis. Upregulated Slr2089 is a membrane-based squalene hopene cyclase (Shc), and evidence has suggested that the Shc protein was related to pH stress (Welander et al., 2009). Upregulated Slr1755 functioned as a NADPH-dependent glycerol-3-phosphate dehydrogenase (GPDH). GPDH catalyzed the conversion of dihydroxyacetone phosphate (DHAP) and NADH to glycerol-3-phosphate (G3P) and NAD^+. G3P is an important precursor for glycerol and glycerolipid synthesis and has been found related to *n*-butanol stress responses in *C. acetobutylicum* (Janssen et al., 2012). Upregulated Ssl2084 is an acyl carrier protein involved in fatty acid biosynthesis. This enzyme has been found involved in hyperosmotic stress in *Synechocystis* (Kanesaki, Suzuki, Allakhverdiev, Mikami, & Murata, 2002).

TCS is a predominant signal transduction mechanism employed by microorganisms to adapt to the complex environments. Besides two CheY proteins (Slr1982 and Sll0039) involved in chemotaxis discussed earlier in Section 3.1, three response regulators (Slr1909, Slr1783, and Sll1708) and one sensory histidine kinase (Slr0484) were also found upregulated by butanol treatment. Interestingly, all four response regulators belong to NarL subfamily of TCS, which were involved in responses to various environmental stresses (Rodriguez-Moya et al., 2010). In addition, a global nitrogen regulator (Sll1423) was also upregulated. The increased expression of nitrogen metabolism during butanol stress may be consistent with the fact that autotrophic growth of cyanobacteria requires the constant assimilation of nitrogen sources (i.e., ammonium) (Espinosa, Forchhammer, & Contreras, 2007).

3.3 *n*-Hexane

Alkanes have also attracted significant attentions in recent years for its better chemical properties like low water solubility and high energy density (Rottig, Wenning, Broker, & Steinbuchel, 2010). In cyanobacteria, by overexpressing aldehyde decarboxylase-encoding gene *orf1593* in *Synechococcus* 7942, Kaiser et al. (2013) achieved a production of alkanes at a level of 2.3 mg/L (Kaiser et al., 2013). More recently, Wang, Liu, and Lu (2013) overexpressed an acyl–acyl carrier protein reductase-encoding gene *sll0209* coupled with an aldehyde-deformylating oxygenase-encoding gene *sll0208* in *Synechocystis* and achieved a production of alkanes at 26 mg/L (Wang et al., 2013). Meanwhile, like other biofuels discussed earlier, alkanes are also toxic to cyanobacteria (Fig. 1).

To address the issue, Liu et al. (2012) carried out an iTRAQ-LC–MS/MS proteomic analysis for two independent cultivations of both control and 0.8% hexane-treated *Synechocystis* samples at 24 and 48 h, each with two replicates. After data filtering using a cutoff of 1.5-fold change and a p-value less than 0.05, 140 and 148 unique proteins were found differentially regulated after *n*-hexane treatments at 24 and 48 h, respectively (Liu et al., 2012). Among them, a total of 164 and 77 proteins were upregulated and downregulated by *n*-hexane treatments, respectively. Pathway enrichment analysis showed that among the regulated proteins, four ABC transporters were matched to the KEGG pathway "ABC transporters" (KO02010) at 48 h. The results showed that under hexane stress, a spectrum of putative transporters, including five transporters involved in bicarbonate transporting (Slr0041, Slr0043, Sll0834, Slr1512, Sll1734), two related with cation transporting (Slr2131, Sll0672), two involved in Na^+ and K^+ transporting (Sll0689, Sll0493), as well as one relative to mercuric transporting (Ssr2857), were found upregulated. Slr0041, Slr0043, and Slr1512 have also been found and discussed under ethanol or *n*-butanol stress (Qiao et al., 2012; Tian et al., 2013). In addition, Sll0493 encodes a NAD^+-binding peripheral membrane protein (KtrA) that played a crucial role in the early phase of cell turgor regulation after hyperosmotic shock in *Synechocystis* by working with a K^+ transporters (KtrB; Slr1509) and KtrE (Slr1508) (Matsuda et al., 2004). Meanwhile, several membranes bound proteins were also found downregulated by *n*-hexane exposure. Though the exact function of these proteins was mostly unknown, the significant changes occurred at the cell membrane level may represent an important resistance strategy against *n*-hexane toxicity, and worth further investigation.

Four different regulatory proteins were upregulated, and two regulatory proteins were downregulated by hexane exposure. Interestingly, upregulated Slr1697 encoding a serine/threonine kinase and upregulated Sll0043 serving as a positive phototaxis hybrid histidine kinase were both required for cell motility and chemotaxis (Kamei, Yoshihara, Yuasa,

Geng, & Ikeuchi, 2003; Kamei, Yuasa, Orikawa, Geng, & Ikeuchi, 2001). Chemotaxis and flagellar motility are essential mechanisms for bacteria to adapt to and survive in environments with various natural stresses (Parnell et al., 2011), thus results suggested that *Synechocystis* cells may adopt the similar mechanism in dealing with the stress caused by *n*-hexane.

The proteomic analysis showed that common stress responses were also observed in the hexane-treated *Synechocystis* cells though the protein species varied between microbial species. For example, consistently with the results from ethanol and *n*-butanol stress response, a heat-shock protein (GrpE, Sll0057) was induced by *n*-hexane (Qiao et al., 2012; Tian et al., 2013). Oxidative stress responses have been reported for cells under treatment of organic solvents as they induced production of ROS. In addition to two responsive glutathione peroxidase, a bacterioferritin comigratory protein (Slr0242), a rehydrin (Slr1198), and a MSR A (Sll1394) were also induced in respond to the oxidative stress caused by *n*-hexane.

4. METABOLIC RESPONSES TO CHEMICALS REVEALED BY PROTEOMICS IN CYANOBACTERIA

4.1 L-Lactate

Lactic acid, widely used in food and pharmaceutical industries and as a building block for biodegradable polymers (Datta & Henry, 2006), can be produced in *Synechocystis* after heterologous expression of a lactate dehydrogenase (LDH) (Angermayr et al., 2014). To explore molecular physiological consequences of a major redirection of carbon in engineered lactate-producing *Synechocystis*, quantitative whole-cell proteomics analyses of ^{14}N-labeled *Synechocystis* mutant SAW041 with lactate biosynthetic pathways (harboring a *ldh* from *Lactococcus lactis*, and a pyruvate kinase from *Enterococcus faecalis*) and its ^{15}N-labeled WT were conducted (Borirak et al., 2015). Subsequent LC–FT-MS/MS analysis identified and quantified 761 and 881 proteins in WT and SAW041, respectively. Using boundary conditions (significance $\alpha = 0.01$ if p-value $< 1.214 \times 10^{-5}$), a total number of 153 proteins showed a significantly altered abundance in mutant SAW041. Interestingly, the enrichment analysis of SAW041 did not reveal specific upregulated pathways but only downregulated pathways including photosynthesis (syn00195), photosynthesis-antenna proteins (syn00196), and glycolysis/gluconeogenesis (syn00010). Consequently, proteins participated in central metabolic pathways, including Pgk, Tal, TktA, Eno, and Gap1, were found to have a decreasing abundance. However, all the ribosomal proteins quantified in this strain were measurably unchanged in their abundance.

In addition, Sll7065, Sll7087, and Sll7090 were found to be upregulated with fold changes greater than two. These proteins were recently determined as CRISPR2- or CRISPR3-system-associated proteins (Scholz, Lange, Hein, Hess, & Backofen, 2013). Yet little was known about the exact functions of these two CRISPR systems in *Synechocystis*, except that the CRISPR3 system is the most abundantly expressed CRISPR system in WT (Scholz et al., 2013). The CRISPR/Cas systems are related with immune systems in bacteria and archaea (Sorek, Lawrence, & Wiedenheft, 2013), which confer resistance to horizontal gene transfer, including phage transduction, transformation, and conjugation (Marraffini & Sontheimer, 2008). As the SAW041 strain has been transformed with a foreign self-replicative plasmid, it remained to be investigated whether the upregulated CRISPR/Cas system was related with function of maintaining genetic stability (Angermayr et al., 2014). Finally, as many proteins that were significantly altered in their abundance in the product-forming strains are either hypothetical proteins or proteins with unknown function, suggesting the high complexity in terms of the metabolic changes and tolerance of the lactate-producing "cell factories."

4.2 3-Hydroxypropionic Acid

3-Hydroxypropionic acid (3-HP) is a nonchiral carboxylic acid that contained a hydroxyl group on its third carbon atom and has a high potential as a platform compound to produce many other chemicals, such as 1,3-propanediol (Kim, Kim, Park, & Seo, 2014). In year 2010, 3-HP was regarded as one of the top value-added chemicals produced from biomass by US Department of Energy (Bozell & Petersen, 2010). After several steps of system optimization, including expression increase of malonyl-CoA reductase (MCR) using different promoters, improved supply of precursor malonyl-CoA and NADPH, a production of 837.18 mg/L 3-HP directly from CO_2 was achieved after 6 days' cultivation of the engineered *Synechocystis* (Wang et al., 2015). However, when compared with engineered *E. coli* systems, the productivity in *Synechocystis* is lower, and further efforts to optimize the production system from both pathway and chassis aspects are still necessary.

Using 3-HP-producing strain SM (with *mcr* gene from *Chloroflexus aurantiacus* with high-capability promoter P_{cpc560}) and its corresponding WT, Wang, Chen, and Zhang (2016) conducted iTRAQ-LC–MS/MS quantitative proteomic analyses, each sample with two replicates. With a cutoff of 1.2-fold change and a p-value < 0.05, 204 and 123 unique proteins were up- and downregulated in the SM strain compared to WT, respectively (Wang et al., 2016). In the 3-HP engineered SM strain, three proteins Sll0629, Ssl0020, and Sll1051 related to photosynthesis were found upregulated. Sll0629 is an important component of photosystem I (Tian et al., 2013), and Ssl0020 is one of the most abundant ferredoxin products in cells (Poncelet, Cassier-Chauvat, Leschelle, Bottin, & Chauvat, 1998). Sll1051-encoding phycocyanobilin lyase CpcF was involved in phycocyanobilin attachment to the subunit of phycocyanin (CpcA) (Fairchild et al., 1992). As no growth arrest was observed in the SM strain, it is speculative that enhanced photosynthesis and energy metabolism may be necessary for the high level of 3-HP production. Consistently, the upregulation of a F0F1 ATP synthase subunit A (Sll1322) and a F0F1 ATP synthase subunit epsilon (Slr1330) was also observed in the SM strain, which could increase ATP supply to the engineered cells for 3-HP biosynthesis. In addition, proteomic analysis also showed that Sll0519, Slr1281, and Slr0844 related to oxidative phosphorylation were upregulated. They all belong to the NAD(P)H: quinone oxidoreductase (NDH-1) family and function to transfer electrons from an electron donor (usually NADH) to quinone to generate a proton motive force used for ATP synthesis (Yagi, Yano, Di Bernardo, & Matsuno-Yagi, 1998).

In 3-HP-producing SM strain, Slr0051 and Slr0899 involved in nitrogen metabolism were upregulated. Carbonic anhydrase Slr0051 could convert CO_2 to HCO_3^-, which could then increase the HCO_3^- concentration in *Synechocystis* (Borirak et al., 2015), while HCO_3^- is a precursor for malonyl-CoA production. Carbonic anhydrase Slr0051 was also found upregulated in ethanol-producing *Synechocystis* as discussed earlier in Section 3.1 (Borirak et al., 2015), suggesting this may be a common response of cyanobacteria to the introduction of foreign pathways. Eight proteins (Sll1746, Sll1807, Slr1678, Ssl3445, Ssr1398, Ssl1784, Ssl2233, and Ssr1399) involved in ribosome synthesis were also upregulated.

In addition, 11 transporters were found upregulated in the SM strain, including Slr1247, Sll0540, Sll0679, Sll0680, Sll1598, Sll0385, Sll1699, Sll0374, Sll1270, Slr0559, and Slr1201. These transporters have previously been found involved in multiprocesses like phosphorus binding and transporting, manganese transporting, cobalt transporting, peptide transporting, and urea transporting (Huertas, Lopez-Maury, Giner-Lamia, Sanchez-Riego, & Florencio, 2014; Kashino et al., 2002; Kobayashi &

Shimizu, 1999; Mikami, Suzuki, & Murata, 2004). Although function for some of these proteins is currently known, the upregulation of Sll1598 and Sll0385 may enhance photosynthesis for better supply of ATP and NADPH and amino acids synthesis into the 3-HP production, respectively.

5. CONCLUSIONS

Proteomics is now nearly 20 years old since originally coined in 1996. The proteomics strategies possess several advantages over genome-based technologies as they directly deal with the functional molecules rather than genetic code or mRNA abundance (Abdallah et al., 2012). In addition, as the accumulation of a protein changes responding to the environment perturbations and stresses is a combination result of transcription, translation, protein turnover, and posttranslational modifications (Abdallah et al., 2012), the response mechanisms revealed by proteomics might be more reliable. In recent years, more advanced proteomics technologies like label-free approaches based on mass spectrometry have been developed, which avoid challenge of incomplete labeling and reduce the analytical cost (Wong & Cagney, 2010). Nevertheless, the low resolution and coverage as well as the accuracy of peptides quantification using proteomics still need to be modified in the future.

In this chapter, we discussed several popular high-throughput proteomics methods and their application to determine metabolic responses of photosynthetic cyanobacteria to biofuels and chemicals, such as ethanol, *n*-butanol, *n*-hexane, lactate, and 3-HP. Together, the proteomic analysis demonstrated that the *Synechocystis* cells employed a combination of common stress response, modifications of cell membrane and envelope, and induction of multiple transporters and cell mobility-related proteins as protection mechanisms against these stresses. All in all, the proteomic studies provide important knowledge for rational design and engineering of cyanobacteria chassis with enhanced tolerance and productivity of biofuels or chemicals in the future.

ACKNOWLEDGMENTS

The research was supported by Grants from National Basic Research Program of China (National "973" program, Project No. 2012CB721101 and No. 2014CB745101), the National High-tech RD Program of China (National "863" program) (No. 2012AA02A707), the Natural Science Foundation of China (NSFC) (No. 31370115 and 31470217), and the Tianjin Municipal Science and Technology Commission (No. 15JCZDJC32500).

REFERENCES

Abdallah, C., Dumas-Gaudot, E., Renaut, J., & Sergeant, K. (2012). Gel-based and gel-free quantitative proteomics approaches at a glance. *International Journal of Plant Genomics, 2012*, 17.

Alsaker, K. V., Spitzer, T. R., & Papoutsakis, E. T. (2004). Transcriptional analysis of spo0A overexpression in *Clostridium acetobutylicum* and its effect on the cell's response to butanol stress. *Journal of Bacteriology, 186*, 1959–1971.

Anfelt, J., Kaczmarzyk, D., Shabestary, K., Renberg, B., Rockberg, J., Nielsen, J., et al. (2015). Genetic and nutrient modulation of acetyl-CoA levels in *Synechocystis* for n-butanol production. *Microbial Cell Factories, 14*, 167.

Angermayr, S. A., van der Woude, A. D., Correddu, D., Vreugdenhil, A., Verrone, V., & Hellingwerf, K. J. (2014). Exploring metabolic engineering design principles for the photosynthetic production of lactic acid by *Synechocystis* sp. PCC6803. *Biotechnology for Biofuels, 7*, 99.

Arenas, F. A., Diaz, W. A., Leal, C. A., Perez-Donoso, J. M., Imlay, J. A., & Vasquez, C. C. (2010). The *Escherichia coli* btuE gene, encodes a glutathione peroxidase that is induced under oxidative stress conditions. *Biochemical and Biophysical Research Communications, 398*, 690–694.

Bantscheff, M., Lemeer, S., Savitski, M. M., & Kuster, B. (2012). Quantitative mass spectrometry in proteomics: Critical review update from 2007 to the present. *Analytical and Bioanalytical Chemistry, 404*, 939–965.

Barbosa, M. F., Yomano, L. P., & Ingram, L. O. (1994). Cloning, sequencing and expression of stress genes from the ethanol-producing bacterium *Zymomonas mobilis*: The groESL operon. *Gene, 148*, 51–57.

Borirak, O., de Koning, L. J., van der Woude, A. D., Hoefsloot, H. C., Dekker, H. L., Roseboom, W., et al. (2015). Quantitative proteomics analysis of an ethanol- and a lactate-producing mutant strain of *Synechocystis* sp. PCC6803. *Biotechnology for Biofuels, 8*, 111.

Bozell, J. J., & Petersen, G. R. (2010). Technology development for the production of biobased products from biorefinery carbohydrates—The US Department of Energy's "Top 10" revisited. *Green Chemistry, 12*, 539–554.

Chandramouli, K., & Qian, P.-Y. (2009). Proteomics: Challenges, techniques and possibilities to overcome biological sample complexity. *Human Genomics and Proteomics: HGP, 2009*, 239204.

Datta, R., & Henry, M. (2006). Lactic acid: Recent advances in products, processes and technologies—A review. *Journal of Chemical Technology & Biotechnology, 81*, 1119–1129.

Deng, M. D., & Coleman, J. R. (1999). Ethanol synthesis by genetic engineering in cyanobacteria. *Applied and Environmental Microbiology, 65*, 523–528.

Dexter, J., & Fu, P. (2009). Metabolic engineering of cyanobacteria for ethanol production. *Energy & Environmental Science, 2*, 857.

Ding, J., Huang, X., Zhang, L., Zhao, N., Yang, D., & Zhang, K. (2009). Tolerance and stress response to ethanol in the yeast *Saccharomyces cerevisiae*. *Applied Microbiology and Biotechnology, 85*, 253–263.

Du, X., & Takagi, H. (2007). N-Acetyltransferase Mpr1 confers ethanol tolerance on *Saccharomyces cerevisiae* by reducing reactive oxygen species. *Applied Microbiology and Biotechnology, 75*, 1343–1351.

Dunlop, M. J. (2011). Engineering microbes for tolerance to next-generation biofuels. *Biotechnology for Biofuels, 4*, 32.

Durre, P. (2007). Biobutanol: An attractive biofuel. *Biotechnology Journal, 2*, 1525–1534.

Espinosa, J., Forchhammer, K., & Contreras, A. (2007). Role of the *Synechococcus* PCC 7942 nitrogen regulator protein PipX in NtcA-controlled processes. *Microbiology, 153*, 711–718.

Fairchild, C. D., Zhao, J., Zhou, J., Colson, S. E., Bryant, D. A., & Glazer, A. N. (1992). Phycocyanin alpha-subunit phycocyanobilin lyase. *Proceedings of the National Academy of Sciences of the United States of America, 89*, 7017–7021.

Figge, R. M., Cassier-Chauvat, C., Chauvat, F., & Cerff, R. (2001). Characterization and analysis of an NAD(P)H dehydrogenase transcriptional regulator critical for the survival of cyanobacteria facing inorganic carbon starvation and osmotic stress. *Molecular Microbiology, 39*, 455–468.

Gao, X., Sun, T., Pei, G., Chen, L., & Zhang, W. (2016). Cyanobacterial chassis engineering for enhancing production of biofuels and chemicals. *Applied Microbiology and Biotechnology, 100*, 3401–3413.

Gao, Z., Zhao, H., Li, Z., Tan, X., & Lu, X. (2012). Photosynthetic production of ethanol from carbon dioxide in genetically engineered cyanobacteria. *Energy & Environmental Science, 5*, 9857.

Graham, R., Graham, C., & McMullan, G. (2007). Microbial proteomics: A mass spectrometry primer for biologists. *Microbial Cell Factories, 6*, 26.

Hihara, Y., Sonoike, K., Kanehisa, M., & Ikeuchi, M. (2003). DNA microarray analysis of redox-responsive genes in the genome of the cyanobacterium *Synechocystis* sp. strain PCC 6803. *Journal of Bacteriology, 185*, 1719–1725.

Huertas, M. J., Lopez-Maury, L., Giner-Lamia, J., Sanchez-Riego, A. M., & Florencio, F. J. (2014). Metals in cyanobacteria: Analysis of the copper, nickel, cobalt and arsenic homeostasis mechanisms. *Life, 4*, 865–886.

Janssen, H., Grimmler, C., Ehrenreich, A., Bahl, H., & Fischer, R. J. (2012). A transcriptional study of acidogenic chemostat cells of *Clostridium acetobutylicum*—Solvent stress caused by a transient n-butanol pulse. *Journal of Biotechnology, 161*, 354–365.

Jin, H., Chen, L., Wang, J., & Zhang, W. (2014). Engineering biofuel tolerance in non-native producing microorganisms. *Biotechnology Advances, 32*, 541–548.

Kaiser, B. K., Carleton, M., Hickman, J. W., Miller, C., Lawson, D., Budde, M., et al. (2013). Fatty aldehydes in cyanobacteria are a metabolically flexible precursor for a diversity of biofuel products. *PloS One, 8*, e58307.

Kamei, A., Yoshihara, S., Yuasa, T., Geng, X., & Ikeuchi, M. (2003). Biochemical and functional characterization of a eukaryotic-type protein kinase, SpkB, in the cyanobacterium, *Synechocystis* sp. PCC 6803. *Current Microbiology, 46*, 296–301.

Kamei, A., Yuasa, T., Orikawa, K., Geng, X. X., & Ikeuchi, M. (2001). A eukaryotic-type protein kinase, SpkA, is required for normal motility of the unicellular Cyanobacterium *Synechocystis* sp. strain PCC 6803. *Journal of Bacteriology, 183*, 1505–1510.

Kaneko, T., Nakamura, Y., Sasamoto, S., Watanabe, A., Kohara, M., Matsumoto, M., et al. (2003). Structural analysis of four large plasmids harboring in a unicellular cyanobacterium, *Synechocystis* sp. PCC 6803. *DNA Research, 10*, 221–228.

Kanesaki, Y., Suzuki, I., Allakhverdiev, S. I., Mikami, K., & Murata, N. (2002). Salt stress and hyperosmotic stress regulate the expression of different sets of genes in *Synechocystis* sp. PCC 6803. *Biochemical and Biophysical Research Communications, 290*, 339–348.

Kashino, Y., Lauber, W. M., Carroll, J. A., Wang, Q., Whitmarsh, J., Satoh, K., et al. (2002). Proteomic analysis of a highly active photosystem II preparation from the cyanobacterium *Synechocystis* sp. PCC 6803 reveals the presence of novel polypeptides. *Biochemistry, 41*, 8004–8012.

Kim, K., Kim, S. K., Park, Y. C., & Seo, J. H. (2014). Enhanced production of 3-hydroxypropionic acid from glycerol by modulation of glycerol metabolism in recombinant *Escherichia coli*. *Bioresource Technology, 156*, 170–175.

Kobayashi, M., & Shimizu, S. (1999). Cobalt proteins. *European Journal of Biochemistry, 261*, 1–9.

Lan, E. I., & Liao, J. C. (2011). Metabolic engineering of cyanobacteria for 1-butanol production from carbon dioxide. *Metabolic Engineering*, *13*, 353–363.

Lan, E. I., & Liao, J. C. (2012). ATP drives direct photosynthetic production of 1-butanol in cyanobacteria. *Proceedings of the National Academy of Sciences of the United States of America*, *109*, 6018–6023.

Lan, E. I., Ro, S. Y., & Liao, J. C. (2013). Oxygen-tolerant coenzyme A-acylating aldehyde dehydrogenase facilitates efficient photosynthetic n-butanol biosynthesis in cyanobacteria. *Energy & Environmental Science*, *6*, 2672.

Liang, C., Zhang, X., Chi, X., Guan, X., Li, Y., Qin, S., et al. (2011). Serine/threonine protein kinase SpkG is a candidate for high salt resistance in the unicellular cyanobacterium *Synechocystis* sp. PCC 6803. *PloS One*, *6*, e18718.

Liu, S. (2014). Ethanol: A smart choice of product in sustainable biorefinery of plant biomass. *Journal of Bioprocess Engineering and Biorefinery*, *3*, 40–49.

Liu, J., Chen, L., Wang, J., Qiao, J., & Zhang, W. (2012). Proteomic analysis reveals resistance mechanism against biofuel hexane in *Synechocystis* sp. PCC 6803. *Biotechnology for Biofuels*, *5*, 68.

Magdeldin, S., Blaser, R. E., Yamamoto, T., & Yates, J. R., 3rd. (2015). Behavioral and proteomic analysis of stress response in zebrafish (Danio rerio). *Journal of Proteome Research*, *14*, 943–952.

Marraffini, L. A., & Sontheimer, E. J. (2008). CRISPR interference limits horizontal gene transfer in staphylococci by targeting DNA. *Science*, *322*, 1843–1845.

Matsuda, N., Kobayashi, H., Katoh, H., Ogawa, T., Futatsugi, L., Nakamura, T., et al. (2004). Na^+-dependent K^+ uptake Ktr system from the cyanobacterium *Synechocystis* sp. PCC 6803 and its role in the early phases of cell adaptation to hyperosmotic shock. *The Journal of Biological Chemistry*, *279*, 54952–54962.

Mikami, K., Suzuki, I., & Murata, N. (2004). Sensors of abiotic stress in *Synechocystis*. In H. Hirt & K. Shinozaki (Eds.), *Plant responses to abiotic stress* (pp. 103–119). Berlin, Heidelberg: Springer Berlin Heidelberg.

Mikkat, S., Hagemann, M., & Schoor, A. (1996). Active transport of glucosylglycerol is involved in salt adaptation of the cyanobacterium *Synechocystis* sp. strain PCC 6803. *Microbiology*, *142*(Pt. 7), 1725–1732.

Neilson, K. A., Mariani, M., & Haynes, P. A. (2011). Quantitative proteomic analysis of cold-responsive proteins in rice. *Proteomics*, *11*, 1696–1706.

Nicolaou, S. A., Gaida, S. M., & Papoutsakis, E. T. (2010). A comparative view of metabolite and substrate stress and tolerance in microbial bioprocessing: From biofuels and chemicals, to biocatalysis and bioremediation. *Metabolic Engineering*, *12*, 307–331.

Nomura, M., Ishitani, M., Takabe, T., Rai, A. K., & Takabe, T. (1995). Synechococcus sp. PCC7942 transformed with *Escherichia coli* bet genes produces glycine betaine from choline and acquires resistance to salt stress. *Plant Physiology*, *107*, 703–708.

Omata, T., Gohta, S., Takahashi, Y., Harano, Y., & Maeda, S. (2001). Involvement of a CbbR homolog in low CO_2-induced activation of the bicarbonate transporter operon in cyanobacteria. *Journal of Bacteriology*, *183*, 1891–1898.

Ong, S. E., Blagoev, B., Kratchmarova, I., Kristensen, D. B., Steen, H., Pandey, A., et al. (2002). Stable isotope labeling by amino acids in cell culture, SILAC, as a simple and accurate approach to expression proteomics. *Molecular & Cellular Proteomics: MCP*, *1*, 376–386.

Parnell, J. J., Callister, S. J., Rompato, G., Nicora, C. D., Pasa-Tolic, L., Williamson, A., et al. (2011). Time-course analysis of the Shewanella amazonensis SB2B proteome in response to sodium chloride shock. *Scientific Reports*, *1*, 25.

Poncelet, M., Cassier-Chauvat, C., Leschelle, X., Bottin, H., & Chauvat, F. (1998). Targeted deletion and mutational analysis of the essential (2Fe-2S) plant-like ferredoxin in *Synechocystis* PCC6803 by plasmid shuffling. *Molecular Microbiology*, *28*, 813–821.

Qiao, J., Wang, J., Chen, L., Tian, X., Huang, S., Ren, X., et al. (2012). Quantitative iTRAQ LC-MS/MS proteomics reveals metabolic responses to biofuel ethanol in cyanobacterial *Synechocystis* sp. PCC 6803. *Journal of Proteome Research*, *11*, 5286–5300.

Ren, Q., Shi, M., Chen, L., Wang, J., & Zhang, W. (2014). Integrated proteomic and metabolomic characterization of a novel two-component response regulator Slr 1909 involved in acid tolerance in Synechocystis sp. PCC 6803. *Journal of Proteomics*, *109*, 76–89.

Rodriguez-Moya, J., Argandona, M., Reina-Bueno, M., Nieto, J. J., Iglesias-Guerra, F., Jebbar, M., et al. (2010). Involvement of EupR, a response regulator of the NarL/FixJ family, in the control of the uptake of the compatible solutes ectoines by the halophilic bacterium *Chromohalobacter salexigens*. *BMC Microbiology*, *10*, 256.

Rottig, A., Wenning, L., Broker, D., & Steinbuchel, A. (2010). Fatty acid alkyl esters: Perspectives for production of alternative biofuels. *Applied Microbiology and Biotechnology*, *85*, 1713–1733.

Savakis, P., & Hellingwerf, K. J. (2014). Engineering cyanobacteria for direct biofuel production from CO. *Current Opinion in Biotechnology*, *33*, 8–14.

Scholz, I., Lange, S. J., Hein, S., Hess, W. R., & Backofen, R. (2013). CRISPR-Cas systems in the cyanobacterium *Synechocystis* sp. PCC6803 exhibit distinct processing pathways involving at least two Cas6 and a Cmr2 protein. *PloS One*, *8*, e56470.

Shoumskaya, M. A., Paithoonrangsarid, K., Kanesaki, Y., Los, D. A., Zinchenko, V. V., Tanticharoen, M., et al. (2005). Identical Hik-Rre systems are involved in perception and transduction of salt signals and hyperosmotic signals but regulate the expression of individual genes to different extents in *Synechocystis*. *The Journal of Biological Chemistry*, *280*, 21531–21538.

Singh, A. K., Bhattacharyya-Pakrasi, M., Elvitigala, T., Ghosh, B., Aurora, R., & Pakrasi, H. B. (2009). A systems-level analysis of the effects of light quality on the metabolism of a cyanobacterium. *Plant Physiology*, *151*, 1596–1608.

Sivagnanam, K., Raghavan, V. G., Shah, M., Hettich, R. L., Verberkmoes, N. C., & Lefsrud, M. G. (2011). Comparative shotgun proteomic analysis of *Clostridium acetobutylicum* from butanol fermentation using glucose and xylose. *Proteome Science*, *9*, 66.

Sorek, R., Lawrence, C. M., & Wiedenheft, B. (2013). CRISPR-mediated adaptive immune systems in bacteria and archaea. *Annual Review of Biochemistry*, *82*, 237–266.

Stanier, R. Y., & Cohen-Bazire, G. (1977). Phototrophic prokaryotes: The cyanobacteria. *Annual Review of Microbiology*, *31*, 225–274.

Stanley, D., Bandara, A., Fraser, S., Chambers, P. J., & Stanley, G. A. (2010). The ethanol stress response and ethanol tolerance of *Saccharomyces cerevisiae*. *Journal of Applied Microbiology*, *109*, 13–24.

Thiel, T. (1988). Phosphate transport and arsenate resistance in the cyanobacterium Anabaena variabilis. *Journal of Bacteriology*, *170*, 1143–1147.

Tian, X., Chen, L., Wang, J., Qiao, J., & Zhang, W. (2013). Quantitative proteomics reveals dynamic responses of *Synechocystis* sp. PCC 6803 to next-generation biofuel butanol. *Journal of Proteomics*, *78*, 326–345.

Tomanek, L. (2011). Environmental proteomics: Changes in the proteome of marine organisms in response to environmental stress, pollutants, infection, symbiosis, and development. *Annual Review of Marine Science*, *3*, 373–399.

Tomas, C. A., Welker, N. E., & Papoutsakis, E. T. (2003). Overexpression of groESL in *Clostridium acetobutylicum* results in increased solvent production and tolerance, prolonged metabolism, and changes in the cell's transcriptional program. *Applied and Environmental Microbiology*, *69*, 4951–4965.

Verma, S. K., & Singh, H. N. (1991). Evidence for energy-dependent copper efflux as a mechanism of Cu2+ resistance in the cyanobacterium *Nostoc calcicola*. *FEMS Microbiology Letters*, *84*, 291–294.

Wang, Y., Chen, L., & Zhang, W. (2016). Proteomic and metabolomic analyses reveal metabolic responses to 3-hydroxypropionic acid synthesized internally in cyanobacterium *Synechocystis* sp. PCC 6803. *Biotechnology for Biofuels,* accepted.

Wang, W., Liu, X., & Lu, X. (2013). Engineering cyanobacteria to improve photosynthetic production of alka(e)nes. *Biotechnology for Biofuels, 6,* 69.

Wang, Y., Shi, M., Niu, X., Zhang, X., Gao, L., Chen, L., et al. (2014). Metabolomic basis of laboratory evolution of butanol tolerance in photosynthetic *Synechocystis* sp. PCC 6803. *Microbial Cell Factories, 13,* 151.

Wang, Y., Sun, T., Gao, X., Shi, M., Wu, L., Chen, L., et al. (2015). Biosynthesis of platform chemical 3-hydroxypropionic acid (3-HP) directly from CO_2 in cyanobacterium *Synechocystis* sp. PCC 6803. *Metabolic Engineering, 34,* 60–70.

Welander, P. V., Hunter, R. C., Zhang, L., Sessions, A. L., Summons, R. E., & Newman, D. K. (2009). Hopanoids play a role in membrane integrity and pH homeostasis in *Rhodopseudomonas palustris* TIE-1. *Journal of Bacteriology, 191,* 6145–6156.

Wong, J. W. H., & Cagney, G. (2010). An overview of label-free quantitation methods in proteomics by mass spectrometry. In J. S. Hubbard & R. A. Jones (Eds.), *Proteome bioinformatics* (pp. 273–283). Totowa, NJ: Humana Press.

Wright, P. C., Noirel, J., Ow, S. Y., & Fazeli, A. (2012). A review of current proteomics technologies with a survey on their widespread use in reproductive biology investigations. *Theriogenology,* 77, 738–765.e752.

Yagi, T., Yano, T., Di Bernardo, S., & Matsuno-Yagi, A. (1998). Procaryotic complex I (NDH-1), an overview. *Biochimica et Biophysica Acta, 1364,* 125–133.

Zhang, W., Culley, D. E., Hogan, M., Vitiritti, L., & Brockman, F. J. (2006). Oxidative stress and heat-shock responses in *Desulfovibrio vulgaris* by genome-wide transcriptomic analysis. *Antonie Van Leeuwenhoek, 90,* 41–55.

Zhang, W., Li, F., & Nie, L. (2010). Integrating multiple 'omics' analysis for microbial biology: Application and methodologies. *Microbiology, 156,* 287–301.

Zhang, Z., Wu, S., Stenoien, D. L., & Pasa-Tolic, L. (2014). High-throughput proteomics. *Annual Review of Analytical Chemistry (Palo Alto, California),* 7, 427–454.

Zhou, J., Zhu, T., Cai, Z., & Li, Y. (2016). From cyanochemicals to cyanofactories: A review and perspective. *Microbial Cell Factories, 15,* 2.

CHAPTER TWENTY

Deep Profiling of Proteome and Phosphoproteome by Isobaric Labeling, Extensive Liquid Chromatography, and Mass Spectrometry

B. Bai*,[1], H. Tan†,[1], V.R. Pagala†, A.A. High†, V.P. Ichhaporia*, L. Hendershot*, J. Peng*,†,[2]

*St. Jude Children's Research Hospital, Memphis, TN, United States
†St. Jude Proteomics Facility, St. Jude Children's Research Hospital, Memphis, TN, United States
[2]Corresponding author: e-mail address: junmin.peng@stjude.org

Contents

Abstract

Mass spectrometry-based proteomics has experienced an unprecedented advance in comprehensive analysis of proteins and posttranslational modifications, with particular technical progress in liquid chromatography coupled to tandem mass spectrometry (LC-MS/MS) and isobaric labeling multiplexing capacity. Here, we introduce a deep proteomics profiling protocol that combines 10-plex tandem mass tag (TMT) labeling with an optimized LC-MS/MS platform to quantitate whole proteome and phosphoproteome. The major steps include protein extraction and digestion, TMT labeling, two-dimensional liquid

[1] Cofirst author.

Methods in Enzymology, Volume 585
ISSN 0076-6879
http://dx.doi.org/10.1016/bs.mie.2016.10.007

chromatography, TiO_2-mediated phosphopeptide enrichment, high-resolution mass spectrometry, and computational data processing. This protocol routinely leads to confident quantification of more than 10,000 proteins and approximately 30,000 phosphosites in mammalian samples. Quality control steps are implemented for troubleshooting and evaluating experimental variation. Such a multiplexed robust method provides a powerful tool for dissecting proteomic signatures at the systems level in a variety of complex samples, ranging from cell culture, animal tissues to human clinical specimens.

1. INTRODUCTION

The development of next-generation sequencing technologies ushers in a new era in studying biological systems and human disease, in which high-throughput measurements of the genome, transcriptome, proteome, metabolome, and other molecular landscapes become possible. In the proteomics field, technical progress over the past years has led to a major advance in mass spectrometry-based quantitative analysis, particularly in the isobaric labeling multiplexing capability combined with extensive liquid chromatography and mass spectrometry, which emerges as a mainstream approach for profiling proteins and posttranslational modifications in highly complex protein samples (Altelaar, Munoz, & Heck, 2013; Azimifar, Nagaraj, Cox, & Mann, 2014; Bensimon, Heck, & Aebersold, 2012; Geyer et al., 2016; Hein et al., 2015; Kim et al., 2014; Wilhelm et al., 2014).

Multiplexed isobaric labeling methods (e.g., iTRAQ and TMT) have significantly improved the throughput of proteomics (Everley, Kunz, McAllister, & Gygi, 2013; McAlister et al., 2012; Pagala et al., 2015), although traditional quantification methods (e.g., label-free quantification and SILAC) have provided successful examples (Bai et al., 2014, 2013; Kim et al., 2014; Thongboonkerd, LaBaer, & Domont, 2014). For example, 10 protein samples can be analyzed together in one batch by the TMT 10-plex reagents. These structurally identical TMT bear the same overall mass, but heavy isotopes are differentially distributed on carbon or nitrogen atoms (Fig. 1). Therefore, each tag generates a unique reporter ion during MS/MS fragmentation, enabling relative quantification of the 10 samples by distinct reporter ions. The TMT strategy is commonly applied in dissecting biological pathways and cellular processes (Churchman et al., 2015; Mertz et al., 2015; Wang et al., 2016).

In addition, significant technical improvement has been made to enhance LC-MS/MS systems. The first-dimensional separation of peptides using LC with high orthogonal resolution is critical in this type of shotgun proteomics (Xu, Duong, & Peng, 2009). High-pH reversed-phase liquid

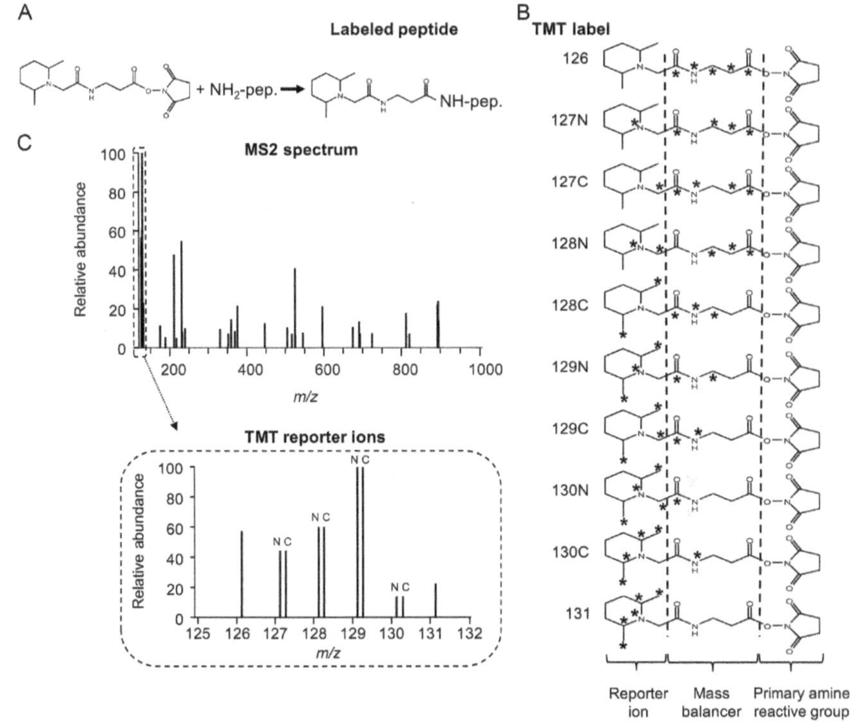

Fig. 1 TMT10-plex reagents. (A) TMT reagents label amine-reactive group (e.g., Lys residues and the amino termini of peptides). (B) The 10 reagents in the TMT10-plex kit (*asterisks* indicate the positions of ^{13}C and ^{15}N heavy isotopes). (C) TMT reporter ions in an MS/MS spectrum. The intensities of these reporters are used for protein quantitation.

chromatography (RPLC) provides better performance than the conventional strong cation-exchange chromatography. Combined with the second dimension of low-pH RPLC, the system improves both analytical dynamic range and protein coverage, yielding impressive results in whole proteome and phosphoproteome analyses (Batth, Francavilla, & Olsen, 2014; Song et al., 2010; Wang, Sun, et al., 2015; Wang, Yang, et al., 2015; Yang, Shen, Camp, & Smith, 2012). For analyzing protein posttranslational modifications, affinity enrichment is critical as modified proteins are normally present in low stoichiometry. Phosphopeptide enrichment by titanium dioxide (TiO_2) has been demonstrated to be a selective, efficient, and reproducible method, and is widely used (Beltran & Cutillas, 2012; Tan et al., 2015; Thingholm & Larsen, 2016). Other technical advances include small C18 particles (1.9 μm) and extended long column (~1 m) (Wang, Sun, et al., 2015; Wang, Yang, et al., 2015). Other notable improvements are new versions of mass spectrometers with rapid scan rates, high sensitivity, and high resolution (Hebert et al., 2014), and revised bioinformatics pipelines for MS data mining (Wang et al., 2014).

Here, we describe a detailed protocol that incorporates the newly updated methodologies with modifications to improve sensitivity and throughput. The protocol includes protein extraction and digestion, TMT 10-plex labeling, basic pH and acid pH RPLC fractionation, phosphopeptide enrichment, high-resolution MS detection, and MS data processing (Fig. 2). When analyzing a range of samples (e.g., cell culture, animal tissues, and clinical specimens), we routinely identify and quantify over 10,000 proteins and ~30,000 phosphosites with high confidence from protein samples (0.1–1 mg of protein per sample). Moreover, we implement several quality control steps for troubleshooting and evaluating experimental variation.

2. SAMPLE PREPARATION BY HARVESTING CELLS AND TISSUES

In this protocol, TMT 10-plex™ isobaric label reagent set (Thermo Scientific, USA) is used for the proteome or phosphoproteome quantification

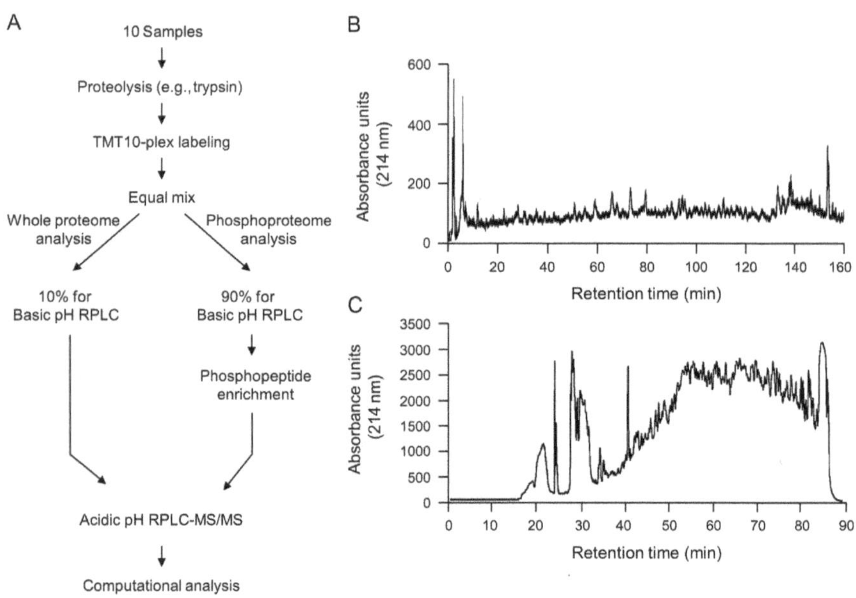

Fig. 2 Workflow of TMT-LC/LC-MS/MS for profiling whole proteome and phosphoproteome. (A) The analysis scheme. (B) Basic pH RPLC fractionation for whole proteome analysis. (C) Basic pH RPLC fractionation for phosphoproteome analysis.

of 10 samples simultaneously. 1 mg protein (100 μg per sample) is recommended as starting material for whole proteome quantification, although the analysis may be performed with as low as 100 μg protein (10 μg per sample). For phosphoproteome quantification, 10 mg of proteins are typically used (1 mg per sample) due to the low stoichiometry of phosphorylation. Approximately 1 mg of proteins can be extracted from 10 million mammalian cells on a 15-cm plate with 80% confluence, while the protein yield of tissues is 5–10% of the weight. Considering that protein content varies in different samples, a pilot experiment is often carried out to examine protein yield. Moreover, the samples should be harvested by the identical manner in the shortest period of time to minimize variation.

1. Suspension cell harvest: collect cells in 15-mL tubes, wash twice with 10 mL ice-cold phosphate-buffered saline (PBS), transfer into 1.5-mL tubes, and centrifuge to remove supernatant completely.
2. Adherent cell harvest: wash twice with 10 mL ice-cold PBS, scrape and collect cells in PBS, transfer to 1.5-mL tubes, and remove PBS by centrifugation. Store cell pellets at −80°C until lysis. Alternatively, the adherent cells can be lysed on-plate by directly adding lysis buffer (see later) on the plate after wash. The on-plate lysis is highly recommended for collecting adherent cell samples during phosphoproteome analysis because it is critical to minimize phosphoproteome perturbation during the harvest, given that phosphorylation is highly sensitive to cellular condition changes. To maintain the desired concentration of lysis buffer, the PBS in the wash steps should be removed completely.
3. Tissue collection: collect tissues in 1.5-mL tubes, weigh, and keep tissues in liquid nitrogen immediately after dissection and store at −80°C. For the 10 samples, the tissue sizes and anatomic regions should be similar to reduce sample heterogeneity. For possible blood contamination, the samples should be rinsed with ice-cold PBS, as highly abundant blood proteins negatively affect protein quantification and downstream data analysis.

3. PROTEIN EXTRACTION, QUALITY CONTROL WESTERN BLOTTING, IN-SOLUTION DIGESTION, AND PEPTIDE PURIFICATION

The lysis buffer used in this protocol contains 50 m*M* HEPES, pH 8.5, 8 *M* urea, and 0.5% sodium deoxycholate. For phosphoproteome analysis, 1 × PhosSTOP phosphatase inhibitor cocktail (Roche) is added. Handling each of the 10 samples, the same way during any step before the pooling of

TMT-labeled samples is critical to reduce variation. Incomplete sample lysis and partial protein denaturation affect protein digestion efficiency. Therefore, it is important to maintain 8 *M* urea concentration by considering that the urea itself occupies volume (100 mg of urea occupies ~73 μL in-solution), and the volume of cell pellets or tissues should also be taken into account as well.

1. Add appropriate amount of lysis buffer to cell pellets or tissues (e.g., 10:1 volume ratio between buffer and samples) to achieve a final protein concentration of 5–10 mg/mL. For on-plate lysis of adherent cells, add lysis buffer on-plate, scrape, and collect the cells to 1.5-mL tubes. Add in ~20% of the final volume of glass beads (1 mm diameter, Sigma).
2. Cap the tubes tightly, lyse the samples in Bullet Blender (Next Advance) at 4°C (speed 8, 30 s on and 5 s off for five cycles) until the samples are homogenized. Alternatively, the samples may be lysed by vortex for 30 s at room temperature (RT) with 30 s cooling in between on ice for ~10 cycles. The exact lysis condition may vary upon samples.
3. Mix the lysate well without centrifugation and make at least three aliquots: two small aliquots (~15 μL) for measuring protein concentration and positive control validation by Western blotting, and the remaining large aliquot for the proteomics analysis.

   Note 1: The insoluble debris is kept in the final aliquot, which allow the digestion of insoluble proteins (Pirmoradian et al., 2013).

   Note 2: One small aliquot of each sample is used to test for predicted protein changes before proteomics analysis that is time consuming and expensive. Perform Western blotting analysis on these samples to confirm positive and negative control proteins that behave the way anticipated. This quality control step also allows the detection of possible mislabeling or other sample issues during sample preparation (Fig. 3).
4. Measure protein concentration by a standard BCA protein assay (Thermo Scientific) or by a Coomassie stained short SDS gel (10%) (Xu et al., 2009) with bovine serum albumin (BSA) as a standard.

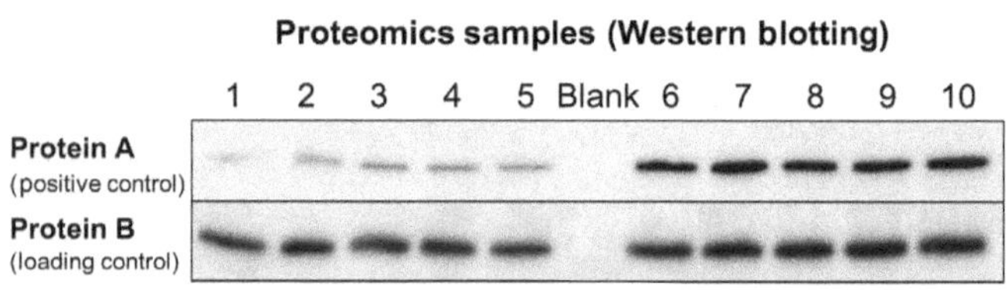

Fig. 3 Quality control analysis of one batch of 10 samples by Western blotting analysis. Protein A is a positive control protein expressed at a higher level in samples 6–10. Protein B is a loading control showing no difference in the samples.

5. Add 10% acetonitrile (ACN) and LysC (Wako) at an enzyme-to-substrate ratio of 1:100 (w/w) to digest proteins under highly denaturing conditions at RT for 2 h (Pagala et al., 2015). Also add dithiothreitol (DTT) to 1 m*M* to reduce disulfide bonds in proteins.
6. Dilute samples to 2 *M* urea with 50 m*M* HEPES (pH 8.5) and further digest samples with trypsin (Promega) at RT for 3 h or overnight. The trypsin to protein ratio is 1:50 (w/w).
7. Further reduce peptides by adding DTT to 1 m*M* for 2 h at RT.
8. Add iodoacetamide (IAA) to 10 m*M* at RT for 30 min in the dark to alkylate Cys-containing peptides.
9. Quench unreacted IAA by adding DTT to 30 m*M* and incubate for 30 min at RT.
10. Terminate the digestion reaction by adding trifluoroacetic acid (TFA) to 1% and check pH by applying a small drop of the solution to pH paper. The pH should be ~2–3.
11. Centrifuge at 20,000 × *g* at RT for 10 min and collect the supernatant.
12. Choose the size of C18 desalting cartridge: either an excessive or limited amount of resin may lead to low recovery of peptides during desalting. The size of C18 cartridge should be selected based on the amount of input material. Cartridge capacities are about 3–5% of the sorbent weight (Villen & Gygi, 2008), alternatively we use 1 μL of C18 resin for every 10 μg of peptides as our criteria (Pagala et al., 2015).
13. Wash cartridge with 5 × bed volumes of methanol, 5 × bed volumes of 60% ACN/0.1% TFA, and condition cartridge with at least of 5 × bed volumes of 0.1% TFA.
14. Load sample on cartridge and wash cartridge with at least 5 × bed volumes of 0.1% TFA, elute with 5 × bed volumes of 60% ACN/0.1% TFA.
15. Dry the eluted peptides by Speedvac and store at −80°C until further analysis.

4. TMT LABELING OF PEPTIDES

Isobaric labeling (e.g., TMT and iTRAQ) has become a versatile technique in relative protein quantification by examining up to 10 biological samples simultaneously in a single run. It provides a fast and easy way for protein quantification with high throughput. To achieve an accurate quantification result, the TMT labeling efficiency should be examined to ensure all samples are fully labeled by TMT. This is a critical step as the amount of TMT label used, alterations in pH, or incorrect protein quantitation can lead to inefficient labeling, and thus negatively alter all downstream results. In addition, a premix ratio test should be performed to determine the correct ratio for equal mixing before pooling for the remainder of the protocol. Again this is extremely important as pipetting errors can occur along the way that will change concentrations. The premix ratio test is the best way to assure that the amount of each of the 10 samples used for the final mix are all at an equal ratio.

1. Resuspend each desalted peptide sample in 50 m*M* HEPES buffer (pH 8.5), keep ~ 1 μg of each unlabeled sample for later TMT labeling efficiency test.

   Note: pH value should be checked by pH paper. Sample may be acidic if not dried completely after desalting which will affect the labeling efficiency.
2. Reconstitute TMT reagents in anhydrous acetonitrile and label peptide samples by following the manufacture's instruction and incubate at RT for 1 h.
3. Test labeling efficiency: desalt ~1 μg of each TMT labeled and unlabeled samples by Ziptips (Millipore) and analyze TMT-labeled samples and unlabeled samples by LC-MS/MS. In a completely TMT-labeled sample, the unlabeled peptide should not be detected in the labeled samples (Fig. 4).
4. If fully labeled, quench the reaction by adding 8 μL of 5% hydroxylamine and incubate for 15 min.
5. Premix ratio test: mix a small aliquot of each TMT-labeled sample, desalt, and analyze by LC-MS/MS to determine the global peptide ratio of all 10 samples.
6. Equally mix 10 samples based on the result of the premix ratio test.

   Note: The premix ratio test determines which of the 10 samples has the lowest amount of peptides. Use this sample as the baseline to adjust the amounts of other nine samples.
7. Desalt the pooled TMT-labeled samples by C18 cartridge to remove byproducts of the quenching reaction.

5. OFFLINE BASIC pH RPLC FRACTIONATION

The number of peptides or proteins that can be identified in one single conventional LC-MS/MS analysis is limited due to sample complexity (dynamic range and number of peptides) even with the latest platform.

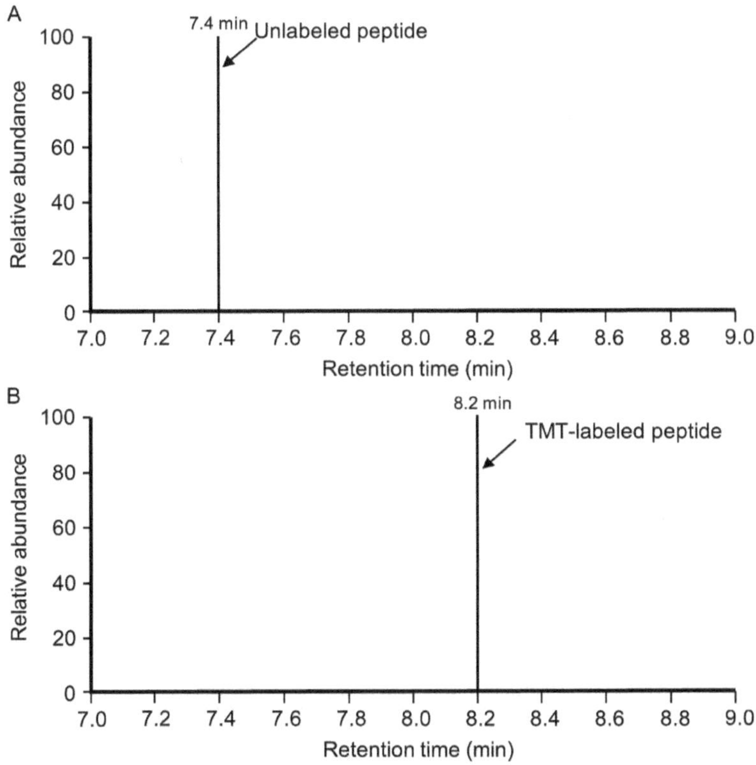

Fig. 4 Labeling efficiency analysis. (A) An unlabeled peptide of strong intensity at retention time of 7.4 min during LC-MS/MS. (B) Complete absence of the same peptide at 7.2 min and the presence of the labeled peptide at 8.2 min, indicating fully TMT labeling.

Several peptide fractionation approaches, such as strong cation-exchange and hydrophilic interaction chromatography, have been applied before standard acidic pH reversed-phase LC-MS/MS analysis to reduce sample complexity. Recently, high pH RPLC has become favored due to its high-resolving power and its salt-free solvent system making it compatible with the downstream acidic pH RPLC-MS/MS analysis (Batth et al., 2014).

1. Set up an XBridge C18 column (4.6 mm × 25 cm, 3.5 μm particle size, Waters) by Agilent 1220 infinity pump for fractionation.
2. Install a 100 μL loop for whole proteome analysis and a 1 mL loop for phosphoproteome analysis. Wash the loop with 3 × loop volume of methanol, water, and buffer A (10 m*M* ammonium formate, pH 8.0) sequentially.
3. Wash the column with 100 μL equally mixed isopropanol, methanol, acetonitrile, and water, and equilibrate the column in 95% buffer A for 1 h.

4. Solubilize the desalted pooled TMT-labeled peptide sample in buffer A. The sample volume should not be bigger than 70% of the loop volume. Verify the sample pH ~8.0. If still acidic then use ammonium hydroxide to adjust pH to 8.0.
5. Fractionate whole proteome sample using the following gradient with buffer B (buffer A plus 90% ACN): 15–20% for 15 min, 20–35% for 100 min, and 35–50% for 30 min. The fraction collector (FC 203B, Gilson) is set to collect fractions every 2 min including loading time. Alternatively, fractionate phosphoproteome sample in 1 h gradient from 15% to 45% buffer B and collect fractions every minute. The flow rate for both runs is 0.4 mL/min.
6. For whole proteome samples, ~40 fractions (every other fraction, a total of 80 fractions) are selected and dried by Speedvac and used for LC-MS/MS analysis. If ultra-deep proteome coverage is required, all 80 fractions are analyzed. For phosphoproteome samples, dry all fractions (~40 fractions in total) for TiO_2 phosphopeptide enrichment.

6. TiO_2 PHOSPHOPEPTIDE ENRICHMENT

TiO_2 phosphopeptide enrichment is dependent on the affinity binding of a phosphate group to TiO_2 through coordinate bonds. The nonspecific binding of carboxyl groups is reduced by the addition of competitors, such as 2,5-dihydroxybenzoic acid, glutamic acid, and lactic acid.

Previously, we presented a refined TiO_2 phosphopeptide enrichment protocol using a low concentration of phosphate as an additive to improve selectivity. Compared with other nonphosphopeptide competitor, phosphate is easy to prepare, effective, and robust while giving similar or higher selectivity. Due to the hydrophilicity of phosphopeptides, there are more phosphopeptides eluted in the early basic pH RPLC fractions compared with later fractions. Therefore, slightly lower concentrations of phosphate and increased amounts of TiO_2 beads are used in early fractions.

1. Reconstitute each dried peptide fraction in binding buffer (65% ACN, 2% TFA, 1 μg/μL).
2. Add KH_2PO_4 to 0.5 mM for first five fractions and 1 mM for other fractions.
3. Weigh TiO_2 beads: 1 mg TiO_2 beads/fraction for first 20 fractions, 0.7 mg/fraction for fractions 21–35 and 0.5 mg/fraction for all remaining fractions.

4. Wash TiO_2 beads twice with washing buffer (65% ACN, 0.1% TFA, 200 μL per mg beads) and mix with the peptide solution of the fractions.
5. Incubate the samples by end-over-end rotation at RT for 20 min.
6. Briefly centrifuge to collect the beads to the bottom of tube, transfer the supernatant to another tube which can be used for consecutive phosphopeptide enrichment if desired.

   Note: The supernatant contains nonphosphopeptides that can be used for whole proteome TMT analysis if desired.
7. Wash phosphopeptide-bound beads twice with wash buffer (200 μL per mg beads), resuspend into 50 μL of wash buffer, and transfer beads to a C18 StageTip (Thermo Fisher Scientific) sitting on the top of a 2-mL centrifuge tube. Collect residual beads in the incubation tube by adding another 50 μL of wash buffer and transferring to the StageTip as well.
8. Centrifuge StageTip to remove the wash buffer, and elute the phosphopeptides with 20 μL of elution buffer (15% NH_4OH, 40% ACN, pH = ~11). The eluents are dried to completeness rapidly.
9. Dissolve the enriched phosphopeptides in 5% formic acid (FA) immediately prior to LC-MS/MS analysis to minimize sample loss.

7. ACIDIC pH RPLC-MS/MS ANALYSIS

During TMT-based quantification, peptide ions are indistinguishable in survey MS spectra but are usually quantified based on the intensity of reporter ions after the peptide ion is fragmented and analyzed in MS/MS spectra. A common limitation is that the TMT report ion ratios may be suppressed due to the coelution of TMT-labeled ions, and the coisolation and cofragmentation of these interfering ions (Ting, Rad, Gygi, & Haas, 2011). Several strategies have been detailed recently to address the ratio suppression effect. One is to narrow ion isolation window (Savitski et al., 2011). In regards to this, mass spectrometers with quadrupole mass filters (e.g., Q Exactive HF and Orbitrap Fusion Lumos) have higher isolation efficiency at narrower mass windows than instruments using waveform-based isolation such as the linear ion trap (e.g., Orbitrap Elite and Velos). Alternatively, MS-based separation, such as gas-phase purification (Wenger et al., 2011) and MultiNotch MS3 method (McAlister et al., 2014), enables more accurate quantifications, but the loss of spectral identifications can be significant. Another approach is to reduce sample complexity by extensive fractionation, including multidimensional LC, small C18 particles, and long gradients (Zhou et al., 2013). In this protocol, we combine offline basic pH RPLC, high-resolving power acidic pH RPLC, and MS/MS quantification method using optimized isolation width to reduce ratio suppression without diminishing the number of identified peptides or proteins.

To sufficiently resolve near-isobaric reporter ions of some 10-plex TMT reagents (6.32 mDa difference), MS/MS resolution must be at least 30,000 at 400 *m/z*. Higher energy collision dissociation (HCD) is used for TMT10-plex reporter ion fragmentation. Electron transfer dissociation (ETD) is not recommended for TMT10-plex reagents because ETD cleavage sites are different from HCD and this result in reporter ion overlap. The mass spectrometer used in this protocol is Q Exactive HF (Thermo Scientific).

1. Pack empty columns (New objective, 75 μm ID for whole proteome sample and 50 μm ID for phosphoproteome samples) with 1.9 μm C18 resin (Dr. Maisch GmbH, Germany) to ~30 cm length. Bed volume of the two columns is 1.3 and 0.6 μL, respectively. Heat the columns at 65°C by a butterfly portfolio heater (Phoenix S&T) to reduce backpressure.
2. Prepare buffer A (5% dimethyl sulfoxide, 0.2% FA) and buffer B (buffer A plus 65% ACN) and wash the columns thoroughly with 95% buffer B and equilibrate in 95% buffer A.
3. Examine the quality of the LC-MS/MS system by two runs using 100 ng of rat brain peptides before analyzing precious TMT-labeled samples. The following parameters should be monitored and evaluated: survey MS signal intensity, quality of MS/MS spectra, peak width, overall retention time, the pressure of LC system as well as the run-to-run variability. Clean, calibrate, and tune MS instrument regularly and use fresh LC buffers to improve the performance of the system.
4. For whole proteome samples: reconstitute the dried peptides in 5% FA, load ~0.2 μg on the column and elute by 15–45% of buffer B gradient in 150 min (~0.3 μL/min flow rate). Adjust the gradient slightly for early and late basic pH RPLC fractions. The mass spectrometer is operated in data-dependent mode with a survey scan in Orbitrap (60,000 resolution, 1×10^6 AGC target and 50 ms maximal ion time) and 20 MS/MS high-resolution scans (60,000 resolution, 1×10^5 AGC target, ~100 ms maximal ion time, 31 HCD normalized collision energy, 0.4 *m/z* isolation window, and 20 s dynamic exclusion).
5. For phosphoproteome samples: resuspend the dried phosphopeptides in 2 μL of 5% FA, load all on the column and elute with 9–35% of buffer B in 200 min and 35–65% buffer B for 40 min (0.15 μL/min flow rate).

Adjust the gradient slightly for early and late basic pH RPLC fractions. Mass spectrometer setting is the same as the earlier except larger isolation window to improve sensitivity (1 *m*/*z* isolation width with 0.3 *m*/*z* offset).

8. MS DATA ANALYSIS

1. Database search: The MS/MS raw files are processed by the SEQUEST algorithm (Eng, McCormack, & Yates, 1994), or the newly developed tag-based hybrid search engine JUMP (Wang et al., 2014). Pattern-based database search (e.g., SEQUEST) and tag-based de novo sequencing (e.g., PEAKS but with limited sensitivity) are common computer algorithms. The JUMP algorithm combines pattern matching and de novo sequencing to score putative peptides, and shows better sensitivity and specificity. The search is typically against the UniProt human database (or other appropriate species specific database) concatenated with a reverse protein sequence decoy database. Briefly, the spectra are converted to mzXML files and then JUMP first performs preprocessing of precursor ions and MS/MS spectra to generate tags and then does MS/MS pattern matching against a database and provides a ranking score (J score). The target-decoy strategy is used to evaluate false discovery rate (FDR) of peptide and protein identification (Peng, Elias, Thoreen, Licklider, & Gygi, 2003). Searches were performed using a 10-ppm mass tolerance for precursor ions and fragment ions, fully tryptic restriction with two maximal missed cleavages, three maximal modification sites, and the assignment of *a*, *b*, and *y* ions. TMT on Lys residues and N-termini (+229.162932 Da) and carbamidomethylation of Cys residues (+57.021 Da) are used for static modifications. Dynamic modifications include Met oxidation (+15.99492 Da) and Ser/Thr/Tyr phosphorylation (+79.96633, only for phosphoproteome data).
2. Data filtering for whole proteome data: To remove false-positive matches and low-quality peptide spectrum matches (PSMs), the assigned peptides are filtered by minimal peptide length (seven amino acid), *m*/*z* accuracy (e.g., 5 ppm), and matching scores (J score and deltaCn). The peptides are then divided into groups according to peptide length, trypticity, and charge state, and then further filtered by matching scores to reduce protein FDR to below 1%. For proteins identified by a single spectral count, higher matching scores are required. If peptides are shared

by multiple members of a protein family, the matched protein members are clustered into one group. According to the parsimony principle, the group is represented by the protein with the highest number of assigned peptides and other proteins matched by unique peptides.

3. Data filtering for phosphoproteome data is similar to whole proteome data except peptide FDR is used as a cutoff as opposed to protein FDR.
4. To evaluate the confidence of assigned phosphosites in peptides, we used the concept of the phosphoRS algorithm (Taus et al., 2011). Another available option to evaluate phosphosite localization is the Ascore algorithm (Beausoleil, Villen, Gerber, Rush, & Gygi, 2006).
5. Quantification: proteins or phosphopeptides are quantified by summing reporter ion counts across all matched PSMs using an in-house program in the JUMP software suite. Briefly, for each identified PSM, TMT reporter ion intensities are extracted and corrected based on isotopic distribution of labeling reagents. For example, the TMT126 reagent generates 91.8% 126 m/z reporter ion, 7.9% 127C m/z reporter ion, and 0.3% of 128C m/z reporter ion. PSMs with zero intensities are recorded but not used for further calculation. PSMs with low intensities are filtered by user-defined thresholds. High noise PSMs are also discarded by percentage of precursor ion in the isolation window of the survey scan (e.g., at least 80%). Then the average of all 10 reporter ions is used to calculate a relative signal between each reporter ion and the average. The relative signals of PSMs are summed for identified proteins or phosphopeptides. Finally, these relative signals are converted to absolute signals by multiplying the averaged reporter ion intensity of the top 3 PSMs in corresponding proteins or phosphopeptides.
6. The protein or phosphopeptide quantification values are exported in Excel for further analysis. In general, the method of analysis of variance (ANOVA) is used to identify proteins or phosphopeptides that are differentially expressed.

9. PROTEOMICS RESULTS AND SELECTED VALIDATION

In the table of proteome data, the following information is exported for each quantified protein: protein accession number, protein annotation, gene name, PSM number, total peptide number, unique peptide number, percentage of coverage, matching scores (J score and deltaJn) of the top scored peptide, and the final reporter ion intensities of each protein. The phosphoproteome data are quantified at phosphopeptide level, including the related protein information (protein accession number, protein annotation, and gene name), the phosphopeptide's PSM number, phosphosite ambiguity score (e.g., from phosphoRS or Ascore), and reporter ion intensities.

To differentiate experimental variance from biological alteration, the biological replicates are compared as a null experiment (Fig. 5). The log2 ratio distribution of the null comparison is closer to zero compared with that of case–control comparison, indicating that the biological alteration of the case–control comparison is larger than experimental variance.

Although the data processing can significantly reduce false positives in the steps of identification and quantification, we cannot completely eliminate false results, and find that there are more false positives in the group of proteins or peptides identified by only one single peptide. Inaccuracy in quantification also exists due to low intensity, errors in data processing (e.g., missing reporter ions due to irregular mass shift), peptide sharing, and ambiguity of phosphosite assignment. Therefore, it is highly recommend that extra validation steps are applied to confirm the MS data. (1) Manually examine the MS/MS spectra of proteins or phosphosite of

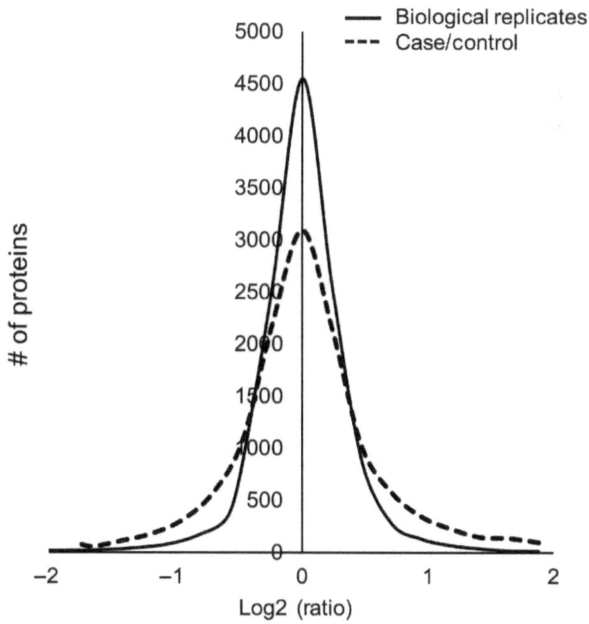

Fig. 5 Analysis of experimental variance and control–case comparison. Distribution of log2 ratios between biological replicates (null experiment) and the control–case comparison.

interest to validate the peptide sequence and TMT reporter ion quantification. (2) Use known protein or phosphosite quantification data to confirm MS results. (3) Verify protein or phosphosite changes by antibody-based approaches (e.g., Western blotting and immunohistochemistry). (4) Chemically synthesize the peptides of interest, and then use them as internal standards to confirm the identification of native peptides. Their MS/MS patterns and retention time during LC–MS/MS should be identical. (5) Verify protein changes by targeted MS strategy approaches. At least one if not multiple methods of validation need to be performed before proceeding with time-consuming biological experiments for testing derived hypothesis.

10. CONCLUSIONS AND DISCUSSION

Here, we describe a high-throughput protocol for quantification of proteins and posttranslational modifications by isobaric labeling. By this routine protocol, from 10 protein samples analyzed in a batch, we analyze over 10,000 proteins and ~30,000 phosphosites with high consistency. In this protocol, phosphopeptides are enriched after labeling to reduce variance. However, for some protein modifications (e.g., ubiquitination), if the TMT conjugation interferes with the enrichment process, the enrichment has to be carried out before the isobaric labeling (Kim et al., 2011). With slight procedure rearrangement, this protocol can be widely used as a general proteomics approach to the study of proteins and multiple modifications. In summary, the protocol here provides a powerful technique to obtain deep protein coverage and alleviate ratio suppression during quantification. Such comprehensive proteomics analysis can be integrated with genomics, transcriptomics, and possibly metabolomics, for comprehensive understanding of biological systems and novel discoveries of molecular mechanisms, biomarkers, and therapeutic targets in human disease (Li et al., 2016; Nesvizhskii, 2014; Wang et al., 2016; Yuan et al., 2014; Zhou et al., 2013).

ACKNOWLEDGMENTS

The authors thank all other members in the Peng lab and St. Jude Proteomics Facility for insightful discussion. This work was partially supported by National Institutes of Health grants R01AG047928 (J.P.), R01AG053987 (J.P.), R01GM114260 (J.P.), and ALSAC (American Lebanese Syrian Associated Charities). The MS analysis was performed in St. Jude Proteomics Facility, partially supported by NIH Cancer Center Support Grant (P30CA021765).

REFERENCES

Altelaar, A. F., Munoz, J., & Heck, A. J. (2013). Next-generation proteomics: Towards an integrative view of proteome dynamics. *Nature Reviews. Genetics*, *14*(1), 35–48.

Azimifar, S. B., Nagaraj, N., Cox, J., & Mann, M. (2014). Cell-type-resolved quantitative proteomics of murine liver. *Cell Metabolism*, *20*(6), 1076–1087.

Bai, B., Chen, P. C., Hales, C. M., Wu, Z., Pagala, V., High, A. A., et al. (2014). Integrated approaches for analyzing U1-70 K cleavage in Alzheimer's disease. *Journal of Proteome Research*, *13*(11), 4526–4534.

Bai, B., Hales, C. M., Chen, P. C., Gozal, Y., Dammer, E. B., Fritz, J. J., et al. (2013). U1 small nuclear ribonucleoprotein complex and RNA splicing alterations in Alzheimer's disease. *Proceedings of the National Academy of Sciences of the United States of America*, *110*(41), 16562–16567.

Batth, T. S., Francavilla, C., & Olsen, J. V. (2014). Off-line high-pH reversed-phase fractionation for in-depth phosphoproteomics. *Journal of Proteome Research*, *13*(12), 6176–6186.

Beausoleil, S. A., Villen, J., Gerber, S. A., Rush, J., & Gygi, S. P. (2006). A probability-based approach for high-throughput protein phosphorylation analysis and site localization. *Nature Biotechnology*, *24*(10), 1285–1292.

Beltran, L., & Cutillas, P. R. (2012). Advances in phosphopeptide enrichment techniques for phosphoproteomics. *Amino Acids*, *43*(3), 1009–1024.

Bensimon, A., Heck, A. J., & Aebersold, R. (2012). Mass spectrometry-based proteomics and network biology. *Annual Review of Biochemistry*, *81*, 379–405.

Churchman, M. L., Low, J., Qu, C., Paietta, E. M., Kasper, L. H., Chang, Y., et al. (2015). Efficacy of retinoids in IKZF1-mutated BCR-ABL1 acute lymphoblastic leukemia. *Cancer Cell*, *28*(3), 343–356.

Eng, J. K., McCormack, A. L., & Yates, J. R. (1994). An approach to correlate tandem mass spectral data of peptides with amino acid sequences in a protein database. *Journal of the American Society for Mass Spectrometry*, *5*(11), 976–989.

Everley, R. A., Kunz, R. C., McAllister, F. E., & Gygi, S. P. (2013). Increasing throughput in targeted proteomics assays: 54-plex quantitation in a single mass spectrometry run. *Analytical Chemistry*, *85*(11), 5340–5346.

Geyer, P. E., Kulak, N. A., Pichler, G., Holdt, L. M., Teupser, D., & Mann, M. (2016). Plasma proteome profiling to assess human health and disease. *Cell Systems*, *2*(3), 185–195.

Hebert, A. S., Richards, A. L., Bailey, D. J., Ulbrich, A., Coughlin, E. E., Westphall, M. S., et al. (2014). The one hour yeast proteome. *Molecular & Cellular Proteomics*, *13*(1), 339–347.

Hein, M. Y., Hubner, N. C., Poser, I., Cox, J., Nagaraj, N., Toyoda, Y., et al. (2015). A human interactome in three quantitative dimensions organized by stoichiometries and abundances. *Cell*, *163*(3), 712–723.

Kim, W., Bennett, E. J., Huttlin, E. L., Guo, A., Li, J., Possemato, A., et al. (2011). Systematic and quantitative assessment of the ubiquitin-modified proteome. *Molecular Cell*, *44*(2), 325–340.

Kim, M. S., Pinto, S. M., Getnet, D., Nirujogi, R. S., Manda, S. S., Chaerkady, R., et al. (2014). A draft map of the human proteome. *Nature*, *509*(7502), 575–581.

Li, Y., Wang, X., Cho, J. H., Shaw, T. I., Wu, Z., Bai, B., et al. (2016). JUMPg: An integrative proteogenomics pipeline identifying unannotated proteins in human brain and cancer cells. *Journal of Proteome Research*, *15*(7), 2309–2320.

McAlister, G. C., Huttlin, E. L., Haas, W., Ting, L., Jedrychowski, M. P., Rogers, J. C., et al. (2012). Increasing the multiplexing capacity of TMTs using reporter ion isotopologues with isobaric masses. *Analytical Chemistry*, *84*(17), 7469–7478.

McAlister, G. C., Nusinow, D. P., Jedrychowski, M. P., Wuhr, M., Huttlin, E. L., Erickson, B. K., et al. (2014). MultiNotch MS3 enables accurate, sensitive, and multiplexed detection of differential expression across cancer cell line proteomes. *Analytical Chemistry*, *86*(14), 7150–7158.

Mertz, J. L., Tan, H., Pagala, V., Bai, B., Chen, P. C., Li, Y., et al. (2015). Sequential elution interactome analysis of the Mind bomb 1 ubiquitin ligase reveals a novel role in dendritic spine outgrowth. *Molecular & Cellular Proteomics*, *14*, 1898–1910.

Nesvizhskii, A. I. (2014). Proteogenomics: Concepts, applications and computational strategies. *Nature Methods*, *11*(11), 1114–1125.

Pagala, V. R., High, A. A., Wang, X., Tan, H., Kodali, K., Mishra, A., et al. (2015). Quantitative protein analysis by mass spectrometry. *Methods in Molecular Biology*, *1278*, 281–305.

Peng, J., Elias, J. E., Thoreen, C. C., Licklider, L. J., & Gygi, S. P. (2003). Evaluation of multidimensional chromatography coupled with tandem mass spectrometry (LC/LC-MS/MS) for large-scale protein analysis: The yeast proteome. *Journal of Proteome Research*, *2*(1), 43–50.

Pirmoradian, M., Budamgunta, H., Chingin, K., Zhang, B., Astorga-Wells, J., & Zubarev, R. A. (2013). Rapid and deep human proteome analysis by single-dimension shotgun proteomics. *Molecular & Cellular Proteomics*, *12*(11), 3330–3338.

Savitski, M. M., Sweetman, G., Askenazi, M., Marto, J. A., Lang, M., Zinn, N., et al. (2011). Delayed fragmentation and optimized isolation width settings for improvement of protein identification and accuracy of isobaric mass tag quantification on Orbitrap-type mass spectrometers. *Analytical Chemistry*, *83*(23), 8959–8967.

Song, C., Ye, M., Han, G., Jiang, X., Wang, F., Yu, Z., et al. (2010). Reversed-phase-reversed-phase liquid chromatography approach with high orthogonality for multidimensional separation of phosphopeptides. *Analytical Chemistry*, *82*(1), 53–56.

Tan, H., Wu, Z., Wang, H., Bai, B., Li, Y., Wang, X., et al. (2015). Refined phosphopeptide enrichment by phosphate additive and the analysis of human brain phosphoproteome. *Proteomics*, *15*(2–3), 500–507.

Taus, T., Kocher, T., Pichler, P., Paschke, C., Schmidt, A., Henrich, C., et al. (2011). Universal and confident phosphorylation site localization using phosphoRS. *Journal of Proteome Research*, *10*(12), 5354–5362.

Thingholm, T. E., & Larsen, M. R. (2016). The use of titanium dioxide for selective enrichment of phosphorylated peptides. *Methods in Molecular Biology*, *1355*, 135–146.

Thongboonkerd, V., LaBaer, J., & Domont, G. B. (2014). Recent advances of proteomics applied to human diseases. *Journal of Proteome Research*, *13*(11), 4493–4496.

Ting, L., Rad, R., Gygi, S. P., & Haas, W. (2011). MS3 eliminates ratio distortion in isobaric multiplexed quantitative proteomics. *Nature Methods*, *8*(11), 937–940.

Villen, J., & Gygi, S. P. (2008). The SCX/IMAC enrichment approach for global phosphorylation analysis by mass spectrometry. *Nature Protocols*, *3*(10), 1630–1638.

Wang, X., Li, Y., Wu, Z., Wang, H., Tan, H., & Peng, J. (2014). JUMP: A tag-based database search tool for peptide identification with high sensitivity and accuracy. *Molecular & Cellular Proteomics*, *13*(12), 3663–3673.

Wang, X., Pandey, A. K., Mulligan, M. K., Williams, E. G., Mozhui, K., Li, Z., et al. (2016). Joint mouse-human phenome-wide association to test gene function and disease risk. *Nature Communications*, 7, 10464.

Wang, H., Sun, S., Zhang, Y., Chen, S., Liu, P., & Liu, B. (2015). An off-line high pH reversed-phase fractionation and nano-liquid chromatography-mass spectrometry method for global proteomic profiling of cell lines. *Journal of Chromatography. B, Analytical Technologies in the Biomedical and Life Sciences*, *974*, 90–95.

Wang, H., Yang, Y., Li, Y., Bai, B., Wang, X., Tan, H., et al. (2015). Systematic optimization of long gradient chromatography mass spectrometry for deep analysis of brain proteome. *Journal of Proteome Research*, *14*(2), 829–838.

Wenger, C. D., Lee, M. V., Hebert, A. S., McAlister, G. C., Phanstiel, D. H., Westphall, M. S., et al. (2011). Gas-phase purification enables accurate, multiplexed proteome quantification with isobaric tagging. *Nature Methods*, *8*(11), 933–935.

Wilhelm, M., Schlegl, J., Hahne, H., Moghaddas Gholami, A., Lieberenz, M., Savitski, M. M., et al. (2014). Mass-spectrometry-based draft of the human proteome. *Nature*, *509*(7502), 582–587.

Xu, P., Duong, D. M., & Peng, J. (2009). Systematical optimization of reverse-phase chromatography for shotgun proteomics. *Journal of Proteome Research*, *8*(8), 3944–3950.

Yang, F., Shen, Y., Camp, D. G., 2nd, & Smith, R. D. (2012). High-pH reversed-phase chromatography with fraction concatenation for 2D proteomic analysis. *Expert Review of Proteomics*, *9*(2), 129–134.

Yuan, Y., Van Allen, E. M., Omberg, L., Wagle, N., Amin-Mansour, A., Sokolov, A., et al. (2014). Assessing the clinical utility of cancer genomic and proteomic data across tumor types. *Nature Biotechnology*, *32*(7), 644–652.

Zhou, F., Lu, Y., Ficarro, S. B., Adelmant, G., Jiang, W., Luckey, C. J., et al. (2013). Genome-scale proteome quantification by DEEP SEQ mass spectrometry. *Nature Communications*, *4*, 2171.

CHAPTER TWENTY-ONE

Glycoprotein Enrichment Analytical Techniques: Advantages and Disadvantages

R. Zhu, L. Zacharias, K.M. Wooding, W. Peng, Y. Mechref[1]
Texas Tech University, Lubbock, TX, United States
[1]Corresponding author: e-mail address: yehia.mechref@ttu.edu

Contents

Methods in Enzymology, Volume 585
ISSN 0076-6879
http://dx.doi.org/10.1016/bs.mie.2016.11.009

Abstract

Protein glycosylation is one of the most important posttranslational modifications. Numerous biological functions are related to protein glycosylation. However, analytical challenges remain in the glycoprotein analysis. To overcome the challenges associated with glycoprotein analysis, many analytical techniques were developed in recent years. Enrichment methods were used to improve the sensitivity of detection, while HPLC and mass spectrometry methods were developed to facilitate the separation of glycopeptides/proteins and enhance detection, respectively. Fragmentation techniques applied in modern mass spectrometers allow the structural interpretation of glycopeptides/proteins, while automated software tools started replacing manual processing to improve the reliability and throughput of the analysis. In this chapter, the current methodologies of glycoprotein analysis were discussed. Multiple analytical techniques are compared, and advantages and disadvantages of each technique are highlighted.

1. PROTEOMICS: POSTTRANSLATIONAL MODIFICATIONS

Proteins perform numerous functions in living organisms, including, but not limited to (i) catalyzing reactions, (ii) replicating DNA, (iii) responding to stimuli, and (iv) transporting molecules (Zhang, Fonslow, Shan, Baek, & Yates, 2013). The importance of identifying the proteins present in cell and tissue samples, so-called proteomics studies, becomes readily apparent with aberrant cells; often times a splice variant or unique posttranslational modification (PTM) lends insight into the mutant phenotype.

The term "proteomics" originated from the combination of "protein" and "genomics" in the 1990s (James, 1997; Wilkins et al., 1996). Proteomics focuses on the characterization and quantification of all the proteins originating in a biological system, including cellular location, splice-variants, interactions, and PTMs. The functional capabilities of proteins are often expanded by PTMs which add to the chemical repertoire of proteins by adding small chemical groups (such as phosphate, simple carbohydrates, acetate, and alkyl groups) or larger molecules (such as ubiquitin, small ubiquitin-like modifier, and complex glycans).

The essential biological functions of protein PTMs require a comprehensive characterization for a complete proteomics analysis. This is a rather daunting task as it has been estimated that the 20,000 genes in the human genome encode approximately 100,000 proteoforms (Gstaiger &

Aebersold, 2009; Smith, Kelleher, & Consortium for Top Down, 2013); to further complicate analysis, over 400 protein PTMs have been identified (Gstaiger & Aebersold, 2009). The accurate proteome and PTM characterization of these complex samples remain a challenge.

One of the most commonly used proteomics strategies is bottom-up or shotgun proteomics (Wolters, Washburn, & Yates, 2001; Yates, 2004); this approach analyzes peptides to gain insight into proteins. This is an indirect measurement of proteins through peptides derived from proteolytic digestion of intact proteins. Trypsin is the most commonly used protease in shotgun proteomics studies because of its high cleavage specificity to the carboxyl side of arginine and lysine (Vandermarliere, Mueller, & Martens, 2013). Endoproteinase Lys-C, endoproteinase Glu-C, and elastase are other commonly used proteases; they are typically employed when tryptic peptides do not yield the information desired (Keil, 1987; Siepen, Keevil, Knight, & Hubbard, 2007). The combination of both highly selective and nonselective proteases improves protein and PTMs identification (MacCoss et al., 2002; Xu, Wong, Kashina, & Yates, 2009).

Peptide mixtures are routinely analyzed with a liquid chromatography sample separation step coupled to a tandem mass spectrometer detector (LC-MS/MS). Briefly, the complex peptide mixture is separated via liquid chromatography immediately followed by mass spectrometry analysis. The isolated peptide ions are fragmented to yield product ions; these product ions are then used to sequence the peptide. A database search with the peptide sequence yields a protein identification. The quantification is generally achieved by relative quantification, such as isotope labeling strategy and label-free quantification.

2. CHALLENGES ASSOCIATED WITH GLYCOPROTEIN ANALYSIS

Glycosylation is one of the most prevalent protein PTMs; it has been estimated that approximately 50% of human proteins are glycosylated (Apweiler, Hermjakob, & Sharon, 1999). Glycans are either covalently bound to asparagine (Asp, N) to form an N-linked glycan or serine (Ser, S) and threonine (Thr, T) residues to form an O-linked glycan (Moremen, Tiemeyer, & Nairn, 2012). Numerous biological functions are related to protein glycosylation, including cell–cell signaling, protein stability, protein localization, and immune response (Helenius & Aebi, 2001; O'Connor &

Imperiali, 1996; Varki, 1993). The importance of protein glycosylation is readily apparent in human diseases such as inflammatory diseases (Campbell, Yu, & Rhodes, 2001; Dube & Bertozzi, 2005; Song, Zhu, Hammoud, & Mechref, 2014), rheumatoid arthritis (Elliott et al., 1997; Smith, Pollacchi, Field, & Watson, 2002), Alzheimer's disease (Botella-Lopez et al., 2006; Wang, Grundke-Iqbal, & Iqbal, 1996), Pompe disease (Dasouki et al., 2014), and cancer (Mechref, Hu, Garcia, & Hussein, 2012; Tsai et al., 2015).

Glycoproteomics, a subset of proteomics focused on glycoproteins, focuses on determining changes in glycoprotein macroheterogeneity (glycosylation site occupancy) and microheterogeneity (different glycoforms attached to the same glycosylation site). Currently, several technical challenges are associated with the analysis of glycoproteins derived from biological samples. First, the highly hydrophilic nature of glycans reduces their ionization efficiency, which gives one the impression that they are not abundant. To overcome this reduced ionization efficiency issue, an enrichment step is usually employed.

Second, glycoproteins commonly have more than one glycosylation sites and/or other PTMs, which greatly complicate the characterization of site-specific glycosylation. In order to acquire macroheterogeneity and microheterogeneity information on glycoproteins, a bottom-up strategy is commonly employed; proteases reduce a complex glycoprotein mixture into a simpler glycopeptide sample. LC-MS/MS is currently the most common and powerful analytical platform for the analysis of protein glycosylation via glycopeptide analysis.

Third, the MS/MS spectra of glycopeptides contain information for both the peptide backbone and the glycan moieties. Several MS/MS fragmentation techniques, often sequentially, are utilized to identify the peptide backbone, glycosylation site, and diagnostic product ions indicative of a glycopeptide.

Fourth, the interpretation of MS/MS spectra requires a significant amount of data processing time and experienced personnel; the proteomics data processing workflow is relatively robust whereas the glycoprotemics data processing is still in its infancy. Accurate and precise software tools for automated data analysis are highly desired. Fig. 1 summarizes the general workflow of glycoproteomics analysis involving (i) cell lysis, (ii) protein extraction, (iii) glycoprotein enrichment, (iv) LC-MS/MS analysis, (v) data analysis, and (vi) validation (Song & Mechref, 2015). Current enrichment techniques, separation platforms, fragmentation methods, and software tools will be reviewed in this chapter.

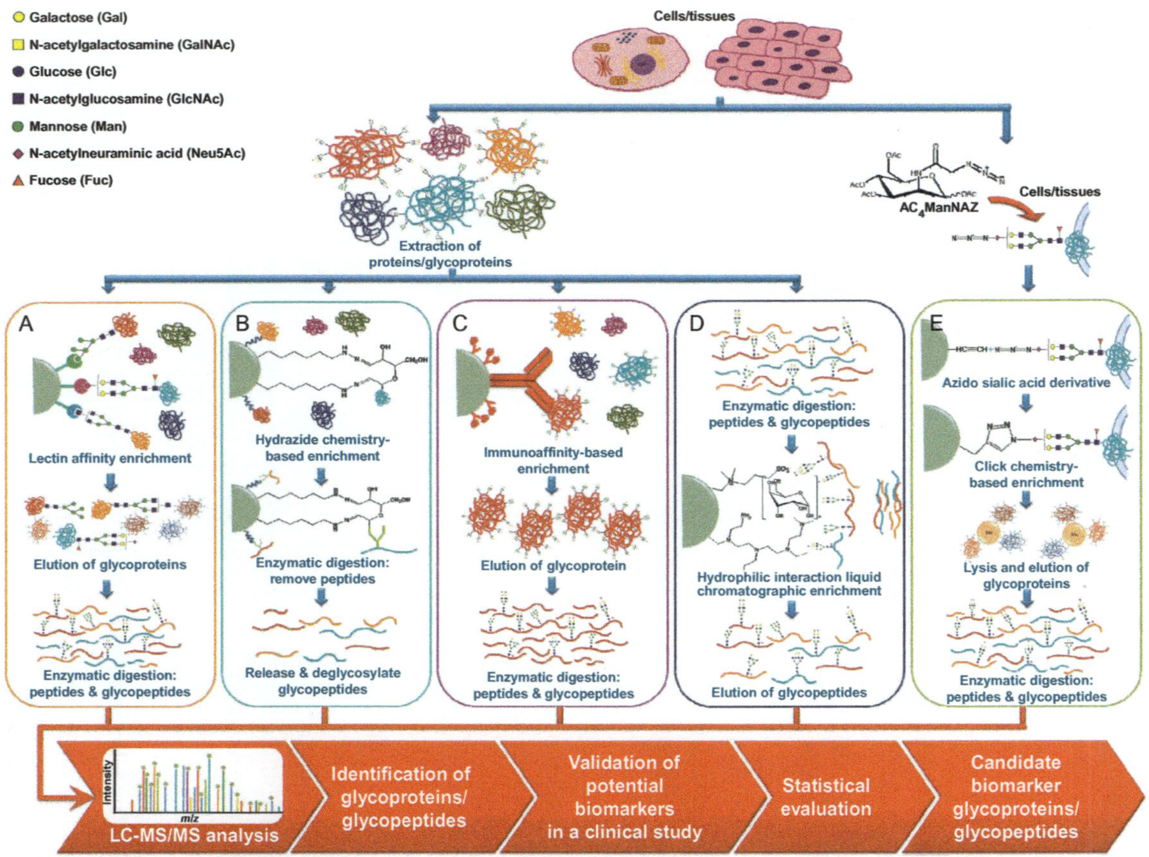

Fig. 1 Flowchart highlighting typical glycoproteomics workflow. Glycopeptide enrichment techniques: (A) lectin affinity, (B) hydrazide chemistry, (C) immunoaffinity, (D) HILIC, and (E) click chemistry. *Figure reproduced with permission from Song, E., & Mechref, Y. (2015). Defining glycoprotein cancer biomarkers by MS in conjunction with glycoprotein enrichment.* Biomarkers in Medicine, *9(9), 835–844. http://dx.doi.org/10.2217/bmm.15.55.*

3. ENRICHMENT TECHNIQUES OF GLYCOPEPTIDES

The low abundance of glycoproteins in biological samples and the poor ionization efficiency, when analyzed by electrospray ionization, remain challenges in protein glycosylation analysis. In order to overcome these challenges, enrichment techniques are often employed with biological specimens (Ongay, Boichenko, Govorukhina, & Bischoff, 2012); these enrichment techniques are reviewed later.

3.1 Lectin Affinity Chromatographic Enrichment

Lectin affinity chromatographic enrichment has been utilized for glycoproteomic analysis of lung (Heo, Lee, Ryoo, Park, & Cho, 2007; Hongsachart et al., 2009), breast (Wang et al., 2008; Yang, Harris, Palmer-Toy, & Hancock, 2006), and liver cancers (Qin et al., 2013; Fig. 1A). Lectins are proteins that specifically bind glycans; the different lectins have different PTMs which differentiate their glycan-binding specificity. During enrichment, lectins bind the various glycopeptides through hydrogen bonding and hydrophilic interactions with saccharide units to retain glycopeptides (Madera, Mann, Mechref, & Novotny, 2008; Mechref, Madera, & Novotny, 2008; Ongay et al., 2012). Lectin affinity chromatography selectively enriches for a single type of glycan structure. For example, Concanavalin A (Con A) selectively binds glycans that possess mannose saccharide units, *Sambucus nigra* agglutinin (SNA) will bind *N*-acetyl neuraminic acid (NeuAc; sialic acid) units, and *Aleuria aurantia* lectin (AAL) will bind fucose units. Table 1 lists a few of the lectin materials that are commercially available. The major disadvantage of lectin separation is that for any single lectin, only a single glycopeptide with a specific saccharide unit can be purified. To overcome this, recent studies have used a combination of lectins for broader glycan enrichment (Yang & Hancock, 2004).

3.2 Covalent Hydrazide Chemistry Enrichment

Hydrazide chemistry enrichment has also been used for glycoprotein purification (Nilsson et al., 2009; Ongay et al., 2012; Zhang, Li, Martin, & Aebersold, 2003; Fig. 1B). In this technique, the glycan moieties on glycopeptides are first oxidized to form aldehydes. The glycans are then covalently immobilized on the solid hydrazide support. Following tryptic digestion, the nonglycosylated peptides are washed off while the glycopeptides are retained. The major disadvantage of hydrazide enrichment is that it only

Table 1 Commercially Available Lectin Materials

Lectin	Source	Glycan Specificity
Aleuria aurantia lectin (AAL)	*Aleuria aurantia* mushrooms	Fucα6GlcNAc
Concanavalin A (Con A)	*Canavalia ensiformis* (Jack Bean) seeds	αMan, αGlc
Galanthus nivalis agglutinin (GNL, GNA)	*Galanthus nivalis* (Snowdrop) bulbs	αMan
Jacalin lectin (Jacalin, AIL)	*Artocarpus integrifolia* (Jackfruit) seeds	Galβ3GlcNAc
Lens culinaris agglutinin (LCA, LcH)	*Lens culinaris* (lentil) seeds	αMan, αGlc
Maackia amurensis lectin I (MAL I, MAL)	*Maackia amurensis* seeds	Galβ4GlcNAc
Peanut agglutinin (PNA)	*Arachis hypogaea* peanuts	Galβ3GlcNAc
Sambucus nigra agglutinin (SNA, EBL)	*Sambucus nigra* (Elderberry) bark	Neu5Acα6Gal/GalNAc
Wheat Germ agglutinin (WGA)	*Triticum vulgaris* (wheat germ)	GlcNAc

Common vendors: Associates of Cape Cod Inc., Biomeda, Biotinylated Lectins, EMD Biosciences, Europa Bioproducts, EY Laboratories, Glycorex, Glygen Corporation, Invitrogen, Prozyme, Sigma-Aldrich, U.S. Biologicals, Vector Laboratories, and Worthington Biochemical Company.
See ref. Halim et al. (2012) for a more thorough review on lectin materials.
Table adapted from Gstaiger, M., & Aebersold, R. (2009). Applying mass spectrometry-based proteomics to genetics, genomics and network biology. *Nature Reviews. Genetics, 10*(9), 617–627. doi:10.1038/nrg2633 with permission.

provides site-specific information and does not allow for determination of the glycan structure due to the oxidative chemical coupling between the glycan and hydrazide solid support after release by PNGase F (Ongay et al., 2012). Hydrazide enrichment has recently been used to identify glycoprotein variations in hepatocellular carcinoma (liver cancer) (Chen et al., 2011; Li, Jiang, et al., 2013), lung (Li, Shah, et al., 2013; Zeng et al., 2010), breast (Whelan et al., 2009), and prostate cancers (Chen, Xi, Tian, Bova, & Zhang, 2013).

3.3 Immunoprecipitation Enrichment

Immunoaffinity columns can also be used to enrich for specific glycoproteins (Fig. 1C). This type of enrichment allows for characterization of site-specific occupancies or glycan saccharide units. Since this method relies on the capability of available antibodies to recognize target glycoproteins, this approach proves to be more expensive than the previous enrichment techniques. Immunoprecipitation has been used to analyze haptoglobin samples from lung cancer (Tsai et al., 2011; Wang, Hincapie, Rejtar, & Karger, 2011).

3.4 Hydrophilic Interaction Liquid Chromatography Enrichment

Hydrophilic interaction liquid chromatography (HILIC) is an enrichment technique that has gained popularity in the last few years (Buszewski & Noga, 2012; Fig. 1D). This enrichment technique is based on the hydrophilic interaction between the cellulose of the stationary phase and the hydroxyl groups of the glycoproteins. Peptide washing occurs in an organic solvent followed by glycopeptide elution with an aqueous buffer. HILIC enrichment is beneficial due to the sufficient removal of competing peptides from a digested biological sample. HILIC enrichment is being developed for both off- and on-line glycoprotein studies.

3.5 Click Chemistry Enrichment

The term "click chemistry" is used to define a reaction process that is quick, easy to employ, results in a product that is easy to purify, and gives high yields (Kolb, Finn, & Sharpless, 2001; Fig. 1E). Click chemistry was initially developed for pharmaceuticals but has since then expanded into other areas of research (Hein, Liu, & Wang, 2008). One such area is in the field of glycoproteomics. Azide-modified glycan residues are inserted into cell membranes during metabolic labeling and then following the pathway of a click chemistry reaction based on a copper-catalyzed azide–alkyne cycloaddition; glycoproteins are bound, and the highly abundant cytosolic proteins are removed. Glycoproteins on the cell surface of prostate cancer have been identified using this enrichment technique (Yang, Nyalwidhe, Guo, Drake, & Semmes, 2011).

3.6 Covalent Boronic Acid Chemistry Enrichment

Boronic acid chemistry is another type of a covalent separation technique that is often used in glycopeptide enrichment processes (Ongay et al., 2012; Fig. 2A). In this method, boronic acid is used to capture 1,2 and 1,3 *cis*-diol groups of glycopeptides by creating a covalent bond and forming a cyclic bromate ester. The noncovalently bound species are then washed away, and the glycopeptides can be further analyzed. The benefit of boronic acid chemistry is the ability to reverse the reaction by a change in pH for the release of the glycopeptide. Boronic acid has been functionalized to a variety

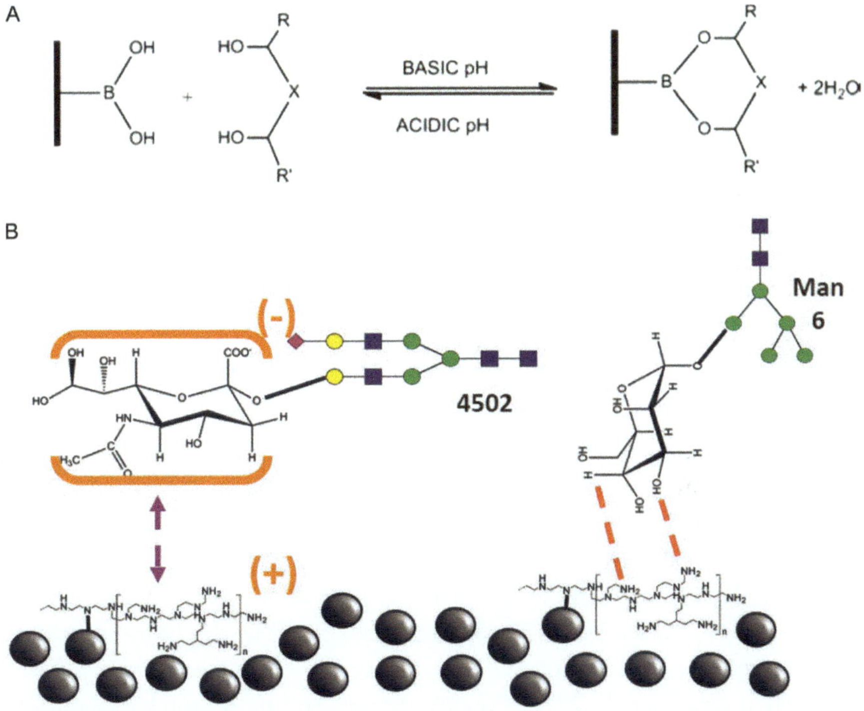

Fig. 2 Boronic acid enrichment and electrostatic repulsion hydrophilic interaction chromatography enrichment. (A) A covalent bond between the stationary media and the glycan hydroxyl groups allows for clean separations. (B) Silica beads functionalized with polyethyleneimine attract uncharged glycans (*right*) while repelling charged glycans (*left*). *(A) Reproduced with permission from Ongay, S., Boichenko, A., Govorukhina, N., & Bischoff, R. (2012). Glycopeptide enrichment and separation for protein glycosylation analysis.* Journal of Separation Science, 35*(18), 2341–2372. doi:10.1002/jssc.201200434. (B) Reproduced with permission from Zacharias, L. G., Hartmann, A. K., Song, E., Zhao, J., Zhu, R., Mirzaei, P., & Mechref, Y. (2016). HILIC and ERLIC enrichment of glycopeptides derived from breast and brain cancer cells.* Journal of Proteome Research, 15, *3624–3634. doi:10.1021/acs.jproteome.6b00429.*

of surfaces for enrichment procedures including, monoliths, mesoporous silica, magnetic particles, or gold nanoparticles (Ongay et al., 2012). Each functionalized material has shown a high specificity for glycopeptides during the enrichment process.

3.7 Electrostatic Repulsion Hydrophilic Interaction Chromatography Enrichment

Glycoprotein enrichment by electrostatic repulsion hydrophilic interaction chromatography (ERLIC) is a technique that takes advantage of the electrostatic charge exhibited by the glycan saccharide units (Zacharias et al., 2016; Fig. 2B). In ERLIC, the stationary phase consists of silica beads with various modified positively charged, functional groups attached. The negative charge on a sialic acid is attracted to the positive functional groups while peptides with positive charges will be repelled. Additionally, a hydrophilic interaction occurs between the glycan and the functional group. The basic, hydrophobic peptides can, therefore, be washed out with an organic solvent (such as acetonitrile, ACN) and the captured glycopeptides can then be eluted with a more polar solvent (such as H_2O/0.5% formic acid) (Hao, Guo, & Sze, 2011).

3.8 Enrichment Methods Comparison

Many studies have compared the efficiency and specificity of different glycopeptide enrichment methods (Calvano, Zambonin, & Jensen, 2008; Ishihara et al., 2011; Lewandrowski, Moebius, Walter, & Sickmann, 2006; Song et al., 2014; Wohlgemuth, Karas, Eichhorn, Hendriks, & Andrecht, 2009; Zacharias et al., 2016; Zhang et al., 2010, 2016; Zhang, Go, & Desaire, 2008). HILIC-based methods result in the highest number of glycopeptide identification when compared to other methods (Zacharias et al., 2016; Zhang et al., 2016; Fig. 3A); however, HILIC and lectin enrichment proved to be less specific than covalent hydrazide chemistry enrichment (Whelan et al., 2009). On the other hand, different enrichment methods exhibit complementarity to others (Song et al., 2014; Fig. 3B). Thaysen-Andersen et al. compared the glycopeptide enrichment methods using a broad spectrum of HILIC materials (ZIC-HILIC, PolyHydroxyethyl A, PolySulfethyl A, TSK Amide-80, and LudgerClean S, TiO_2), graphitized carbon (Hypersil and LudgerClean EB10), and lectin affinity chromatography (ConA). The comparison showed that a comprehensive glycopeptide profile is achieved by any of the approaches (Thaysen-Andersen, Mysling, & Hojrup, 2009).

4. HIGH-PERFORMANCE LIQUID CHROMATOGRAPHY IN PROTEOMICS AND GLYCOPROTEOMICS

High-performance liquid chromatography (HPLC) is the most commonly used separation technique in shotgun proteomic and glycoproteomic analysis because of its high resolving power, high reproducibility, and compatibility with mass spectrometry; HPLC is considered as the most efficient separation technique for large-scale peptide/glycopeptide analysis.

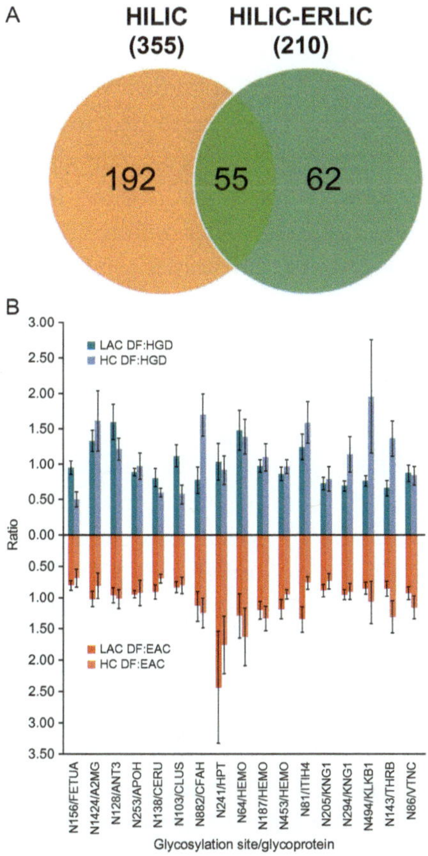

Fig. 3 Comparison of different glycopeptide enrichment methods. (A) The Venn diagram depicts the number of identified unique glycosylation sites for HILIC and HILIC-ERLIC enrichment methods (shown *bolded* in the parentheses is the total number of glycopeptides identified in each case) and of those, which glycosylation sites were unique to the individual methods. (B) Comparisons of MRM quantitation for 17 common glycosylation site between lectin affinity and hydrazide chemistry enrichments in terms of ratio between different groups of biological samples. *DF*, disease-free; *EAC*, esophageal adenocarcinoma; *HGD*, high-grade dysplasia. *(A) Reproduced with permission from Zacharias, L. G., Hartmann, A. K., Song, E., Zhao, J., Zhu, R., Mirzaei, P., & Mechref, Y. (2016). HILIC and ERLIC enrichment of glycopeptides derived from breast and brain cancer cells.* Journal of Proteome Research, 15, *3624–3634. doi:10.1021/acs.jproteome.6b00429. (B) Reproduced with permission from Song, E., Zhu, R., Hammoud, Z. T., & Mechref, Y. (2014). LC-MS/MS quantitation of esophagus disease blood serum glycoproteins by enrichment with hydrazide chemistry and lectin affinity chromatography.* Journal of Proteome Research, 13*(11), 4808–4820. doi:10.1021/pr500570m.*

Glycopeptide separation remains a challenge because of the intrinsic dichotomy between the hydrophobic component (peptide backbone) and the hydrophilic component (glycan). To address this, a wide range of chromatography approaches have been employed in proteomics and glycoproteomics studies, including reversed-phase liquid chromatography (RPLC), HILIC, and porous graphitized carbon LC (PGC-LC). Despite all of the improvements in sample separation, the separation of glycopeptides remains a challenge because of the isomeric heterogeneity of the glycan moieties.

4.1 Reversed-Phase Liquid Chromatography

RPLC utilizes a nonpolar stationary phase and a polar mobile phase. In the past several decades, RPLC with a C18 stationary phase has been widely used in peptides separation because of the relatively high hydrophobicity of peptides (Zhang et al., 2013). Meanwhile, large-scale glycopeptide separations have relied almost exclusively on C18-RPLC (Anonsen, Vik, Egge-Jacobsen, & Koomey, 2012; Halim, Nilsson, Ruetschi, Hesse, & Larson, 2012; Halim, Ruetschi, Larson, & Nilsson, 2013; Yin et al., 2013). Since RPLC is the standard separation system employed by proteomics laboratories, its compatibility with other lines of proteomic research permits the wide use of RPLC in glycopeptide separation (Thaysen-Andersen & Packer, 2014). The retention of glycopeptides in RPLC is directly related to the hydrophobicity of the molecules. In a glycopeptide molecule, the glycan moiety contributes hydrophilicity, while the peptide backbone contributes hydrophobicity. Therefore, the hydrophobicity of a glycopeptide is generally lower than its peptide backbone. As a result, those hydrophilic glycopeptides containing relatively larger glycan moieties and smaller peptide moieties might not be retained by C18. Moreover, the hydrophobicity variation of glycan isomers is very limited. Therefore, RPLC offers poor glycan isomer differentiation (Hua et al., 2012). This limits the use of RPLC in the glycoproteomics analysis.

4.2 Hydrophilic Interaction Liquid Chromatography

HILIC allows analytes to interact with hydrophilic stationary phases such as bare silica particles, amine-, hydroxy-, amide-bonded, or zwitterionic (ZIC-HILIC) particles by applying a relative nonpolar mobile phase (Alpert, 1990). The capability of separating glycopeptides using HILIC has been shown in several studies (Zauner, Deelder, & Wuhrer, 2011;

Zauner, Koeleman, Deelder, & Wuhrer, 2010). Since the mobile phase of HILIC contains high organic solvents, which might lead to solubility problems of glycopeptide mixture, the use of HILIC in large-scale glycopeptide separation is limited. On the other hand, HILIC has been proved as the most efficient enrichment method for glycopeptides as abovementioned (Mysling, Palmisano, Hojrup, & Thaysen-Andersen, 2010).

4.3 Porous Graphitized Carbon-LC

PGC was developed as a very insoluble and stable support for HPLC in the 1980s. PGC-LC was first used in the early 1990s to separate glycans isomers (Davies et al., 1993). The combined retention mechanism allows PGC-LC to become one of the major methods to achieve structural isomeric separation of detached glycans (Wuhrer, Deelder, & Hokke, 2005). As an alternative separation technique, PGC-LC has also been applied to glycopeptide analysis (An, Peavy, Hedrick, & Lebrilla, 2003; Hua et al., 2012; Nwosu et al., 2013); however, the retention of glycopeptides on PGC might be too strong with vast and hydrophobic peptide moieties especially those glycopeptides with highly sialylated glycan moieties (Alley, Mechref, & Novotny, 2009b; Thaysen-Andersen, Wilkinson, Payne, & Packer, 2011). In order to overcome this issue, nonspecific proteases are used to generate glycopeptides with small peptide moieties in several works (An et al., 2003; Froehlich et al., 2011; Hua et al., 2013; Stavenhagen, Plomp, & Wuhrer, 2015). Nonspecific proteases such as pronase digest glycoproteins into glycopeptides with small peptide backbone (<4–6 amino acids). These glycopeptides have better chance to elute from PGC-LC and generate more useful MS/MS spectra than those with large peptide backbones. On the other hand, since the nonspecific protease will generate multiple peptide backbone of one glycosylation site simultaneously, this quantitative application of this strategy is limited. In addition, compared to the separation of native glycans isomeric selectivity of glycopeptides was also compromised even with a small peptide backbone (Hua et al., 2013). The abovementioned commonly used HPLC separation techniques are limited to its natural disadvantages in separating glycopeptides. Therefore, the combination of these techniques is applied to glycomic studies (Lam et al., 2011; Liu et al., 2014; Zhao et al., 2014). Most of the works have been performed on a PGC and C18-RPLC platform (Fig. 4; Zhao, Song, Zhu, & Mechref, 2016). The C18-RPLC is majorly used for glycopeptide separation with the PGC component used for purification (Liu et al., 2014) or a secondary dimension of

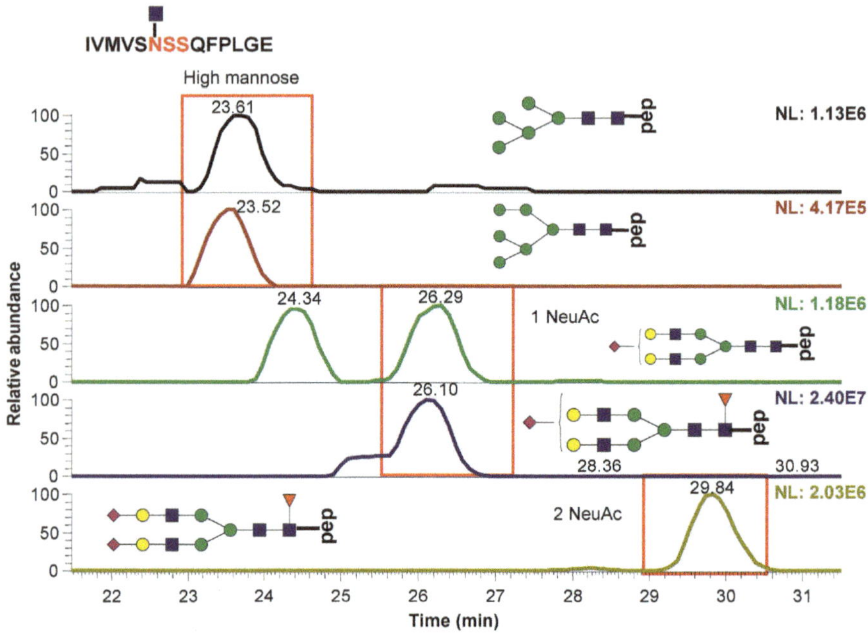

Fig. 4 Separation of glycopeptides derived from porcine thyroglobulin (PTG) with C18 analytical column. The same peptide backbone is associated with multiple glycan residues highlighting the microheterogeneity at this glycosylation site. *Reproduced with permission from Zhao, J., Song, E., Zhu, R., & Mechref, Y. (2016). Parallel data acquisition of in-source fragmented glycopeptides to sequence the glycosylation sites of proteins.* Electrophoresis, 37*(11), 1420–1430. doi:10.1002/elps.201500562.*

separation (Stavenhagen et al., 2015; Zhao et al., 2014). The combination of different stationary phases with complementary retention mechanisms has been proved to advance glycopeptide analysis.

5. ION MOBILITY-MASS SPECTROMETRY

Ion mobility-mass spectrometry (IM-MS) has gained the attention of numerous researchers because of its unique ability to separate ions based on size, shape, or charge in addition to being coupled to a high mass accuracy time-of-flight mass analyzer. There are several different types of ion mobility that have been utilized for glycopeptide analysis. Traveling wave ion mobility-mass spectrometry (TWIMS-MS) (Giles et al., 2004), the most popular commercially available IM-MS instrument, has been used to thoroughly characterize an IgG1 mAb's glycosylation heterogeneity profile with glycopeptides of each glycoform being readily identified (Olivova, Chen, Chakraborty, & Gebler, 2008). Li et al. found that TWIMS could separate peptides from glycopeptides into distinct trend lines that could be used for predicting glycosylation status of other peptides; they also noted that ion mobility reduced chemical noise to allow for the detection of lower abundant ions (Li, Bendiak, Siems, Gang, & Hill, 2013). TWIMS-MS has also been used to distinguish epimeric glycopeptides derived from Muc 2 (Both et al., 2014; Fig. 5A). Two isomeric glycopeptides, same peptide backbone but differing only in the attachment of either α-GlcNAc or α-GalNAc, were partially separated with TWIMS-MS; interestingly, multiple overlapping conformers were identified for each glycopeptide. The authors confirmed the identity of each glycopeptide by using CID-IMS-MS (Fig. 5B); the diagnostic oxonium ions were separated with TWIMS-MS, highlighting the utility and importance of characterizing product ions.

High-field asymmetric wave ion mobility spectrometry (FAIMS) has also been used to rapidly separate coeluting isomeric O-linked glycopeptides, differing only in glycosylation site (Fig. 5C). The coeluting peptides were separated by altering the correction voltage applied. Confirmation of the glycosylation sites was confirmed with electron transfer dissociation (ETD) (Fig. 5D and E); the two glycosylation sites are readily identified by their diagnostic product ions (Creese & Cooper, 2012). The combination of FAIMS to separate isobaric peptides and ETD to confirm the glycosylation site has proven to be quite useful in analyzing complex glycopeptide mixtures.

6. CAPILLARY ELECTROPHORESIS

Capillary electrophoresis (CE) has long been employed for the rapid and reproducible separation and analysis of fluorescently labeled glycans and intact glycoproteins. CE has been employed for the rapid separation of glycopeptides before analysis by mass spectrometry. Gimenez, Ramos-Hernan, Benavente, Barbosa, and Sanz-Nebot (2012) developed a method utilizing CE coupled to ESI-orthogonal accelerating time-of-flight mass spectrometry (CE-ESI-oaTOF-MS) to rapidly analyze *N*- and *O*-glycopeptides derived from erythropoietin (EPO), a commonly used glycoprotein hormone abused in athletic competitions. Their method also discovered a novel sulfated sialoform glycopeptide.

Lew et al. developed a rapid method for therapeutic antibody quality control utilizing CE-MS to analyze the tryptic digest of Trastuzumab; the

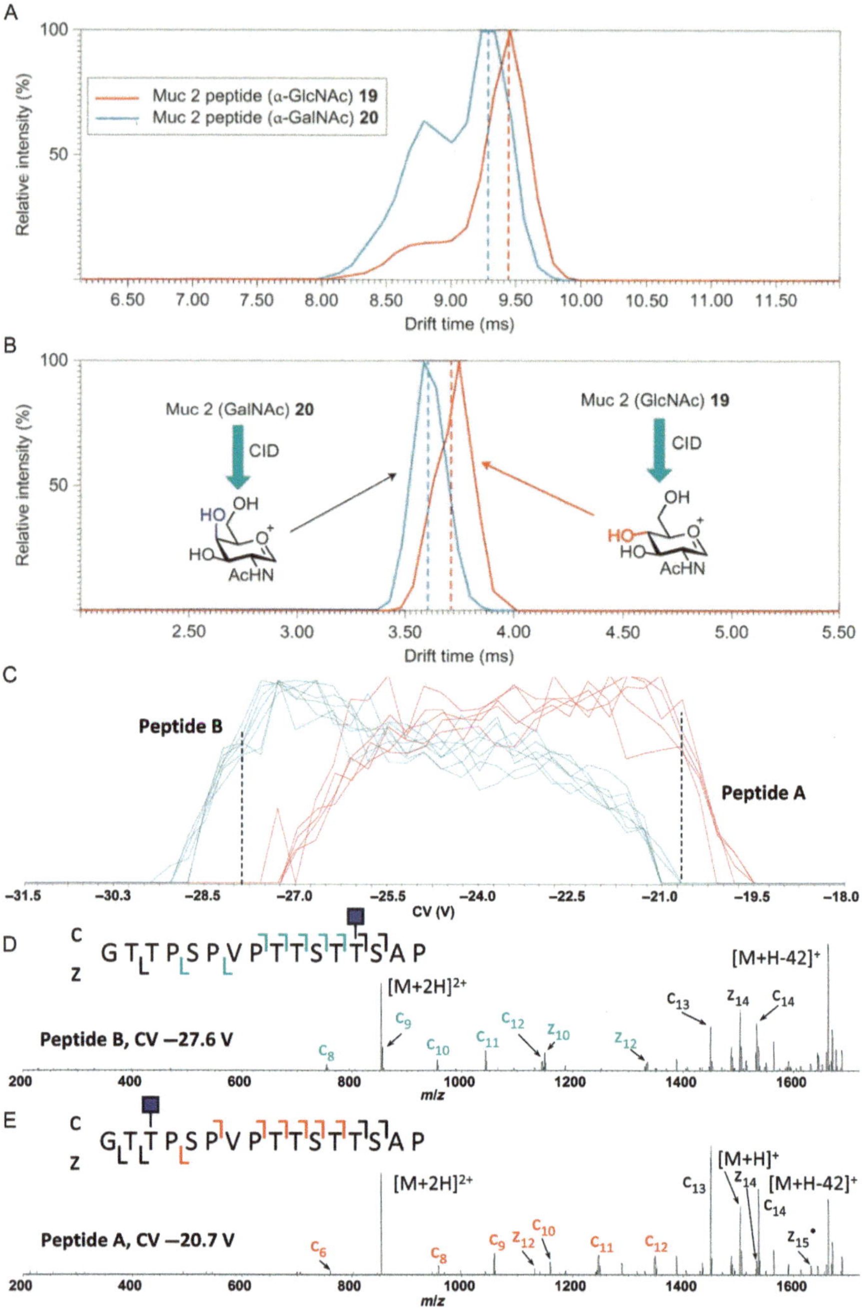

Fig. 5 Separation and analysis of glycopeptides by ion mobility-mass spectrometry. (A) traveling wave ion mobility mass spectrometry (TWIMS) identified multiple conformers of the isobaric Muc2 glycopeptides (PTTTPITTTTTVTPTPTPTGTQT with GalNAc 19 and GlcNac 20). TWIMS was not able to differentiate between the two intact glycopeptides; (B) however, after CID, TWIMS-MS could distinguish the diagnostic oxonium ions from each Mucin glycopeptide. (C) High-field asymmetric wave ion mobility mass spectrometry (FAIMS-MS) separation of two isobaric O-linked glycopeptides, differing only in glycan site attachment. (D, E) Glycopeptide identity confirmed with ETD MS^2; diagnostic c and z ions are indicated in *blue* and *red*.

relative abundance of 14-glycoforms on one glycopeptide were readily attained (Lew, Gallegos-Perez, Fonslow, Lies, & Guttman, 2015; Fig. 6).

Barroso et al. coupled CE-TOF-MS and μLC-TOF-MS for a clinical research application; their method was able to detect the relative glycoform abundance of human Transferrin Tf glycopeptide to predict the alcohol dependence of individuals utilizing a small serum sample (Barroso, Gimenez, Benavente, Barbosa, & Sanz-Nebot, 2013).

7. CHARACTERIZATION OF INTACT GLYCOPEPTIDES

In recent years, proteomic and glycoproteomic analysis has benefited from the technological advances of new mass spectrometry techniques. The identification of glycopeptides heavily relies on tandem MS (MS/MS). Informative and complementary fragments created by various fragmentation methods permit the confident interpretation of intact glycopeptides. Three useful tandem MS (MS/MS) techniques are widely used, namely collision-induced dissociation (CID), higher energy collision dissociation (HCD), and ETD.

7.1 Collision-Induced Dissociation

CID-MS/MS is the most common and robust fragmentation method used in peptide analysis. It has also been widely used in glycopeptide analysis. During CID process, ions are transmitted through lenses and ion optics from electrospray ion source. In linear ion trap instruments, ions are trapped, isolated, fragmented through resonant-excitation, and then scanned by electron multipliers. Since glycosidic linkages of glycoforms are more labile than amide bonds of peptide backbone, a typical CID-MS/MS spectrum of glycopeptides is dominated by glycan fragmentation patterns with no or less peptide backbone sequence information (Huddleston, Bean, & Carr, 1993; Mayampurath, Wu, Segu, Mechref, & Tang, 2011). The resulting fragments of glycopeptides are named as B- and Y-type ions (Domon & Costello, 1988).

7.2 Higher Energy Collision Dissociation

HCD is a beam-type fragmentation has been demonstrated on LIT instruments and on hybrid instruments with dedicated collision cells (Bereman et al., 2012; Olsen et al., 2007; Ting et al., 2009). HCD yields informative low mass fragment of glycopeptides. Specifically, it creates fragmentation

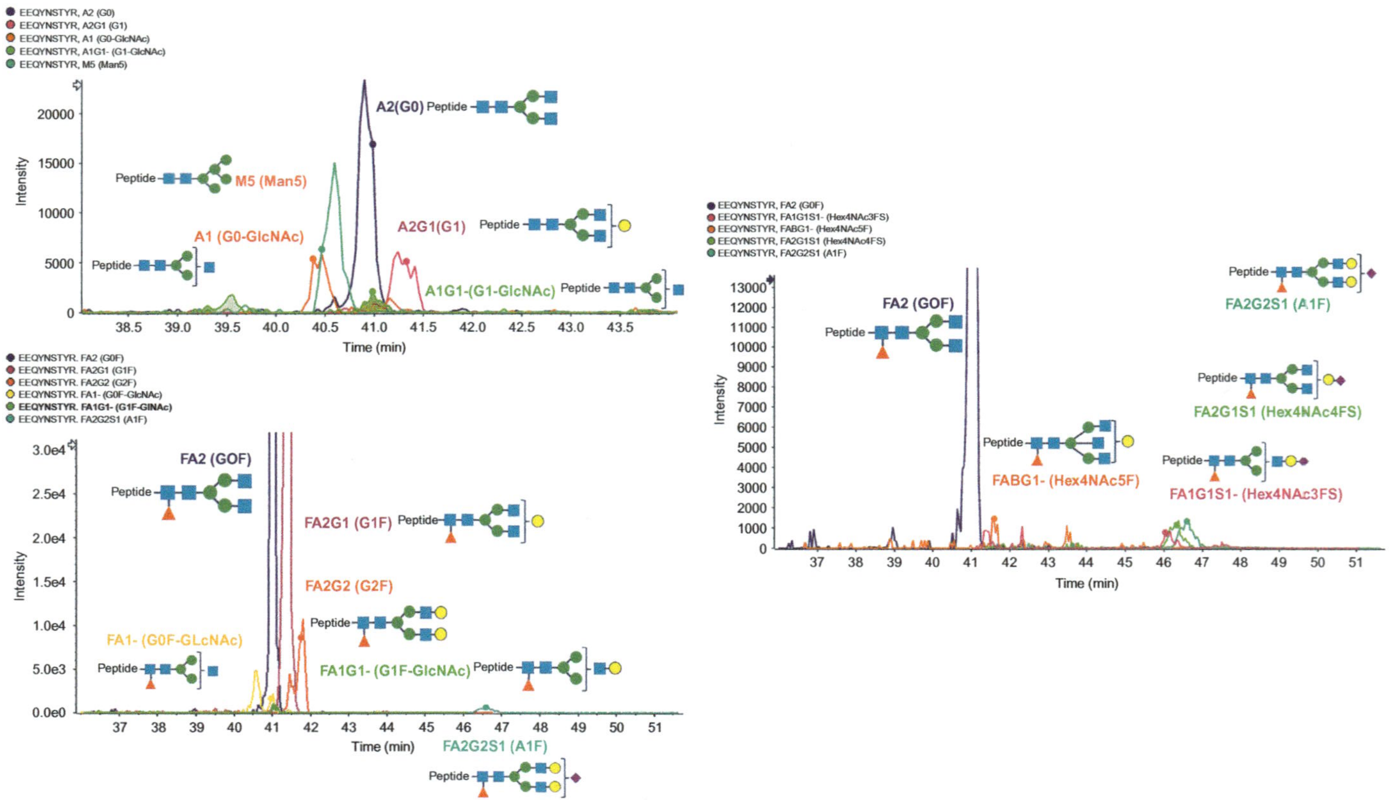

Fig. 6 Capillary electrophoresis mass spectrometry (CE-MS) of therapeutic antibody. CE-MS quantified 14-glycoforms on one glycopeptide from Trastuzumab. *Reproduced with permission from Lew, C., Gallegos-Perez, J. L., Fonslow, B., Lies, M., & Guttman, A. (2015). Rapid level-3 characterization of therapeutic antibodies by capillary electrophoresis electrospray ionization mass spectrometry.* Journal of Chromatographic Science, 53*(3), 443–449. doi:10.1093/chromsci/bmu229.*

pattern of peptides or diagnostic ions of glycan moieties with high intensities at low *m/z*. The diagnostic ions are also referred as oxonium ions (diagnostic ions), such as *m/z* 138.0551 (HexNAc-2H_2O-CH_2O), 204.0875 (HexNAc), 274.0976 (NeuAc-H_2O), 366.1395 (Hex + HexNAc), 512.1977 (Hex + HexNAc + dHex), and 657.2353 (NeuAc + Hex + HexNAc) (Mayampurath et al., 2011; Segu, Hammad, & Mechref, 2010). The appearance of oxonium ions usually indicates the presence of a glycopeptide. The effective characterization of glycopeptides can be performed with the use of oxonium ions in HCD and the glycan fragmentations in CID.

7.3 Electron Transfer Dissociation

Alternatively, ETD, as a relatively new invented fragmentation scheme can also be applied to glycopeptide analysis. ETD induces random fragmentation along the peptide backbone of a glycopeptide creating mostly c- and z-type ions (Stephenson & McLuckey, 1998; Syka, Coon, Schroeder, Shabanowitz, & Hunt, 2004). ETD has been recently employed to better identify the glycosylation sites of peptides. However, the major drawback of ETD is that the efficiency is largely dependent on *m/z* values of glycopeptides, thus limiting its use in large-scale glycoproteomic analysis (Alley, Mechref, & Novotny, 2009a; Wu, Huhmer, Hao, & Karger, 2007).

8. CHARACTERIZATION OF INTACT GLYCOPROTEINS

The previous sections all employed a bottom-up proteomics approach involving a protease, usually trypsin, to reduce the large intact protein into smaller peptides before LC-MS/MS analysis. Top-down mass spectrometry eliminates the protease step resulting in the analysis of intact proteins; proteins are fragmented inside the instrument utilizing different dissociation techniques such as ECD, ETD, and UVPD. Top-down mass spectrometry has several advantages over the conventional bottom-up approach in that protein heterogeneity is preserved; that is different PTM combinations are preserved whereas the bottom-up approach loses that connectivity information. Top-down experiments require high mass accuracy and so are performed exclusively on FT-based mass analyzer such as FTICR and orbitrap instruments. Middle-down mass spectrometry is similar to the bottom-up approach in that proteins are proteolyzed into smaller peptides but differ in peptide size; middle-down peptides are considerably smaller as they employ a more restrictive enzyme such as IdeS.

Tran et al. combined a top-down and middle-down approach to determine the glycosylation profile of IgG and IgG-fusion protein (Tran et al., 2016; Fig. 7). They were able to identify the heterogeneous glycosylation on the Fc portion; glycan heterogeneity matched oligosaccharide profiling. This approach yielded combinatorial information regarding the

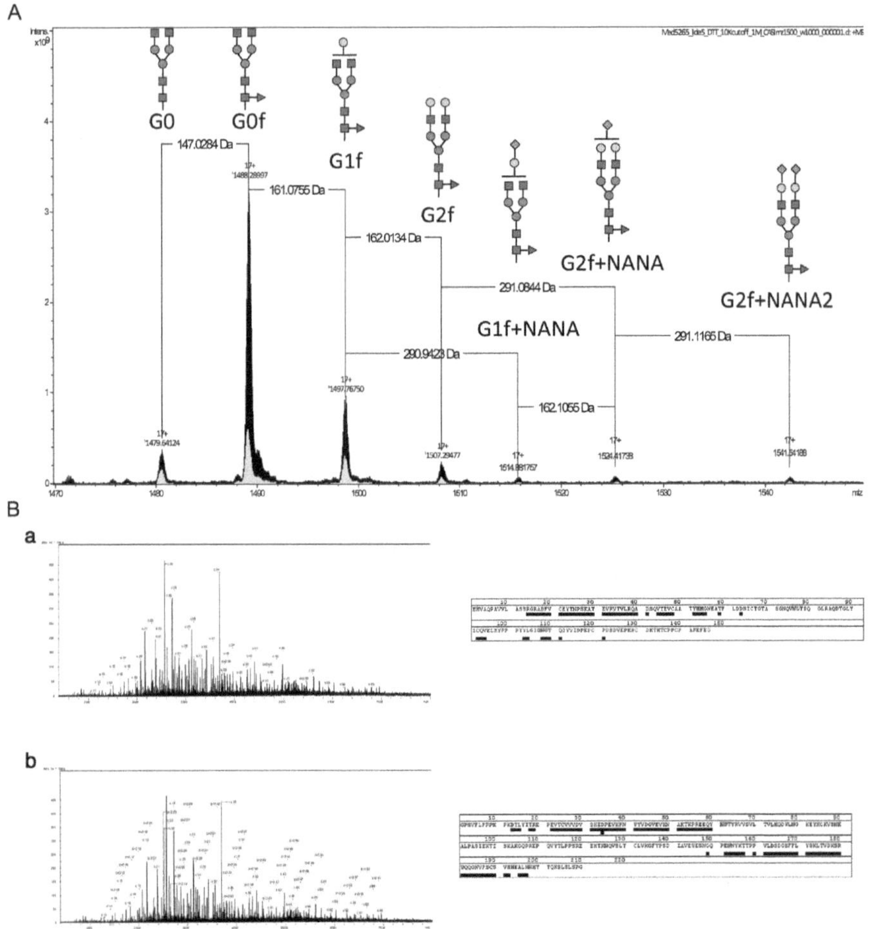

Fig. 7 Combined top-down and middle-down proteomics approach to identifying IgG glycosylation profile. (A) Top-down native mass spectrometry of intact IgG to readily identify glycoforms. (B) Middle-down mass spectrometry to identify peptide backbone and any posttranslational modifications. *Reproduced with permission from Tran, B. Q., Barton, C., Feng, J., Sandjong, A., Yoon, S. H., Awasthi, S., …, Goo, Y. A. (2016). Comprehensive glycosylation profiling of IgG and IgG-fusion proteins by top-down MS with multiple fragmentation techniques.* Journal of Proteomics, 134, *93–101. doi:10.1016/j.jprot.2015.10.021.*

glycosylation sites that could not have been attained with either a bottom-up approach or an oligosaccharide mapping approach.

Nagel et al. employed a top-down approach to rapidly characterize the different sequence polymorphisms (SNPs) as well as assess relative abundances of phosphorylation and glycosylation on fibrinogen in a single experiment; this was done to accurately screen patients for possible cardiovascular risks including stroke (Nagel & Meyer, 2014). They were able to identify primary sequence variants in their clinical samples while also determining the phosphorylation and glycosylation status of each fibrinogen subunit.

Yang et al. employed a native-MS top-down mass spectrometry approach to identify 59 proteoforms of chicken ovalbumin including multiple phosphorylations and 45 different glycans (Yang, Barendregt, Kamerling, & Heck, 2013). This approach revealed native-MS could rapidly identify the heterogeneity in glycoprotein sample. The authors highlight the advantage of little sample consumption but note that this method cannot yield glycan connectivity information, only intact mass measurements.

Currently, most top-down mass spectrometry approaches aimed at characterizing glycoproteins rely on existing glycan databases to assign a glycan structure to a proteoform. Native-MS and top-down mass spectrometry are well situated to assess the macroheterogeneity of glycoproteins but struggle with identifying the sites of glycosylation on proteins.

9. SOFTWARE TOOLS FOR MS/MS-BASED GLYCOPEPTIDE CHARACTERIZATION

The combination of state-of-the-art mass spectrometers and the accompanying bioinformatics tools has allowed proteomics studies to evolve from single protein identification studies to identifying changes in entire proteomes. The data generated from these entire proteome studies would be impractical to interpret manually but manageable with currently available software. Software for bottom-up proteomics experiments is quite robust in determining peptide sequence and consequently protein identity (Hu, Khatri, & Zaia, 2016); however, accurate quantitation is still an open question being addressed by numerous researchers (Blein-Nicolas & Zivy, 2016).

Glycoproteomics differs from routine proteomics studies in that the glycan cannot be treated as a simple PTM, like methylation or phosphorylation, because of the enormous variance of glycan structures. Glycoproteomics studies are generally focused on identifying the different glycans present on glycopeptide, identifying the peptide sequence and consequently the protein's identity, and determining both the glycan and glycoprotein abundance. To further complicate analysis, more than one glycan can be identified with the same peptide sequence; this is referred to as glycan micro-heterogeneity. Currently, several fragmentation techniques are routinely employed to aid in the identification of glycan and peptide. CID is the most common technique for fragmenting peptides in a bottom-up proteomics approach but it does not work well with glycopeptides; CID fragments the glycan while leaving the peptide predominantly intact. This has been overcome by using a combination of CID and ETD in a data-dependent acquisition experiment (Singh, Zampronio, Creese, & Cooper, 2012); CID yields glycan product ions, including the glycopeptide diagnostic oxonium ions, while ETD selectively yields peptide product ions. The data generated from the abovementioned fragmentation techniques can then be analyzed with bioinformatics tools.

The glycoproteomics bioinformatics field is not as well established as the older proteomics field. In addition to the previously mentioned difficulties, glycopeptides with similar glycans tend to coelute. Numerous tools encompassing different strategies have been developed for various glycoproteomic studies to address these issues. The current software tools annotate glycopeptides utilize precursor ion mass and fragmentation patterns matching with known glycopeptides. For example, Peptoonist (Goldberg et al., 2007) combines precursor and fragments information to annotate glycopeptides. On the other hand, GlypID (Wu et al., 2010) utilizes the coelution of *N*-glycopeptide on RPLC, thus combines coeluted *N*-glycopeptide ions from full scans and fragmentation from CID-MS/MS scans to identify putative glycoforms associated with the same glycosylation site. In order to improve the confidence of glycopeptide identification, software tools are combined with multiple LC-MS/MS analysis of the same sample (Toghi Eshghi, Shah, Yang, Li, & Zhang, 2015). In this type of experiment, a portion of glycopeptide is deglycosylated first and analyzed to create an experimental peptide back bone database, while another part of the sample is subjected to LC-MS as intact glycopeptide. The experimental database is further utilized by the software tool to identify glycopeptides confidently.

He et al. developed Glycomaster DB to analyze glycopeptides extracted from the human urinary proteome and fragmented with both HCD and ETD (He, Xin, Shan, Lajoie, & Ma, 2014). They were able to identify multiple N-linked glycan forms on the same glycosylation site but were not able to identify any O-linked glycopeptides. Cheng et al. utilized their own program to identify 2200+ unique *N*-glycopeptides assigned to 1700+ site-specific *N*-glycans on 450+ glycosylation sites in HEK 293T cells; their low false-discovery rate (FDR) indicates a robust identification algorithm for the glycopeptide (Cheng et al., 2014).

Zhu et al. developed GlycoPep Evaluator (GPE) to improve upon existing FDR approaches; GPE generates decoy glycopeptides de novo for every target peptide (Zhu, Su, Go, & Desaire, 2014). Zhang et al. developed for the specific task of identifying O-linked glycosylations in which Gal or Glc-Gal are attached to hydroxylysine residues commonly found in collagen (Zhang et al., 2015); this type of glycopeptide is not commonly identified with most software packages. The authors proceeded to identify more glycopeptides in collagen, including two additional glycosylation sites than what had been previously reported.

GlycoSeq was implemented to determine glycopeptides derived from two breast cancer cell lines (Yu et al., 2016); the software program utilizes a heuristic iterated glycan sequencing algorithm coupled to biological glycosidic linkage information to readily identify glycans. Jansen et al. created LaCytools to improve glycopeptide identification in therapeutic monoclonal antibodies; a focus on quality control measurements including signal to noise, isotope pattern quality, and mass accuracy resulted in a powerful software suite (Jansen et al., 2016). Jansen developed MassyTools to improve glycopeptide data generated from MALDI-MS resulting in the improved calibration and quantitation over existing softwares (Jansen et al., 2015). Park et al. developed a robust software package, Integrated GlycoProteome Analyzer (I-GPA), to analyze and map *N*-glycoproteomes; their method identified 600+ *N*-glycopeptides and quantified ~600 *N*-glycopeptides from human plasma with an astonishing 0% false-positive discovery rate manually verified (Park et al., 2016). Lih et al. developed, MAGIC-web, a web server dedicated to targeted and untargeted N-linked glycan analysis (Lih et al., 2016). Zeng et al. developed pGlyco to identify *N*-glycans from HCD-MS/MS and CID-MS/MS data sets; a novel target decoy method was implemented to estimate FDR of glycan identification (Zeng et al., 2016).

10. CONCLUDING REMARKS AND FUTURE DIRECTIONS

In recent years, significant development and improvement of analytical methods have been implemented to glycoproteomics study.

Glycopeptides enrichment using lectin affinity, hydrazide chemistry, click chemistry, immunoprecipitation, boronic acid chemistry, HILIC, and ERLIC allow for sensitive detection of glycopeptides. LC, CE, and IM-MS improve the separation of different glycopeptides and increase the identification of microheterogeneity and macroheterogeneity of glycoproteins. Multiple mass spectrometry-based dissociation mechanisms aid with the structural elucidation of glycopeptides. In addition to advancements in instrumentation, the development of automated annotation software tools for glycopeptide identification and quantitation has significantly improved the analytical throughput.

Despite all of the recent advancements, various hurdles still remain in the analysis of glycopeptides. A comprehensive enrichment technique unique to all glycopeptides has not been developed resulting in an inherent compromise; HILIC-based separation is able to enrich and allow for the identification of the highest number of glycopeptides, but it is less specific than hydrazide chemistry enrichment. Accurate glycopeptide structural elucidation results from the use of complementary tandem MS mechanisms to yield accurate information about the peptide backbone (ETD), glycan structure (CID), and oxonium ions (HCD). Although modern mass spectrometers are able to perform CID, HCD, and ETD simultaneously, the sensitivity and throughput loss is still considerable. In addition to enrichment and instrumentation short comings, the false-positive discovery rate of automated software tools remains a challenge. There is much room for improvement in developing a comprehensive protocol that will yield accurate and quantitative information on all glycopeptides in samples.

ACKNOWLEDGMENTS

This work was supported by a grant from the Cancer Prevention Institute of Texas (CPRIT, RP130624) and partially by an NIH grant (1R01GM112490).

REFERENCES

Alley, W. R., Jr., Mechref, Y., & Novotny, M. V. (2009a). Characterization of glycopeptides by combining collision-induced dissociation and electron-transfer dissociation mass spectrometry data. *Rapid Communications in Mass Spectrometry*, *23*(1), 161–170. http://dx.doi.org/10.1002/rcm.3850.

Alley, W. R., Jr., Mechref, Y., & Novotny, M. V. (2009b). Use of activated graphitized carbon chips for liquid chromatography/mass spectrometric and tandem mass spectrometric analysis of tryptic glycopeptides. *Rapid Communications in Mass Spectrometry*, *23*(4), 495–505. http://dx.doi.org/10.1002/rcm.3899.

Alpert, A. J. (1990). Hydrophilic-interaction chromatography for the separation of peptides, nucleic acids and other polar compounds. *Journal of Chromatography*, *499*, 177–196. Retrieved from http://www.ncbi.nlm.nih.gov/pubmed/2324207.

An, H. J., Peavy, T. R., Hedrick, J. L., & Lebrilla, C. B. (2003). Determination of N-glycosylation sites and site heterogeneity in glycoproteins. *Analytical Chemistry*, *75*(20), 5628–5637. Retrieved from http://www.ncbi.nlm.nih.gov/pubmed/14710847.

Anonsen, J. H., Vik, A., Egge-Jacobsen, W., & Koomey, M. (2012). An extended spectrum of target proteins and modification sites in the general O-linked protein glycosylation system in Neisseria gonorrhoeae. *Journal of Proteome Research*, *11*(12), 5781–5793. http://dx.doi.org/10.1021/pr300584x.

Apweiler, R., Hermjakob, H., & Sharon, N. (1999). On the frequency of protein glycosylation, as deduced from analysis of the SWISS-PROT database. *Biochimica et Biophysica Acta*, *1473*(1), 4–8. Retrieved from http://www.ncbi.nlm.nih.gov/pubmed/10580125.

Barroso, A., Gimenez, E., Benavente, F., Barbosa, J., & Sanz-Nebot, V. (2013). Analysis of human transferrin glycopeptides by capillary electrophoresis and capillary liquid chromatography-mass spectrometry. Application to diagnosis of alcohol dependence. *Analytica Chimica Acta*, *804*, 167–175. http://dx.doi.org/10.1016/j.aca.2013.09.044.

Bereman, M. S., Canterbury, J. D., Egertson, J. D., Horner, J., Remes, P. M., Schwartz, J., … MacCoss, M. J. (2012). Evaluation of front-end higher energy collision-induced dissociation on a benchtop dual-pressure linear ion trap mass spectrometer for shotgun proteomics. *Analytical Chemistry*, *84*(3), 1533–1539. http://dx.doi.org/10.1021/ac203210a.

Blein-Nicolas, M., & Zivy, M. (2016). Thousand and one ways to quantify and compare protein abundances in label-free bottom-up proteomics. *Biochimica et Biophysica Acta*, *1864*(8), 883–895. http://dx.doi.org/10.1016/j.bbapap.2016.02.019.

Botella-Lopez, A., Burgaya, F., Gavin, R., Garcia-Ayllon, M. S., Gomez-Tortosa, E., Pena-Casanova, J., … Saez-Valero, J. (2006). Reelin expression and glycosylation patterns are altered in Alzheimer's disease. *Proceedings of the National Academy of Sciences of the United States of America*, *103*(14), 5573–5578. http://dx.doi.org/10.1073/pnas.0601279103.

Both, P., Green, A. P., Gray, C. J., Sardzik, R., Voglmeir, J., Fontana, C., … Eyers, C. E. (2014). Discrimination of epimeric glycans and glycopeptides using IM-MS and its potential for carbohydrate sequencing. *Nature Chemistry*, *6*(1), 65–74. http://dx.doi.org/10.1038/nchem.1817.

Buszewski, B., & Noga, S. (2012). Hydrophilic interaction liquid chromatography (HILIC)—A powerful separation technique. *Analytical and Bioanalytical Chemistry*, *402*(1), 231–247. http://dx.doi.org/10.1007/s00216-011-5308-5.

Calvano, C. D., Zambonin, C. G., & Jensen, O. N. (2008). Assessment of lectin and HILIC based enrichment protocols for characterization of serum glycoproteins by mass spectrometry. *Journal of Proteomics*, *71*(3), 304–317. http://dx.doi.org/10.1016/j.jprot.2008.06.013.

Campbell, B. J., Yu, L. G., & Rhodes, J. M. (2001). Altered glycosylation in inflammatory bowel disease: A possible role in cancer development. *Glycoconjugate Journal*, *18*(11-12), 851–858.

Chen, R., Tan, Y., Wang, M., Wang, F., Yao, Z., Dong, L., … Zou, H. (2011). Development of glycoprotein capture-based label-free method for the high-throughput screening of differential glycoproteins in hepatocellular carcinoma. *Molecular & Cellular Proteomics*, *10*(7). http://dx.doi.org/10.1074/mcp.M110.006445. M110.006445.

Chen, J., Xi, J., Tian, Y., Bova, G. S., & Zhang, H. (2013). Identification, prioritization, and evaluation of glycoproteins for aggressive prostate cancer using quantitative glycoproteomics and antibody-based assays on tissue specimens. *Proteomics*, *13*(15), 2268–2277. http://dx.doi.org/10.1002/pmic.201200541.

Cheng, K., Chen, R., Seebun, D., Ye, M., Figeys, D., & Zou, H. (2014). Large-scale characterization of intact N-glycopeptides using an automated glycoproteomic method. *Journal of Proteomics*, *110*, 145–154. http://dx.doi.org/10.1016/j.jprot.2014.08.006.

Creese, A. J., & Cooper, H. J. (2012). Separation and identification of isomeric glycopeptides by high field asymmetric waveform ion mobility spectrometry. *Analytical Chemistry*, *84*(5), 2597–2601. http://dx.doi.org/10.1021/ac203321y.

Dasouki, M., Jawdat, O., Almadhoun, O., Pasnoor, M., McVey, A. L., Abuzinadah, A., … Dimachkie, M. M. (2014). Pompe disease: Literature review and case series. *Neurologic Clinics, 32*(3), 751–776. http://dx.doi.org/10.1016/j.ncl.2014.04.010. ix.

Davies, M. J., Smith, K. D., Carruthers, R. A., Chai, W., Lawson, A. M., & Hounsell, E. F. (1993). Use of a porous graphitised carbon column for the high-performance liquid chromatography of oligosaccharides, alditols and glycopeptides with subsequent mass spectrometry analysis. *Journal of Chromatography, 646*(2), 317–326. Retrieved from http://www.ncbi.nlm.nih.gov/pubmed/8408434.

Domon, B., & Costello, C. (1988). A systematic nomenclature for carbohydrate fragmentations in FAB-MS/MS spectra of glycoconjugates. *Glycoconjugate Journal, 5*(4), 397–409. http://dx.doi.org/10.1007/bf01049915.

Dube, D. H., & Bertozzi, C. R. (2005). Glycans in cancer and inflammation—Potential for therapeutics and diagnostics. *Nature Reviews. Drug Discovery, 4*(6), 477–488. http://dx.doi.org/10.1038/nrd1751.

Elliott, M. A., Elliott, H. G., Gallagher, K., McGuire, J., Field, M., & Smith, K. D. (1997). Investigation into the concanavalin A reactivity, fucosylation and oligosaccharide microheterogeneity of alpha 1-acid glycoprotein expressed in the sera of patients with rheumatoid arthritis. *Journal of Chromatography. B, Biomedical Sciences and Applications, 688*(2), 229–237.

Froehlich, J. W., Barboza, M., Chu, C., Lerno, L. A., Jr., Clowers, B. H., Zivkovic, A. M., … Lebrilla, C. B. (2011). Nano-LC-MS/MS of glycopeptides produced by nonspecific proteolysis enables rapid and extensive site-specific glycosylation determination. *Analytical Chemistry, 83*(14), 5541–5547. http://dx.doi.org/10.1021/ac2003888.

Giles, K., Pringle, S. D., Worthington, K. R., Little, D., Wildgoose, J. L., & Bateman, R. H. (2004). Applications of a travelling wave-based radio-frequency-only stacked ring ion guide. *Rapid Communications in Mass Spectrometry, 18*(20), 2401–2414. http://dx.doi.org/10.1002/rcm.1641.

Gimenez, E., Ramos-Hernan, R., Benavente, F., Barbosa, J., & Sanz-Nebot, V. (2012). Analysis of recombinant human erythropoietin glycopeptides by capillary electrophoresis electrospray-time of flight-mass spectrometry. *Analytica Chimica Acta, 709*, 81–90. http://dx.doi.org/10.1016/j.aca.2011.10.028.

Goldberg, D., Bern, M., Parry, S., Sutton-Smith, M., Panico, M., Morris, H. R., & Dell, A. (2007). Automated N-glycopeptide identification using a combination of single- and tandem-MS. *Journal of Proteome Research, 6*(10), 3995–4005. http://dx.doi.org/10.1021/pr070239f.

Gstaiger, M., & Aebersold, R. (2009). Applying mass spectrometry-based proteomics to genetics, genomics and network biology. *Nature Reviews. Genetics, 10*(9), 617–627. http://dx.doi.org/10.1038/nrg2633.

Halim, A., Nilsson, J., Ruetschi, U., Hesse, C., & Larson, G. (2012). Human urinary glycoproteomics; attachment site specific analysis of N- and O-linked glycosylations by CID and ECD. *Molecular & Cellular Proteomics, 11*(4). http://dx.doi.org/10.1074/mcp.M111.013649. M111 013649.

Halim, A., Ruetschi, U., Larson, G., & Nilsson, J. (2013). LC-MS/MS characterization of O-glycosylation sites and glycan structures of human cerebrospinal fluid glycoproteins. *Journal of Proteome Research, 12*(2), 573–584. http://dx.doi.org/10.1021/pr300963h.

Hao, P., Guo, T., & Sze, S. K. (2011). Simultaneous analysis of proteome, phospho- and glycoproteome of rat kidney tissue with electrostatic repulsion hydrophilic interaction chromatography. *PloS One, 6*(2). e16884, http://dx.doi.org/10.1371/journal.pone.0016884.

He, L., Xin, L., Shan, B., Lajoie, G. A., & Ma, B. (2014). GlycoMaster DB: Software to assist the automated identification of N-linked glycopeptides by tandem mass spectrometry. *Journal of Proteome Research, 13*(9), 3881–3895. http://dx.doi.org/10.1021/pr401115y.

Hein, C. D., Liu, X. M., & Wang, D. (2008). Click chemistry, a powerful tool for pharmaceutical sciences. *Pharmaceutical Research*, *25*(10), 2216–2230. http://dx.doi.org/10.1007/s11095-008-9616-1.

Helenius, A., & Aebi, M. (2001). Intracellular functions of N-linked glycans. *Science*, *291*(5512), 2364–2369. Retrieved from http://www.ncbi.nlm.nih.gov/pubmed/11269317.

Heo, S. H., Lee, S. J., Ryoo, H. M., Park, J. Y., & Cho, J. Y. (2007). Identification of putative serum glycoprotein biomarkers for human lung adenocarcinoma by multilectin affinity chromatography and LC-MS/MS. *Proteomics*, 7(23), 4292–4302. http://dx.doi.org/10.1002/pmic.200700433.

Hongsachart, P., Huang-Liu, R., Sinchaikul, S., Pan, F. M., Phutrakul, S., Chuang, Y. M., … Chen, S. T. (2009). Glycoproteomic analysis of WGA-bound glycoprotein biomarkers in sera from patients with lung adenocarcinoma. *Electrophoresis*, *30*(7), 1206–1220. http://dx.doi.org/10.1002/elps.200800405.

Hu, H., Khatri, K., & Zaia, J. (2016). Algorithms and design strategies towards automated glycoproteomics analysis. *Mass Spectrometry Reviews*, http://dx.doi.org/10.1002/mas.21487 (Epub ahead of print).

Hua, S., Hu, C. Y., Kim, B. J., Totten, S. M., Oh, M. J., Yun, N., … An, H. J. (2013). Glyco-analytical multispecific proteolysis (Glyco-AMP): A simple method for detailed and quantitative Glycoproteomic characterization. *Journal of Proteome Research*, *12*(10), 4414–4423. http://dx.doi.org/10.1021/pr400442y.

Hua, S., Nwosu, C. C., Strum, J. S., Seipert, R. R., An, H. J., Zivkovic, A. M., … Lebrilla, C. B. (2012). Site-specific protein glycosylation analysis with glycan isomer differentiation. *Analytical and Bioanalytical Chemistry*, *403*(5), 1291–1302. http://dx.doi.org/10.1007/s00216-011-5109-x.

Huddleston, M. J., Bean, M. F., & Carr, S. A. (1993). Collisional fragmentation of glycopeptides by electrospray ionization LC/MS and LC/MS/MS: Methods for selective detection of glycopeptides in protein digests. *Analytical Chemistry*, *65*(7), 877–884.

Ishihara, T., Fukuda, I., Morita, A., Takinami, Y., Okamoto, H., Nishimura, S., & Numata, Y. (2011). Development of quantitative plasma N-glycoproteomics using label-free 2-D LC-MALDI MS and its applicability for biomarker discovery in hepatocellular carcinoma. *Journal of Proteomics*, *74*(10), 2159–2168. http://dx.doi.org/10.1016/j.jprot.2011.06.010.

James, P. (1997). Protein identification in the post-genome era: The rapid rise of proteomics. *Quarterly Reviews of Biophysics*, *30*(4), 279–331. Retrieved from http://www.ncbi.nlm.nih.gov/pubmed/9634650.

Jansen, B. C., Falck, D., de Haan, N., Hipgrave Ederveen, A. L., Razdorov, G., Lauc, G., & Wuhrer, M. (2016). LaCyTools: A targeted liquid chromatography-mass spectrometry data processing package for relative quantitation of glycopeptides. *Journal of Proteome Research*, *15*(7), 2198–2210. http://dx.doi.org/10.1021/acs.jproteome.6b00171.

Jansen, B. C., Reiding, K. R., Bondt, A., Hipgrave Ederveen, A. L., Palmblad, M., Falck, D., & Wuhrer, M. (2015). MassyTools: A high-throughput targeted data processing tool for relative quantitation and quality control developed for glycomic and glycoproteomic MALDI-MS. *Journal of Proteome Research*, *14*(12), 5088–5098. http://dx.doi.org/10.1021/acs.jproteome.5b00658.

Keil, B. (1987). Proteolysis data bank: Specificity of alpha-chymotrypsin from computation of protein cleavages. *Protein Sequences & Data Analysis*, *1*(1), 13–20. Retrieved from http://www.ncbi.nlm.nih.gov/pubmed/3447153.

Kolb, H. C., Finn, M. G., & Sharpless, K. B. (2001). Click chemistry: Diverse chemical function from a few good reactions. *Angewandte Chemie (International Ed. in English)*, *40*(11), 2004–2021.

Lam, M. P., Lau, E., Siu, S. O., Ng, D. C., Kong, R. P., Chiu, P. C., … Chu, I. K. (2011). Online combination of reversed-phase/reversed-phase and porous graphitic carbon liquid chromatography for multicomponent separation of proteomics and glycoproteomics samples. *Electrophoresis*, *32*(21), 2930–2940. http://dx.doi.org/10.1002/elps.201100092.

Lew, C., Gallegos-Perez, J. L., Fonslow, B., Lies, M., & Guttman, A. (2015). Rapid level-3 characterization of therapeutic antibodies by capillary electrophoresis electrospray ionization mass spectrometry. *Journal of Chromatographic Science*, *53*(3), 443–449. http://dx.doi.org/10.1093/chromsci/bmu229.

Lewandrowski, U., Moebius, J., Walter, U., & Sickmann, A. (2006). Elucidation of N-glycosylation sites on human platelet proteins: A glycoproteomic approach. *Molecular & Cellular Proteomics*, *5*(2), 226–233. http://dx.doi.org/10.1074/mcp.M500324-MCP200.

Li, H., Bendiak, B., Siems, W. F., Gang, D. R., & Hill, H. H., Jr. (2013). Ion mobility-mass correlation trend line separation of glycoprotein digests without deglycosylation. *International Journal for Ion Mobility Spectrometry*, *16*(2), 105–115. http://dx.doi.org/10.1007/s12127-013-0127-3.

Li, X., Jiang, J., Zhao, X., Wang, J., Han, H., Zhao, Y., … Qian, X. (2013). N-glycoproteome analysis of the secretome of human metastatic hepatocellular carcinoma cell lines combining hydrazide chemistry, HILIC enrichment and mass spectrometry. *PloS One*, *8*(12). e81921. http://dx.doi.org/10.1371/journal.pone.0081921.

Li, Q. K., Shah, P., Li, Y., Aiyetan, P. O., Chen, J., Yung, R., … Zhang, H. (2013). Glycoproteomic analysis of bronchoalveolar lavage (BAL) fluid identifies tumor-associated glycoproteins from lung adenocarcinoma. *Journal of Proteome Research*, *12*(8), 3689–3696.

Lih, T. M., Choong, W. K., Chen, C. C., Cheng, C. W., Lin, H. N., Chen, C. T., … Sung, T. Y. (2016). MAGIC-web: A platform for untargeted and targeted N-linked glycoprotein identification. *Nucleic Acids Research*, *44*(W1), W575–W580. http://dx.doi.org/10.1093/nar/gkw254.

Liu, J., Wang, F., Zhu, J., Mao, J., Liu, Z., Cheng, K., … Zou, H. (2014). Highly efficient N-glycoproteomic sample preparation by combining C(18) and graphitized carbon adsorbents. *Analytical and Bioanalytical Chemistry*, *406*(13), 3103–3109. http://dx.doi.org/10.1007/s00216-014-7716-9.

MacCoss, M. J., McDonald, W. H., Saraf, A., Sadygov, R., Clark, J. M., Tasto, J. J., … Yates, J. R., 3rd. (2002). Shotgun identification of protein modifications from protein complexes and lens tissue. *Proceedings of the National Academy of Sciences of the United States of America*, *99*(12), 7900–7905. http://dx.doi.org/10.1073/pnas.122231399.

Madera, M., Mann, B., Mechref, Y., & Novotny, M. V. (2008). Efficacy of glycoprotein enrichment by microscale lectin affinity chromatography. *Journal of Separation Science*, *31*(14), 2722–2732. http://dx.doi.org/10.1002/jssc.200800094.

Mayampurath, A. M., Wu, Y., Segu, Z. M., Mechref, Y., & Tang, H. (2011). Improving confidence in detection and characterization of protein N-glycosylation sites and microheterogeneity. *Rapid Communications in Mass Spectrometry*, *25*(14), 2007–2019. http://dx.doi.org/10.1002/rcm.5059.

Mechref, Y., Hu, Y., Garcia, A., & Hussein, A. (2012). Identifying cancer biomarkers by mass spectrometry-based glycomics. *Electrophoresis*, *33*(12), 1755–1767. http://dx.doi.org/10.1002/elps.201100715.

Mechref, Y., Madera, M., & Novotny, M. V. (2008). Glycoprotein enrichment through lectin affinity techniques. *Methods in Molecular Biology*, *424*, 373–396. http://dx.doi.org/10.1007/978-1-60327-064-9_29.

Moremen, K. W., Tiemeyer, M., & Nairn, A. V. (2012). Vertebrate protein glycosylation: Diversity, synthesis and function. *Nature Reviews. Molecular Cell Biology*, *13*(7), 448–462. http://dx.doi.org/10.1038/nrm3383.

Mysling, S., Palmisano, G., Hojrup, P., & Thaysen-Andersen, M. (2010). Utilizing ion-pairing hydrophilic interaction chromatography solid phase extraction for efficient glycopeptide enrichment in glycoproteomics. *Analytical Chemistry*, *82*(13), 5598–5609. http://dx.doi.org/10.1021/ac100530w.

Nagel, T., & Meyer, B. (2014). Simultaneous characterization of sequence polymorphisms, glycosylation and phosphorylation of fibrinogen in a direct analysis by LC-MS. *Biochimica et Biophysica Acta*, *1844*(12), 2284–2289. http://dx.doi.org/10.1016/j.bbapap.2014.09.021.

Nilsson, J., Ruetschi, U., Halim, A., Hesse, C., Carlsohn, E., Brinkmalm, G., & Larson, G. (2009). Enrichment of glycopeptides for glycan structure and attachment site identification. *Nature Methods*, *6*(11), 809–811. http://dx.doi.org/10.1038/nmeth.1392.

Nwosu, C. C., Huang, J., Aldredge, D. L., Strum, J. S., Hua, S., Seipert, R. R., & Lebrilla, C. B. (2013). In-gel nonspecific proteolysis for elucidating glycoproteins: A method for targeted protein-specific glycosylation analysis in complex protein mixtures. *Analytical Chemistry*, *85*(2), 956–963. http://dx.doi.org/10.1021/ac302574f.

O'Connor, S. E., & Imperiali, B. (1996). Modulation of protein structure and function by asparagine-linked glycosylation. *Chemistry & Biology*, *3*(10), 803–812. Retrieved from http://www.ncbi.nlm.nih.gov/pubmed/8939697.

Olivova, P., Chen, W., Chakraborty, A. B., & Gebler, J. C. (2008). Determination of N-glycosylation sites and site heterogeneity in a monoclonal antibody by electrospray quadrupole ion-mobility time-of-flight mass spectrometry. *Rapid Communications in Mass Spectrometry*, *22*(1), 29–40. http://dx.doi.org/10.1002/rcm.3330.

Olsen, J. V., Macek, B., Lange, O., Makarov, A., Horning, S., & Mann, M. (2007). Higher-energy C-trap dissociation for peptide modification analysis. *Nature Methods*, *4*(9), 709–712. http://dx.doi.org/10.1038/nmeth1060.

Ongay, S., Boichenko, A., Govorukhina, N., & Bischoff, R. (2012). Glycopeptide enrichment and separation for protein glycosylation analysis. *Journal of Separation Science*, *35*(18), 2341–2372. http://dx.doi.org/10.1002/jssc.201200434.

Park, G. W., Kim, J. Y., Hwang, H., Lee, J. Y., Ahn, Y. H., Lee, H. K., … Yoo, J. S. (2016). Integrated GlycoProteome Analyzer (I-GPA) for automated identification and quantitation of site-specific N-glycosylation. *Scientific Reports*, *6*, 21175. http://dx.doi.org/10.1038/srep21175.

Qin, X., Chen, Q., Sun, C., Wang, C., Peng, Q., Xie, L., … Li, S. (2013). High-throughput screening of tumor metastatic-related differential glycoprotein in hepatocellular carcinoma by iTRAQ combines lectin-related techniques. *Medical Oncology*, *30*(1), 420. http://dx.doi.org/10.1007/s12032-012-0420-8.

Segu, Z. M., Hammad, L. A., & Mechref, Y. (2010). Rapid and efficient glycoprotein identification through microwave-assisted enzymatic digestion. *Rapid Communications in Mass Spectrometry*, *24*(23), 3461–3468. http://dx.doi.org/10.1002/rcm.4774.

Siepen, J. A., Keevil, E. J., Knight, D., & Hubbard, S. J. (2007). Prediction of missed cleavage sites in tryptic peptides aids protein identification in proteomics. *Journal of Proteome Research*, *6*(1), 399–408. http://dx.doi.org/10.1021/pr060507u.

Singh, C., Zampronio, C. G., Creese, A. J., & Cooper, H. J. (2012). Higher energy collision dissociation (HCD) product ion-triggered electron transfer dissociation (ETD) mass spectrometry for the analysis of N-linked glycoproteins. *Journal of Proteome Research*, *11*(9), 4517–4525. http://dx.doi.org/10.1021/pr300257c.

Smith, L. M., Kelleher, N. L., & Consortium for Top Down, P. (2013). Proteoform: A single term describing protein complexity. *Nature Methods*, *10*(3), 186–187. http://dx.doi.org/10.1038/nmeth.2369.

Smith, K. D., Pollacchi, A., Field, M., & Watson, J. (2002). The heterogeneity of the glycosylation of alpha-1-acid glycoprotein between the sera and synovial fluid in rheumatoid arthritis. *Biomedical Chromatography*, *16*(4), 261–266. http://dx.doi.org/10.1002/bmc.158.

Song, E., & Mechref, Y. (2015). Defining glycoprotein cancer biomarkers by MS in conjunction with glycoprotein enrichment. *Biomarkers in Medicine*, *9*(9), 835–844. http://dx.doi.org/10.2217/bmm.15.55.

Song, E., Zhu, R., Hammoud, Z. T., & Mechref, Y. (2014). LC-MS/MS quantitation of esophagus disease blood serum glycoproteins by enrichment with hydrazide chemistry and lectin affinity chromatography. *Journal of Proteome Research*, *13*(11), 4808–4820. http://dx.doi.org/10.1021/pr500570m.

Stavenhagen, K., Plomp, R., & Wuhrer, M. (2015). Site-specific protein N- and O-glycosylation analysis by a C18-porous graphitized carbon-liquid chromatography-electrospray ionization mass spectrometry approach using pronase treated glycopeptides. *Analytical Chemistry*, *87*(23), 11691–11699. http://dx.doi.org/10.1021/acs.analchem.5b02366.

Stephenson, J. L., Jr., & McLuckey, S. A. (1998). Simplification of product ion spectra derived from multiply charged parent ions via ion/ion chemistry. *Analytical Chemistry*, *70*(17), 3533–3544.

Syka, J. E., Coon, J. J., Schroeder, M. J., Shabanowitz, J., & Hunt, D. F. (2004). Peptide and protein sequence analysis by electron transfer dissociation mass spectrometry. *Proceedings of the National Academy of Sciences of the United States of America*, *101*(26), 9528–9533. http://dx.doi.org/10.1073/pnas.0402700101.

Thaysen-Andersen, M., Mysling, S., & Hojrup, P. (2009). Site-specific glycoprofiling of N-linked glycopeptides using MALDI-TOF MS: Strong correlation between signal strength and glycoform quantities. *Analytical Chemistry*, *81*(10), 3933–3943. http://dx.doi.org/10.1021/ac900231w.

Thaysen-Andersen, M., & Packer, N. H. (2014). Advances in LC-MS/MS-based glycoproteomics: Getting closer to system-wide site-specific mapping of the N- and O-glycoproteome. *Biochimica et Biophysica Acta*, *1844*(9), 1437–1452. http://dx.doi.org/10.1016/j.bbapap.2014.05.002.

Thaysen-Andersen, M., Wilkinson, B. L., Payne, R. J., & Packer, N. H. (2011). Site-specific characterisation of densely O-glycosylated mucin-type peptides using electron transfer dissociation ESI-MS/MS. *Electrophoresis*, *32*(24), 3536–3545. http://dx.doi.org/10.1002/elps.201100294.

Ting, L., Cowley, M. J., Hoon, S. L., Guilhaus, M., Raftery, M. J., & Cavicchioli, R. (2009). Normalization and statistical analysis of quantitative proteomics data generated by metabolic labeling. *Molecular & Cellular Proteomics*, *8*(10), 2227–2242. http://dx.doi.org/10.1074/mcp.M800462-MCP200.

Toghi Eshghi, S., Shah, P., Yang, W., Li, X., & Zhang, H. (2015). GPQuest: A spectral library matching algorithm for site-specific assignment of tandem mass spectra to intact N-glycopeptides. *Analytical Chemistry*, *87*(10), 5181–5188. http://dx.doi.org/10.1021/acs.analchem.5b00024.

Tran, B. Q., Barton, C., Feng, J., Sandjong, A., Yoon, S. H., Awasthi, S., … Goo, Y. A. (2016). Comprehensive glycosylation profiling of IgG and IgG-fusion proteins by top-down MS with multiple fragmentation techniques. *Journal of Proteomics*, *134*, 93–101. http://dx.doi.org/10.1016/j.jprot.2015.10.021.

Tsai, H. Y., Boonyapranai, K., Sriyam, S., Yu, C. J., Wu, S. W., Khoo, K. H., … Chen, S. T. (2011). Glycoproteomics analysis to identify a glycoform on haptoglobin associated with lung cancer. *Proteomics*, *11*(11), 2162–2170. http://dx.doi.org/10.1002/pmic.201000319.

Tsai, T. H., Song, E., Zhu, R., Di Poto, C., Wang, M., Luo, Y., … Ressom, H. W. (2015). LC-MS/MS-based serum proteomics for identification of candidate biomarkers for hepatocellular carcinoma. *Proteomics*, *15*(13), 2369–2381. http://dx.doi.org/10.1002/pmic.201400364.

Vandermarliere, E., Mueller, M., & Martens, L. (2013). Getting intimate with trypsin, the leading protease in proteomics. *Mass Spectrometry Reviews*, *32*(6), 453–465. http://dx.doi.org/10.1002/mas.21376.

Varki, A. (1993). Biological roles of oligosaccharides: All of the theories are correct. *Glycobiology*, *3*(2), 97–130. Retrieved from http://www.ncbi.nlm.nih.gov/pubmed/8490246.

Wang, Y., Ao, X., Vuong, H., Konanur, M., Miller, F. R., Goodison, S., & Lubman, D. M. (2008). Membrane glycoproteins associated with breast tumor cell progression identified by a lectin affinity approach. *Journal of Proteome Research*, 7(10), 4313–4325. http://dx.doi.org/10.1021/pr8002547.

Wang, J. Z., Grundke-Iqbal, I., & Iqbal, K. (1996). Glycosylation of microtubule-associated protein tau: An abnormal posttranslational modification in Alzheimer's disease. *Nature Medicine*, *2*(8), 871–875.

Wang, D., Hincapie, M., Rejtar, T., & Karger, B. L. (2011). Ultrasensitive characterization of site-specific glycosylation of affinity-purified haptoglobin from lung cancer patient plasma using 10 μm i.d. porous layer open tubular liquid chromatography-linear ion trap collision-induced dissociation/electron transfer dissociation mass spectrometry. *Analytical Chemistry*, *83*(6), 2029–2037. http://dx.doi.org/10.1021/ac102825g.

Whelan, S. A., Lu, M., He, J., Yan, W., Saxton, R. E., Faull, K. F., … Chang, H. R. (2009). Mass spectrometry (LC-MS/MS) site-mapping of N-glycosylated membrane proteins for breast cancer biomarkers. *Journal of Proteome Research*, *8*(8), 4151–4160. http://dx.doi.org/10.1021/pr900322g.

Wilkins, M. R., Pasquali, C., Appel, R. D., Ou, K., Golaz, O., Sanchez, J. C., … Hochstrasser, D. F. (1996). From proteins to proteomes: Large scale protein identification by two-dimensional electrophoresis and amino acid analysis. *Biotechnology (N.Y.)*, *14*(1), 61–65. Retrieved from http://www.ncbi.nlm.nih.gov/pubmed/9636313.

Wohlgemuth, J., Karas, M., Eichhorn, T., Hendriks, R., & Andrecht, S. (2009). Quantitative site-specific analysis of protein glycosylation by LC-MS using different glycopeptide-enrichment strategies. *Analytical Biochemistry*, *395*(2), 178–188. http://dx.doi.org/10.1016/j.ab.2009.08.023.

Wolters, D. A., Washburn, M. P., & Yates, J. R., 3rd. (2001). An automated multidimensional protein identification technology for shotgun proteomics. *Analytical Chemistry*, *73*(23), 5683–5690. Retrieved from http://www.ncbi.nlm.nih.gov/pubmed/11774908.

Wu, S. L., Huhmer, A. F., Hao, Z., & Karger, B. L. (2007). On-line LC-MS approach combining collision-induced dissociation (CID), electron-transfer dissociation (ETD), and CID of an isolated charge-reduced species for the trace-level characterization of proteins with post-translational modifications. *Journal of Proteome Research*, *6*(11), 4230–4244. http://dx.doi.org/10.1021/pr070313u.

Wu, Y., Mechref, Y., Klouckova, I., Mayampurath, A., Novotny, M. V., & Tang, H. (2010). Mapping site-specific protein N-glycosylations through liquid chromatography/mass spectrometry and targeted tandem mass spectrometry. *Rapid Communications in Mass Spectrometry*, *24*(7), 965–972. http://dx.doi.org/10.1002/rcm.4474.

Wuhrer, M., Deelder, A. M., & Hokke, C. H. (2005). Protein glycosylation analysis by liquid chromatography-mass spectrometry. *Journal of Chromatography. B, Analytical Technologies in the Biomedical and Life Sciences*, *825*(2), 124–133. http://dx.doi.org/10.1016/j.jchromb.2005.01.030.

Xu, T., Wong, C. C., Kashina, A., & Yates, J. R., 3rd. (2009). Identification of N-terminally arginylated proteins and peptides by mass spectrometry. *Nature Protocols*, *4*(3), 325–332. http://dx.doi.org/10.1038/nprot.2008.248.

Yang, Y., Barendregt, A., Kamerling, J. P., & Heck, A. J. (2013). Analyzing protein micro-heterogeneity in chicken ovalbumin by high-resolution native mass spectrometry exposes qualitatively and semi-quantitatively 59 proteoforms. *Analytical Chemistry*, *85*(24), 12037–12045. http://dx.doi.org/10.1021/ac403057y.

Yang, Z., & Hancock, W. S. (2004). Approach to the comprehensive analysis of glycoproteins isolated from human serum using a multi-lectin affinity column. *Journal of*

Chromatography. A, 1053(1-2), 79–88. Retrieved from http://www.ncbi.nlm.nih.gov/pubmed/15543974.

Yang, Z., Harris, L. E., Palmer-Toy, D. E., & Hancock, W. S. (2006). Multilectin affinity chromatography for characterization of multiple glycoprotein biomarker candidates in serum from breast cancer patients. *Clinical Chemistry, 52*(10), 1897–1905. http://dx.doi.org/10.1373/clinchem.2005.065862.

Yang, L., Nyalwidhe, J. O., Guo, S., Drake, R. R., & Semmes, O. J. (2011). Targeted identification of metastasis-associated cell-surface sialoglycoproteins in prostate cancer. *Molecular & Cellular Proteomics, 10*(6). http://dx.doi.org/10.1074/mcp.M110.007294.M110.007294.

Yates, J. R., 3rd. (2004). Mass spectral analysis in proteomics. *Annual Review of Biophysics and Biomolecular Structure, 33*, 297–316. http://dx.doi.org/10.1146/annurev.biophys.33.111502.082538.

Yin, X., Bern, M., Xing, Q., Ho, J., Viner, R., & Mayr, M. (2013). Glycoproteomic analysis of the secretome of human endothelial cells. *Molecular & Cellular Proteomics, 12*(4), 956–978. http://dx.doi.org/10.1074/mcp.M112.024018.

Yu, C. Y., Mayampurath, A., Zhu, R., Zacharias, L., Song, E., Wang, L., … Tang, H. (2016). Automated glycan sequencing from tandem mass spectra of N-linked glycopeptides. *Analytical Chemistry, 88*(11), 5725–5732. http://dx.doi.org/10.1021/acs.analchem.5b04858.

Zacharias, L. G., Hartmann, A. K., Song, E., Zhao, J., Zhu, R., Mirzaei, P., & Mechref, Y. (2016). HILIC and ERLIC enrichment of glycopeptides derived from breast and brain cancer cells. *Journal of Proteome Research, 15*(10), 3624–3634. http://dx.doi.org/10.1021/acs.jproteome.6b00429.

Zauner, G., Deelder, A. M., & Wuhrer, M. (2011). Recent advances in hydrophilic interaction liquid chromatography (HILIC) for structural glycomics. *Electrophoresis, 32*(24), 3456–3466. http://dx.doi.org/10.1002/elps.201100247.

Zauner, G., Koeleman, C. A., Deelder, A. M., & Wuhrer, M. (2010). Protein glycosylation analysis by HILIC-LC-MS of Proteinase K-generated N- and O-glycopeptides. *Journal of Separation Science, 33*(6-7), 903–910. http://dx.doi.org/10.1002/jssc.200900850.

Zeng, X., Hood, B. L., Sun, M., Conrads, T. P., Day, R. S., Weissfeld, J. L., … Bigbee, W. L. (2010). Lung cancer serum biomarker discovery using glycoprotein capture and liquid chromatography mass spectrometry. *Journal of Proteome Research, 9*(12), 6440–6449. http://dx.doi.org/10.1021/pr100696n.

Zeng, W. F., Liu, M. Q., Zhang, Y., Wu, J. Q., Fang, P., Peng, C., … Yang, P. (2016). pGlyco: A pipeline for the identification of intact N-glycopeptides by using HCD- and CID-MS/MS and MS3. *Scientific Reports, 6*, 25102. http://dx.doi.org/10.1038/srep25102.

Zhang, Y., Fonslow, B. R., Shan, B., Baek, M. C., & Yates, J. R., 3rd. (2013). Protein analysis by shotgun/bottom-up proteomics. *Chemical Reviews, 113*(4), 2343–2394. http://dx.doi.org/10.1021/cr3003533.

Zhang, Y., Go, E. P., & Desaire, H. (2008). Maximizing coverage of glycosylation heterogeneity in MALDI-MS analysis of glycoproteins with up to 27 glycosylation sites. *Analytical Chemistry, 80*(9), 3144–3158. http://dx.doi.org/10.1021/ac702081a.

Zhang, H., Guo, T., Li, X., Datta, A., Park, J. E., Yang, J., … Sze, S. K. (2010). Simultaneous characterization of glyco- and phosphoproteomes of mouse brain membrane proteome with electrostatic repulsion hydrophilic interaction chromatography. *Molecular & Cellular Proteomics, 9*(4), 635–647. http://dx.doi.org/10.1074/mcp.M900314-MCP200.

Zhang, H., Li, X. J., Martin, D. B., & Aebersold, R. (2003). Identification and quantification of N-linked glycoproteins using hydrazide chemistry, stable isotope labeling and mass spectrometry. *Nature Biotechnology, 21*(6), 660–666. http://dx.doi.org/10.1038/nbt827.

Zhang, C., Ye, Z., Xue, P., Shu, Q., Zhou, Y., Ji, Y., … Yang, F. (2016). Evaluation of different N-glycopeptide enrichment methods for N-glycosylation sites mapping in mouse brain. *Journal of Proteome Research*, *15*(9), 2960–2968. http://dx.doi.org/10.1021/acs.jproteome.6b00098.

Zhang, Y., Yu, C. Y., Song, E., Li, S. C., Mechref, Y., Tang, H., & Liu, X. (2015). Identification of glycopeptides with multiple hydroxylysine O-glycosylation sites by tandem mass spectrometry. *Journal of Proteome Research*, *14*(12), 5099–5108. http://dx.doi.org/10.1021/acs.jproteome.5b00299.

Zhao, J., Song, E., Zhu, R., & Mechref, Y. (2016). Parallel data acquisition of in-source fragmented glycopeptides to sequence the glycosylation sites of proteins. *Electrophoresis*, *37*(11), 1420–1430. http://dx.doi.org/10.1002/elps.201500562.

Zhao, Y., Szeto, S. S., Kong, R. P., Law, C. H., Li, G., Quan, Q., … Chu, I. K. (2014). Online two-dimensional porous graphitic carbon/reversed phase liquid chromatography platform applied to shotgun proteomics and glycoproteomics. *Analytical Chemistry*, *86*(24), 12172–12179. http://dx.doi.org/10.1021/ac503254t.

Zhu, Z., Su, X., Go, E. P., & Desaire, H. (2014). New glycoproteomics software, GlycoPep evaluator, generates decoy glycopeptides de novo and enables accurate false discovery rate analysis for small data sets. *Analytical Chemistry*, *86*(18), 9212–9219. http://dx.doi.org/10.1021/ac502176n.

CHAPTER TWENTY-TWO

Quantitative Glycomics: A Combined Analytical and Bioinformatics Approach

L. Veillon, S. Zhou, Y. Mechref[1]
Texas Tech University, Lubbock, TX, United States
[1]Corresponding author: e-mail address: yehia.mechref@ttu.edu

Contents

Abstract

Glycosylation is one of the most common and essential protein modifications. Glycans conjugated to biomolecules modulate the function of such molecules through both direct recognition of glycan structures and indirect mechanisms that involve the control of protein turnover rates, stability, and conformation. The biological attributes of glycans in numerous biological processes and implications in a number of diseases highlight the necessity for comprehensive characterization of protein glycosylation. This chapter reviews cutting-edge methods and tools developed to facilitate quantitative glycomics. This chapter highlights the different methods employed for the release and purification of glycans from biological samples. The most effective labeling methods developed for sensitive quantitative glycomics are also described and discussed. The chromatographic approaches that have been used effectively in glycomics are also highlighted.

Methods in Enzymology, Volume 585
ISSN 0076-6879
http://dx.doi.org/10.1016/bs.mie.2016.11.006

1. PROTEIN GLYCOSYLATION

Glycosylation is one of the most common and essential protein modifications. More than half of proteins are glycosylated (Apweiler, Hermjakob, & Sharon, 1999). Glycans conjugated to biomolecules modulate the function of such molecules through both direct recognition of glycan structures and indirect mechanisms that involve the control of protein turnover rates, stability, and conformation. The biological attributes of glycans in numerous biological processes and implications in a number of diseases highlight the necessity for comprehensive characterization of protein glycosylation.

1.1 Definition of Glycosylation

Proteins are a biomolecule biosynthesized utilizing 20 natural amino acids. Shortly after or during the biosynthesis of proteins, several modifications may occur to amino acid residues. These modifications are referred to as posttranslational modification (PTM). PTM chemically or enzymatically alters specific amino acid residues, which modify the properties and functions of proteins as a consequence. Among more than 200 types of PTMs, acetylation, methylation, phosphorylation, ubiquitination, lipidation, proteolysis, and glycosylation occur most commonly. Glycosylation refers to modifications that commonly add carbohydrate molecules to asparagine and serine or threonine in the polypeptide backbone of proteins. This process is distinct from other common PTMs because of two facts, first is the molecular weight shift by glycosylation is much larger (usually greater than 800 Da) than other PTMs (phosphorylation adds 80 Da, acetylation adds 42 Da, and methylation adds 14 Da to the modified amino acid) and the second is the uncertainty of the added carbohydrate structure, since this process is a template-free process. This structural heterogeneity contributes to the diverse biofunctions of glycoconjugates, while simultaneously increasing the difficulty of achieving a comprehensive characterization of such protein modification.

1.2 Classification of Protein Glycosylation

Based on the occurrence site and core structure, glycosylation can be classified as N-linked glycosylation or O-linked glycosylation. N-linked glycosylation occurs on asparagine in a subsequence of Asn-Xxx-Ser/Thr, where

Xxx refers to any amino acid except proline. There are several steps in the biosynthesis of N-glycans. N-glycosylation is initiated on the cytosolic surface of the endoplasmic reticulum (ER) membrane, where a lipid precursor acts as the carrier for immature glycans (Kornfeld & Kornfeld, 1985). In this process, glycans composed of two *N*-acetylglucosamine and five mannose residues are synthesized and anchored to the lipid precursor. After which, the oligosaccharide is flipped to the luminal side of the membrane, where four mannose and three glucose units are added to the glycan structure. The formation of this structure indicates that the glycan is ready to be transferred from the lipid precursor to the asparagine on the polypeptide chain by oligosaccharyltransferase (Lizak, Gerber, Numao, Aebi, & Locher, 2011). Terminal glucose and mannose are subsequently trimmed in the ER (Helenius & Aebi, 2004). When the glycoprotein reaches the Golgi complex, glycans are remodeled into diverse structures by different glycosidases and glycosyltransferases, while the core structure of the N-glycan that consists of two *N*-acetylglucosamine and three mannose units is always kept untouched (Rabouille et al., 1995; Stanley, 2011). Monosaccharides that commonly compose N-glycans are *N*-acetylglucosamine, *N*-acetylgalactosamine, mannose, galactose, fucose, and neuraminic acids. The biosynthesis of O-linked glycans is less complex than that of N-glycans. Synthesis starts by direct modification of polypeptide chain Ser/Thr residues in the Golgi apparatus (Clausen & Bennett, 1996; Ten Hagen, Fritz, & Tabak, 2003). However, O-glycan structures are also highly differentiated because of trimming and addition by various glycosidases and glycosyltransferases (Gill, Clausen, & Bard, 2011).

N-glycans can be classified into high mannose, complex, and hybrid types according to their structures. High mannose refers to glycans that only have mannose residues beyond the core structure, while the complex type has other monosaccharides. Hybrid glycans have biantennary structures with characteristics of high mannose and complex type on each branch. Owing to the unique biological functions of fucose and sialic acid, glycans can also be classified as fucosylated and sialylated types when they contain corresponding monosaccharides.

1.3 Glycan Biofunctions and Relationships to Diseases

Glycosylation is a type of PTM that adds significant mass to the polypeptide backbone, sometimes glycans cover a major portion of the protein surface. Hence, it is not surprising that glycosylation has significant impacts on protein structure and properties. The functions of protein glycosylation can be observed in a variety of ways. First, glycans may interfere with protein folding (Freeze, Esko, & Parodi, 2009; Molinari, 2007); during the protein folding process, glycans act as quality control and checkpoints for protein folding and determine whether the protein can be folded in the luminal side of the ER or be degraded in the cytosolic side (Helenius & Aebi, 2004; Lederkremer, 2009). Second, glycans influence the stability and solubility of folded proteins; glycans play roles in protecting polypeptides from proteolysis by hindering the accessibility of proteases to peptide bonds (Ellgaard, Molinari, & Helenius, 1999; Means & Desrosiers, 2000; Trombetta & Helenius, 1998). Moreover, glycans mediate protein–protein interactions. Glycans with certain structures may also act as binding targets for pathogens and microbes.

Although only 1–2% of the genome is responsible for encoding glycan-related enzymes (Lowe & Marth, 2003), glycan patterns and abundance are altered significantly in various human diseases such as skin diseases (Vabres et al., 1998), diabetes (Miura & Endo, 2016), rheumatoid arthritis (Gornik & Lauc, 2008), liver diseases (Mehta, Herrera, & Block, 2015), and numerous cancers (Cheresh, Reisfeld, & Varki, 1984; Dall'Olio & Chiricolo, 2001; Kohla, Stockfleth, & Schauer, 2002; Laubli, Stevenson, Varki, Varki, & Borsig, 2006; Seales et al., 2005; Stevenson, Choi, & Varki, 2005; Varki & Varki, 2002). Glycans have been found to play roles in cell recognition and adhesion (Bucior, Scheuring, Engel, & Burger, 2004), these capabilities are responsible for cancer progression and migration. Various studies have been conducted to investigate the relationships between glycan expression and cancer progression, and several structures of glycans have been identified as putative biomarkers for tumor development (Varki, Kannagi, & Toole, 2009).

The importance of obtaining a comprehensive understanding of protein glycosylation can be concluded from the examples previously discussed. Hence, developing sensitive and reliable strategies for glycomic analysis are decidedly necessary. In the following sections, different glycomic platforms will be reviewed and compared.

2. SAMPLE HANDLING IN GLYCOMICS

Glycans that are linked to proteins are generally oriented such that they either extend outward from the protein surface or are embraced inside of the folded protein. Regardless of their placement on the protein molecule, sample treatments are always required for their analysis. In the field of glycomics, glycans are typically released from heat-denatured proteins or peptides prior to analysis. In the following sections, a comprehensive list of the methods utilized to release glycans is provided.

2.1 Glycan Release

Although glycans are analyzed together with the peptide backbone in glycoproteomics, studies to acquire site-specific glycosylation information are not common, largely because they require sophisticated tandem mass spectrometers and complex data processing efforts. Releasing glycans from glycoproteins benefits their analysis by increasing sensitivity and providing simplicity. Glycan-releasing methods differ between N-linked and O-linked glycans, and can be classified into chemical- and enzymatic-based reactions.

The most commonly utilized N-glycan releasing approach is cleavage of glycans from asparagine by enzymatic digestion with peptide *N*-glycosidase F (PNGase F), which can be isolated from *Flavobacterium meningosepticum* (Plummer & Tarentino, 1991). PNGase F is capable of selectively and efficiently releasing all N-linked glycans from peptide backbones except for those containing core α1–3-linked fucose, which usually do not exist in mammalian glycans. In the case of plants and insects, PNGase A should be employed to release glycans with core α1–3-linked fucose (Takahashi & Nishibe, 1978; Tretter, Altmann, & Marz, 1991). However, PNGase A is not sensitive to sialylated glycans. After PNGase F treatment, asparagine in the peptide backbone is converted into aspartic acid, making identification of N-glycosylation sites possible. Additional enzymes, such as *endo*-F1, -F2, -F3, and -H, are also capable of releasing N-glycans (Trimble & Tarentino, 1991). Unlike PNGase F, the *endo*-F series, and *endo*-H only cleave N-glycans containing particular structures and leave one *N*-acetylglucosamine on the peptide backbone. These specific N-glycan-releasing enzymes assist addressing specific scientific questions, such as selecting or depleting high-mannose glycans in complex samples. N-glycans can also be released by chemical reactions; however, chemical cleavage relies on high temperature and pH, which tends to prompt glycan cleavages and formation of by-products.

There are no universal and efficient enzymes for releasing O-linked glycans. For example, *endo*-α-*N*-acetylgalactosaminidase only cleaves O-glycans with a core one structure (Huang & Aminoff, 1972). Therefore, in the case of O-glycans, chemical release methods are more efficient.

Reductive β-elimination is the most common method of releasing O-glycans (Carlson, 1968). This reaction can be achieved by mixing a weak base (e.g., ammonium hydroxide) with an O-glycoprotein prior to an overnight incubation at 60°C in a water bath. However, side reactions during this procedure are inevitable, and site information is lost. Several studies have sought to optimize the reaction conditions for β-elimination. For example, in one recent study, pyrazolone analogs (BEP) were utilized, and the reaction was accelerated by microwave radiation (Furukawa et al., 2015). BEP improved reaction selectivity and reduced the frequency of peeling reactions commonly observed in conventional β-elimination methods. The assistance of a microwave also allowed reaction time to be reduced from 16 to 2 h. The different methods available for releasing glycans are summarized in Table 1.

It is worth mentioning that although most glycosidase enzymes work with intact native protein, enzymatic digestion efficiency is much lower compared to protocols in which glycoproteins are denatured prior to digestion. For example, the effectiveness of PNGase F digestion is 100 times higher when denaturation is conducted to pretreat protein samples. The benefits of effective protein denaturation include accelerating enzymatic digestion, reducing enzyme consumption, and ensuring complete release of all glycans, especially when the analyte is a complex biological sample such as serum or tissue. Thermal denaturation and detergent denaturation are two commonly utilized strategies. Although detergents improve protein extraction from complex biological samples and ensure the disruption of 3D protein structures, most detergents are not MS friendly and are difficult to remove. Hence, thermal denaturation or MS-compatible detergents such as RapiGest™ are more widely used in MS-based glycomics.

2.2 Glycan Purification

After enzymatic digestion, glycans are in solutions containing a mix of proteins, peptides, salts, and possibly detergents. Therefore, purification is necessary to eliminate inference from the sample matrix in the derivatization and analytical procedures that follow. Protein precipitation is commonly utilized to remove proteins from sample mixtures; high concentration methanol, ethanol, or aqueous acetone solutions are added to samples and incubation at −80°C follows, the precipitated proteins can then be isolated by high-speed centrifugation. Dialysis or filtration methods, based on the molecular weight difference between glycans and their contaminants, can be used to separate glycans from buffer salts or detergents.

Table 1 Release of Glycans From Glycoproteins Is Attained Through Different Methods

N-glycan	
Chemical Release	
Using anhydrous hydrazine at 95°C	Takasaki, Mizuochi, and Kobata (1982)
Oxidation using sodium hypochlorite	Song et al. (2016)
Enzymatic Release	
PNGase F purified from *Flavobacterium meningosepticum*	Takahashi and Nishibe (1978) and Tarentino, Gomez, and Plummer (1985)
PNGase A purified from almond *Prunus amygdalus* var. *dulcis*	Takahashi and Nishibe (1978) and Tretter et al. (1991)
Endoglycosidase F1 purified from *Elizabethkingia miricola*	Trimble and Tarentino (1991)
Endoglycosidase F2 purified from *Elizabethkingia miricola*	Trimble and Tarentino (1991)
Endoglycosidase F3 purified from *Elizabethkingia miricola*	Trimble and Tarentino (1991)
Endoglycosidase H purified from *Streptomyces plicatus*	Tarentino and Maley (1974) and Tarentino, Plummer, and Maley (1974)
Endo-β-*N*-acetylglucosaminidase (*Endo*-M) purified from *Mucor hiemalis*	Kadowaki et al. (1990)
O-glycan	
Chemical Release	
Using anhydrous hydrazine at 60°C	Patel et al. (1993)
Alkaline-β-elimination	Carlson (1968)
Nonreductive β-elimination	Hulsmeier et al. (2002)
Oxidation using sodium hypochlorite	Song et al. (2016)
Enzymatic Release	
Endo-α-*N*-acetylgalactosaminidase	Huang and Aminoff (1972)
Enzymatic/Chemical Release	
Pronase followed by solid-phase permethylation	Goetz, Novotny, andMechref (2009)

Solid-phase extraction (SPE) is another efficient method for extracting glycans from digested mixtures according to differences in hydrophobicity. Commonly used SPE materials are C18 and charcoal, both of which have high selectivity for glycans. The affinity of charcoal for carbohydrate structures enables the separation of different types of glycans, which has been demonstrated by its use for glycan fractionation in some studies. However, glycans are not completely trapped by or eluted out of SPE stationary phase materials which results in sample loss, the major drawback of SPE purification.

2.3 Glycan Derivatization

Most glycomic analysis strategies require derivatization of released glycans. The advantages of derivatizing glycans can be reflected in the following aspects. Unlike proteins or peptides, native glycans are not suitable for optical detection methods, labeling glycans with tags containing chromophores or fluorophores can make glycans detectable by UV or laser-induced fluorescence (LIF). Native glycans, especially acidic species, are hydrophilic and have low proton affinity, resulting in low ionization efficiency in positive mode MS analysis. Increased hydrophobicity through derivatization improves the detection sensitivity of glycans in MS analysis. Furthermore, native sialylated and fucosylated glycans are not stable during ionization and MS analysis. Hence, derivatization is necessary in order to stabilize these glycan structures. A variety of derivatization methods that are employed to simultaneously improve glycan detection and separation efficiency in capillary electrophoresis (CE) and various types of LC will be discussed in the following sections.

2.3.1 Derivatization Through Reductive Amination

The unique chemical properties of the reducing end of glycans make it an ideal site for the attachment of labeling tags. The most commonly utilized reducing end labeling reaction is reductive amination. The reaction occurs between an aldehyde on the glycan reducing end and an amine from the labeling tag (Fig. 1A). The intermediate imine is quickly reduced in the presence of a reducing reagent in the reaction mixture. The reaction is usually accomplished in dimethyl sulfoxide (Bigge et al., 1995), tetrahydrofuran (Evangelista, Guttman, & Chen, 1996), or methanol (Anumula, 1994) with the presence of acids such as acetic acid and citric acid (Evangelista et al., 1996; Szabo, Guttman, Rejtar, & Karger, 2010). Sodium cyanoborohydride is most frequently used as the reducing reagent in this reaction, although

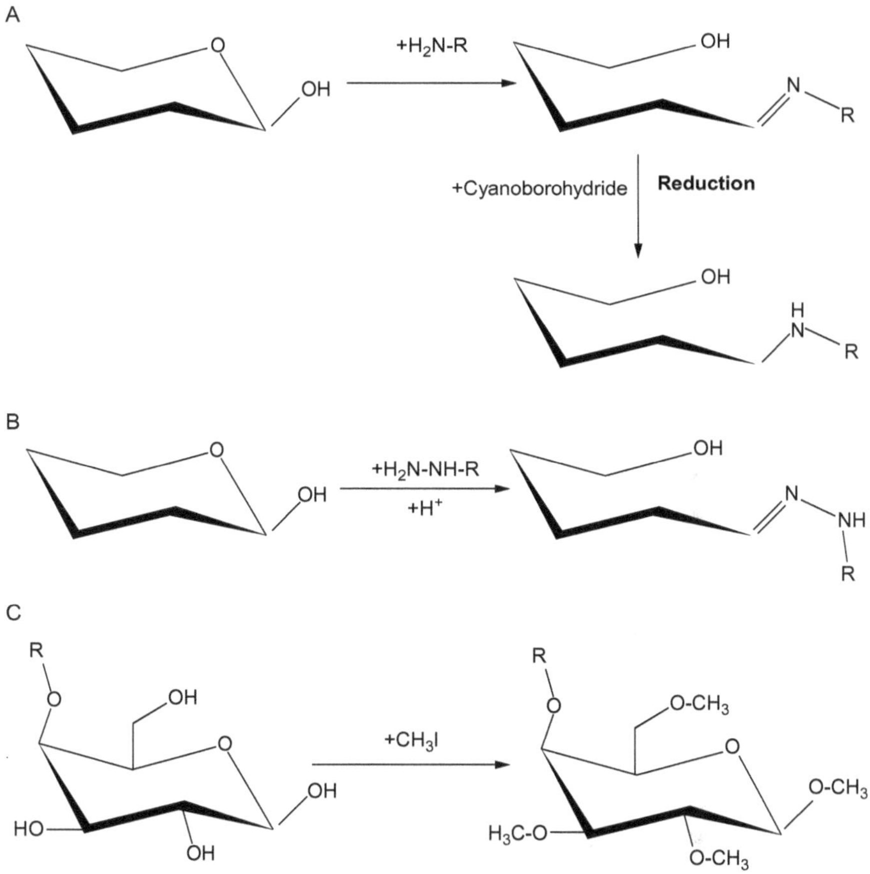

Fig. 1 Derivatization by (A) reductive amination, (B) hydrazide derivatization, and (C) permethylation.

recently 2-picoline-borane was reported as a nontoxic reducing reagent for glycan labeling (Ruhaak, Steenvoorden, Koeleman, Deelder, & Wuhrer, 2010). Common labeling reagents are anthranilic acid (2-AA) (Anumula, 1994), 2-aminobenzamide (2-AB; Bigge et al., 1995), and 2-aminopyridine (2-AP; Hase, Ikenaka, & Matsushima, 1978), all of which have fluorophores enabling LIF detection after HPLC separation. 9-Aminopyrene-1,4,6-trisulfonic acid (APTS) (Guttman, Chen, & Evangelista, 1996) and 8-aminonaphthalene-1,3,6-trisulfonic acid (ANTS) (Chiesa & Horvath, 1993) are regularly utilized to induce fixed charges to glycans so that they can be separated by CE. An additional function of these labeling reagents is to improve ionization efficiency and glycan stability in both matrix-assisted laser desorption ionization (MALDI) and electrospray ionization (ESI). Several other labeling reagents have also been reported for quantitative glycomics, including 2-aminoacridone (AMAC) (Charlwood, Birrell, Organ, & Camilleri, 1999), 4-aminobenzonitrile (ABN) (Schwaiger, Oefner, Huber, Grill, & Bonn, 1994), 3-aminobenzoic acid (3-AA) (Nakajima, Oda, Kinoshita, Masuko, & Kakehi, 2002), and 3-aminobenzamide (3-AB) (Kakehi, Funakubo, Suzuki, Oda, & Kitada, 1999).

Reductive amination requires excess labeling and reducing reagents, and a clean-up step is necessary to purify labeled glycans. Commonly utilized purification strategies are SPE with different stationary phase materials (e.g., C18, HILIC, and porous graphitized carbon (PGC)), paper chromatography (Bigge et al., 1995), filtration (Hase, 1994; Prater, Anumula, & Hutchins, 2007), dialysis, and precipitation (Pabst et al., 2009). Unfortunately, the introduction of an additional purification step reduces sample preparation throughput and creates the potential for sample loss.

2.3.2 Derivatization Through the Hydrazide Reaction

The hydrazide reaction is another widely applied glycan reducing end labeling method. Hydrazide on labeling reagents interacts with the reducing end aldehyde of glycans, after which the resulting CN double bond is left in place. A positive aspect of this reaction is that the employment of toxic reducing reagents is avoided (Fig. 1B), this renders a postlabeling purification step optional and makes hydrazide labeling more convenient in comparison with reductive amination. 2,4-Dinitrophenylhydrazine (Karamanos, Tsegenidis, & Antonopoulos, 1987), pyrenebutyric acid hydrazide (Sugahara, Amano, & Irimura, 2003), and 1-alkoxyamino-1-deoxy alditols (Vondeyn, York, Albersheim, & Darvill, 1990) have been reported to label glycans such that their derivatives are compatible with optical detection methods. Additionally, research involving phenylhydrazine has exhibited a 100% labeling yield (Lattova & Perreault, 2003a), demonstrating the efficiency and reliability of hydrazide derivatization for glycan analysis.

Despite the potential for optical detection, most researchers utilizing hydrazide glycan derivatization are employing MS for glycomic profiling. The function of hydrazide in these cases is to induce a permanent charge so that both neutral and acidic glycans can be ionization with similar efficiency in MALDI and ESI. 2-Hydrazino-*N*,*N*,*N*-trimethyl-2-*oxo*-ethanaminium chloride (Girard's T reagent) is a well-established reagent for improvement of ionization efficiency in positive mode (Gil, Kim, &

Kim, 2008; Gouw, Burgers, Trikoupis, & Terlouw, 2002; Naven & Harvey, 1998). An essential characteristic for evaluating hydrazide reagents are their ability to boost sensitivity. Walker and coworkers, in the laboratory of David C. Muddiman, have made comparisons among five labeling reagents: 5-phenylpetane hydrazide, 3-phenylpropane hydrazide, 2-phenylacethydrazide, 4-phenylbutane hydrazide, and 4-phenylbenzohydrazide (Phenyl2-GPN), in order to evaluate their ionization efficiency enhancement capability. The results indicated a linear relationship between ion abundance and tag hydrophobicity with Phenyl2-GPN, the most hydrophobic of the labels, being the most effective with a 10% higher sensitivity (Fig. 2). Other reported hydrazide labeling reagents for glycomics include biotinyl-L-3-(2-naphthyl)-alanine hydrazide (Walker, Lilley, Enamorado, Comins, & Muddiman, 2011; Walker, Papas, Comins, & Muddiman, 2010), keratan sulfates (Zhang et al., 2008), cyanine dyes (Kameyama et al., 2004), and 4-phenyl pyridine (Bereman, Comins, & Muddiman, 2010).

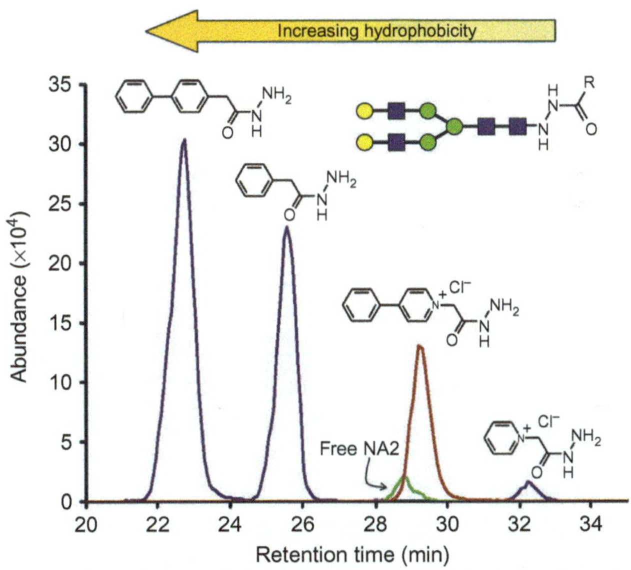

Fig. 2 Extracted ion chromatogram of the NA2 glycan labeled with hydrazide tags, demonstrating the relationship between hydrophobicity and ionization efficiency. *Reprinted and modified from Walker, S. H., Papas, B. N., Comins, D. L., & Muddiman, D. C. (2010). Interplay of permanent charge and hydrophobicity in the electrospray ionization of glycans.* Analytical Chemistry, *82(15), 6636–6642. http://dx.doi.org/10.1021/Ac101227a, with permission. Copyright (2010) American Chemical Society.*

Hydrazide glycan interactions have also been utilized for recognizing and capturing glycans (Leteux et al., 1998; Yang & Zhang, 2012). A solid-phase glycan isolation method using hydrazide beads was reported (Yang & Zhang, 2012) to achieve more specific glycan binding than conventional glycan isolation methods, which rely on the hydrophobicity difference between glycan and sample matrix. The formed hydrazone bond can be cleaved in an acidic elution solution so that glycans can be recovered from the hydrazide beads. A negative aspect of the hydrazide labeling approach is that hydrazide derivatized glycans are less stable than reductive amination labeled glycans, due to the lability of the hydrazone bond in acidic or basic solutions.

2.3.3 Derivatization of Glycan Sialic Acids

Unlike the two reducing end labeling methods discussed earlier, the derivatization method discussed in this section is focused on the sialic acids of glycans, which are commonly observed as terminal residues. Because sialic acid is the most labile monosaccharide found in glycans and is commonly found to exhibit significant changes in disease progression, it has attracted attention from a large number of researchers. 4-(4,6-Dimethoxy-1,3,5-triazin-2-yl)-4-methylmorpholinium (DMT-MM) is often utilized to stabilize sialic acids by amidation, especially in MALDI-MS where in source decay regularly occurs (Sekiya, Wada, & Tanaka, 2005). Moreover, DMT-MM can act as a catalyst to differentiate α2–3-linked and α2–6-linked sialic acids when proper reaction conditions are employed (Wheeler, Domann, & Harvey, 2009). In this reaction, α2–3-linked acids are converted into lactones while α2–6-linked sialic acids are converted into methyl esters, resulting in a 13 Da mass difference. These mass differences can be resolved using MS to reveal the number of differently linked sialic acids in a glycan (Alley & Novotny, 2010; Harvey, Royle, Radcliffe, Rudd, & Dwek, 2008; Montesino et al., 2012; Tousi, Bones, Hancock, & Hincapie, 2013). To date, this is still one of the most efficient sialic acid linkage isomer identification methods.

2.3.4 Permethylation

Permethylation is another commonly used glycomic derivatization method (Fig. 1C). It was specifically developed for MS-based glycomics because permethylation only improves glycan stability and ionization efficiency during MS analysis, no chromophores or fluorophores are introduced by this derivatization method. Another difference between permethylation and the reducing end labeling methods described earlier is that permethylation is a multiple site modification method. Hence, the proton affinity improvement by permethylation is much higher than single site derivatization methods. However, multiple site modification brings with it the challenge of managing derivatization efficiency. The overall incomplete derivatization ratio increases exponentially with the number of modification sites. Thus, selecting a proper permethylation protocol is critical for achieving efficient permethylation. Glycan permethylation was first reported by Ciucanu and Kerek (1984), DMSO, solid sodium hydroxide, and iodomethane were utilized in this protocol. The procedure was optimized by the same group in 2003 in order to reduce oxidative degradation in the reaction (Ciucanu & Costello, 2003). In 2005, Kang and Mechref proposed a solid-phase permethylation protocol. In this protocol, conventional in solution reaction was replaced by a reaction in spin columns of capillaries packed with sodium hydroxide beads (Kang, Mechref, Klouckova, & Novotny, 2005; Kang, Mechref, & Novotny, 2008). This solid-phase permethylation protocol reduced reaction time, improved derivatization efficiency for sialylated glycans, and exhibited higher throughput compared with in solution protocols (Fig. 3).

2.3.5 Glycosylamine Derivatization

Recently reagents for the rapid labeling of released N-glycans have been investigated. In contrast to traditional reductive amination approaches, which tag the aldehyde termini of glycans after they are hydrolyzed from their glycosylamine forms, these reagents quickly label glycans as glycosylamines. Cook, Bullock, and Sullivan (2012) initially reported a labeling method based on a 2-AB analog where glycans were tagged prior to the elimination of water from the reducing end by a urea linked aminobenzamide (Cook et al., 2012). ProZyme (Hayward, CA) has since marketed a kit termed GlykoPrep® Rapid-Reductive-Amination™ 2-AB labeling module that utilizes this strategy. The product is reported to provide highly fluorescent labeling, with no loss of sialic acid, with an efficiency of greater than 85%. Although this technique improved the speed of the labeling procedure, its application was limited to fluorescent detection. The Waters Corp. (Milford, MA) has since introduced a labeling reagent termed RapiFluor-MS™ that imparts both fluorescent properties and enhanced ionization efficiency for mass spectrometric analysis (Lauber et al., 2015). The RapiFluor-MS™ reagent is composed of a quinolone fluorophore, an N-hydroxysuccinimide carbamate group, for tagging, and a basic tertiary amine that provides improved ionization efficiency. The reagent is sold in a kit that reports the preparation of glycans for analysis in less than 30 min

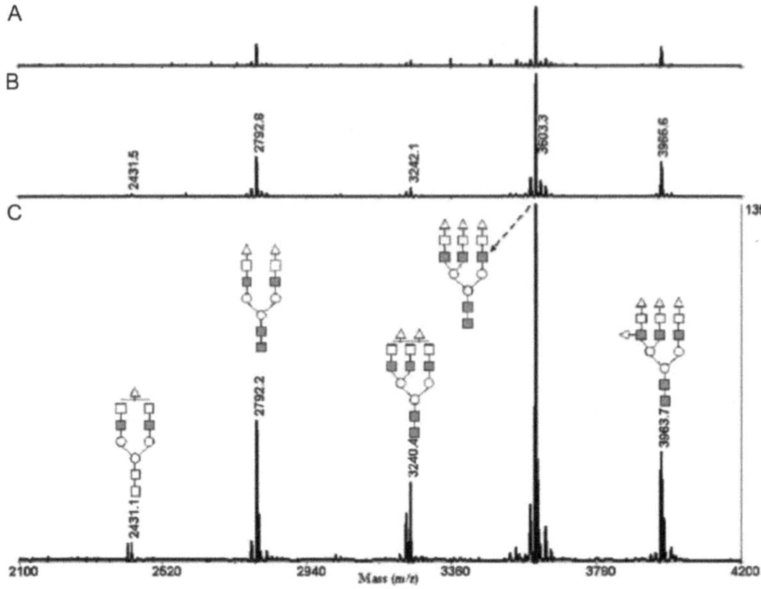

Fig. 3 MALDI-TOF-MS spectra of permethylated N-glycans from fetuin, demonstrating the advantage of solid-phase permethylation utilizing a microspin column (B) or fused-silica capillary (C) compared to the conventional in solution method (A). *Reprinted and modified from Kang, P., Mechref, Y., Klouckova, I., & Novotny, M. V. (2005). Solid-phase permethylation of glycans for mass spectrometric analysis.* Rapid Communications in Mass Spectrometry, 19*(23), 3421–3428. http://dx.doi.org/10.1002/rcm.2210, with permission. Copyright (2005) John Wiley and Sons.*

through the use of a rapid PNGase F deglycosylation protocol involving *Rapi*Gest SF surfactant and a HILIC μElution SPE purification step. This method was recently utilized for the comparative glycomic analysis of two biosimilar monoclonal antibodies and the highly sensitive detection of released N-glycans by mass spectrometry was reported (Fang et al., 2016).

3. QUANTITATIVE GLYCOMICS PLATFORMS

Reliable quantitative analysis of glycans is essential for understanding glycan biofunctions and discovering potential biomarkers for diseases. Numerous strategies for quantitative glycomics have been developed in recent decades. While glycomics is marching toward high sensitivity, high reliability, and high throughput, the remarkable complexity of the glycome still presents a number of challenges that must be addressed. In the following sections, the strategies currently employed in the field of quantitative glycomics are discussed.

3.1 MS-Based Glycomics

Since the invention of soft ionization techniques, such as MALDI and ESI, mass spectrometry has become one of the most powerful bioanalytical platforms available. The rapid development of high resolution mass spectrometry (HRMS) in recent decades has enabled precise glycan identification through molecular weight determination. Tandem MS assists glycan structure elucidation. Direct analysis by MS skips the glycan separation step, significantly accelerating the analysis speed, allowing quantitative glycomic profiling to be accomplished in seconds. However, competitive ionization is not negligible when analyte composition is complex. Therefore, direct MS analysis is most often utilized for relatively simple glycan samples or glycan structural elucidation studies.

MALDI-MS is one of the direct MS glycomic analysis strategies. Native or derivatized glycans are mixed with matrix and then ionized by a laser beam. The selection of matrix is a critical factor that influences mass spectrum quality. 2,5-Dihydroxybenzoic acid (2,5-DHB) is currently the most suitable matrix for glycan analysis (Stahl, Steup, Karas, & Hillenkamp, 1991). Sodium salts are often spiked into the matrix to prompt the formation of singly charged sodium adducts. Compared with ESI, the most pronounced advantage of MALDI is its capability to generate ions with a simplified charge state and adduct form, thus concentrating ion species while avoiding complicated spectra.

Many studies have reported improved sensitivity when comatrix chemicals have been spiked into solution with 2,5-DHB. For example, 2-hydroxy-5-methoxybenzoic acid (Karas et al., 1993), 1-hydroxyisoquinoline (Mohr, Bornsen, & Widmer, 1995), base spermine (Mechref & Novotny, 1998), and aniline (Snovida, Chen, & Perreault, 2006) are all reported as efficient comatrices for stabilizing glycans and improving ionization efficiency. The derivatization methods discussed earlier are necessary for MALDI-MS glycan analysis due to the fact that native glycans are fragile during ionization.

MALDI-MS has been utilized for the discovery of many glycomic biomarkers for a variety of diseases, such as hepatocellular carcinoma (HCC) (Ressom et al., 2008), breast cancer (Kyselova et al., 2008), prostate cancer (de Leoz et al., 2008), and ovarian cancer (Barkauskas et al., 2009) as well as the therapeutic antibody trastuzumab (Matsumoto et al., 2009). Glycan analytes derived from serum, tissue, or cell lines are all compatible with MALDI-MS analyses. Therefore, this technique is well suited for the identification of putative glycan biomarkers of disease.

ESI-MS is another major direct MS analysis strategy. It can be utilized in both positive and negative ionization modes for glycomic analysis. However, the performance of this type of analysis, when applied to complex biological samples, is limited due to the high dynamic range of glycans. Direct infusion ESI-MS is most commonly used for glycan structural elucidation studies, where only one or a few glycans are subjected to analysis (Harvey, 2005a, 2005b; Pfenninger, Karas, Finke, & Stahl, 2002a, 2002b).

3.2 LC-MS-Based Glycomics

Although direct MS analysis facilitates precise glycan identification and reliable quantification for glycans in high abundance, separation techniques are necessary when only optical detection devices are available, or minor glycans are masked due to competitive ionization in direct MS analysis. Furthermore, deconvolution of the fragmentation spectra of isomeric structures is greatly simplified by the elimination of coelution through chromatographic separation. Various off-line and on-line separation techniques, for both native and derivatized glycans, have been developed and applied in quantitative glycomics and are discussed herein.

3.2.1 Offline Separation and Fractionation

Several early studies combined LC separation with MALDI-MS analysis (Catala, Howe, Hucko, Rose, & Thannhauser, 2011; Lochnit & Geyer, 2004; Sparbier et al., 2006), where LC fractions were spotted on MALDI plates and subjected to MS analysis. In this method, a complex mixture of glycans is fractionated to reduce competitive ionization, which facilitates the identification of minor glycan species. In a recent study, MALDI-MS, LC-MALDI-MS, and LC-ESI-MS were compared side by side for the analysis of N-glycans derived from human blood serum (Hu & Mechref, 2012). According to the results, the LC-MALDI-MS method identified 12 more glycans than direct MALDI-MS, indicating the necessity of utilizing a separation technique prior to MS analysis. However, LC-ESI-MS provided the highest glycan identification number and sensitivity, demonstrating the advantage of coupling on-line separation to MS.

Offline glycan fractionation can also be achieved using SPE spin columns (Addeo et al., 1978; Packer, Lawson, Jardine, & Redmond, 1998). Lebrilla and coworkers have been using PGC SPE prefractionation in their sample preparation protocol for years. In this method, neutral glycans can be separated from acidic glycans to minimize suppression effects (Chu et al., 2009; Ninonuevo et al., 2006; Ruhaak, Miyamoto, & Lebrilla, 2013).

3.2.2 Glycan Separation by Hydrophilic Interaction Chromatography (HILIC)

Native glycans are highly hydrophilic molecules that contain hydroxyl groups capable of hydrogen bonding. HILIC-LC provides highly efficient separation of native and reducing end labeled glycans. The retention mechanism of this chromatographic technique is primarily based on the formation of hydrogen bonds with the hydrophilic stationary phase; ionic interactions also play a role in the separation (Fig. 4). Commonly used HILIC materials for glycan separation are amide-coated silica, amine-coated silica, and zwitterionic (ZIC) HILIC phases (Alpert et al., 1994; Takegawa et al., 2006; Wuhrer, de Boer, & Deelder, 2009). LC elution solutions are most commonly composed of water and acetonitrile with the addition of ammonium formate (5–50 m*M*) to increase ionic strength. The gradients generally begin with a high organic concentration and native or reducing end labeled

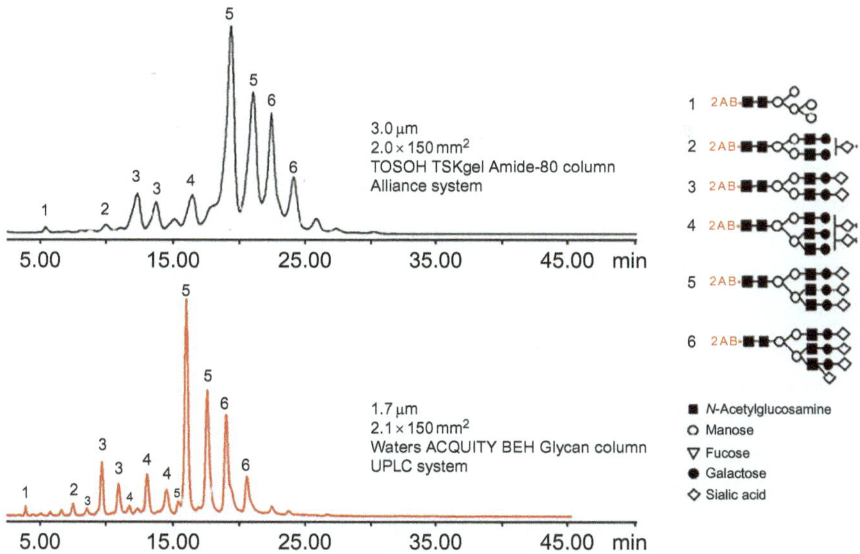

Fig. 4 HILIC separation of 2-AB-labeled glycans. *Reprinted and modified from Zauner, G., Deelder, A. M., & Wuhrer, M. (2011). Recent advances in hydrophilic interaction liquid chromatography (HILIC) for structural glycomics.* Electrophoresis, 32*(24), 3456–3466. http://dx.doi.org/10.1002/elps.201100247, with permission.*

glycans are eluted as the percentage of water increases. Smaller and neutral glycans elute first, while larger and sialylated glycans are retained by the column for a longer period of time. HILIC-LC is a separation technique that is sensitive to structural isomerism, enabling the partial separation of compositional or linkage glycan isomers.

HILIC-LC can be coupled with UV or fluorescence detection for glycomic analysis. As mentioned earlier, this strategy requires native glycans to be derivatized with chromophore or fluorophore containing tags. 2-AA (Benet & Austin, 2011; Lonardi, Deelder, Wuhrer, & Balog, 2012; Royle et al., 2008), 2-AB (Kalay et al., 2012; Kozak, Royle, Gardner, Fernandes, & Wuhrer, 2012; Mauko et al., 2012), and PA (Takahashi, 1996) are frequently used in HILIC-fluorescence (FL) glycomic profiling. In contrast to MS detection strategies, optical detection methods are not capable of identifying the sequence of glycans; glycan identification is therefore based solely on retention time. Owing to retention times varying over time and between instruments, the introduction of an internal standard is necessary for retention time calibration. Dextrin and dextran, which are composed of a series of repeating glucose units, are often used as retention time references with the concept of glycan units (GUs) employed. Hence, the glycan retention times in HILIC-LC can be represented as GUs rather than a precise measure of time. The most comprehensive HILIC-FL glycomic strategy was developed by Rudd and coworkers, who established the GlycoBase and autoGU databases (Campbell, Royle, Radcliffe, Dwek, & Rudd, 2008; Royle et al., 2008) in which more than 350 N-glycan structures can be found. Similarly, databases for O-glycans have also been developed (Royle et al., 2002).

Automatic data processing systems are available for HILIC-FL profiling of 2-AB-labeled glycans with the assistance of databases, making rapid and high-throughput glycomic analysis for scores of samples possible. This platform has been widely used for biomarker discovery studies involving diseases such as ovarian (Saldova et al., 2007), lung (Arnold et al., 2011), and breast cancer (Abd Hamid et al., 2008; Saldova, Reuben, Abd Hamid, Rudd, & Cristofanilli, 2011).

Common flow rates and mobile phases of HILIC-LC permit convenient coupling of HILIC separation to MS analysis through ESI. As a more informative detection platform, MS can provide precise glycan identification and detailed glycan structural information can be obtained using tandem MS (Wuhrer et al., 2009). Additionally, native or reduced glycans can be subjected to HILIC-MS analysis without the need for fluorescence labeling.

Owing to the isomeric separation capability of HILIC, it has been coupled to tandem MS for the structural analysis of glycans. Efforts were made to minimize the physical size of the HILIC separation platform, to increase its sensitivity, leading to the development of nanoLC (Clarke, Harmon, & DeFelippis, 2009; Wuhrer, Koeleman, Hokke, & Deelder, 2004) and chip-based (Staples et al., 2009) HILIC systems. The detection limit of these miniaturized systems can be as low as femtomoles (Wuhrer, Koeleman, Deelder, & Hokke, 2004; Wuhrer, Koeleman, Hokke, et al., 2004).

3.2.3 Glycan Separation by Reversed-Phase Liquid Chromatography (RPLC)

One of the most obvious advantages of utilizing RPLC to separate glycans is that RPLC is the separation technique of choice for proteomics and metabolomics. Therefore no adjustment is required for chromatography columns or mobile phases when a laboratory has multiple research projects. The retention mechanism on RPLC-based C18 stationary phase columns relies on the hydrophobicity of analyte molecules. Hence, derivatization is required in order to achieve sufficient retention of glycans (Wuhrer, Deelder, & Hokke, 2005). Choices for reducing end labeling tags for glycomic RPLC separation include 2-AB (Chen & Flynn, 2007; Wing et al., 2001), PA (Yamada, Tsukamoto, Sasaki, Yagyu, & Takahashi, 1997), 1-phenyl-3-methyl-5-pyrazolone (PMP) (Saba, Shen, Jamieson, & Perreault, 1999, 2001), phenylhydrazine (Lattova & Perreault, 2003b), 2-amino-5-bromopyridine (Li & Kinzer, 2003), and ANTS (Saba et al., 2001). After derivatization, the tags and the glycan itself influence retention time in concert. When the hydrophobicity of a tag is greater than that of the glycan, retention time decreases with increased numbers of monosaccharides in the glycan; this relationship reverses when the hydrophobicity of a tag is less than the glycan (Pabst et al., 2009).

Permethylation is another widely used derivatization method for RPLC glycan separation. After permethylation, glycan hydrophobicity increases significantly. Additionally, all hydrophobic groups are directly related to the glycan structure, thus making the retention of permethylated glycans on C18 stationary phase more predictable. Isomeric separations can be achieved using RPLC for permethylated O-glycans (Fig. 5) (Hanisch & Muller, 2009). However, when the analyte is an N-glycan, which is larger on average, the isomeric separation capability of RPLC decreases (Fig. 6) (Dong, Zhou, & Mechref, 2016). Furthermore, peak broadening can often be observed for highly branched N-glycan structures. In this case, high

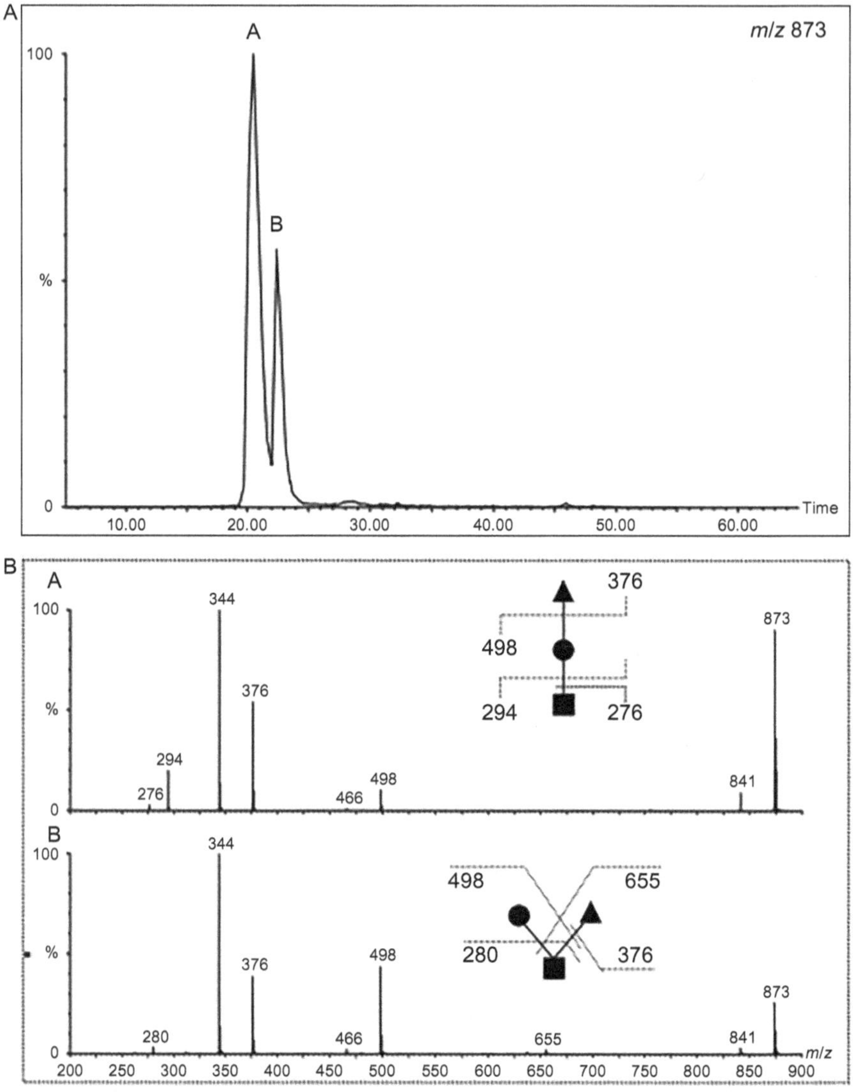

Fig. 5 (A) RPLC separation of permethylated isomeric O-glycan trisaccharides. (B) MS/MS spectra of isomers in peaks A and B (*upper panel*). *Reprinted and modified from Hanisch, F. G., & Muller, S. (2009). Analysis of methylated O-glycan alditols by reversed-phase NanoLC coupled CAD-ESI mass spectrometry.* Carbohydrate Microarrays: Methods and Protocols, 534, *107–115. http://dx.doi.org/10.1007/978-1-59745-022-5_8, with permission.*

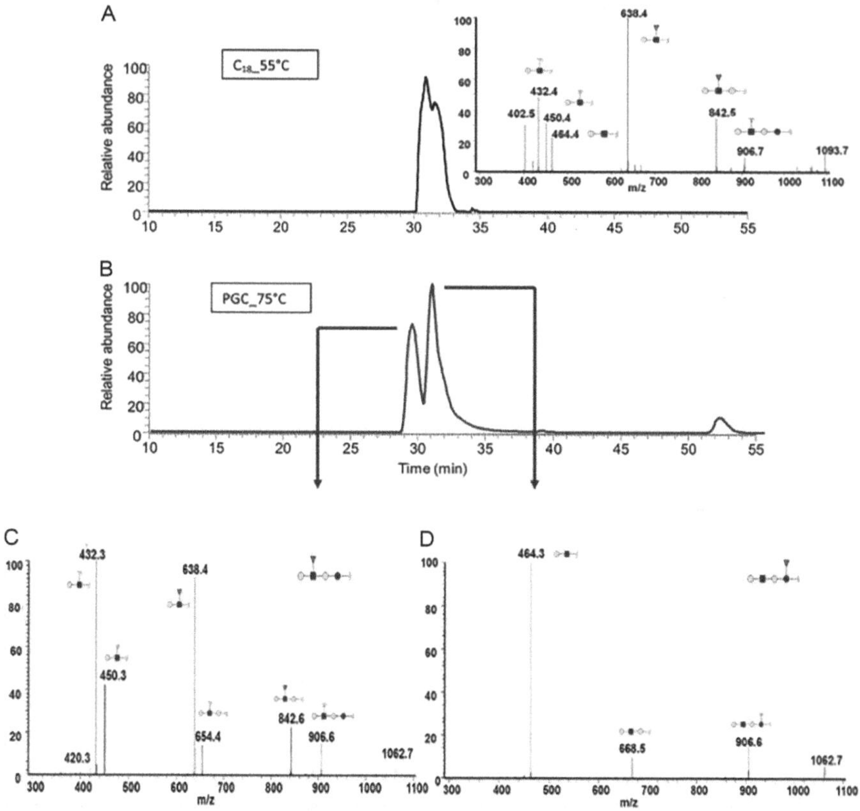

Fig. 6 EIC of fucosylated Gal-GlcNAc-lactose isomers determined in human milk using (A) C_{18} and (B) PGC columns. Tandem MS of the two isomers associated with the free oligosaccharide structures are depicted in (C) and (D). *Reprinted and modified from Dong, X., Zhou, S., & Mechref, Y. (2016). LC-MS/MS analysis of permethylated free oligosaccharides and N-glycans derived from human, bovine, and goat milk samples.* Electrophoresis, 37*(11), 1532–1548. http://dx.doi.org/10.1002/elps.201500561, with permission.*

separation temperatures may be employed to increase peak resolution. Nanoscale RPLC systems have also been developed for increased sensitivity. Online purification may be achieved by utilizing a C18 trap in a 10 port valve system (Desantos-Garcia, Khalil, Hussein, Hu, & Mechref, 2011). Studies have demonstrated improved sensitivity when off-line liquid–liquid extraction is replaced by on-line purification. This strategy can be utilized for quantitative glycomic profiling of serum, the surface of tissues and cell lines to study a variety of diseases (Alley et al., 2012; Hu, Zhou, Khalil, Renteria, & Mechref, 2013; Mechref et al., 2012; Tsai et al., 2014).

3.2.4 Glycan Separation by PGC Liquid Chromatography

PGC was first utilized as an SPE material for glycan purification in 1998 (Packer et al., 1998) and was subsequently applied to LC-ESI-MS in 1999 (Thomsson, Karlsson, & Hansson, 1999). Among currently utilized LC stationary phases, PGC exhibits the greatest capacity for isomeric separation, despite the separation mechanism not being completely understood. Currently, retention is believed to be caused by a combination of hydrophobic and ionic interactions (Ruhaak, Deelder, & Wuhrer, 2009). Native and reducing end labeled glycans are most commonly subjected to PGC separation. However, the elution of acidic glycans on PGC may be problematic. Pabst and coworkers have systematically investigated the separation of acidic glycans on PGC. The results indicated that elution buffer ionic strength is critical for elution of acidic glycans (Pabst & Altmann, 2008).

PGC separation coupled with UV detection or MS analysis has been widely applied for quantitative glycomic analysis. For example, a chip-based PGC nanoLC-ESI-MS technique was developed by the Lebrilla group. The method stability was investigated, and the results indicated that analyses repeated on different days had a coefficient of variation of 4%, calculated from log 10 transformed peak intensities (Ruhaak, Taylor, et al., 2013). This strategy was successfully applied to the analysis of human and bovine milk (Nwosu et al., 2012; Wu, Grimm, German, & Lebrilla, 2011), dry blood spots (Ruhaak, Miyamoto, Kelly, & Lebrilla, 2012), plasma from lung cancer patients (Ruhaak, Nguyen, et al., 2013), prostate cancer, and lymphoproliferative cancer cell lines (Hua et al., 2014).

In 2007, permethylated glycans were also reported to be separated by PGC-LC (Costello, Contado-Miller, & Cipollo, 2007). In this study, isomeric separations of high-mannose glycans were observed, and detailed structures were interpreted from MS2 spectra. However, peak shapes were not smooth, and baseline resolution was not achieved. Moreover, sialylated glycans were not efficiently separated by the method employed. In a study more recently conducted, in our laboratory, the optimal conditions for PGC separation of permethylated glycans were determined. Improved resolution, similar to that attained for native glycans, was achieved while more informative MS2 spectra were acquired, prompted by glycan permethylation (Fig. 6) (Dong et al., 2016).

3.2.5 Glycan Separation by CE

CE has been described in-depth in recent reviews (Campa, Coslovi, Flamigni, & Rossi, 2006; Mechref, Hu, Desantos-Garcia, Hussein, &

Tang, 2013; Suzuki, 2013; Zamfir & Peter-Katalinic, 2004). The method involves the use of capillaries with 50–100 μm bore sizes in high electric fields (750 V/cm). Short separation times and small injection volumes are typical. Thus, highly sensitive detection methods must be employed. One such method is LIF following reductive amination labeling with fluorophores such as APTS (Chen & Evangelista, 1998; Evangelista et al., 1996), utilizing reducing agents such as cyanoborohydride (Evangelista et al., 1996) or 2-picoline-borane (Ruhaak et al., 2010). CE coupled to LIF has been successfully applied for the analysis of APTS-labeled N-glycans derived from a number of purified glycoproteins (Chen & Evangelista, 1998; Evangelista et al., 1996; Guttman et al., 1996; Guttman & Pritchett, 1995; Ma & Nashabeh, 1999; Mechref, Muzikar, & Novotny, 2005; Suzuki, Muller, Guttman, & Karger, 1997) as well as human serum from chronic hepatitis C patients (Callewaert, Geysens, Molemans, & Contreras, 2001; Callewaert et al., 2004; Vanderschaeghe, Debruyne, Van Vlierberghe, Callewaert, & Delanghe, 2009; Vanderschaeghe et al., 2009). In addition to LIF, CE has also been successfully conjugated to tandem MS for the quantification of glycans, which has been discussed in previous reviews (Campa et al., 2006; Mechref et al., 2013; Zamfir & Peter-Katalinic, 2004). Both on-line and off-line coupling of CE to MS instrumentation have been utilized. In the case of off-line coupling, MALDI-MS was initially employed to analyze APTS-labeled N-glycans from ribonuclease B (Suzuki et al., 1997). On-line coupling entails the use of CE-ESI-MS systems, which have been utilized for the analysis of mannooligosaccharide caps in mycobacterial lipoarabinomannan (Monsarrat, Brando, Condouret, Nigou, & Puzo, 1999) and phosphorylated APTS-labeled N-glycans from cellobiohydrolase I (Sandra, Devreese, Van Beeumen, Stals, & Claeyssens, 2004).

3.2.6 Glycan Separation by Ion Mobility Spectroscopy-Mass Spectrometry (IMS-MS)

IMS separates gas-phase ions, typically generated by ESI, based on their collisional cross sections which are approximately proportionate to molecular shape and size (Clowers, Dwivedi, Steiner, Hill, & Bendiak, 2005; Clowers & Hill, 2005). Owing to the isomeric separation capabilities of IMS, the technique has been gaining popularity in the field of glycan analysis. The advantages of having an additional gas-phase separation during IMS-MS have been demonstrated by the characterization of N-glycan isomers from RNase B, fetuin, and gp120 (Fenn & McLean, 2009; Harvey et al., 2012, 2013; Zhu, Lee, Valentine, Reilly, & Clemmer, 2012). IMS has also been applied to the analysis of glycans in complex biological matrices; for example, the profiling of total serum glycans from individuals with liver cancer or cirrhosis of the liver has been reported (Isailovic et al., 2008), as well as the comparison of N-glycans from the serum of patients with Barrett's esophagus, high-grade dysplasia, and esophageal adenocarcinoma with healthy control individuals (Gaye et al., 2012). The applications of IMS for the analysis of glycans in the context of high-throughput and high-resolution studies have recently been reviewed (Gray et al., 2016).

3.3 MS-Based Quantitation Methods

Although MS is a powerful glycan identification and structure elucidation tool, its performance in glycomic quantitation is not as stable as optical detection methods. This is due to the nature of ionization and detection of MS instruments. First, both MALDI and ESI do not ionize all analyte molecules; the ionization efficiency is influenced by many factors. For MALDI, the uncertainty of crystal quality, spot layer thickness, and salt concentration in the sample results in the fluctuation of ionization efficiency. For ESI, the needle inner diameter, sheath gas flow, and analyte solution composition all influence the stability of ionization. Especially in the case of nanoESI, which is the common interface between nanoLC or chip/capillary-based separation techniques and MS, ionization fluctuation can be as high as 30%.

A multiinstitute glycomic profiling study was proposed by the Human Proteome Organization (HUPO) and the Human Disease Glycomics/Proteome Initiative (HGPI) in 2007 (Wada et al., 2007). While the final conclusion, of the study, was that MS-based analysis appears to be an efficient method for identification and quantitation of oligosaccharides in glycomic studies. Variation was also observed in the data, indicating the necessity of efforts to improve the reliability of MS quantitation. Currently, spiking internal standards and multiplexing through the use of stable-isotope labeling represent two major approaches for increasing the reliability of analyses.

3.3.1 Internal Standard

The necessity of spiking internal standards is well recognized in MALDI-MS glycomic profiling. An internal standard can effectively normalize results from different runs. Maltoheptaose and other carbohydrates with repeating units of glucose have been utilized as internal standards (Bereman et al., 2010; Snovida, Rak-Banville, & Perreault, 2008; Szabo et al., 2010; Walker, Budhathoki-Uprety, Novak, & Muddiman, 2011). The selection criteria for internal standards are that their molecular weight should be close to the average molecular weight of the glycans being analyzed and they should have chemical properties similar to glycans so they can be derivatized using the same methods applied to the glycans of interest. Recently, iGlycoMab was reported as an internal standard for normalizing variation observed in glycan release, sample preparation, and MS analysis. iGlycoMab is a purified murine IgG2b monoclonal antibody that has all glycan nitrogen atoms replaced by the stable-isotope ^{15}N. Hence, the glycan internal standard does not have a molecular weight that overlaps with analyte glycans ($\sim$4 Da mass difference is induced by ^{15}N on N-acetylglucosamine). The coefficient of variation of model glycoprotein and cell line glycan analysis was found to decrease when results were normalized by iGlycoMab (Fig. 7), demonstrating its utility (Zhou, Tello, et al., 2016).

3.3.2 Multiplexing Through Derivatization

As mentioned earlier, variation between different runs is the primary source of uncertainty when reliable analysis for multiple samples is required, for example, biomarker discovery. Hence, combining multiple samples in one MS analysis functions to eliminate a great deal of uncertainty. Multiplexed MS analysis also provides the additional advantage of saving time. Total analysis time can be reduced by a factor of the number of multiplexed samples. In order to resolve identical glycan structures derived from different multiplexed samples in a single MS analysis, molecular weight differences must be generated. The most convenient way to introduce different masses to glycans is through derivatization. Atoms on the derivatization reagents can be replaced with stable isotopes such as deuterium (D), ^{13}C, or ^{15}N. Glycans labeled with standard reagents are commonly referred to as "light" and those labeled with the stable isotope reagents are termed "heavy." It is worth mentioning that the mass difference between light- and heavy-labeled glycans should be larger than 4 Da. Otherwise, the overlapping natural isotopic distribution of light-labeled glycans will interfere with the intensity of heavy-labeled glycans.

Reductive amination is a single site derivatization method, therefore the mass difference between heavy and light reagents utilized should be larger than 4 Da. Commonly used stable-isotope replacement patterns include introducing four deuteriums or six ^{13}C atoms into an aromatic ring. These patterns are widely used for 2-AA (Hitchcock, Costello, & Zaia, 2006; Hitchcock, Yates, Shortkroff, Costello, & Zaia, 2006), aniline (Lawrence et al., 2008; Xia, Feasley, Sachdev, Smith, & Cummings, 2009), and PA

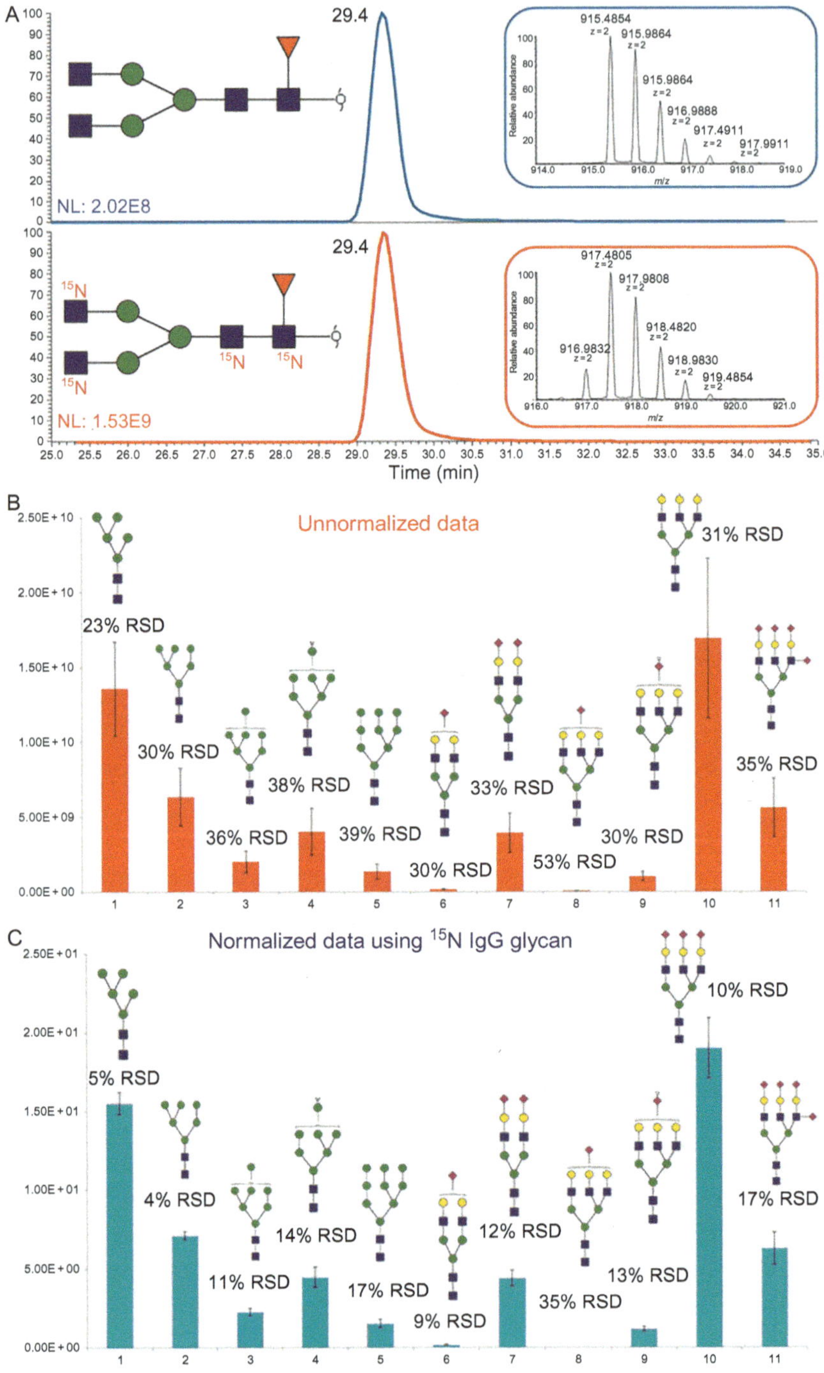

Fig. 7 See legend on opposite page.

(Yuan et al., 2005). However, due to the limitation of the labeling reagent size, these strategies only allow for the simultaneous analysis of two samples per run. To overcome this, larger tags have been developed to enable the inclusion of more glycan samples per analysis. Bowman and Zaia introduced a tetraplex stable-isotope coded tagging strategy (Bowman & Zaia, 2007), where four tags (d_0, d_4, d_8, and d_{12}) with a 4 Da mass difference between them were utilized to analyze glycans derived from model glycoproteins mixed in a predetermined ratio. The results demonstrated the reliability of glycomic quantitation using this type of tag. The tag strategy was then applied for the analysis of glycans derived from human, ovine, bovine, and baboon samples (Bowman & Zaia, 2010).

Heavy/light labeling can also be achieved through permethylation. Because permethylation is a multiple site derivatization method, the rule that the mass difference between tags should be equal to or greater than 4 Da can be ignored. The first report of multiplexing through permethylation, introduced by Mechref, Novotny, and coworkers in 2007, was termed comparative glycomic mapping (C-GlycoMAP) (Kang, Mechref, Kyselova, Goetz, & Novotny, 2007). The permethylation reagents CH_3I and CD_3I were used for labeling glycans released from RNaseB and the spectra generated by MALDI-MS exhibited reliable quantitation ratios from 0.125 to 6. This strategy was subsequently extended by introducing an additional iodomethane derivative, CH_2DI, allowing triplex analysis to be achieved (Hu, Desantos-Garcia, & Mechref, 2013). In this study, glycans released from human blood serum of patients with esophageal cancer, Barret's disease, and high-grade dysplasia were analyzed. Quantitation by isobaric labeling (QUIBL) represents an additional type of permethylation that incorporates multiplexing (Atwood et al., 2008). $^{13}CH_3I$ and $^{12}CH_2DI$ are utilized for the isobaric labeling of glycans, resulting in a mass difference of 0.002922 Da between each permethylation site. The use of a high resolution mass spectrometer is required for the isolation of such similar masses. When utilized for the analysis of N-glycans from a model glycoprotein and

Fig. 7 (A) EIC and MS spectra of N-glycans released from unlabeled IgG standard and ^{15}N iGlycoMab. *Bar graphs* representing intensity variation of biological ($n=3$) analyses of N-glycans released from model glycoprotein mixtures (B) without normalization and (C) with normalization using ^{15}N-glycans derived from iGlycoMab. *Reprinted and modified from Zhou, S., Tello, N., Harvey, A., Boyes, B., Orlando, R., & Mechref, Y. (2016). Reliable LC-MS quantitative glycomics using iGlycoMab stable isotope labeled glycans as internal standards.* Electrophoresis, 37*(11), 1489–1497. http://dx.doi.org/10.1002/elps.201600013, with permission.*

human serum, a dynamic range of two orders of magnitude was exhibited. O-glycans from mucin were also quantified using this strategy (Botelho et al., 2008). When applying a permethylation-based multiplexing strategy on an RPLC-MS platform, it is important to note that the retention times are modified when deuterium iodomethane is used. This occurs as a result of the hydrophobicity of hydrogen being greater than deuterium, the relationship of retention times is $CH_3 > CH_2D > CHD2 > CD_3$ on a C18 column. This phenomenon does not extend to carbon, as no difference in RPLC retention time is observed between carbon and ^{13}C iodomethane derivatized glycans. Although multiplexing techniques reduce variation, ESI efficiency changes throughout an LC-MS analysis. This temporal variation may not be eliminated when multiplexed samples have significantly different retention times. The isobaric permethylation reagents, $^{13}CH_3I$ and $^{12}CH_3I$, have been utilized for LC-MS analysis of N-glycans released from embryos of *Drosophila melanogaster* at various temporal stages of development (Aoki et al., 2007).

3.3.3 Metabolic Isotopic Labeling

Metabolic labeling presents an alternative to derivatization reagents for the incorporation of stable isotopes into glycan structures. This strategy allows for simplified sample preparation and selection of analytical platforms due to the elimination of potential issues stemming from derivatization reagents. In this case, even native glycans can be subjected to multiplex analysis. Isotopic detection of aminosugars with glutamine (IDAWG) is a metabolic glycan labeling strategy introduced by Orlando et al. (2009). The labeling mechanism employed in the strategy is as follows: ^{15}N-Gln was spiked into cell culture media to act as the nitrogen source for the biosynthesis of UDP-GlcNAc, which could be further converted to UDP-GalNAc and CMP-Neu5Ac. As a result, all GlcNAc, Neu5Ac, and GalNAc in cell glycans were ^{15}N labeled. The nitrogen replacement efficiency of this method was demonstrated to be higher than 95%. Another metabolic labeling strategy was proposed by Breidenbacha et al. (2010). In their approach, $^{13}C_6$-GlcNAc was introduced to cell secreted glycoproteins by an exogenous GlcNAc salvage pathway.

Stable isotopes can also be introduced into glycans through PNGase F digestion in ^{18}O-water; this method is known as glycan reducing end ^{18}O labeling (GREOL) (Zhang, Wang, Tang, & Yang, 2011). During the procedure, oxygen atoms of glycan reducing ends are replaced by ^{18}O and a 2 Da mass difference is generated. The quantitative capability of this method was verified using ovalbumin glycans; it was additionally applied to glycans released from the serum of patients with HCC. The linear response range of this method was determined to be two orders of magnitude or higher.

3.3.4 Multiplexing Using a Reporter Ion in Tandem MS

Full MS-based multiplexing always relies on the criteria of a mass difference of at least 4 Da between heavy- and light-labeled molecules. This can significantly limit the maximum number of samples that can be analyzed together. Additionally, full MS-based multiplexing can be complicated by tags resulting in different retention times and isotopic labeling resulting in identical glycan structures having different m/z values. In order to overcome these drawbacks, a multiplex method employing reporter ions in MS/MS was developed. This strategy was accomplished by utilizing a series of reducing end labeling tags with reporter ion portions that differed by mass. A mass balancer, or normalizer, was included in the tags to ensure they were isobaric. Hence, the same glycans with different multiplex tags were identical during LC separation and full MS analysis; this reduced the complexity of LC-MS data collected during the analysis of multiple samples. Owing to the fact that the reporter ions were small, $m/z = 100 - 200$ Da, the influence of isotopic peaks was negligible. In this strategy, the mass gap between adjacent reporter ions can be 1 Da or less, which contributes to its potential capacity for multiplex analysis.

AminoxyTMT (Hahne et al., 2012; Zhong et al., 2015), iART (Yang et al., 2013), and QUANTITY (Yang et al., 2015) are all examples of multiplexing reagents for quantitative glycomics. AminoxyTMT is a derivative of the TMT reagent developed for proteomics. A six-plex kit of aminoxyTMT is commercially available. However, reporter ion yield has been reported to be an issue for aminoxyTMT because glycosidic bonds are weaker than the bonds linking the reporter ions. This issue can be overcome by conducting MS^3 on Y1 ions or by modifying fragmentation patterns through sodium adducts (Fig. 8) (Zhou, Hu, et al., 2016). Quantitation error was reported to be controlled, within 10%, when an optimized LC-MS/MS method was utilized to facilitate the formation of sodium adducts. The iART tetra-plex reagent is reported to have issues with low reporter ion yield, similar to what has been observed with aminoxyTMT. On the other hand, QUANTITY (Fig. 9) is a tetra-plex reagent that results in high reporter ion yield without the help of sodium or metal ion adducts. This strategy was applied to the quantitative analysis of glycans released from

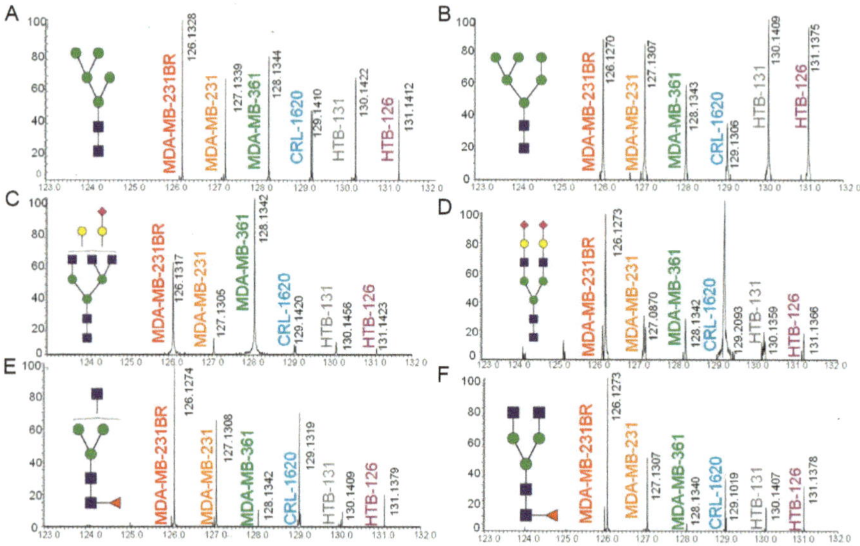

Fig. 8 Tandem mass spectra of different aminoxyTMT-labeled glycans derived from various cell lines. Tandem mass spectra of several glycan structures are shown in (A)–(F). *Reprinted and modified from Zhou, S., Hu, Y., Veillon, L., Snovida, S. I., Rogers, J. C., Saba, J., & Mechref, Y. (2016). Quantitative LC-MS/MS glycomic analysis of biological samples using AminoxyTMT.* Analytical Chemistry, 88*(15), 7515–7522. http://dx.doi.org/10.1021/acs.analchem.6b00465, with permission.*

three bioengineered Chinese hamster ovary (CHO) cell lines (Yang et al., 2015).

3.4 Data Processing Software for Glycomics

Omics analyses are always based on the analysis of a vast number of samples, therefore reliable and high-throughput data processing software is necessary. While this is also the case for glycomics, data processing software for glycomics is relatively under developed compared with proteomics software. MultiGlycan is software that was specifically developed for LC-MS analysis of permethylated glycans, where an algorithm was employed for deconvolution in order to reduce false negatives (Fig. 10) (Hu, Zhou, Yu, Tang, & Mechref, 2015). Additionally, MultiGlycan is capable of analyzing permethylated glycan data with the functionality of enabling the normalization of reagent purity and isotopic distribution differences.

SimGlycan (Apte & Meitei, 2010; Meitei, Apte, Snovida, Rogers, & Saba, 2015) (Fig. 11) is an additional software package that allows the identification and quantitation of glycans derivatized using several methods, in

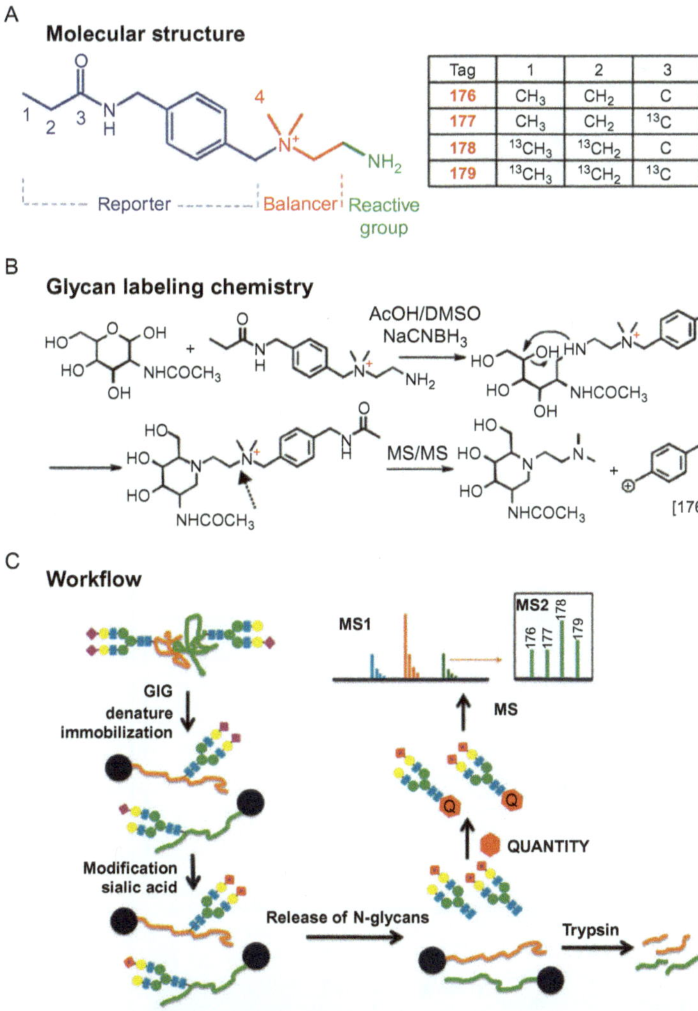

Fig. 9 QUANTITY isobaric tandem mass tags for glycan labeling and quantitation. *Reprinted and modified from Yang, S., Wang, M. Y., Chen, L. J., Yin, B. J., Song, G. Q., Turko, I. V., … Li, S. W. (2015). QUANTITY: An isobaric tag for quantitative glycomics.* Scientific Reports, 5. *doi: Artn 17585, with permission.*

addition to permethylation, including aminoxyTMT labeling. Furthermore, a library has been built for the automated analysis of analytes separated on different LC columns. It represents one of the most powerful software solutions for quantitative glycan data analysis and structural elucidation based on MS/MS spectra.

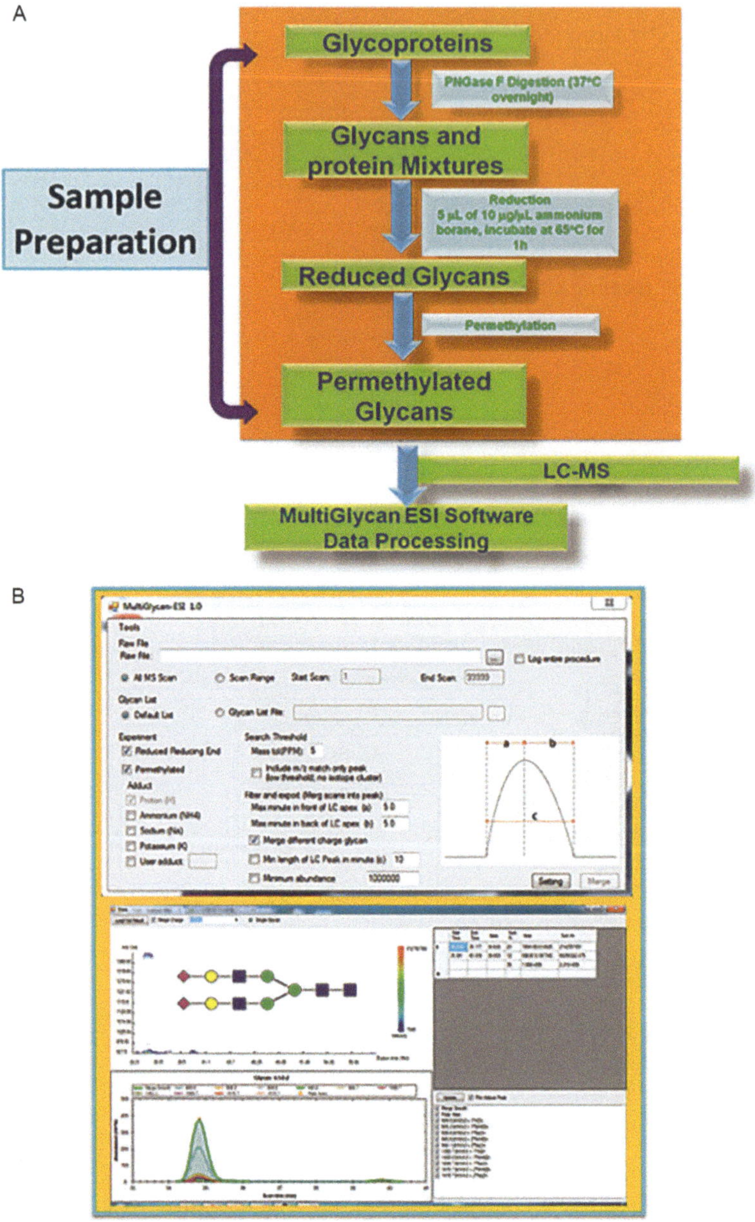

Fig. 10 Workflow of MultiGlycan-ESI-based automatic quantitation strategy: (A) permethylated N-glycan preparation workflow and (B) software annotation and quantitation method. *Reprinted and modified from Hu, Y., Zhou, S., Yu, C. Y., Tang, H., & Mechref, Y. (2015). Automated annotation and quantitation of glycans by liquid chromatography/electrospray ionization mass spectrometric analysis using the MultiGlycan-ESI computational tool.* Rapid Communications in Mass Spectrometry, *29(1), 135–142. http://dx.doi.org/10.1002/rcm.7093, with permission.*

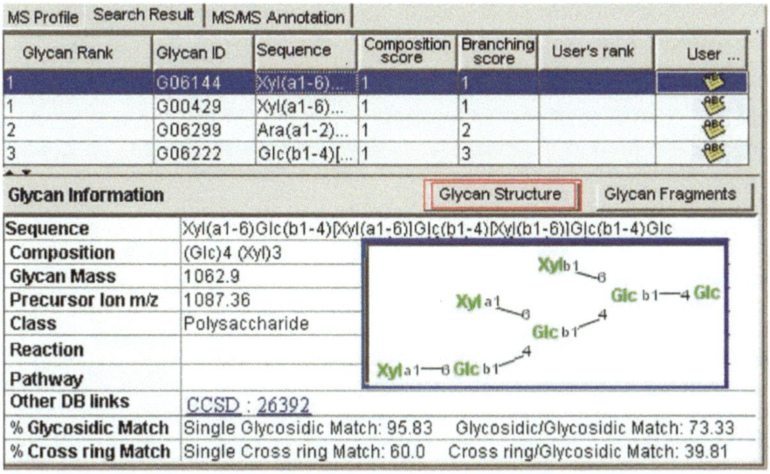

Fig. 11 Glycan identified through SimGlycan MS/MS database search. *Reprinted and modified from Apte, A., & Meitei, N. S. (2010). Bioinformatics in glycomics: Glycan characterization with mass spectrometric data using SimGlycan.* Carbohydrate Microarrays: Methods and Protocols, 600, *269–281. http://dx.doi.org/10.1007/978-1-60761-454-8_19, with permission.*

GlycoWorkbench (Ceroni et al., 2008; Damerell et al., 2012, 2015) possesses the function of predicting fragment ions for glycans; Cartoonist (Goldberg, Sutton-Smith, Paulson, & Dell, 2005) provides automated annotation of MALDI-MS analyzed glycans; software like Glycomod (Cooper, Gasteiger, & Packer, 2001), GlycoQuest, Glyco-Peakfinder (Maass, Ranzinger, Geyer, von der Lieth, & Geyer, 2007), and SysBioWare (Vakhrushev, Dadimov, & Peter-Katalinic, 2009) are also employed in glycomic data processing.

ACKNOWLEDGMENTS

This work was supported by an NIH grant (1R01GM112490-01) and a grant from the Cancer Prevention and Research Institute of Texas (RP130624).

REFERENCES

Abd Hamid, U. M., Royle, L., Saldova, R., Radcliffe, C. M., Harvey, D. J., Storr, S. J., … Rudd, P. M. (2008). A strategy to reveal potential glycan markers from serum glycoproteins associated with breast cancer progression. *Glycobiology*, *18*(12), 1105–1118.

Addeo, F., Soulier, S., Pelissier, J. P., Chobert, J. M., Mercier, J. C., & Ribadeaudumas, B. (1978). Preparation and fractionation of goat kappa-casein—Analysis of glycan and peptide components. *Journal of Dairy Research*, *45*(2), 191–196.

Alley, W. R., Jr., & Novotny, M. V. (2010). Glycomic analysis of sialic acid linkages in glycans derived from blood serum glycoproteins. *Journal of Proteome Research*, *9*(6), 3062–3072. http://dx.doi.org/10.1021/pr901210r.

Alley, W. R. J., Vasseur, J. A., Goetz, J. A., Svoboda, M., Mann, B. F., Matei, D. E., … Novotny, M. V. (2012). N-linked glycan structures and their expressions change in the blood sera of ovarian cancer patients. *Journal of Proteome Research*, *11*, 2282–2300.

Alpert, A. J., Shukla, M., Shukla, A. K., Zieske, L. R., Yuen, S. W., Ferguson, M. A. J., … Orlando, R. (1994). Hydrophilic-interaction chromatography of complex carbohydrates. *Journal of Chromatography A*, *676*(1), 191–202. http://dx.doi.org/10.1016/0021-9673(94)00467-6.

Anumula, K. R. (1994). Quantitative determination of monosaccharides in glycoproteins by high-performance liquid chromatography with highly sensitive fluorescence detection. *Analytical Biochemistry*, *220*(2), 275–283. http://dx.doi.org/10.1006/abio.1994.1338.

Aoki, K., Perlman, M., Lim, J.-M., Cantu, R., Wells, L., & Tiemeyer, M. (2007). Dynamic developmental elaboration of N-linked glycan complexity in the Drosophila melanogaster embryo. *The Journal of Biological Chemistry*, *282*, 9127–9142.

Apte, A., & Meitei, N. S. (2010). Bioinformatics in glycomics: Glycan characterization with mass spectrometric data using SimGlycan. *Carbohydrate Microarrays: Methods and Protocols*, *600*, 269–281. http://dx.doi.org/10.1007/978-1-60761-454-8_19.

Apweiler, R., Hermjakob, H., & Sharon, N. (1999). On the frequency of protein glycosylation, as deduced from analysis of the SWISS-PROT database. *Biochimica et Biophysica Acta*, *1473*(1), 4–8.

Arnold, J. N., Saldova, R., Galligan, M. C., Murphy, T. B., Mimura-Kimura, Y., Telford, J. E., … Rudd, P. M. (2011). Novel glycan biomarkers for the detection of lung cancer. *Journal of Proteome Research*, *10*(4), 1755–1764.

Atwood, J. A., Cheng, L., Alvarez-Manilla, G., Warren, N. L., York, W. S., & Orlando, R. (2008). Quantitation by isobaric labeling: Applications to glycomics. *Journal of Proteome Research*, 7(1), 367–374. http://dx.doi.org/10.1021/pr070476i.

Barkauskas, D. A., An, H. J., Kronewitter, S. R., de Leoz, M. L., Chew, H. K., White, R. W. D., … Rocke, D. M. (2009). Detecting glycan cancer biomarkers in serum samples using MALDI FT-ICR mass spectrometry data. *Bioinformatics*, *25*(2), 251–257. http://dx.doi.org/10.1093/bioinformatics/btn610.

Benet, T., & Austin, S. (2011). On-line cleanup for 2-aminobenzamide-labeled oligosaccharides. *Analytical Biochemistry*, *414*(1), 166–168. http://dx.doi.org/10.1016/j.ab.2011.03.002.

Bereman, M. S., Comins, D. L., & Muddiman, D. C. (2010). Increasing the hydrophobicity and electrospray response of glycans through derivatization with novel cationic hydrazides. *Chemical Communications*, *46*(2), 237–239. http://dx.doi.org/10.1039/B915589a.

Bigge, J. C., Patel, T. P., Bruce, J. A., Goulding, P. N., Charles, S. M., & Parekh, R. B. (1995). Nonselective and efficient fluorescent labeling of glycans using 2-amino benzamide and anthranilic acid. *Analytical Biochemistry*, *230*(2), 229–238. http://dx.doi.org/10.1006/abio.1995.1468.

Botelho, J. C., Atwood, J. A., III, Cheng, L., Alvarez-Manilla, G., York, W. S., & Orlando, R. (2008). Quantification by isobaric labeling (QUIBL) for the comparative glycomic study of O-linked glycans. *International Journal of Mass Spectrometry*, *278*, 137–142.

Bowman, M. J., & Zaia, J. (2007). Tags for the stable isotopic labeling of carbohydrates and quantitative analysis by mass spectrometry. *Analytical Chemistry*, *79*(15), 5777–5784. http://dx.doi.org/10.1021/ac070581b.

Bowman, M. J., & Zaia, J. (2010). Comparative glycomics using a tetraplex stable-isotope coded tag. *Analytical Chemistry*, *82*(7), 3023–3031. http://dx.doi.org/10.1021/ac100108w.

Breidenbacha, M. A., Gallagherb, J. E. G., Kingc, D. S., Smarta, B. P., Wua, P., & Bertozzia, C. R. (2010). Targeted metabolic labeling of yeast N-glycans with unnatural sugars. *Proceedings of the National Academy of Sciences of the United States of America*, *107*, 3988–3993.

Bucior, I., Scheuring, S., Engel, A., & Burger, M. M. (2004). Carbohydrate-carbohydrate interaction provides adhesion force and specificity for cellular recognition. *The Journal of Cell Biology*, *165*(4), 529–537. http://dx.doi.org/10.1083/jcb.200309005.

Callewaert, N., Geysens, S., Molemans, F., & Contreras, R. (2001). Ultrasensitive profiling and sequencing of N-linked oligosaccharides using standard DNA-sequencing equipment. *Glycobiology*, *11*(4), 275–281.

Callewaert, N., Van Vlierberghe, H., Van Hecke, A., Laroy, W., Delanghe, J., & Contreras, R. (2004). Noninvasive diagnosis of liver cirrhosis using DNA sequencer-based total serum protein glycomics. *Nature Medicine*, *10*(4), 429–434.

Campa, C., Coslovi, A., Flamigni, A., & Rossi, M. (2006). Overview on advances in capillary electrophoresis-mass spectrometry of carbohydrates: A tabulated review. *Electrophoresis*, *27*(11), 2027–2050. http://dx.doi.org/10.1002/elps.200500960.

Campbell, M. P., Royle, L., Radcliffe, C. M., Dwek, R. A., & Rudd, P. M. (2008). GlycoBase and autoGU: Tols for HPLC-based glycan analysis. *Bioinformatics*, *24*(9), 1214–1216. http://dx.doi.org/10.1093/bioinformatics/btn090.

Carlson, D. M. (1968). Structures and immunochemical properties of oligosaccharides isolated from pig submaxillary mucins. *The Journal of Biological Chemistry*, *243*(3), 616–626.

Catala, C., Howe, K. J., Hucko, S., Rose, J. K. C., & Thannhauser, T. W. (2011). Towards characterization of the glycoproteome of tomato (Solanum lycopersicum) fruit using Concanavalin A lectin affinity chromatography and LC-MALDI-MS/MS analysis. *Proteomics*, *11*(8), 1530–1544. http://dx.doi.org/10.1002/pmic.201000424.

Ceroni, A., Maass, K., Geyer, H., Geyer, R., Dell, A., & Haslam, S. M. (2008). GlycoWorkbench: A tool for the computer-assisted annotation of mass spectra of glycans. *Journal of Proteome Research*, 7(4), 1650–1659. http://dx.doi.org/10.1021/pr7008252.

Charlwood, J., Birrell, H., Organ, A., & Camilleri, P. (1999). A chromatographic and mass spectrometric strategy for the analysis of oligosaccharides: Determination of the glycan structures in porcine thyroglobulin. *Rapid Communications in Mass Spectrometry*, *13*(8), 716–723. http://dx.doi.org/10.1002/(SICI)1097-0231(19990430)13:8<716::AID-RCM547>3.0.CO;2-C10.1002/(SICI)1097-0231(19990430)13:8<716::AID-RCM547>3.0.CO;2-C.

Chen, F. T., & Evangelista, R. A. (1998). Profiling glycoprotein N-linked oligosaccharide by capillary electrophoresis. *Electrophoresis*, *19*(15), 2639–2644. http://dx.doi.org/10.1002/elps.1150191512.

Chen, X., & Flynn, G. C. (2007). Analysis of N-glycans from recombinant immunoglobulin G by on-line reversed-phase high-performance liquid chromatography/mass spectrometry. *Analytical Biochemistry*, *370*(2), 147–161. http://dx.doi.org/10.1016/j.ab.2007.08.012.

Cheresh, D. A., Reisfeld, R. A., & Varki, A. P. (1984). O-acetylation of disialoganglioside GD3 by human melanoma cells creates a unique antigenic determinant. *Science*, *225*(4664), 844–846.

Chiesa, C., & Horvath, C. (1993). Capillary zone electrophoresis of malto-oligosaccharides derivatized with 8-aminonaphthalene-1,3,6-trisulfonic acid. *Journal of Chromatography*, *645*(2), 337–352.

Chu, C. S., Ninonuevo, M. R., Clowers, B. H., Perkins, P. D., An, H. J., Yin, H. F., … Lebrilla, C. B. (2009). Profile of native N-linked glycan structures from human serum using high performance liquid chromatography on a microfluidic chip and time-of-flight mass spectrometry. *Proteomics*, *9*(7), 1939–1951. http://dx.doi.org/10.1002/pmic.200800249.

Ciucanu, I., & Costello, C. E. (2003). Elimination of oxidative degradation during the per-O-methylation of carbohydrates. *Journal of the American Chemical Society, 125*(52), 16213–16219. http://dx.doi.org/10.1021/ja035660t.

Ciucanu, I., & Kerek, F. (1984). A simple and rapid method for the permethylation of carbohydrates. *Carbohydrate Research, 131*(2), 209–217.

Clarke, A., Harmon, B., & DeFelippis, M. R. (2009). Analysis of 3-(acetylamino)-6-aminoacridine-derivatized oligosaccharides from recombinant monoclonal antibodies by liquid chromatography-mass spectrometry. *Analytical Biochemistry, 390*, 209–211.

Clausen, H., & Bennett, E. P. (1996). A family of UDP-GalNAc: Polypeptide N-acetylgalactosaminyl-transferases control the initiation of mucin-type O-linked glycosylation. *Glycobiology, 6*(6), 635–646.

Clowers, B. H., Dwivedi, P., Steiner, W. E., Hill, H. H., Jr., & Bendiak, B. (2005). Separation of sodiated isobaric disaccharides and trisaccharides using electrospray ionization-atmospheric pressure ion mobility-time of flight mass spectrometry. *Journal of the American Society for Mass Spectrometry, 16*(5), 660–669. http://dx.doi.org/10.1016/j.jasms.2005.01.010.

Clowers, B. H., & Hill, H. H., Jr. (2005). Mass analysis of mobility-selected ion populations using dual gate, ion mobility, quadrupole ion trap mass spectrometry. *Analytical Chemistry*, 77(18), 5877–5885. http://dx.doi.org/10.1021/ac050700s.

Cook, K. S., Bullock, K., & Sullivan, T. (2012). Development and qualification of an antibody rapid deglycosylation method. *Biologicals, 40*(2), 109–117. http://dx.doi.org/10.1016/j.biologicals.2011.12.010.

Cooper, C. A., Gasteiger, E., & Packer, N. H. (2001). GlycoMod—A software tool for determining glycosylation compositions from mass spectrometric data. *Proteomics, 1*(2), 340–349. http://dx.doi.org/10.1002/1615-9861(200102)1:2<340::AID-PROT340->3.0.CO;2-B.

Costello, C. E., Contado-Miller, J. M., & Cipollo, J. F. (2007). A glycomics platform for the analysis of permethylated oligosaccharide alditols. *Journal of the American Society for Mass Spectrometry, 18*(10), 1799–1812. http://dx.doi.org/10.1016/j.jasms.2007.07.016.

Dall'Olio, F., & Chiricolo, M. (2001). Sialyltransferases in cancer. *Glycoconjugate Journal, 18*(11–12), 841–850.

Damerell, D., Ceroni, A., Maass, K., Ranzinger, R., Dell, A., & Haslam, S. M. (2012). The GlycanBuilder and GlycoWorkbench glycoinformatics tools: Updates and new developments. *Biological Chemistry, 393*(11), 1357–1362. http://dx.doi.org/10.1515/hsz-2012-0135.

Damerell, D., Ceroni, A., Maass, K., Ranzinger, R., Dell, A., & Haslam, S. M. (2015). Annotation of glycomics MS and MS/MS spectra using the GlycoWorkbench software tool. *Carbohydrate Microarrays: Methods and Protocols, 1273*, 3–15. http://dx.doi.org/10.1007/978-1-4939-2343-4_1.

de Leoz, M. L. A., An, H. J., Kronewitter, S., Kim, J., Beecroft, S., Vinall, R., … Lebrilla, C. (2008). Glycomic approach for potential biomarkers on prostate cancer: Profiling of N-linked glycans in human sera and pRNS cell lines. *Disease Markers, 25*(4–5), 243–258.

Desantos-Garcia, J. L., Khalil, S. I., Hussein, A., Hu, Y., & Mechref, Y. (2011). Enhanced sensitivity of LC-MS analysis of permethylated N-glycans through online purification. *Electrophoresis, 32*(24), 3516–3525. http://dx.doi.org/10.1002/elps.201100378.

Dong, X., Zhou, S., & Mechref, Y. (2016). LC-MS/MS analysis of permethylated free oligosaccharides and N-glycans derived from human, bovine, and goat milk samples. *Electrophoresis, 37*(11), 1532–1548. http://dx.doi.org/10.1002/elps.201500561.

Ellgaard, L., Molinari, M., & Helenius, A. (1999). Setting the standards: Quality control in the secretory pathway. *Science, 286*(5446), 1882–1888.

Evangelista, R. A., Guttman, A., & Chen, F. T. (1996). Acid-catalyzed reductive amination of aldoses with 8-aminopyrene-1,3,6-trisulfonate. *Electrophoresis*, *17*(2), 347–351. http://dx.doi.org/10.1002/elps.1150170210.

Fang, J., Doneanu, C., Alley, W. R., Jr., Yu, Y. Q., Beck, A., & Chen, W. (2016). Advanced assessment of the physicochemical characteristics of Remicade(R) and Inflectra(R) by sensitive LC/MS techniques. *MAbs*, *8*(6), 1021–1034. http://dx.doi.org/10.1080/19420862.2016.1193661.

Fenn, L. S., & McLean, J. A. (2009). Simultaneous glycoproteomics on the basis of structure using ion mobility-mass spectrometry. *Molecular BioSystems*, *5*(11), 1298–1302. http://dx.doi.org/10.1039/b909745g.

Freeze, H. H., Esko, J. D., & Parodi, A. J. (2009). Glycans in glycoprotein quality control. In A. Varki, R. D. Cummings, J. D. Esko, H. H. Freeze, P. Stanley, C. R. Bertozzi, G. W. Hart, & M. E. Etzler (Eds.), *Essentials of glycobiology*. (2nd ed.). Cold Spring Harbor, NY: Cold Spring Harbor Laboratory Press.

Furukawa, J., Piao, J., Yoshida, Y., Okada, K., Yokota, I., Higashino, K., … Shinohara, Y. (2015). Quantitative O-glycomics by microwave-assisted beta-elimination in the presence of pyrazolone analogues. *Analytical Chemistry*, *87*(15), 7524–7528. http://dx.doi.org/10.1021/acs.analchem.5b02155.

Gaye, M. M., Valentine, S. J., Hu, Y., Mirjankar, N., Hammoud, Z. T., Mechref, Y., … Clemmer, D. E. (2012). Ion mobility-mass spectrometry analysis of serum N-linked glycans from esophageal adenocarcinoma phenotypes. *Journal of Proteome Research*, *11*(12), 6102–6110.

Gil, G. C., Kim, Y. G., & Kim, B. G. (2008). A relative and absolute quantification of neutral N-linked oligosaccharides using modification with carboxymethyl trimethylammonium hydrazide and matrix-assisted laser desorption/ionization time-of-flight mass spectrometry. *Analytical Biochemistry*, *379*(1), 45–59. http://dx.doi.org/10.1016/j.ab.2008.04.039.

Gill, D. J., Clausen, H., & Bard, F. (2011). Location, location, location: New insights into O-GalNAc protein glycosylation. *Trends in Cell Biology*, *21*(3), 149–158. http://dx.doi.org/10.1016/j.tcb.2010.11.004.

Goetz, J. A., Novotny, M. V., & Mechref, Y. (2009). Enzymatic/chemical release of O-glycans allowing MS analysis at high sensitivity. *Analytical Chemistry*, *81*(23), 9546–9552. http://dx.doi.org/10.1021/ac901363h.

Goldberg, D., Sutton-Smith, M., Paulson, J., & Dell, A. (2005). Automatic annotation of matrix-assisted laser desorption/ionization N-glycan spectra. *Proteomics*, *5*(4), 865–875. http://dx.doi.org/10.1002/pmic.200401071.

Gornik, O., & Lauc, G. (2008). Glycosylation of serum proteins in inflammatory diseases. *Disease Markers*, *25*(4–5), 267–278.

Gouw, J. W., Burgers, P. C., Trikoupis, M. A., & Terlouw, J. K. (2002). Derivatization of small oligosaccharides prior to analysis by matrix-assisted laser desorption/ionization using glycidyltrimethylammonium chloride and Girard's reagent T. *Rapid Communications in Mass Spectrometry*, *16*(10), 905–912. http://dx.doi.org/10.1002/Rcm.654.

Gray, C. J., Thomas, B., Upton, R., Migas, L. G., Eyers, C. E., Barran, P. E., & Flitsch, S. L. (2016). Applications of ion mobility mass spectrometry for high throughput, high resolution glycan analysis. *Biochimica et Biophysica Acta*, *1860*(8), 1688–1709. http://dx.doi.org/10.1016/j.bbagen.2016.02.003.

Guttman, A., Chen, F. T., & Evangelista, R. A. (1996). Separation of 1-aminopyrene-3,6,8-trisulfonate-labeled asparagine-linked fetuin glycans by capillary gel electrophoresis. *Electrophoresis*, *17*(2), 412–417. http://dx.doi.org/10.1002/elps.1150170221.

Guttman, A., & Pritchett, T. (1995). Capillary gel electrophoresis separation of high-mannose type oligosaccharides derivatized by 1-aminopyrene-3,6,8-trisulfonic acid. *Electrophoresis*, *16*(10), 1906–1911.

Hahne, H., Neubert, P., Kuhn, K., Etienne, C., Bomgarden, R., Rogers, J. C., & Kuster, B. (2012). Carbonyl-reactive tandem mass tags for the proteome-wide quantification of N-linked glycans. *Analytical Chemistry*, *84*(8), 3716–3724. http://dx.doi.org/10.1021/ac300197c.

Hanisch, F. G., & Muller, S. (2009). Analysis of methylated O-glycan alditols by reversed-phase NanoLC coupled CAD-ESI mass spectrometry. *Carbohydrate Microarrays: Methods and Protocols*, *534*, 107–115. http://dx.doi.org/10.1007/978-1-59745-022-5_8.

Harvey, D. J. (2005a). Fragmentation of negative ions from carbohydrates: Part 1. Use of nitrate and other anionic adducts for the production of negative ion electrospray spectra from N-linked carbohydrates. *Journal of the American Society for Mass Spectrometry*, *16*(5), 622–630. http://dx.doi.org/10.1016/j.jasms.2005.01.004.

Harvey, D. J. (2005b). Fragmentation of negative ions from carbohydrates: Part 3. Fragmentation of hybrid and complex N-linked glycans. *Journal of the American Society for Mass Spectrometry*, *16*(5), 647–659. http://dx.doi.org/10.1016/j.jasms.2005.01.006.

Harvey, D. J., Royle, L., Radcliffe, C. M., Rudd, P. M., & Dwek, R. A. (2008). Structural and quantitative analysis of N-linked glycans by matrix-assisted laser desorption ionization and negative ion nanospray mass spectrometry. *Analytical Biochemistry*, *376*(1), 44–60. http://dx.doi.org/10.1016/j.ab.2008.01.025.

Harvey, D. J., Scarff, C. A., Crispin, M., Scanlan, C. N., Bonomelli, C., & Scrivens, J. H. (2012). MALDI-MS/MS with traveling wave ion mobility for the structural analysis of N-linked glycans. *Journal of the American Society for Mass Spectrometry*, *23*(11), 1955–1966. http://dx.doi.org/10.1007/s13361-012-0425-8.

Harvey, D. J., Scarff, C. A., Edgeworth, M., Crispin, M., Scanlan, C. N., Sobott, F., … Scrivens, J. H. (2013). Travelling wave ion mobility and negative ion fragmentation for the structural determination of N-linked glycans. *Electrophoresis*, *34*(16), 2368–2378. http://dx.doi.org/10.1002/elps.201200669.

Hase, S. (1994). High-performance liquid chromatography of pyridylaminated saccharides. *Methods in Enzymology*, *230*, 225–237.

Hase, S., Ikenaka, T., & Matsushima, Y. (1978). Structure analyses of oligosaccharides by tagging of the reducing end sugars with a fluorescent compound. *Biochemical and Biophysical Research Communications*, *85*(1), 257–263.

Helenius, A., & Aebi, M. (2004). Roles of N-linked glycans in the endoplasmic reticulum. *Annual Review of Biochemistry*, *73*, 1019–1049. http://dx.doi.org/10.1146/annurev.biochem.73.011303.073752.

Hitchcock, A. M., Costello, C. E., & Zaia, J. (2006). Glycoform quantification of chondroitin/dermatan sulfate using an LC/MS/MS platform. *Biochemistry*, *45*, 2350–2361.

Hitchcock, A. M., Yates, K. E., Shortkroff, S., Costello, C. E., & Zaia, J. (2006). Optimized extraction of glycosaminoglycans from normal and osteoarthritic cartilage for glycomics profiling. *Glycobiology*, *17*, 25–35.

Hu, Y., Desantos-Garcia, J. L., & Mechref, Y. (2013). Comparative glycomic profiling of isotopically permethylated N-glycans by LC-ESI-MS. *Rapid Communications in Mass Spectrometry*, *27*, 865–877.

Hu, Y. L., & Mechref, Y. (2012). Comparing MALDI-MS, RP-LC-MALDI-MS and RP-LC-ESI-MS glycomic profiles of permethylated N-glycans derived from model glycoproteins and human blood serum. *Electrophoresis*, *33*(12), 1768–1777. http://dx.doi.org/10.1002/elps.201100703.

Hu, Y., Zhou, S., Khalil, S. I., Renteria, C. L., & Mechref, Y. (2013). Glycomic profiling of tissue sections by LC-MS. *Analytical Chemistry*, *85*(8), 4074–4079. http://dx.doi.org/10.1021/ac400106x.

Hu, Y., Zhou, S., Yu, C. Y., Tang, H., & Mechref, Y. (2015). Automated annotation and quantitation of glycans by liquid chromatography/electrospray ionization mass spectrometric analysis using the MultiGlycan-ESI computational tool. *Rapid Communications in Mass Spectrometry*, *29*(1), 135–142. http://dx.doi.org/10.1002/rcm.7093.

Hua, S., Saunders, M., Dimapasoc, L. M., Jeong, S. H., Kim, B. J., Kim, S., … An, H. J. (2014). Differentiation of cancer cell origin and molecular subtype by plasma membrane N-glycan profiling. *Journal of Proteome Research*, *13*(2), 961–968. http://dx.doi.org/10.1021/pr400987f.

Huang, C. C., & Aminoff, D. (1972). Enzymes that destroy blood group specificity. V. The oligosaccharase of Clostridium perfringens. *The Journal of Biological Chemistry*, *247*(21), 6737–6742.

Hulsmeier, A. J., Gehrig, P. M., Geyer, R., Sack, R., Gottstein, B., Deplazes, P., & Kohler, P. (2002). A major Echinococcus multilocularis antigen is a mucin-type glycoprotein. *The Journal of Biological Chemistry*, *277*(8), 5742–5748. http://dx.doi.org/10.1074/jbc.M107161200.

Isailovic, D., Kurulugama, R. T., Plasencia, M. D., Stokes, S. T., Kyselova, Z., Goldman, R., … Clemmer, D. E. (2008). Profiling of human serum glycans associated with liver cancer and cirrhosis by IMS-MS. *Journal of Proteome Research*, 7(3), 1109–1117.

Kadowaki, S., Yamamoto, K., Fujisaki, M., Izumi, K., Tochikura, T., & Yokoyama, T. (1990). Purification and characterization of a novel fungal endo-beta-N-acetylglucosaminidase acting on complex oligosaccharides of glycoproteins. *Agricultural and Biological Chemistry*, *54*(1), 97–106.

Kakehi, K., Funakubo, T., Suzuki, S., Oda, Y., & Kitada, Y. (1999). 3-Aminobenzamide and 3-aminobenzoic acid, tags for capillary electrophoresis of complex carbohydrates with laser-induced fluorescent detection. *Journal of Chromatography. A*, *863*(2), 205–218.

Kalay, H., Ambrosini, M., van Berkel, P. H. C., Parren, P. W. H. I., van Kooyk, Y., & Vallejo, J. J. G. (2012). Online nanoliquid chromatography-mass spectrometry and nanofluorescence detection for high-resolution quantitative N-glycan analysis. *Analytical Biochemistry*, *423*(1), 153–162. http://dx.doi.org/10.1016/j.ab.2012.01.015.

Kameyama, A., Kaneda, Y., Yamanaka, H., Yoshimine, H., Narimatsu, H., & Shinohara, Y. (2004). Detection of oligosaccharides labeled with cyanine dyes using matrix-assisted laser desorption/ionization mass spectrometry. *Analytical Chemistry*, *76*(15), 4537–4542. http://dx.doi.org/10.1021/Ac049897z.

Kang, P., Mechref, Y., Klouckova, I., & Novotny, M. V. (2005). Solid-phase permethylation of glycans for mass spectrometric analysis. *Rapid Communications in Mass Spectrometry*, *19*(23), 3421–3428. http://dx.doi.org/10.1002/rcm.2210.

Kang, P., Mechref, Y., Kyselova, Z., Goetz, J. A., & Novotny, M. V. (2007). Comparative glycomic mapping through quantitative permethylation and stable-isotope labeling. *Analytical Chemistry*, *79*, 6064–6073.

Kang, P., Mechref, Y., & Novotny, M. V. (2008). High-throughput solid-phase permethylation of glycans prior to mass spectrometry. *Rapid Communications in Mass Spectrometry*, *22*, 721–734.

Karamanos, N. K., Tsegenidis, T., & Antonopoulos, C. A. (1987). Analysis of neutral sugars as dinitrophenyl-hydrazones by high-performance liquid-chromatography. *Journal of Chromatography*, *405*, 221–228. http://dx.doi.org/10.1016/S0021-9673(01)81764-7.

Karas, M., Ehring, H., Nordhoff, E., Stahl, B., Strupat, K., Hillenkamp, F., … Krebs, B. (1993). Matrix-assisted laser-desorption ionization mass-spectrometry with additives to 2,5-dihydroxybenzoic acid. *Organic Mass Spectrometry*, *28*(12), 1476–1481. http://dx.doi.org/10.1002/oms.1210281219.

Kohla, G., Stockfleth, E., & Schauer, R. (2002). Gangliosides with O-acetylated sialic acids in tumors of neuroectodermal origin. *Neurochemical Research*, *27*(7–8), 583–592.

Kornfeld, R., & Kornfeld, S. (1985). Assembly of asparagine-linked oligosaccharides. *Annual Review of Biochemistry*, *54*, 631–664. http://dx.doi.org/10.1146/annurev.bi.54.070185.003215.

Kozak, R. P., Royle, L., Gardner, R. A., Fernandes, D. L., & Wuhrer, M. (2012). Suppression of peeling during the release of O-glycans by hydrazinolysis. *Analytical Biochemistry*, *423*(1), 119–128. http://dx.doi.org/10.1016/j.ab.2012.01.002.

Kyselova, Z., Mechref, Y., Kang, P., Goetz, J. A., Dobrolecki, L. E., Sledge, G. W., … Novotny, M. V. (2008). Breast cancer diagnosis and prognosis through quantitative measurements of serum glycan profiles. *Clinical Chemistry*, *54*(7), 1166–1175. http://dx.doi.org/10.1373/clinchem.2007.087148.

Lattova, E., & Perreault, H. (2003a). Labelling saccharides with phenylhydrazine for electrospray and matrix-assisted laser desorption-ionization mass spectrometry. *Journal of Chromatography. B, Analytical Technologies in the Biomedical and Life Sciences*, *793*(1), 167–179.

Lattova, E., & Perreault, H. (2003b). Profiling of N-linked oligosaccharides using phenylhydrazine derivatization and mass spectrometry. *Journal of Chromatography. A*, *1016*(1), 71–87.

Lauber, M. A., Brousmiche, D. W., Hua, Z., Koza, S. M., Guthrie, E., Magnelli, P., … Fountain, K. J. (2015). *Rapid preparation of released N-glycans for HILIC analysis using a novel fluorescence and MS-active labeling reagent [Application Note]*. Retrieved from Milford, MA: Waters Corp.

Laubli, H., Stevenson, J. L., Varki, A., Varki, N. M., & Borsig, L. (2006). L-selectin facilitation of metastasis involves temporal induction of Fut7-dependent ligands at sites of tumor cell arrest. *Cancer Research*, *66*(3), 1536–1542. http://dx.doi.org/10.1158/0008-5472.can-05-3121.

Lawrence, R., Olson, S. K., Steele, R. E., Wang, L., Warrior, R., Cummings, R. D., & Esko, J. D. (2008). Evolutionary differences in glycosaminoglycan fine structure detected by quantitative glycan reductive isotope labeling. *The Journal of Biological Chemistry*, *283*, 33674–33684.

Lederkremer, G. Z. (2009). Glycoprotein folding, quality control and ER-associated degradation. *Current Opinion in Structural Biology*, *19*(5), 515–523. http://dx.doi.org/10.1016/j.sbi.2009.06.004.

Leteux, C., Childs, R. A., Chai, W. G., Stoll, M. S., Kogelberg, H., & Feizi, T. (1998). Biotinyl-L-3-(2-naphthyl)-alanine hydrazide derivatives of N-glycans: Versatile solid-phase probes for carbohydrate-recognition studies. *Glycobiology*, *8*(3), 227–236. http://dx.doi.org/10.1093/glycob/8.3.227.

Li, M., & Kinzer, J. A. (2003). Structural analysis of oligosaccharides by a combination of electrospray mass spectrometry and bromine isotope tagging of reducing-end sugars with 2-amino-5-bromopyridine. *Rapid Communications in Mass Spectrometry*, *17*(13), 1462–1466. http://dx.doi.org/10.1002/rcm.1064.

Lizak, C., Gerber, S., Numao, S., Aebi, M., & Locher, K. P. (2011). X-ray structure of a bacterial oligosaccharyltransferase. *Nature*, *474*(7351), 350–355. http://dx.doi.org/10.1038/nature10151.

Lochnit, G., & Geyer, R. (2004). An optimized protocol for nano-LC-MALDI-TOF-MS coupling for the analysis of proteolytic digests of glycoproteins. *Biomedical Chromatography*, *18*(10), 841–848. http://dx.doi.org/10.1002/bmc.399.

Lonardi, E., Deelder, A. M., Wuhrer, M., & Balog, C. I. A. (2012). Microarray technology using glycans extracted from natural sources for serum antibody fluorescent detection. *Carbohydrate Microarrays: Methods and Protocols*, *808*, 285–302. http://dx.doi.org/10.1007/978-1-61779-373-8_20.

Lowe, J. B., & Marth, J. D. (2003). A genetic approach to mammalian glycan function. *Annual Review of Biochemistry*, *72*, 643–691. http://dx.doi.org/10.1146/annurev.biochem.72.121801.161809.

Ma, S., & Nashabeh, W. (1999). Carbohydrate analysis of a chimeric recombinant monoclonal antibody by capillary electrophoresis with laser-induced fluorescence detection. *Analytical Chemistry*, *71*(22), 5185–5192.

Maass, K., Ranzinger, R., Geyer, H., von der Lieth, C. W., & Geyer, R. (2007). "Glyco-peakfinder"—De novo composition analysis of glycoconjugates. *Proteomics*, 7(24), 4435–4444. http://dx.doi.org/10.1002/pmic.200700253.

Matsumoto, K., Shimizu, C., Arao, T., Andoh, M., Katsumata, N., Kohno, T., ... Fujiwara, Y. (2009). Identification of predictive biomarkers for response to trastuzumab using plasma FUCA activity and N-glycan identified by MALDI-TOF-MS. *Journal of Proteome Research*, *8*(2), 457–462. http://dx.doi.org/10.1021/pr800655p.

Mauko, L., Lacher, N. A., Pelzing, M., Nordborg, A., Haddad, P. R., & Hilder, E. F. (2012). Comparison of ZIC-HILIC and graphitized carbon-based analytical approaches combined with exoglycosidase digestions for analysis of glycans from monoclonal antibodies. *Journal of Chromatography B: Analytical Technologies in the Biomedical and Life Sciences*, *911*, 93–104. http://dx.doi.org/10.1016/j.jchromb.2012.10.043.

Means, R. E., & Desrosiers, R. C. (2000). Resistance of native, oligomeric envelope on simian immunodeficiency virus to digestion by glycosidases. *Journal of Virology*, *74*(23), 11181–11190.

Mechref, Y., Hu, Y., Desantos-Garcia, J. L., Hussein, A., & Tang, H. (2013). Quantitative glycomics strategies. *Molecular & Cellular Proteomics*, *12*(4), 874–884. http://dx.doi.org/10.1074/mcp.R112.026310.

Mechref, Y., Hu, Y., Garcia, A., Zhou, S., Desantos-Garcia, J. L., & Hussein, A. (2012). Defining putative glycan cancer biomarkers by MS. *Bioanalysis*, *4*, 2457–2469.

Mechref, Y., Muzikar, J., & Novotny, M. V. (2005). Comprehensive assessment of N-glycans derived from a murine monoclonal antibody: A case for multimethodological approach. *Electrophoresis*, *26*(10), 2034–2046. http://dx.doi.org/10.1002/elps.200410345.

Mechref, Y., & Novotny, M. V. (1998). Matrix-assisted laser desorption ionization mass spectrometry of acidic glycoconjugates facilitated by the use of spermine as a co-matrix. *Journal of the American Society for Mass Spectrometry*, *9*(12), 1293–1302. http://dx.doi.org/10.1016/S1044-0305(98)00106-8.

Mehta, A., Herrera, H., & Block, T. (2015). Glycosylation and liver cancer. *Advances in Cancer Research*, *126*, 257–279. http://dx.doi.org/10.1016/bs.acr.2014.11.005.

Meitei, N. S., Apte, A., Snovida, S. I., Rogers, J. C., & Saba, J. (2015). Automating mass spectrometry-based quantitative glycomics using aminoxy tandem mass tag reagents with SimGlycan. *Journal of Proteomics*, *127*(Pt. A), 211–222. http://dx.doi.org/10.1016/j.jprot.2015.05.015.

Miura, Y., & Endo, T. (2016). Glycomics and glycoproteomics focused on aging and age-related diseases—Glycans as a potential biomarker for physiological alterations. *Biochimica et Biophysica Acta*, *1860*(8), 1608–1614. http://dx.doi.org/10.1016/j.bbagen.2016.01.013.

Mohr, M. D., Bornsen, K. O., & Widmer, H. M. (1995). Matrix-assisted laser-desorption ionization mass-spectrometry—Improved matrix for oligosaccharides. *Rapid Communications in Mass Spectrometry*, *9*(9), 809–814. http://dx.doi.org/10.1002/rcm.1290090919.

Molinari, M. (2007). N-glycan structure dictates extension of protein folding or onset of disposal. *Nature Chemical Biology*, *3*(6), 313–320. http://dx.doi.org/10.1038/nchembio880.

Monsarrat, B., Brando, T., Condouret, P., Nigou, J., & Puzo, G. (1999). Characterization of mannooligosaccharide caps in mycobacterial lipoarabinomannan by capillary electrophoresis/electrospray mass spectrometry. *Glycobiology*, *9*(4), 335–342.

Montesino, R., Calvo, L., Vallin, A., Rudd, P. M., Harvey, D. J., & Cremata, J. A. (2012). Structural characterization of N-linked oligosaccharides on monoclonal antibody Nimotuzumab through process development. *Biologicals*, *40*(4), 288–298. http://dx.doi.org/10.1016/j.biologicals.2012.04.005.

Nakajima, K., Oda, Y., Kinoshita, M., Masuko, T., & Kakehi, K. (2002). Time-resolved fluorometric analysis of carbohydrates labeled with amino-aromatic compounds by reductive amination. *Analyst*, *127*(7), 972–976.

Naven, T. J. P., & Harvey, D. J. (1998). Cationic derivatization of oligosaccharides with Girard's T reagent for improved performance in matrix-assisted laser desorption/ionization and electrospray mass spectrometry. *Rapid Communications in Mass Spectrometry*, *10*(7), 829–834.

Ninonuevo, M. R., Park, Y., Yin, H. F., Zhang, J. H., Ward, R. E., Clowers, B. H., … Lebrilla, C. B. (2006). A strategy for annotating the human milk glycome. *Journal of Agricultural and Food Chemistry*, *54*(20), 7471–7480. http://dx.doi.org/10.1021/jf0615810.

Nwosu, C. C., Aldredge, D. L., Lee, H., Lerno, L. A., Zivkovic, A. M., German, J. B., & Lebrilla, C. B. (2012). Comparison of the human and bovine milk N-glycome via high-performance microfluidic chip liquid chromatography and tandem mass spectrometry. *Journal of Proteome Research*, *11*(5), 2912–2924. http://dx.doi.org/10.1021/pr300008u.

Orlando, R., Lim, J. M., Atwood, J. A., 3rd, Angel, P. M., Fang, M., Aoki, K., … Wells, L. (2009). IDAWG: Metabolic incorporation of stable isotope labels for quantitative glycomics of cultured cells. *Journal of Proteome Research*, *8*(8), 3816–3823. http://dx.doi.org/10.1021/pr8010028.

Pabst, M., & Altmann, F. (2008). Influence of electrosorption, solvent, temperature, and ion polarity on the performance of LC-ESI-MS using graphitic carbon for acidic oligosaccharides. *Analytical Chemistry*, *80*(19), 7534–7542. http://dx.doi.org/10.1021/ac801024r.

Pabst, M., Kolarich, D., Poltl, G., Dalik, T., Lubec, G., Hofinger, A., & Altmann, F. (2009). Comparison of fluorescent labels for oligosaccharides and introduction of a new post-labeling purification method. *Analytical Biochemistry*, *384*(2), 263–273. http://dx.doi.org/10.1016/j.ab.2008.09.041.

Packer, N. H., Lawson, M. A., Jardine, D. R., & Redmond, J. W. (1998). A general approach to desalting oligosaccharides released from glycoproteins. *Glycoconjugate Journal*, *15*(8), 737–747. http://dx.doi.org/10.1023/A:1006983125913.

Patel, T., Bruce, J., Merry, A., Bigge, C., Wormald, M., Jaques, A., & Parekh, R. (1993). Use of hydrazine to release in intact and unreduced form both N- and O-linked oligosaccharides from glycoproteins. *Biochemistry*, *32*(2), 679–693.

Pfenninger, A., Karas, M., Finke, B., & Stahl, B. (2002a). Structural analysis of underivatized neutral human milk oligosaccharides in the negative ion mode by nano-electrospray MS(n) (Part 1: Methodology). *Journal of the American Society for Mass Spectrometry*, *13*(11), 1331–1340. http://dx.doi.org/10.1016/S1044-0305(02)00645-1.

Pfenninger, A., Karas, M., Finke, B., & Stahl, B. (2002b). Structural analysis of underivatized neutral human milk oligosaccharides in the negative ion mode by nano-electrospray MS(n) (Part 2: Application to isomeric mixtures). *Journal of the American Society for Mass Spectrometry*, *13*(11), 1341–1348. http://dx.doi.org/10.1016/S1044-0305(02)00646-3.

Plummer, T. H., Jr., & Tarentino, A. L. (1991). Purification of the oligosaccharide-cleaving enzymes of Flavobacterium meningosepticum. *Glycobiology*, *1*(3), 257–263.

Prater, B. D., Anumula, K. R., & Hutchins, J. T. (2007). Automated sample preparation facilitated by PhyNexus MEA purification system for oligosaccharide mapping of glycoproteins. *Analytical Biochemistry*, *369*(2), 202–209. http://dx.doi.org/10.1016/j.ab.2007.07.011.

Rabouille, C., Hui, N., Hunte, F., Kieckbusch, R., Berger, E. G., Warren, G., & Nilsson, T. (1995). Mapping the distribution of Golgi enzymes involved in the construction of complex oligosaccharides. *Journal of Cell Science*, *108*(Pt. 4), 1617–1627.

Ressom, H. W., Varghese, R. S., Goldman, L., An, Y., Loffredo, C. A., Abdel-Hamid, M., … Goldman, R. (2008). Analysis of MALDI-TOF mass spectrometry data for discovery of peptide and glycan biomarkers of hepatocellular carcinoma. *Journal of Proteome Research*, 7(2), 603–610.

Royle, L., Campbell, M. P., Radcliffe, C. M., White, D. M., Harvey, D. J., Abrahams, J. L., ... Dwek, R. A. (2008). HPLC-based analysis of serum N-glycans on a 96-well plate platform with dedicated database software. *Analytical Biochemistry*, *376*(1), 1–12. http://dx.doi.org/10.1016/j.ab.2007.12.012.

Royle, L., Mattu, T. S., Hart, E., Langridge, J. I., Merry, A. H., Murphy, N., ... Rudd, P. M. (2002). An analytical and structural database provides a strategy for sequencing O-glycans from microgram quantities of glycoproteins. *Analytical Biochemistry*, *304*(1), 70–90. http://dx.doi.org/10.1006/abio.2002.5619.

Ruhaak, L. R., Deelder, A. M., & Wuhrer, M. (2009). Oligosaccharide analysis by graphitized carbon liquid chromatography-mass spectrometry. *Analytical and Bioanalytical Chemistry*, *394*(1), 163–174. http://dx.doi.org/10.1007/s00216-009-2664-5.

Ruhaak, L. R., Miyamoto, S., Kelly, K., & Lebrilla, C. B. (2012). N-Glycan profiling of dried blood spots. *Analytical Chemistry*, *84*(1), 396–402. http://dx.doi.org/10.1021/ac202775t.

Ruhaak, L. R., Miyamoto, S., & Lebrilla, C. B. (2013). Developments in the identification of glycan biomarkers for the detection of cancer. *Molecular & Cellular Proteomics*, *12*(4), 846–855. http://dx.doi.org/10.1074/mcp.R112.026799.

Ruhaak, L. R., Nguyen, U. T., Stroble, C., Taylor, S. L., Taguchi, A., Hanash, S. M., ... Miyamoto, S. (2013). Enrichment strategies in glycomics-based lung cancer biomarker development. *Proteomics Clinical Applications*, 7(9–10), 664–676. http://dx.doi.org/10.1002/prca.201200131.

Ruhaak, L. R., Steenvoorden, E., Koeleman, C. A., Deelder, A. M., & Wuhrer, M. (2010). 2-picoline-borane: A non-toxic reducing agent for oligosaccharide labeling by reductive amination. *Proteomics*, *10*(12), 2330–2336. http://dx.doi.org/10.1002/pmic.200900804.

Ruhaak, L. R., Taylor, S. L., Miyamoto, S., Kelly, K., Leiserowitz, G. S., Gandara, D., ... Kim, K. (2013). Chip-based nLC-TOF-MS is a highly stable technology for large-scale high-throughput analyses. *Analytical and Bioanalytical Chemistry*, *405*(14), 4953–4958. http://dx.doi.org/10.1007/s00216-013-6908-z.

Saba, J. A., Shen, X., Jamieson, J. C., & Perreault, H. (1999). Effect of 1-phenyl-3-methyl-5-pyrazolone labeling on the fragmentation behavior of asialo and sialylated N-linked glycans under electrospray ionization conditions. *Rapid Communications in Mass Spectrometry*, *13*(8), 704–711. http://dx.doi.org/10.1002/(SICI)1097-0231(19990430)13:8<704::AID-RCM543>3.0.CO;2-V10.1002/(SICI)1097-0231(19990430)13:8<704::AID-RCM543>3.0.CO;2-V.

Saba, J. A., Shen, X., Jamieson, J. C., & Perreault, H. (2001). Investigation of different combinations of derivatization, separation methods and electrospray ionization mass spectrometry for standard oligosaccharides and glycans from ovalbumin. *Journal of Mass Spectrometry*, *36*(5), 563–574. http://dx.doi.org/10.1002/jms.158.

Saldova, R., Reuben, J. M., Abd Hamid, U. M., Rudd, P. M., & Cristofanilli, M. (2011). Levels of specific serum N-glycans identify breast cancer patients with higher circulating tumor cell counts. *Annals of Oncology*, *22*(5), 1113–1119.

Saldova, R., Royle, L., Radcliffe, C. M., Abd Hamid, U. M., Evans, R., Arnold, J. N., ... Rudd, P. M. (2007). Ovarian cancer is associated with changes in glycosylation in both acute-phase proteins and IgG. *Glycobiology*, *17*(12), 1344–1356.

Sandra, K., Devreese, B., Van Beeumen, J., Stals, I., & Claeyssens, M. (2004). The Q-Trap mass spectrometer, a novel tool in the study of protein glycosylation. *Journal of the American Society for Mass Spectrometry*, *15*(3), 413–423. http://dx.doi.org/10.1016/j.jasms.2003.11.003.

Schwaiger, H., Oefner, P. J., Huber, C., Grill, E., & Bonn, G. K. (1994). Capillary zone electrophoresis and micellar electrokinetic chromatography of 4-aminobenzonitrile carbohydrate derivatives. *Electrophoresis*, *15*(7), 941–952.

Seales, E. C., Jurado, G. A., Brunson, B. A., Wakefield, J. K., Frost, A. R., & Bellis, S. L. (2005). Hypersialylation of beta1 integrins, observed in colon adenocarcinoma, may contribute to cancer progression by up-regulating cell motility. *Cancer Research*, *65*(11), 4645–4652. http://dx.doi.org/10.1158/0008-5472.can-04-3117.

Sekiya, S., Wada, Y., & Tanaka, K. (2005). Derivatization for stabilizing sialic acids in MALDI-MS. *Analytical Chemistry*, 77(15), 4962–4968. http://dx.doi.org/10.1021/ac050287o.

Snovida, S. I., Chen, V. C., & Perreault, H. (2006). Use of a 2,5-dihydroxybenzoic acid/aniline MALDI matrix for improved detection and on-target derivatization of glycans: A preliminary report. *Analytical Chemistry*, *78*(24), 8561–8568. http://dx.doi.org/10.1021/ac061375r.

Snovida, S. I., Rak-Banville, J. M., & Perreault, H. (2008). On the use of DHB/aniline and DHB/N, N-dimethylaniline matrices for improved detection of carbohydrates: Automated identification of oligosaccharides and quantitative analysis of sialylated glycans by MALDI-TOF mass spectrometry. *Journal of the American Society for Mass Spectrometry*, *19*(8), 1138–1146. http://dx.doi.org/10.1016/j.jasms.2008.04.033.

Song, X., Ju, H., Lasanajak, Y., Kudelka, M. R., Smith, D. F., & Cummings, R. D. (2016). Oxidative release of natural glycans for functional glycomics. *Nature Methods*, *13*(6), 528–534. http://dx.doi.org/10.1038/nmeth.3861.

Sparbier, K., Asperger, A., Kessler, I., Koch, S., Shi, G., Wenzel, T., & Kostrzewa, M. (2006). Identification and characterization of glycoproteins in human serum by means of glyco-specific magnetic bead separation and LCMALDI analysis. *Molecular & Cellular Proteomics*, *5*(10), S134.

Stahl, B., Steup, M., Karas, M., & Hillenkamp, F. (1991). Analysis of neutral oligosaccharides by matrix-assisted laser desorption—Ionization mass-spectrometry. *Analytical Chemistry*, *63*(14), 1463–1466. http://dx.doi.org/10.1021/Ac00014a022.

Stanley, P. (2011). Golgi glycosylation. *Cold Spring Harbor Perspectives in Biology*, *3*(4), 1–13. http://dx.doi.org/10.1101/cshperspect.a005199.

Staples, G. O., Bowman, M. J., Costello, C. E., Hitchcock, A. M., Lau, J. M., Leymarie, N., … Zaia, J. (2009). A chip-based amide-HILIC LC/MS platform for glycosaminoglycan glycomics profiling. *Proteomics*, *9*(3), 686–695. http://dx.doi.org/10.1002/pmic.200701008.

Stevenson, J. L., Choi, S. H., & Varki, A. (2005). Differential metastasis inhibition by clinically relevant levels of heparins—Correlation with selectin inhibition, not antithrombotic activity. *Clinical Cancer Research*, *11*(19 Pt. 1), 7003–7011. http://dx.doi.org/10.1158/1078-0432.ccr-05-1131.

Sugahara, D., Amano, J., & Irimura, T. (2003). Fluorescence labeling of oligosaccharides useful in the determination of molecular interactions. *Analytical Sciences*, *19*(1), 167–169. http://dx.doi.org/10.2116/analsci.19.167.

Suzuki, S. (2013). Recent developments in liquid chromatography and capillary electrophoresis for the analysis of glycoprotein glycans. *Analytical Sciences*, *29*(12), 1117–1128.

Suzuki, H., Muller, O., Guttman, A., & Karger, B. L. (1997). Analysis of 1-aminopyrene-3,6,8-trisulfonate-derivatized oligosaccharides by capillary electrophoresis with matrix-assisted laser desorption/ionization time-of-flight mass spectrometry. *Analytical Chemistry*, *69*(22), 4554–4559.

Szabo, Z., Guttman, A., Rejtar, T., & Karger, B. L. (2010). Improved sample preparation method for glycan analysis of glycoproteins by CE-LIF and CE-MS. *Electrophoresis*, *31*(8), 1389–1395. http://dx.doi.org/10.1002/elps.201000037.

Takahashi, N. (1996). Three-dimensional mapping of N-linked oligosaccharides using anion-exchange, hydrophobic and hydrophilic interaction modes of high-performance liquid chromatography. *Journal of Chromatography A*, *720*(1–2), 217–225. http://dx.doi.org/10.1016/0021-9673(95)00328-2.

Takahashi, N., & Nishibe, H. (1978). Some characteristics of a new glycopeptidase acting on aspartylglycosylamine linkages. *Journal of Biochemistry*, *84*(6), 1467–1473.
Takasaki, S., Mizuochi, T., & Kobata, A. (1982). Hydrazinolysis of asparagine-linked sugar chains to produce free oligosaccharides. *Methods in Enzymology*, *83*, 263–268.
Takegawa, Y., Deguchi, K., Ito, H., Keira, T., Nakagawa, H., & Nishimura, S. I. (2006). Simple separation of isomeric sialylated N-glycopeptides by a zwitterionic type of hydrophilic interaction chromatography. *Journal of Separation Science*, *29*(16), 2533–2540. http://dx.doi.org/10.1002/jssc.200600133.
Tarentino, A. L., Gomez, C. M., & Plummer, T. H., Jr. (1985). Deglycosylation of asparagine-linked glycans by peptide:N-glycosidase F. *Biochemistry*, *24*(17), 4665–4671.
Tarentino, A. L., & Maley, F. (1974). Purification and properties of an endo-beta-N-acetylglucosaminidase from Streptomyces griseus. *The Journal of Biological Chemistry*, *249*(3), 811–817.
Tarentino, A. L., Plummer, T. H., Jr., & Maley, F. (1974). The release of intact oligosaccharides from specific glycoproteins by endo-beta-N-acetylglucosaminidase H. *The Journal of Biological Chemistry*, *249*(3), 818–824.
Ten Hagen, K. G., Fritz, T. A., & Tabak, L. A. (2003). All in the family: The UDP-GalNAc: polypeptide N-acetylgalactosaminyltransferases. *Glycobiology*, *13*(1), 1R–16R. http://dx.doi.org/10.1093/glycob/cwg007.
Thomsson, K. A., Karlsson, N. G., & Hansson, G. C. (1999). Liquid chromatography-electrospray mass spectrometry as a tool for the analysis of sulfated oligosaccharides from mucin glycoproteins. *Journal of Chromatography. A*, *854*(1–2), 131–139.
Tousi, F., Bones, J., Hancock, W. S., & Hincapie, M. (2013). Differential chemical derivatization integrated with chromatographic separation for analysis of isomeric sialylated N-glycans: A nano-hydrophilic interaction liquid chromatography-MS platform. *Analytical Chemistry*, *85*(17), 8421–8428. http://dx.doi.org/10.1021/ac4018007.
Tretter, V., Altmann, F., & Marz, L. (1991). Peptide-N4-(N-acetyl-beta-glucosaminyl) asparagine amidase F cannot release glycans with fucose attached alpha 1–3 to the asparagine-linked N-acetylglucosamine residue. *European Journal of Biochemistry*, *199*(3), 647–652.
Trimble, R. B., & Tarentino, A. L. (1991). Identification of distinct endoglycosidase (endo) activities in Flavobacterium meningosepticum: Endo F1, endo F2, and endo F3. Endo F1 and endo H hydrolyze only high mannose and hybrid glycans. *The Journal of Biological Chemistry*, *266*(3), 1646–1651.
Trombetta, E. S., & Helenius, A. (1998). Lectins as chaperones in glycoprotein folding. *Current Opinion in Structural Biology*, *8*(5), 587–592.
Tsai, T. H., Wang, M., Di Poto, C., Hu, Y., Zhou, S., Zhao, Y., … Ressom, H. W. (2014). LC-MS profiling of N-glycans derived from human serum samples for biomarker discovery in hepatocellular carcinoma. *Journal of Proteome Research*, *13*(11), 4859–4868.
Vabres, P., Sevin, C., Amoric, J. C., Odievre, M. H., Saudubray, J. M., & de Prost, Y. (1998). Skin manifestations of protein glycosylation deficiency, the CDG (carbohydrate deficient glycoprotein) type 1 syndrome. *Annales de Dermatologie et de Vénéréologie*, *125*(10), 715–716.
Vakhrushev, S. Y., Dadimov, D., & Peter-Katalinic, J. (2009). Software platform for high-throughput glycomics. *Analytical Chemistry*, *81*(9), 3252–3260. http://dx.doi.org/10.1021/ac802408f.
Vanderschaeghe, D., Debruyne, E., Van Vlierberghe, H., Callewaert, N., & Delanghe, J. (2009). Analysis of gamma-globulin mobility on routine clinical CE equipment: Exploring its molecular basis and potential clinical utility. *Electrophoresis*, *30*(15), 2617–2623. http://dx.doi.org/10.1002/elps.200900054.
Vanderschaeghe, D., Laroy, W., Sablon, E., Halfon, P., Van Hecke, A., Delanghe, J., & Callewaert, N. (2009). GlycoFibroTest is a highly performant liver fibrosis biomarker

derived from DNA sequencer-based serum protein glycomics. *Molecular & Cellular Proteomics*, *8*(5), 986–994. http://dx.doi.org/10.1074/mcp.M800470-MCP200.

Varki, A., Kannagi, R., & Toole, B. P. (2009). Glycosylation changes in cancer. In A. Varki, R. D. Cummings, J. D. Esko, H. H. Freeze, P. Stanley, C. R. Bertozzi, G. W. Hart, & M. E. Etzler (Eds.), *Essentials of glycobiology*. (2nd ed.). Cold Spring Harbor, NY: Cold Spring Harbor Laboratory Press.

Varki, N. M., & Varki, A. (2002). Heparin inhibition of selectin-mediated interactions during the hematogenous phase of carcinoma metastasis: Rationale for clinical studies in humans. *Seminars in Thrombosis and Hemostasis*, *28*(1), 53–66. http://dx.doi.org/10.1055/s-2002-20564.

Vondeyn, W., York, W. S., Albersheim, P., & Darvill, A. G. (1990). 1-Alkoxyamino-1-deoxy alditols, useful UV-absorbing derivatives of neutral and acidic oligosaccharides. *Carbohydrate Research*, *201*(1), 135–144. http://dx.doi.org/10.1016/0008-6215(90)84230-R.

Wada, Y., Azadi, P., Costello, C. E., Dell, A., Dwek, R. A., Geyer, H., … Taniguchi, N. (2007). Comparison of the methods for profiling glycoprotein glycans—HUPO Human Disease Glycomics/Proteome Initiative multi-institutional study. *Glycobiology*, *17*(4), 411–422. http://dx.doi.org/10.1093/glycob/cwl086.

Walker, S. H., Budhathoki-Uprety, J., Novak, B. M., & Muddiman, D. C. (2011). Stable-isotope labeled hydrophobic hydrazide reagents for the relative quantification of N-linked glycans by electrospray ionization mass spectrometry. *Analytical Chemistry*, *83*(17), 6738–6745. http://dx.doi.org/10.1021/ac201376q.

Walker, S. H., Lilley, L. M., Enamorado, M. F., Comins, D. L., & Muddiman, D. C. (2011). Hydrophobic derivatization of N-linked glycans for increased ion abundance in electrospray ionization mass spectrometry. *Journal of the American Society for Mass Spectrometry*, *22*(8), 1309–1317. http://dx.doi.org/10.1007/s13361-011-0140-x.

Walker, S. H., Papas, B. N., Comins, D. L., & Muddiman, D. C. (2010). Interplay of permanent charge and hydrophobicity in the electrospray ionization of glycans. *Analytical Chemistry*, *82*(15), 6636–6642. http://dx.doi.org/10.1021/Ac101227a.

Wheeler, S. F., Domann, P., & Harvey, D. J. (2009). Derivatization of sialic acids for stabilization in matrix-assisted laser desorption/ionization mass spectrometry and concomitant differentiation of alpha(2 − > 3)- and alpha(2 − > 6)-isomers. *Rapid Communications in Mass Spectrometry*, *23*(2), 303–312. http://dx.doi.org/10.1002/rcm.3867.

Wing, D. R., Garner, B., Hunnam, V., Reinkensmeier, G., Andersson, U., Harvey, D. J., … Butters, T. D. (2001). High-performance liquid chromatography analysis of ganglioside carbohydrates at the picomole level after ceramide glycanase digestion and fluorescent labeling with 2-aminobenzamide. *Analytical Biochemistry*, *298*(2), 207–217. http://dx.doi.org/10.1006/abio.2001.5393.

Wu, S., Grimm, R., German, J. B., & Lebrilla, C. B. (2011). Annotation and structural analysis of sialylated human milk oligosaccharides. *Journal of Proteome Research*, *10*(2), 856–868. http://dx.doi.org/10.1021/pr101006u.

Wuhrer, M., de Boer, A. R., & Deelder, A. M. (2009). Structural glycomics using hydrophilic interaction chromatography (Hilic) with mass spectrometry. *Mass Spectrometry Reviews*, *28*(2), 192–206. http://dx.doi.org/10.1002/mas.20195.

Wuhrer, M., Deelder, A. M., & Hokke, C. H. (2005). Protein glycosylation analysis by liquid chromatography-mass spectrometry. *Journal of Chromatography. B, Analytical Technologies in the Biomedical and Life Sciences*, *825*(2), 124–133. http://dx.doi.org/10.1016/j.jchromb.2005.01.030.

Wuhrer, M., Koeleman, C. A., Deelder, A. M., & Hokke, C. H. (2004a). Normal-phase nanoscale liquid chromatography-mass spectrometry of underivatized oligosaccharides at low-femtomole sensitivity. *Analytical Chemistry*, *76*(3), 833–838. http://dx.doi.org/10.1021/ac034936c.

Wuhrer, M., Koeleman, C. A., Hokke, C. H., & Deelder, A. M. (2004b). Nano-scale liquid chromatography-mass spectrometry of 2-aminobenzamide-labeled oligosaccharides at low femtomole sensitivity. *International Journal of Mass Spectrometry*, *232*, 51–57.

Xia, B., Feasley, C. L., Sachdev, G. P., Smith, D. F., & Cummings, R. D. (2009). Glycan reductive isotope labeling for quantitative glycomics. *Analytical Biochemistry*, *387*, 162–170.

Yamada, E., Tsukamoto, Y., Sasaki, R., Yagyu, K., & Takahashi, N. (1997). Structural changes of immunoglobulin G oligosaccharides with age in healthy human serum. *Glycoconjugate Journal*, *14*(3), 401–405.

Yang, S., Wang, M. Y., Chen, L. J., Yin, B. J., Song, G. Q., Turko, I. V., ... Li, S. W. (2015). QUANTITY: An isobaric tag for quantitative glycomics. *Scientific Reports*. *5*. http://dx.doi.org/10.1038/Srep17585. Artn 17585.

Yang, S., Yuan, W., Yang, W. M., Zhou, J. Y., Harlan, R., Edwards, J., ... Zhang, H. (2013). Glycan analysis by isobaric aldehyde reactive tags and mass spectrometry. *Analytical Chemistry*, *85*(17), 8188–8195. http://dx.doi.org/10.1021/ac401226d.

Yang, S., & Zhang, H. (2012). Solid-phase glycan isolation for glycomics analysis. *Proteomics Clinical Applications*, *6*(11–12), 596–608. http://dx.doi.org/10.1002/prca.201200045.

Yuan, J., Hashii, N., Kawasaki, N., Itoh, S., Kawanishi, T., & Hayakawa, T. (2005). Isotope tag method for quantitative analysis of carbohydrates by liquid chromatography–mass spectrometry. *Journal of Chromatography. A*, *1067*, 145–152.

Zamfir, A., & Peter-Katalinic, J. (2004). Capillary electrophoresis-mass spectrometry for glycoscreening in biomedical research. *Electrophoresis*, *25*(13), 1949–1963. http://dx.doi.org/10.1002/elps.200405825.

Zhang, Y., Iwamoto, T., Radke, G., Kariya, Y., Suzuki, K., Conrad, A. H., ... Conrad, G. W. (2008). On-target derivatization of keratan sulfate oligosaccharides with pyrenebutyric acid hydrazide for MALDI-TOF/TOF-MS. *Journal of Mass Spectrometry*, *43*(6), 765–772. http://dx.doi.org/10.1002/jms.1373.

Zhang, W., Wang, H., Tang, H., & Yang, P. (2011). Endoglycosidase-mediated incorporation of 18O into glycans for relative glycan quantitation. *Analytical Chemistry*, *83*(12), 4975–4981. http://dx.doi.org/10.1021/ac200753e.

Zhong, X. F., Chen, Z. W., Snovida, S., Liu, Y., Rogers, J. C., & Li, L. J. (2015). Capillary electrophoresis-electrospray ionization-mass spectrometry for quantitative analysis of glycans labeled with multiplex carbonyl-reactive tandem mass tags. *Analytical Chemistry*, *87*(13), 6527–6534. http://dx.doi.org/10.1021/acs.analchem.5b01835.

Zhou, S., Hu, Y., Veillon, L., Snovida, S. I., Rogers, J. C., Saba, J., & Mechref, Y. (2016). Quantitative LC-MS/MS glycomic analysis of biological samples using AminoxyTMT. *Analytical Chemistry*, *88*(15), 7515–7522. http://dx.doi.org/10.1021/acs.analchem.6b00465.

Zhou, S., Tello, N., Harvey, A., Boyes, B., Orlando, R., & Mechref, Y. (2016). Reliable LC-MS quantitative glycomics using iGlycoMab stable isotope labeled glycans as internal standards. *Electrophoresis*, *37*(11), 1489–1497. http://dx.doi.org/10.1002/elps.201600013.

Zhu, F., Lee, S., Valentine, S. J., Reilly, J. P., & Clemmer, D. E. (2012). Mannose7 glycan isomer characterization by IMS-MS/MS analysis. *Journal of the American Society for Mass Spectrometry*, *23*(12), 2158–2166. http://dx.doi.org/10.1007/s13361-012-0491-y.

AUTHOR INDEX

Note: Page numbers followed by "*f*" indicate figures, and "*t*" indicate tables.

B

D

E

F

G

H

K

L

M

N

Q

R

S

W

X

Y

SUBJECT INDEX

Note: Page numbers followed by "*f*" indicate figures, and "*t*" indicate tables.

N

O

P